Current Topics in the Chemistry of Boron

Current Topics in the Chemistry of Boron

Edited by

George W. Kabalka

*The University of Tennessee,
Knoxville, Tennessee, USA*

ROYAL
SOCIETY OF
CHEMISTRY

The Proceedings of the Eighth International Meeting on Boron Chemistry, held at the University of Tennessee, Knoxville, Tennessee, USA on July 11–15 1993.

Special Publication No. 143

ISBN 0-85186-535-6

A catalogue record for this book is available from the British Library

Published by The Royal Society of Chemistry,
Thomas Graham House, The Science Park, Milton Road,
Cambridge CB4 4WF, UK

Printed and bound by Bookcraft (Bath) Ltd.

Preface

The Eighth International Meeting on Boron Chemistry [IMEBORON VIII] was held in Knoxville, Tennessee, USA from July 11-15, 1993 on the University of Tennessee Campus. IMEBORON meetings are held triennially and are attended by scientists interested in every aspect of boron chemistry. IMEBORON VIII, which was sponsored by the International Union of Pure and Applied Chemistry, attracted over 200 participants from 20 countries making it the largest IMEBORON meeting held to date.

The scientific presentations covered all aspects of boron chemistry including synthetic methods, molecular structure, bonding theory, mechanistic principles, medical applications, and material science. The presentations have been organized into six chapters in which the subject matter has been arranged so that related materials are presented as a group. The first chapter, *Organoborane Chemistry*, focuses on the chemistry of organoboranes and includes their preparation and reactions. The second chapter, *Chiral Organoboranes in Synthesis*, extends the coverage of the organic reactions of boron into the realm of chiral synthesis. Chapter three, *Medical Applications of Boron*, focuses on the use of both organic and inorganic boron reagents in the medical arena. The fourth chapter, *Carborane Chemistry*, details the chemistry of a variety of carborane molecules. Chapter five, *Metallaborane Chemistry*, presents a summary of the current studies involving the metallic derivatives of a variety of borane and carborane reagents. Chapter six, *Heteroborane Derivatives and Complex Borohydrides*, brings together a number of studies which are focused on the theoretical, analytical, and general chemistry of boron hydride reagents.

The IMEBORON VIII lectures are presented in this treatise. The plenary lectures are presented early in each chapter; these are generally followed by the somewhat shorter invited lectures and, finally, the contributed lectures. In some instances, the order of presentation has been modified in an effort to present related chemistry in a logical fashion. The chapters contain summaries of the current research efforts of essentially all of the major boron research groups in the world. These reports present a "snapshot" of state-of-the-art boron chemistry. It is hoped that this book will serve as a catalyst to young scientists interested in a career in boron chemistry while providing fresh insights to those readers actively involved in the world of boron chemistry.

George W. Kabalka, Ph.D.
Chairman - IMEBORON VIII

Contents

Organoborane Chemistry

Medical Applications of Boron

Carborane Chemistry

Metallaborane Chemistry

ORGANOBORANE CHEMISTRY

New Synthetic Transformations via Organoboron Compounds

Akira Suzuki

DEPARTMENT OF APPLIED CHEMISTRY, FACULTY OF ENGINEERING, HOKKAIDO UNIVERSITY, SAPPORO 060, JAPAN

INTRODUCTION

The cross-coupling reactions of various organoboranes with a number of organic halides in the presence of a catalytic amount of palladium complexes and bases were reported to give versatile and useful synthetic methods for conjugated alkadienes and alkenynes, arylated alkenes, 1,4-alkadienes, allylic benzenes, α,β-unsaturated carboxylic acids, and 2,4-alkadienoates (ref. 1). Thereafter, a modified method for the synthesis of conjugated alkadienes, syntheses of stereodefined trisubstituted alkenes, benzo-fused heteroaromatic compounds, and α,β-unsaturated ketones were presented (ref. 2). Although the palladium-catalyzed cross-coupling reactions of 1-alkenyl- and arylboranes with organic halides proceed readily and stereo- and regioselectively to give expected coupling products in high yields, organoboranes with alkyl groups on boron were not used successfully for the coupling under similar conditions. Recently, it was found that the reaction between 9-alkyl-9-BBN derivatives and 1-halo-1-alkenes or haloarenes in the presence of dichloro[1,1'-bis(diphenylphosphino)ferrocene]palladium(II) and base, such as sodium hydroxide, potassium carbonate, and phosphate gives the corresponding alkenes or arenes in excellent yields (ref. 2). By using the reaction, cycloalkenes, benzo-fused cycloalkenes, and exocyclic alkenes are readily synthesized (ref. 2). The palladium-catalyzed coupling reactions of 1-alkenyl-, aryl-, and alkylboron compounds with aryl or 1-alkenyl triflates, instead of organic halides, take place with ease to give expected coupling products in high yields under mild conditions (ref. 2). Most recently, we have found new synthetic methods related to such palladium-catalyzed cross-coupling reactions, which are described in this review.

CARBONYLATIVE CROSS-COUPLING OF 9-ALKYL-9-BBN DERIVATIVES WITH 1-HALO-1-ALKENES (ref. 3) AND IODOALKANES (ref. 4)

Carbonylative cross-coupling of organometallic compounds with organic halides is reported to give a method for the synthesis of ketones. Among a variety of organometallic reagents, a method using organoboron compounds was first examined by Kojima and his coworkers (ref. 5) for the synthesis of alkyl aryl ketones. Due to the low nucleophilicity of alkyl group on boron, the transmetalation between organoboranes and acylpalladium(II) species generated in the catalytic cycle is anticipated to be a step of retarding in the coupling of organoboron compounds. They found that bis(acetylacetonato)zinc(II) accelerates the transmetalation step and the palladium-catalyzed carbonylative coupling of organoboranes with aryl and benzyl halides gives the corresponding ketones. However, when we applied the procedure in the reaction with 1-halo-1-alkenes, it was recognized that no expected α,β-unsaturated ketones are obtained. Thus, we have reinvestigated such a carbonylative cross-coupling reaction, and observed that the coupling of 9-alkyl-9-BBN derivatives (9-R-9-BBN) with carbon monoxide and 1-iodo-1-alkenes is effectively induced with K_3PO_4 at room temperature in the presence of $Pd(PPh_3)_4$ or $PdCl_2(PPh_3)_2$ (eq. 1).

$$R-B\diamondsuit \quad + \quad CO \quad + \quad \underset{R^1}{X}\diagdown\diagup\underset{}{\overset{H}{\diagdown}}R^2 \quad \xrightarrow[K_3PO_4]{Pd(PPh_3)_4} \quad R\overset{O}{\overset{\|}{C}}\underset{R^1}{\diagdown}\diagup\overset{H}{\diagdown}R^2 \qquad (1)$$

$$68 - 99 \%$$

The reaction is considered to proceed via a pathway similar to that of the palladium-catalyzed carbonylative coupling reaction of other organometallics which involves (a) oxidative addition of haloalkenes to Pd(0) complex, (b) insertion of carbon monoxide to give acylpalladium(II) halide, (c) transfer of alkyl group on boron to acylpalladium(II) halide with the aid of K_3PO_4, and (d) reductive elimination to ketone. One advantage of this ketone synthesis is to be able to use it for the synthesis of ketones with various functional groups, because of the tolerant character of organoboranes.

In the course of the studies, we have discovered that 9-alkyl-9-BBN derivatives react with iodoalkanes under a carbon monoxide pressure in the presence of K_3PO_4 and a catalytic amount of $Pd(PPh_3)_4$ yielding unsymmetrical dialkyl ketones. Although the reaction takes place slowly under dark, the irradiation of light, especially visible, accelerates the rate of coupling to give products in high yields (eq. 2, ref. 4). The irradiation of UV or a high carbon monoxide pressure

$$R-B\diamondsuit \quad + \quad CO \quad + \quad I-R' \quad \xrightarrow[Pd(PPh_3)_4 / K_3PO_4]{hv} \quad R-\overset{O}{\overset{\|}{C}}-R' \qquad (2)$$

R = primary   R' = primary, secondary, and tertiary   65 - 76 %

gives no satisfactory results. A particularly interesting transformation is observed in the case of reaction with 1-iodo-5-hexene (eq. 3).

$$CH_2=CH(CH_2)_4-I \quad + \quad CO \quad + \quad 9\text{-octyl-9-BBN} \quad \xrightarrow[Pd(PPh_3)_4 / K_3PO_4]{hv}$$

$$CH_2=CH(CH_2)_4-\overset{O}{\overset{\|}{C}}-(CH_2)_7CH_3 \quad + \quad \diamondsuit-CH_2-\overset{O}{\overset{\|}{C}}-(CH_2)_7CH_3 \qquad (3)$$

$$\text{1, 4 \%} \qquad\qquad\qquad\qquad \text{2, 60 \%}$$

The reaction with 9-octyl-9-BBN under the standard carbonylative coupling conditions does not give the expected ketone (**1**) as the major product; instead, 1-cyclopentyl-2-decanone (**2**) is obtained in 60% yield. The transformation is considered to occur by prior isomerization of the hexenyl iodide to (iodomethyl)cyclopentane followed by cross-coupling of this latter iodide (ref. 6).

The use of alkyl halides for the palladium-catalyzed cross-coupling reaction was regarded to be difficult, because of the slow rate of oxidative addition step and the β-elimination from oxidative adducts. The present study may open a window to utilize alkyl halides as organic electrophiles in the coupling reaction. Indeed, iodoalkanes have been revealed to couple with 9-alkyl-9-BBN derivatives to yield the corresponding alkanes, such a reaction of which will be discussed next.

ALKYL-ALKYL CROSS-COUPLING OF 9-ALKYL-9-BBN DERIVATIVES WITH IODOALKANES POSSESSING β-HYDROGENS (ref. 7)

Although a wide variety of organic electrophiles, such as aryl, 1-alkenyl, benzyl, allyl, and 1-alkynyl halides, have been efficiently utilized for the palladium-catalyzed cross-coupling reactions with various organometallic reagents, it has been considered that such reactions cannot be extended to alkyl halides with sp^3-carbon having β-hydrogens due to the slow rate of oxidative addition of alkyl halides to palladium(0) complexes and the fast β-hydride elimination from σ-alkylpalladium intermediates in the catalytic cycle. Thus, the use of alkyl halides as coupling partners is a challenging problem in several recent publications. Castle and Widdowson (ref. 8) reported recently that Pd(dppf), formed *in situ* by the reduction of PdCl$_2$(dppf) with DIBAL, effectively catalyzes the cross-coupling of iodoalkanes with Grignard reagents. However, Yuan and Scott (ref. 9) published thereafter that the reaction reported by Castle and Widdowson provides exclusively reduction products of alkyl halides instead of coupling products.

In the case of organoboron compounds, we have found that such a reaction of usual iodoalkanes with 9-alkyl-9-BBN derivatives proceeds readily in the presence of Pd(PPh$_3$)$_4$ and K$_3$PO$_4$ in dioxane to give the corresponding coupling products in fairly good yields (eq. 4, ref. 7).

$$\text{R-B} \quad + \quad \text{I-R'} \quad \xrightarrow[\text{K}_3\text{PO}_4 \text{ / dioxane}]{\text{Pd(PPh}_3)_4} \quad \text{R-R'} \qquad (4)$$

R, R' = alkyl   55 - 71 %

In order to examine the effect of organometallic reagents in the coupling reaction, a molar amount of 1-iododecane is allowed to react with various butylmetal reagents under the conditions which we used above, [Pd(PPh$_3$)$_4$-K$_3$PO$_4$/dioxane] and the Widdowson's conditions [PdCl$_2$(dppf)/THF]. Among the reagents we examined, the 9-alkyl-9-BBN derivatives are only effective for the coupling reaction. Neither tributylborane nor lithium tetrabutylborate as a boron reagent is suitable for the reaction. Other metal reagents, such as Mg, Zn, Al, Sn, Zr, and Hg, give no satisfactory results at all. In all of these experiments, decane and decene caused by β-hydride elimination are commonly obtained.

IMINOCARBONYLATIVE CROSS-COUPLING BETWEEN HALOARENES, *t*-BuNC, AND 9-ALKYL-9-BBN DERIVATIVES. SYNTHESIS OF ALKYL ARYL KETONES (ref. 10)

In the previous section, mention was made of the carbonylative cross-coupling reaction between organic halides, carbon monoxide, and organoboron compounds in the presence of Pd catalyst and base to give ketones. Although isocyanides are isoelectronic with carbon monoxide and, hence, might be expected to exhibit a similar feature for insertion reactions, only a few reports on the cross-coupling were published. The difficulty of use of isocyanides for the coupling reaction is mainly due to their tendency to cause multiple insertions to transition metal complexes which leads to polyisocyanides. In order to make clear this point, we have tried the reaction with *t*-butylisocyanide, and discovered that 9-alkyl-9-BBN derivatives react with *t*-butylisocyanide to form relatively stable complexes which readily participate in the cross-coupling reaction catalyzed by palladium complex and base. The results are successfully applied to the iminocarbonylative cross-coupling of 9-alkyl-9-BBN compounds with haloarenes under mild conditions (eq. 5, ref. 10). The protonolysis of intermediates, ketimines, gives the corresponding alkyl aryl ketones in good yields. The coupling is highly specific with 9-alkyl-9-BBN and

$$R\text{-B}\diagdown \quad + \quad t\text{-BuNC} \quad + \quad X\text{-Ar} \quad \xrightarrow[K_3PO_4]{Pd(PPh_3)_4} \quad R\overset{\overset{\displaystyle NBu^t}{\|}}{\text{-C-}}Ar \quad \xrightarrow{H_3O^+} \quad R\overset{\overset{\displaystyle O}{\|}}{\text{-C-}}Ar \qquad (5)$$

$$64 - 97\ \%$$

iodoarenes. The reaction of trialkylboranes with iodobenzene and t-BuNC under similar conditions does not provide the expected ketone. Quite low yields are obtained by using bromobenzene in place of iodobenzene.

SYNTHESIS OF STERICALLY HINDERED BIARYLS VIA CROSS-COUPLING OF ARYLBORONIC ACIDS OR THEIR ESTERS WITH HALOARENES (ref. 11)

Previously, we reported a simple and versatile method for the synthesis of unsymmetrical biaryls via the cross-coupling of arylboronic acids with haloarenes (ref. 12) in the presence of palladium catalyst and base. Then, many articles with respect to the application of this reaction were published. Although our original procedure (ref. 12) using $Pd(PPh_3)_4$ and aqueous Na_2CO_3 in benzene at 80 °C was found to work effectively for the most of arylboronic acids, it was pointed out that sterically hindered (ref. 13) or electron-withdrawing group substituted (ref. 14) arylboronic acids never provide satisfactory results owing to steric hindrance or competitive hydrolytic deboration. Consequently, we have attempted to reinvestigate the coupling reaction of such sterically hindered aryboronic acids having *ortho*-substituents or functional groups which accelerate the hydrolytic deboration.

The cross-coupling reaction of mesitylboronic acid with aryl halides was known to take place only slowly due to the steric hindrance of two *ortho*-methyl groups (ref. 13). Thus, we examined the reaction of mesitylboronic acid and its esters with iodobenzene at 80 °C in the presence of 2 mol% of $Pd(PPh_3)_4$ and 1.5 equivalents of various bases (eq. 6). The results are summarized in Table 1.

$$\text{—}\diagdown \diagup \text{—B(OR)}_2 \quad + \quad X\text{-Ar} \quad \xrightarrow[base]{Pd(PPh_3)_4} \quad \text{—}\diagdown \diagup \text{—Ar} \quad + \quad \text{—}\diagdown \diagup \qquad (6)$$

3a : R = H   **3b** : R = Bu   **3c** : (OR)$_2$ = -O(CH$_2$)$_3$O-   **4**

Our previous conditions [$Pd(PPh_3)_4$/Na_2CO_3/benzene-H_2O, ref. 12] and the modified conditions by Gronowitz [$Pd(PPh_3)_4$/Na_2CO_3/DME-H_2O, ref. 14] are not satisfactory for the coupling of mesitylboronic acid, and the reaction is not completed even after 2 days. Although the side reactions such as homocoupling are negligibly small, the formation of mesitylene obtained by hydrolytic deboronation increases slowly with the reaction time. On the other hand, the addition of stronger bases, e.g., aqueous NaOH or Ba(OH)$_2$, both in benzene and DME exerts remarkable effect on acceleration of the rate of coupling. By using such conditions, mesitylboronic acid couples with iodobenzene within 4 h to give the corresponding biaryl in a quantitative yield. For such hindered arylboronic acids, an alternative procedure employing the boronic esters and anhydrous bases is developed.

2-Formyl group on arylboronic acids is known to accelerate the rate of hydrolytic deboration (ref. 14). Indeed, the coupling of 2-formylphenylboronic acid (**5a**) with 2-iodo-toluene at 80 °C using Na_2CO_3 in DME/H_2O gives only a 54% yield of the corresponding biaryl (**6**) accompanying benzaldehyde (39%), as shown in eq. 7. The aprotic conditions are desirable for such boronic acids

Table 1. Reaction of mesitylboronic acid with iodobenzene

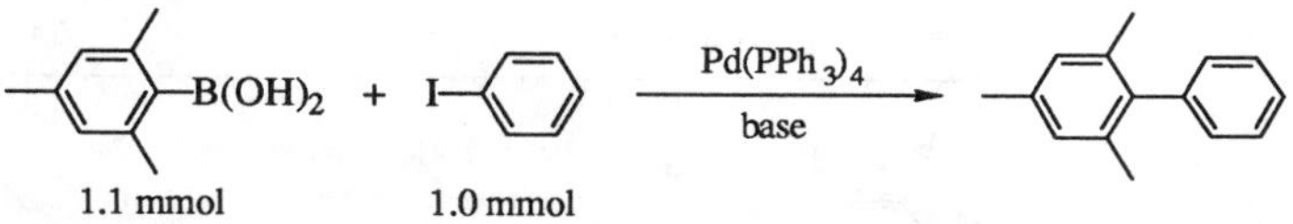

| | 1.1 mmol | | 1.0 mmol | | | |

Solvent	Base	Yield (%)[a]		
		8 h	24 h	48 h
Benzene / H_2O	Na_2CO_3	25 (6)	77 (12)	85 (26)
	$Ba(OH)_2$	92 (13)		
	TlOH	91 (20)		
DME / H_2O	Na_2CO_3	50 (1)	66 (2)	83 (7)
	K_3PO_4	70 (0)	83 (3)	
	$Ba(OH)_2$	99 (2)		

[a] Mesitylene yields are shown in parentheses.

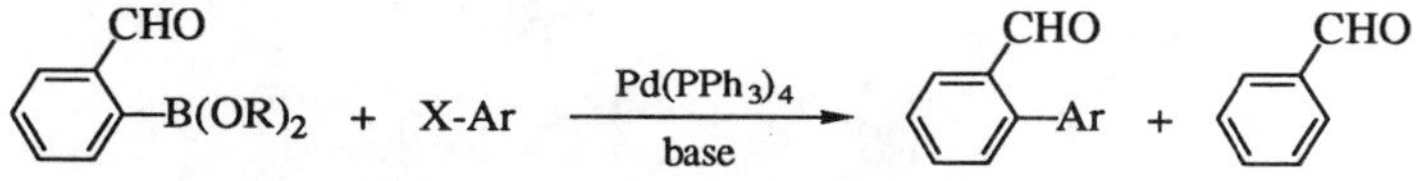

5a : R = H   5b : $(OR)_2$ = $-O(CH_2)_3O-$ 6   (7)

sensitive to aqueous base. Thus, the trimethylene glycol ester of 2-formylphenylboronic acid (5b) readily couples with 2-iodotoluene at 100 °C in the presence of K_3PO_4 in DMF in a yield of 89%, although less than 10% of benzaldehyde is still accompanied. In Table 2 the representative results are summarized. The combination of the present reaction and the cross-coupling with triflates (ref. 2), which is demonstrated in eq. 8, gives the unsymmetrically substituted diarylbenzene in a high yield.

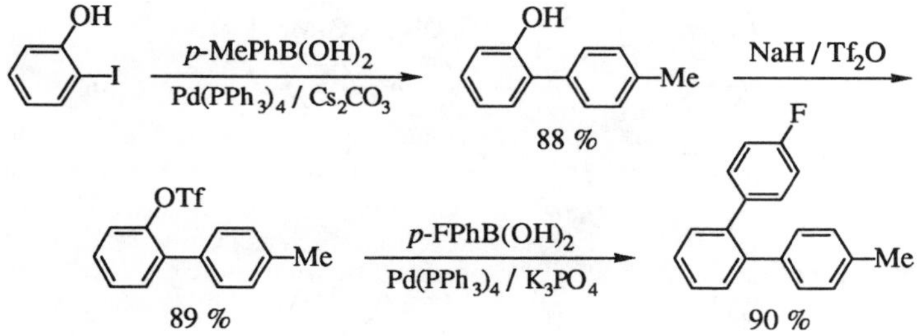

CROSS-COUPLING OF ENOL ACETATES OF α-BROMOKETONES WITH 1-ALKENYL-, ARYL-, OR ALKYLBORON COMPOUNDS (ref. 15)

The cross-coupling reaction of alkyl enol ethers or enol acetates of α-bromo carbonyl compounds with representative organoboranes occurs smoothly to give enol ethers or acetates in good yields, which are readily converted into carbonyl compounds by protonolysis (eq. 9, ref. 15).

Some synthetic applications of the reaction will be shown below (ref. 15).

Table 2. Synthesis of hindered biaryls

Boronic acid	Halide	Method	Product	Yield (%)[a]
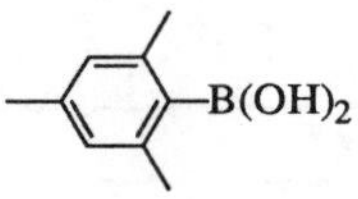 (mesityl)–B(OH)₂	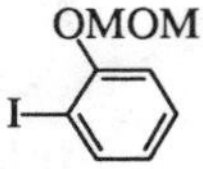 I–(OMOM-phenyl)	A	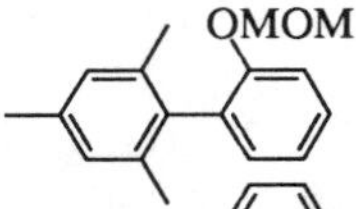 (product)	95
	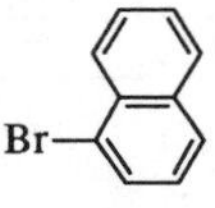 Br–(naphthyl)	A	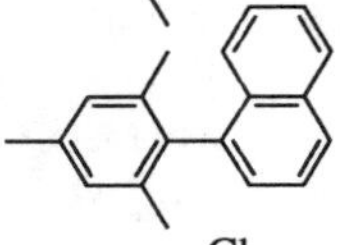 (product)	(86)
	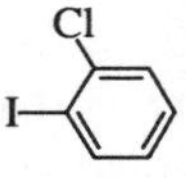 I–(Cl-phenyl)	A	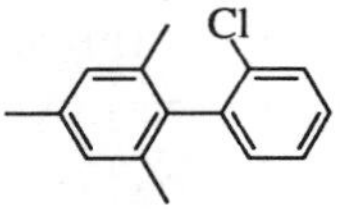 (product)	94
	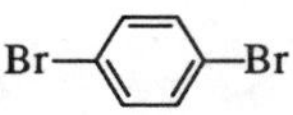 Br–(phenyl)–Br	A[b]	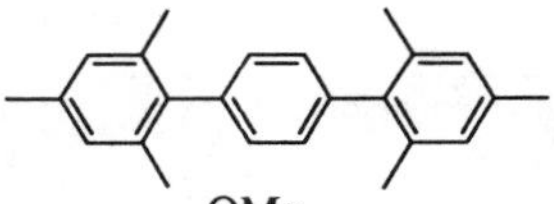 (product)	(75)
	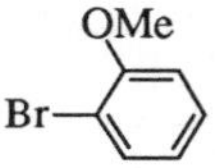 Br–(OMe-phenyl)	A	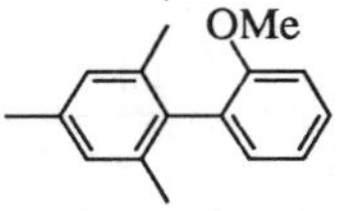 (product)	80
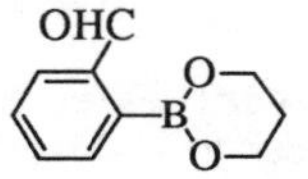 (OHC-phenyl-B(1,3-dioxane))	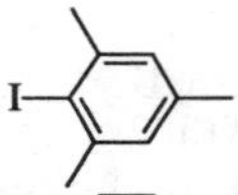 I–(dimethyl-phenyl)	B	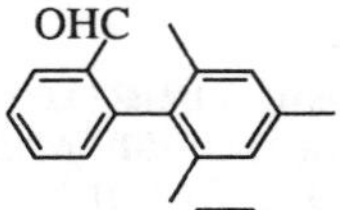 (product, OHC)	(73)
	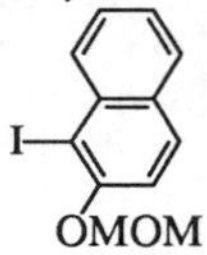 I–(OMOM-naphthyl)	B	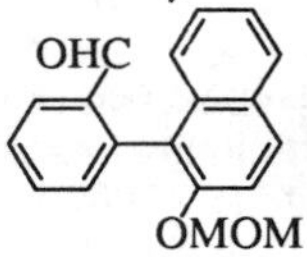 (product, OHC)	(85)
	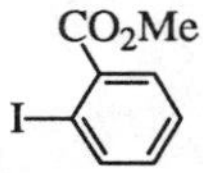 I–(CO₂Me-phenyl)	B	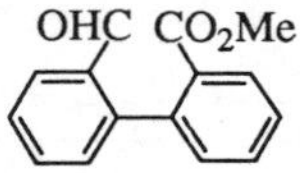 (product, OHC CO₂Me)	63
	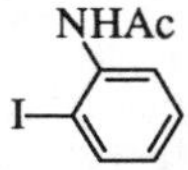 I–(NHAc-phenyl)	B[c]	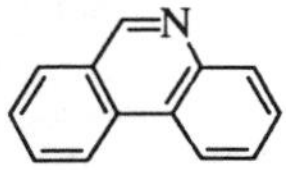 (product, =N)	79
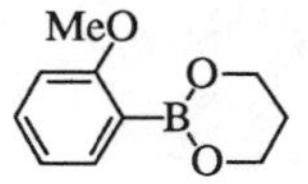 (MeO-phenyl-B(1,3-dioxane))	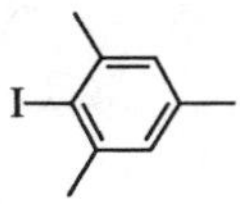 I–(dimethyl-phenyl)	B	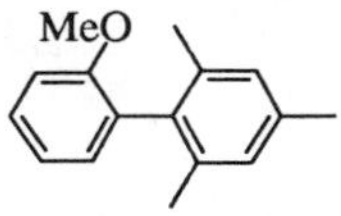 (product, MeO)	99
	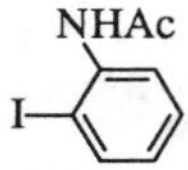 I–(NHAc-phenyl)	B	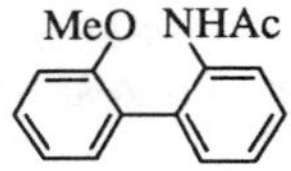 (product, MeO NHAc)	95

[a] GLC yields are based on the aryl halides employed, and isolated yields are in parentheses.
[b] Mesitylboronic acid of 2.2 equiv. was used.   [c] K_3PO_4 of 2.5 equiv. was used.
Method A : $Pd(PPh_3)_4$ (2 mol%) and $Ba(OH)_2$ (1.5 equiv.) in DME-H_2O at 80 °C.
Method B : $Pd(PPh_3)_4$ (2 mol%) and K_3PO_4 (1.5 equiv.) in DMF at 100 °C.

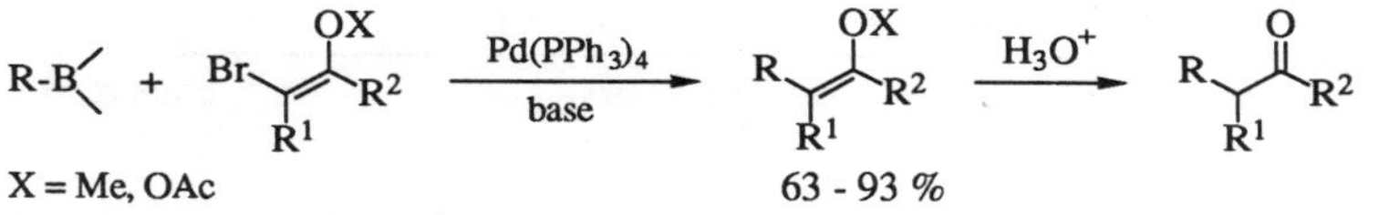

$$R\text{-}B\diagdown + \underset{R^1}{\overset{OX}{\underset{|}{Br\diagup}}}\text{-}R^2 \xrightarrow[\text{base}]{Pd(PPh_3)_4} \underset{R^1}{\overset{OX}{R\diagup}}\text{-}R^2 \xrightarrow{H_3O^+} \underset{R^1}{\overset{O}{R\diagup}}\text{-}R^2 \qquad (9)$$

X = Me, OAc

63 - 93 %

1. 9-BBN

2. Br⎯CH=C(OAc)Ph / Pd(PPh₃)₄ / K₂CO₃

3. KOH / MeOH

4. HCl / H₂O

78 %

$CH_3(CH_2)_5CH=CH_2$

1. 9-BBN

2. (Z)-BrCH=CHOEt / Pd(PPh₃)₄ / K₃PO₄

3. HCl / H₂O

$CH_3(CH_2)_8CHO$

62 %

9-BBN

Br⎯CH=C(OAc)Pr^i / H

Pd(PPh₃)₄ / K₂CO₃

67 %

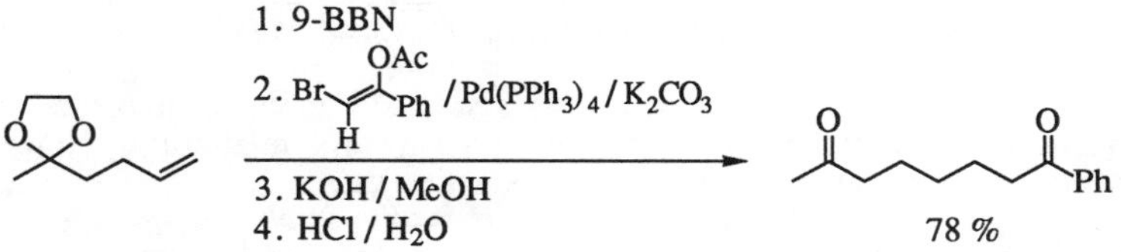

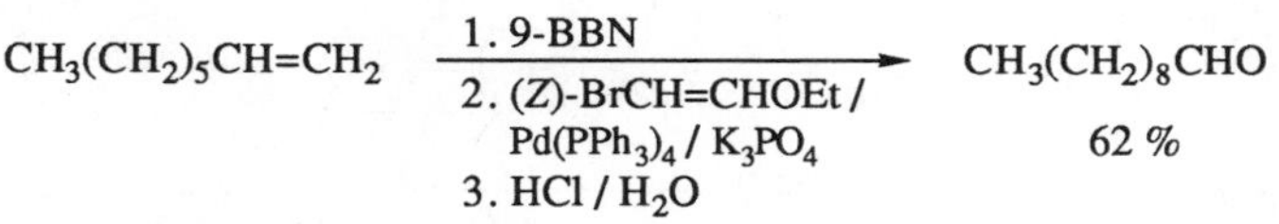

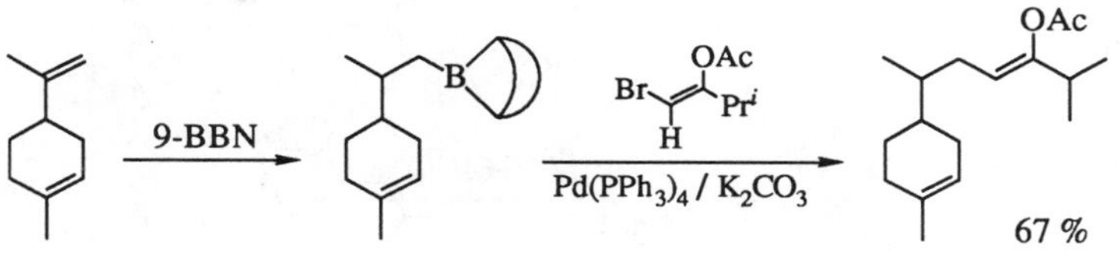

SYNTHESIS OF FUNCTIONALIZED ORGANOTIN COMPOUNDS VIA CROSS-COUPLING OF ARYL OR 1-ALKENYL HALIDES WITH 9-(ω-STANNYLALKYL)-9-BBN DERIVATIVES (ref. 16)

Organotin compounds have been attracting currently chemists' attention because of their importance as synthetic intermediates. In connection with our interest in the palladium-catalyzed transformation of organoboron compounds, we have attempted to develop a new route to such organotin reagents based on the coupling of 9-(ω-stannylalkyl)-9-BBN with organic halides (eq. 10, ref. 16). As the catalyst,

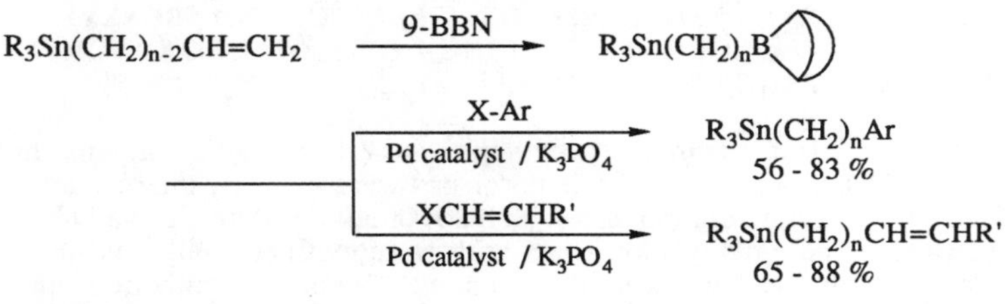

$$R_3Sn(CH_2)_{n-2}CH=CH_2 \xrightarrow{\text{9-BBN}} R_3Sn(CH_2)_nB\text{<}$$

$$\xrightarrow[\text{Pd catalyst / } K_3PO_4]{\text{X-Ar}} \begin{array}{c} R_3Sn(CH_2)_nAr \\ 56 - 83 \% \end{array}$$

$$\xrightarrow[\text{Pd catalyst / } K_3PO_4]{\text{XCH=CHR'}} \begin{array}{c} R_3Sn(CH_2)_nCH=CHR' \\ 65 - 88 \% \end{array} \qquad (10)$$

$PdCl_2(dppf)$ in DMF or $Pd(PPh_3)_4$ in dioxane in the presence of K_3PO_4 gives good results. It is known that enones with ω-stannyl groups are readily destannylated in the presence of acid catalyst to afford carbocyclization products, as shown below.

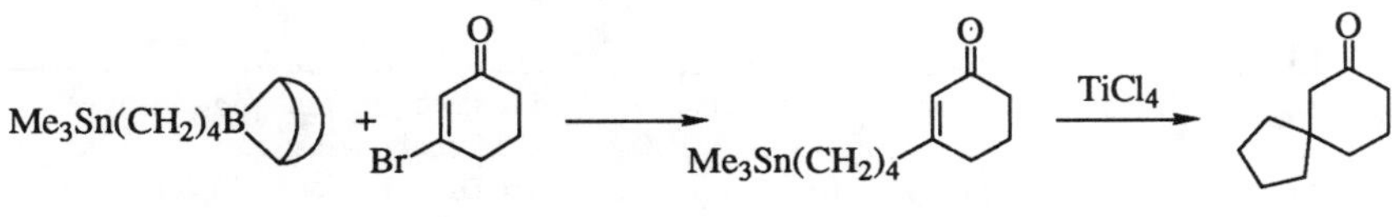

$Me_3Sn(CH_2)_4B\text{<}$ + [bromocyclohexenone] ⟶ $Me_3Sn(CH_2)_4$[cyclohexenone] $\xrightarrow{TiCl_4}$ [spiro ketone]

$$Me_3Sn(CH_2)_2B \text{⟨⟩} + \underset{Br}{\overset{O}{\bigwedge}} \longrightarrow Me_3Sn(CH_2)_2 \overset{O}{\bigwedge} \overset{CF_3COOH}{\longrightarrow} \triangle \overset{O}{\bigwedge}$$

STEREOSELECTIVE SYNTHESIS OF EXOCYCLIC ALKENES VIA HYDROBORATION-COUPLING SEQUENCE (ref. 17)

The synthesis of exocyclic alkenes (ref. 18) is of great importance in natural product synthesis, because there are many biologically and medicinally interesting compounds with such structures, e.g., carbacyclin.

Bromoalkadienes (**7**), the key precursors for the synthesis, were prepared by the procedure as indicated in eq. 11. The palladium-catalyzed intramolecular cross-coupling of **8** gives exocyclic alkenes (eq. 11, ref. 17). Five- and six-membered

$$(11)$$

51 - 83 %

exocyclic derivatives can be easily obtained by such hydroboration-coupling sequence. The rate of hydroboration of alkenes is known to be sterically and electronically affected with substituents on alkenes. Especially, halogen groups give remarkable effect on reducing the rate of hydroboration. Consequently, the reaction of 9-BBN occurs chemoselectively at the unhalogenated C=C bond to give the borane intermediate (**8**).

STEREOSELECTIVE SYNTHESIS OF ALLYLIC AND BENZYLIC BORONATES VIA CROSS-COUPLING OF (DIALKOXYBORYL)-METHYLZINC REAGENTS (refs. 19 and 22)

From the viewpoint of synthetic methodology, the synthesis of allylic boronates has been desired (ref. 20). A variety of allylic boronates were made by the reaction of allylic lithium or magnesium compounds with halo- or alkoxyboron derivatives, or by the alkylation of (halomethyl)boronates with stereodefined 1-alkenyllithium reagents. Most recently, we have been interested in developing a synthetic method of such stereodefined allylic boronates (**11**) via the coupling reaction between (dialkoxyboryl)methylzinc compounds (**10**) and 1-halo-1-alkenes (eq. 12, ref. 19).

$$(RO)_2BCH_2I \overset{\text{"Zn"}}{\longrightarrow} (RO)_2BCH_2ZnI \overset{\underset{R^2}{\overset{R^1}{X \bigwedge} R^3}}{\underset{\text{Pd catalyst}}{\longrightarrow}} (RO)_2B \underset{R^2}{\overset{R^1}{\bigwedge} R^3} \qquad (12)$$

9   **10**   **11**

9a : (RO)$_2$ = -OCMe$_2$CMe$_2$O-   **9b** : (RO)$_2$ = -OCH$_2$CMe$_2$CH$_2$O-   54 - 76 %

Literature survey revealed a few reports on the formation of stable borylmethylmetal reagents which can be utilized for the synthesis of allylic boronates through the cross-coupling reaction. However, the recent success of a direct preparation of borylmethylzinc compounds (**10**) from (iodomethyl)boronates (**9**) and activated zinc (ref. 21) prompted us to use them for the coupling reaction. We have found that not only the pinacol derivative (**9a**) reported by Knochel (ref. 21) but also the 2,2-dimethyl-1,3-propanediol derivative (**9b**) can be synthesized readily in almost quantitative yields. The solution of zinc reagents (**9a** and **9b**) is sufficiently stable to be stored in a cold room.

The cross-coupling reaction of **10** with 1-halo-1-alkenes in the presence of palladium catalyst proceeds through complete retention of configuration of haloalkenes to give stereodefined allylic boronates in good to moderate yields. The potential versatility of such coupling reactions is demonstrated in eqs. 13 and 14.

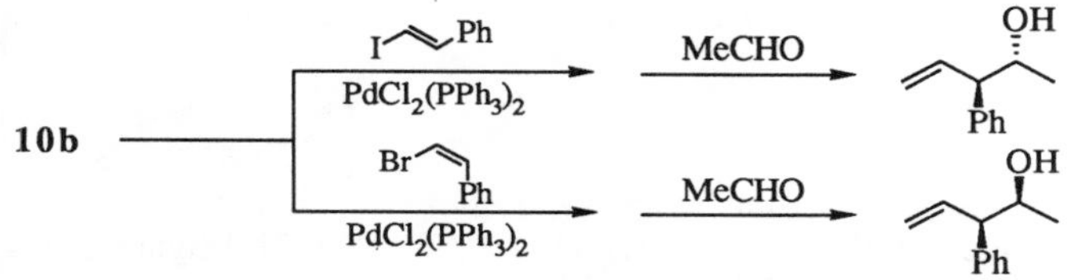

$$10b \quad (13)$$

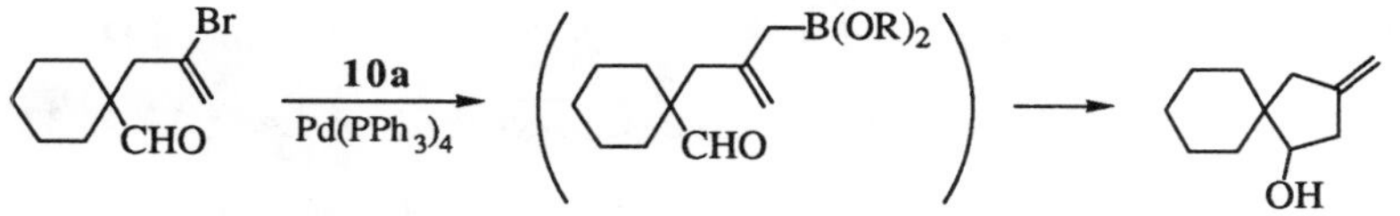

$$\quad (14)$$

The coupling of **10b** with iodoarenes produces the corresponding benzylic boronates in high yields. Benzylic boronates with acyl groups at the *ortho* position readily undergo the 1,5-rearrangement to the carbonyl oxygen giving the *o*-quinodimethane derivatives under thermal or photochemical conditions. The reaction provides benzo-fused cycloalkanes by trapping with dienophiles, one of such examples of which is illustrated in eq. 15 (ref. 22).

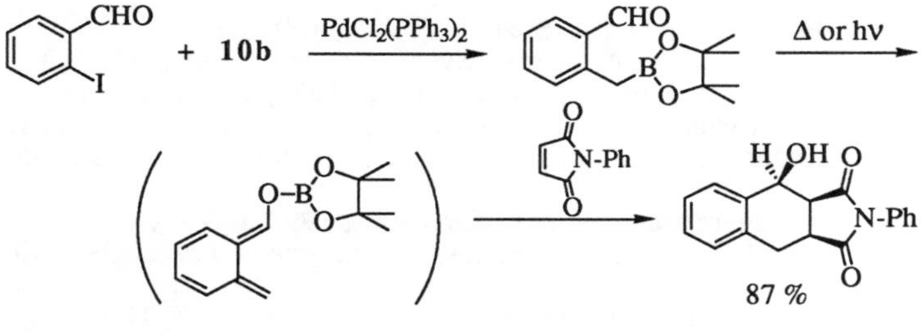

$$\quad (15)$$

87 %

CONCLUSION

In summarizing, we wish to emphasize the usefulness of the coupling reaction in organic synthesis. Although we explored many reactions between organoboranes and organic halides, the problems which should be investigated still remain. The outline of the successful cross-coupling reactions is shown in Scheme 1.

1-Alkynylboron derivatives have never been used in the reaction, because they are unstable under basic conditions. However, as we have perceived most recently a hopeful sign that such compounds react with organic halides under mild basic conditions, the successful result might be reported in the near future.

Scheme 1

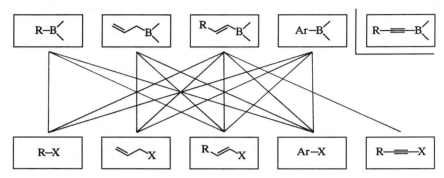

REFERENCES

1. A. Suzuki, *Pure & Appl. Chem.*, 1985, **57**, 1749.
2. A. Suzuki, *Pure & Appl. Chem.*, 1991, **63**, 419; T. Oh-e, N. Miyaura, and A. Suzuki, *J. Org. Chem.*, 1993, **58**, 2201.
3. T. Ishiyama, N. Miyaura, and A. Suzuki, *Bull. Chem. Soc. Jpn.*, 1991, **64**, 1999.
4. T. Ishiyama, N. Miyaura, and A. Suzuki, *Tetrahedron Lett.*, 1991, **32**, 6923.
5. Y. Wakita, T. Yasunaga, M. Akita, and M. Kojima, *J. Organomet. Chem.*, 1986, **301**, C17.
6. D. P. Curran and D. Kim, *Tetrahedron Lett.*, 1986, **27**, 5821; M. Newcomb and D. P. Curran, *Acc. Chem. Res.*, 1988, **21**, 206.
7. T. Ishiyama, S. Abe, N. Miyaura, and A. Suzuki, *Chem. Lett.*, 1992, 691.
8. P. L. Castle and D. A. Widdowson, *Tetrahedron Lett.*, 1986, **27**, 6013.
9. K. Yuan and W. J. Scott, *Tetrahedron Lett.*, 1989, **30**, 4779.
10. T. Ishiyama, T. Oh-e, N. Miyaura, and A. Suzuki, *Tetrahedron Lett.*, 1992, **33**, 4465.
11. T. Watanabe, N. Miyaura, and A. Suzuki, *Synlett.*, 1992, 207.
12. N. Miyaura, T. Yanagi, and A. Suzuki, *Synth. Commun.*, 1981, **11**, 513.
13. W. J. Thompson and J. Gaudino, *J. Org. Chem.*, 1984, **49**, 5237.
14. S. Gronowitz, A-B. Honfeldt, and Y. Yang, *Chem. Scripta*, 1988, **28**, 281.
15. S. Abe, N. Miyaura, and A. Suzuki, *Bull. Chem. Soc. Jpn.*, 1992, **65**, 2863.
16. T. Ishiyama, N. Miyaura, and A. Suzuki, *Synlett.*, 1991, 687.
17. N. Miyaura, M. Ishikawa, and A. Suzuki, *Tetrahedron Lett.*, 1992, **33**, 2571.
18. D. K. Hutchinson and P. L. Fuchs, *J. Am. Chem. Soc.*, 1987, **109**, 4755; M. Sodeoka, S. Satoh, and M. Shibasaki, *ibid.*, 1988, **110**, 4823.
19. T. Watanabe, N. Miyaura, and A. Suzuki, *J. Organomet. Chem.*, 1993, **444**, C1.
20. R. W. Hoffmann, *Angew. Chem. Int. Ed. Engl.*, 1982, **21**, 555; W. R. Roush and R. L. Halterman, *J. Am. Chem. Soc.*, 1986, **108**, 294; H. C. Brown and K. S. Bhat, *ibid.*, 1986, **108**, 293; H. C. Brown and M. V. Rangaishenvi, *Tetrahedron Lett.*, 1990, **31**, 7113.
21. P. Knochel, *J. Am. Chem. Soc.*, 1990, **112**, 7431; M. J. Rozema, A. R. Sidduri, and P. Knochel, *J. Org. Chem.*, 1992, **57**, 1956.
22. G. Kanai, N. Miyaura, and A. Suzuki, *Chem. Lett.*, 1993, 845.

Some Chemistry of Hindered Organoboranes

Andrew Pelter

UNIVERSITY COLLEGE OF SWANSEA, SWANSEA, SA2 9PP, UK

Abstract. Boron stabilised carbanions can be made through the use of highly hindered diarylboranes. Some reactions of the anions so produced are discussed. A spin-off of this work is the production of some stable mono- and diarylboranes which hold great promise as reagents of general application. In a similar fashion, highly hindered and very stable alkyldiarylhydroborates are now available for selective reductions.

I entered the field of hindered diarylorganoboranes due to my interest in extending the chemistry of organoboranes to the chemistry of boron stabilised carbanions. In Table 1 are shown the results of calculations of the stabilisation energies of alkyl [1,2] and alkenyl carbanions [3] adjacent to first row elements. In each case the stabilisation, shown pictorially in Figure 1 for the case of boron, peaks at boron. The stabilisation by a carbonyl group is calculated [2] to be comparable to the stabilisation by an isoelectronic boron atom, and on this basis boron stabilised carbanions should be intermediates of synthetic significance in organic chemistry.

Table 1

Calculated Stabilisation Energies (kJ mol^{-1}) for $CH_2=C-X^{(-)}$ and $X-CH_2^{(-)}$

X	BeH	BH$_2$	CH$_3$	NH$_2$	OH	F
S.E. for $X-CH_2^{(-)}$	-133	-229	+24	+8	-32	-65
S.E. for $CH_2=C-X^{(-)}$	-99	-175	+5	-27	-80	-108

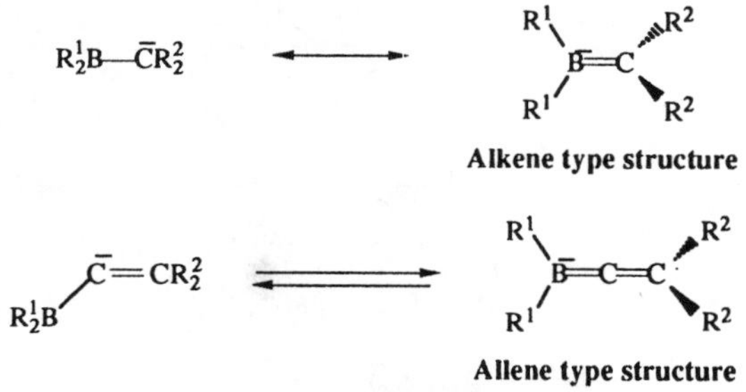

Alkene type structure

Allene type structure

Figure 1

I decided to use the base induced α-deprotonation of organoboranes to produce the required carbanions.   This meant that the formation of organoborates had to be discouraged either electronically or sterically (or both) and in order to take advantage of the rich chemistry of organoboranes a steric hindrance approach was first adopted [4].   Initial experiments soon showed that both the organoboranes and the bases had to be highly hindered if success were to be achieved, and we therefore developed the reagents and methodology shown in Figure 2.   With minor modifications the methodology shown for the dimesitylboryl series applies to both the other sets of compounds.

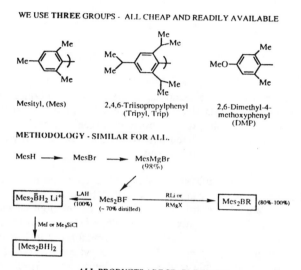

Figure 2

The reactions of bases with alkyldimesitylboranes were then investigated, and a typical study is shown in Figure 3.   In practice mesityllithium is our base of choice as it is more generally applicable than lithium dicyclohexyl-amide and gives only an innocuous hydrocarbon as side product. Mesityllithium is readily available and is of general application as a very hindered base without β-hydrogen atoms.

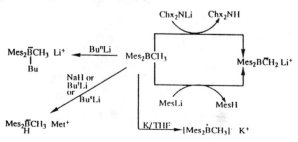

Figure 3

With boron stabilised carbanions readily available, it was now possible to study their physical and chemical properties. An X-ray study [5] of Mes_2BCH_2Li showed a shortening of the C-B bond to that approximating to the length calculated for a C=B, whilst n.m.r. studies showed that the rotational barrier around the C-B bond of the anions approximates to that of the isoelectronic B-N bond [6].

The chemical studies were initiated with a study of the alkylation of dimesitylboryl stabilised carbanions, the results of which are shown in Figure 4 [4].

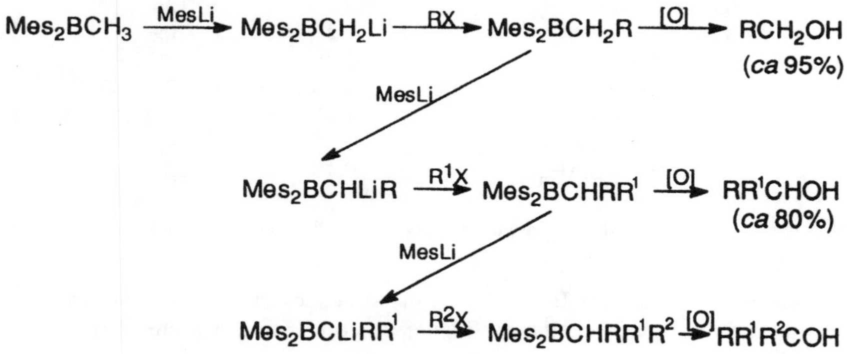

$Mes_2BCH_3 \xrightarrow{MesLi} Mes_2BCH_2Li \xrightarrow{RX} Mes_2BCH_2R \xrightarrow{[O]} RCH_2OH$ (*ca* 95%)

$\xrightarrow{MesLi}$

$Mes_2BCHLiR \xrightarrow{R^1X} Mes_2BCHRR^1 \xrightarrow{[O]} RR^1CHOH$ (*ca* 80%)

$\xrightarrow{MesLi}$

$Mes_2BCLiRR^1 \xrightarrow{R^2X} Mes_2BCHRR^1R^2 \xrightarrow{[O]} RR^1R^2COH$

Figure 4

These reactions provide general syntheses of primary and secondary alcohols in good yields. However, although the production of *tert*-alkyl-dimesitylboranes was surprisingly easy, the release of the *tert*-alkyl group by oxidation became very difficult, even under forcing conditions. For this reason we developed the $(DMP)_2BR$ series in which electrophilic attack is switched from the extremely hindered boron atom to the aromatic group, as shown in equation 1 [7].

$$(DMP)_2BR + 2MeOH \xrightarrow{cat. HX} (MeO)_2BR + 2DMPH \qquad (1)$$

This solvolysis is an extremely easy process, (Table 2), and consequently the DMP group provides a link between the highly hindered boranes and dialkoxyboranes from which boron stabilised carbanions cannot be made [8]. The product alkyldimethoxyborane can then, of course, be readily oxidised to the corresponding alcohol or otherwise manipulated.

Table 2

Methanolysis of Diarylboranes (0.08M) with HCl(0.08M), 1 equiv) in Methanol at 50°C.

Exp.	Compound	Time (h) for removal of 1 or 2 aryl groups	
		1	2
1	DMP_2BMe	0.10	1.1
2	Mes_2BMe	36.4	48(1.15)[a]
3	DMP_2BEt	0.22	1.7
4	DMP_2BBu	0.24	1.6
5	DMP_2Oct	0.30	3.2
6	DMP_2B-2-Pent	1.0	17
7	DMP_2B-3-Dec	6.2	136
8	DMP_2BBu^t	7.2	109
9	Mes_2BBu^t	190	200(1.01)[a]
10	DMP_2B-3-Me-3-Hex	12.6	179
11	$DMP_2BCH=CHBu$	1.1	19
12	$DMP_2BC(Et)=CHEt$	16^b	-
13	$DMP_2BCH_2SiMe_3$	0.15	5.0

[a]Reaction stopped at time shown, with mole equiv. of MDPH released in brackets.
[b]CF_3SO_3H(0.08M) used.

The reactions of Mes_2BCH_2Li with metal halides provide a series of metallated dimesitylborylmethanes (Figure 5), each with a rich chemistry of its own [9].

$$Mes_2BCH_3 \xrightarrow[\text{LiNChx}_2]{\text{MesLi or}} Mes_2BCH_2Li \xrightarrow{R_xMetX} Mes_2BCH_2MetR_x$$

$$Mes_2BF + LiCH_2MetR_x \longrightarrow Mes_2BCH_2MetR_x$$

Compound	m.p.(°C)
$Mes_2CH_2SiMe_3$	72
Mes_2BCH_2SPh	113
$Mes_2BCH_2SnMe_3$	52
$Mes_2BCH_2SnBu_3$	oil
$Mes_2BCH_2SnPh_3$	124
$Mes_2BCH_2PbPh_3$	122
$(Mes_2BCH_2)_2Hg$	180

Figure 5

The reactions with epoxides show that steric factors dominate the position of the attack of the carbanions (Figures 6,7).   Overall the reaction gives ready access to 1,3-diols [10] (equation 2).

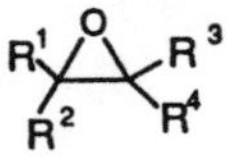

$$\underset{R^2 \quad R^4}{R^1 \overset{O}{\triangle} R^3} + Mes_2B\bar{C}HR\ Li^+ \longrightarrow \xrightarrow{[O]} \begin{array}{c} RCH(OH)C\ R^1R^2C(OH)R^3R^4 \\ + \\ RCH(OH)C\ R^3R^4C(OH)R^1R^2 \end{array} \qquad (2)$$

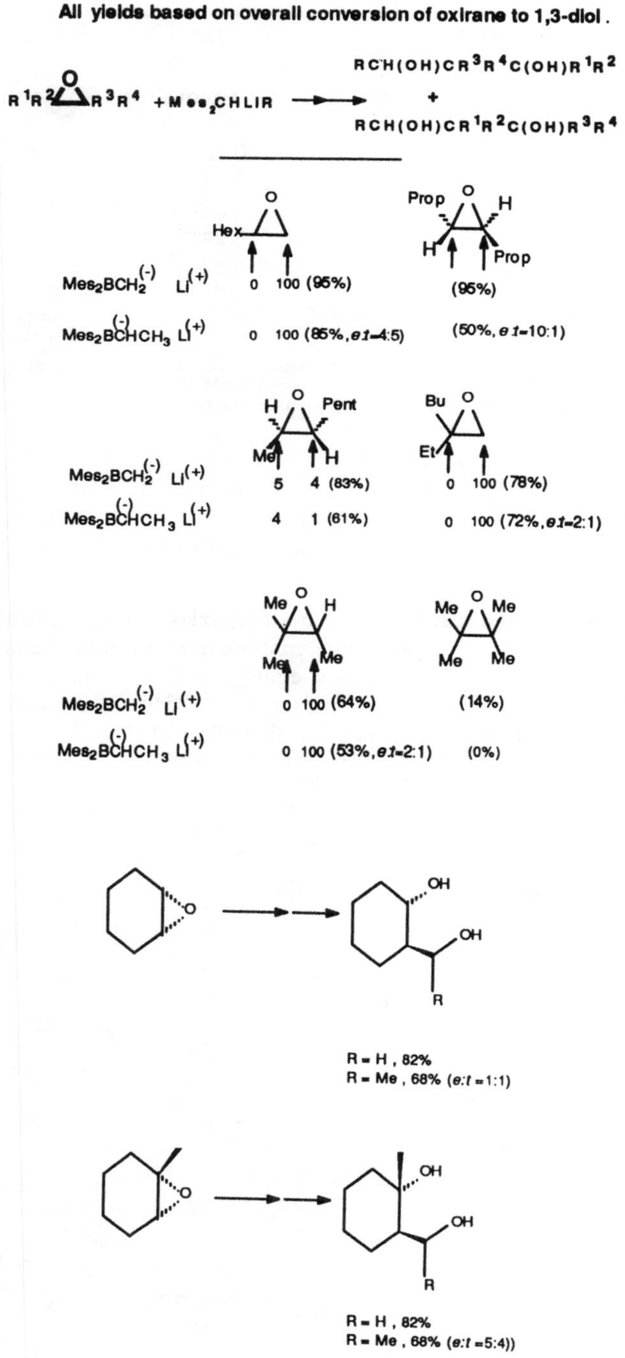

All yields based on overall conversion of oxirane to 1,3-diol .

$$R^1R^2\!\!\!\triangle\!\!\!R^3R^4 + Mes_2CHLiR \longrightarrow \begin{array}{c} RCH(OH)CR^3R^4C(OH)R^1R^2 \\ + \\ RCH(OH)CR^1R^2C(OH)R^3R^4 \end{array}$$

Mes₂BCH₂⁽⁻⁾ Li⁽⁺⁾	0 100 (95%)	(95%)
Mes₂BCHCH₃⁽⁻⁾ Li⁽⁺⁾	0 100 (85%, e.t.=4:5)	(50%, e.t.=10:1)

Mes₂BCH₂⁽⁻⁾ Li⁽⁺⁾	5 4 (83%)	0 100 (78%)
Mes₂BCHCH₃⁽⁻⁾ Li⁽⁺⁾	4 1 (61%)	0 100 (72%, e.t.=2:1)

Mes₂BCH₂⁽⁻⁾ Li⁽⁺⁾	0 100 (64%)	(14%)
Mes₂BCHCH₃⁽⁻⁾ Li⁽⁺⁾	0 100 (53%, e.t.=2:1)	(0%)

R = H , 82%
R = Me , 68% (e:t =1:1)

R = H , 82%
R = Me , 68% (e:t =5:4))

Figure 6

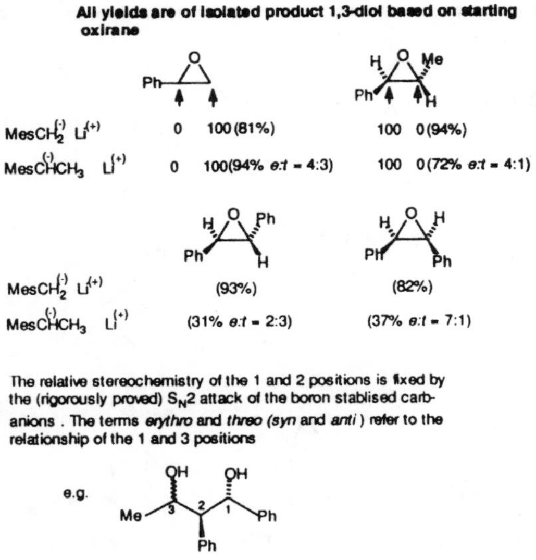
All yields are of isolated product 1,3-diol based on starting oxirane
MesCH2(-) Li(+)
0 100(81%)
100 0(94%)
MesCH(-)CH3 Li(+)
0 100(94% e:t = 4:3)
100 0(72% e:t = 4:1)
MesCH2(-) Li(+)
(93%)
(82%)
MesCH(-)CH3 Li(+)
(31% e:t = 2:3)
(37% e:t = 7:1)
The relative stereochemistry of the 1 and 2 positions is fixed by the (rigorously proved) SN2 attack of the boron stablised carb-anions . The terms erythro and threo (syn and anti) refer to the relationship of the 1 and 3 positions
e.g.

Figure 7

Steric control of the reactions of dimesitylboryl stabilised carbanions is further illustrated by the reactions of the anion derived from allyldimesitylborane (Figure 8). In all cases attack is entirely at C-3, to yield the corresponding *trans*-alkenylborane. This leads to very efficient three carbon homologation reactions as well as to a lactol synthesis [11].

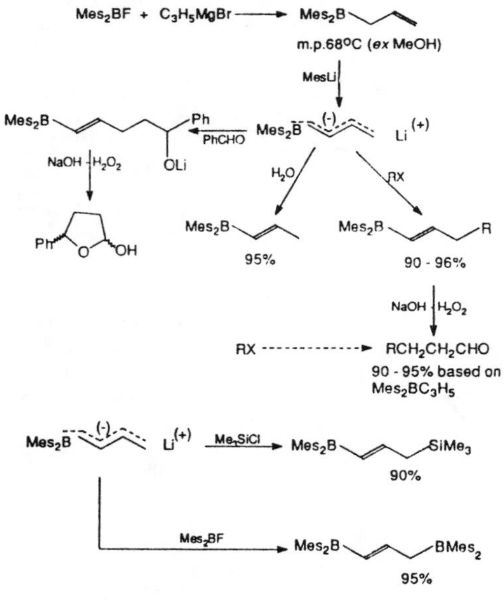
Mes2BF + C3H5MgBr
m.p.68°C (ex MeOH)
MesLi
PhCHO
NaOH H2O2
H2O
RX
95%
90 - 96%
RCH2CH2CHO
90 - 95% based on Mes2BC3H5
Me3SiCl
90%
Mes2BF
95%

Figure 8

Perhaps the most important reactions that we have discovered in this area involve the condensations of boron stabilised carbanions with aldehydes and ketones. The reaction expected was the production of alkenes according to equation 3.

$$Mes_2\overline{B}CHR\ Li^+ + R^1COR^2 \longrightarrow R^1R^2C=CHR + Mes_2OLi \quad (3)$$

We first [12] examined the reactions with symmetrical diaryl ketones so that no stereochemical or enolisation problems would arise. Our results were very satisfactory and are shown in Table 3. They show that a variety of carbanions can be used successfully (exp. 1-3) and that even labile alkenes such as benzofulvene can be prepared by this method (exp. 4,5).

Table 3

The reactions of Mes₂BCHLiR with aromatic ketones

Exp.	Ketone	R	Alkene Product[a]	Conditions	Yield (%)[b]
1	Ph_2CO	H	$Ph_2C=CH_2$	24h, 20°C	75(95)
2	"	CH_3	$Ph_2C=CHCH_3$	1h, 0°C; 12h, 20°C	70(90)
3	"	C_7H_{15}	$Ph_2C=CHC_7H_{15}$	1h, 0°C; 12h, 20°C	70(9))
4	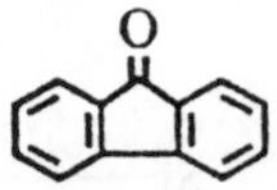	H	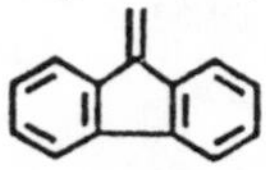	12h; 20°C[c]	80
5	"	CH_3	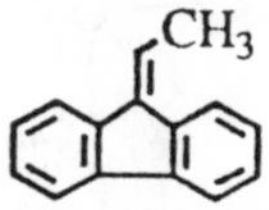	1h,0°C; 12h, 20°C[c]	90

[a]Alkenes either compared directly with authentic samples or fully characterised. [b]Isolated yields of purified products (g.c. yields). [c]Initial product stirred in $CHCl_3$ for 12h/20°C. [d]Initial products stirred in $CHCl_3$ for 3h/20°C.

Benzaldehyde was then examined as a representative aromatic aldehyde. Under the same conditions as used for the ketones no alkene was produced. Lowering the temperature gave rise to some alkene but always less than 50% and always accompanied by ketone, $PhCOCH_2R$, and alcohol $PhCHOHCH_2R$. Our explanation for these products is given in Figure 9.

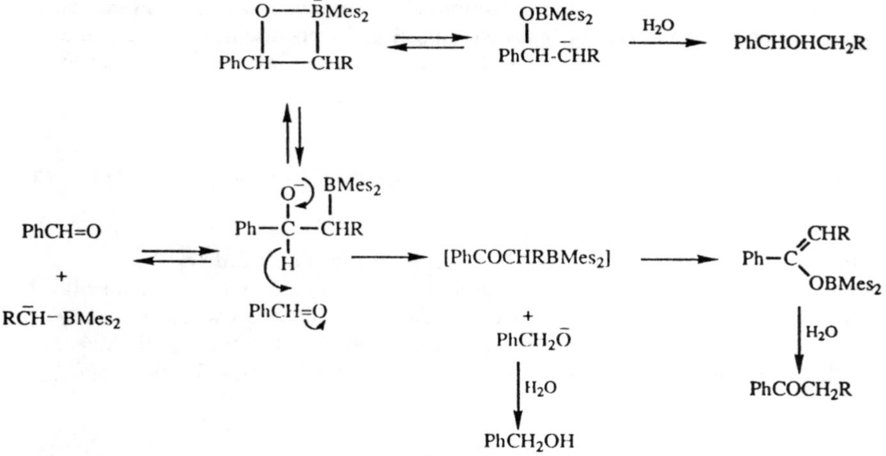

Figure 9

Despite the low yields it was encouraging that the alkene was almost entirely *E*-alkene, in contrast to both the Wittig [13] and Peterson [14] reactions. We had in fact hoped that the reaction would be stereoselective to give *E*-alkenes on the basis of the shorter B-O and B-C bond lengths in a presumed cyclic intermediate as compared with the corresponding P-O and P-C and Si-C bond lengths (Figure 10).

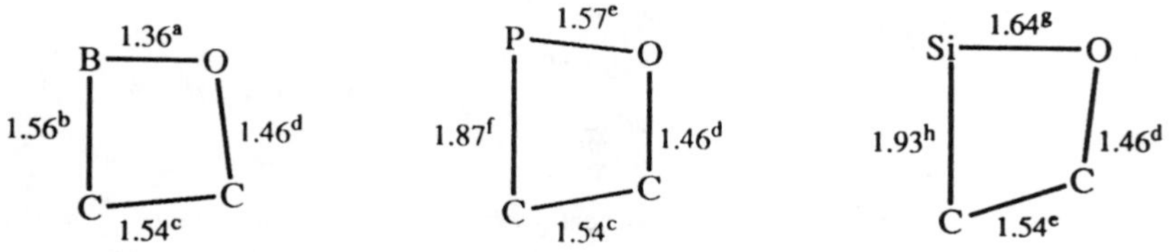

a) in $B(OH)_3$, $B(OMe)_3$; b) in Me_3B ; c) in CH_3-CH_3 ; d) in oxetane ;
e) in $(HO)_3P=O$; f) in Me_3P ; g) in $(Me_3Si)_2O$, h) in Me_4Si.
All bond lengths in Å units, roughly to scale, ignoring bond angles.

Figure 10

In order to gain some insight into the structure of the intermediate that yields alkene, it was decided to carry out some low temperature oxidations using alkaline hydrogen peroxide. This reaction always proceeds with retention [15] and the structure of the 1,2-diol produced would then correlate directly with that of the intermediate. We had expected that the cyclic intermediates that give *E*-alkenes would also give rise to *threo*-1,2-diols. In fact, we obtained the *erythro* isomers!!

It also turned out that **this is an efficient and unique process which gives diols even in those cases that gave no alkene at all** [16] equation 4 (Table 4).

$$Mes_2BCHLiR \ + \ ArCHO \xrightarrow[\text{ii , [O]}]{\text{i , -110°C}} \ \text{erythro-ArCHOHCHOHR} \qquad (4)$$

Table 4
Production of 1,2-Diols.

Entry	Ar	R	Yield % (h, temp)[a]	e:t[b]
1	Ph	Me	84(1h, -120°C)	92:8
2	p-MeC$_6$H$_4$	Me	80(3h, -78°C)	91:9
3	p-ClC$_6$H$_4$	Me	86(1h, -78°C)	91:9
4	p-MeOC$_6$H$_4$	Me	84(6h, -78°C)	>99:<1[c]
5	p-NO$_2$C$_6$H$_4$	Me	78(8h, -120°C)	98:2
6	2,4,6-triMeC$_6$H$_2$	Me	54(6^{1}/$_2$h, - 78°C)	>99:1[c]
7	Ph	Hept^n	80(5h, -78°C)	92:8

[a]All yields are of isolated, characterised diols. [b]Determined by h.p.l.c., ^{13}C n.m.r., ^{1}H n.m.r.
[c]No *threo*-product could be detected by any method.

To explain these results we must assume firstly that the initial condensation is highly stereoselective and secondly that the intermediates in this reaction are not oxaboratetanes that undergo *syn* elimination but instead are acyclic and undergo *anti* elimination (Figure 11).   An alternative, though unlikely explanation is that the B-C bond is oxidised with inversion of configuration.

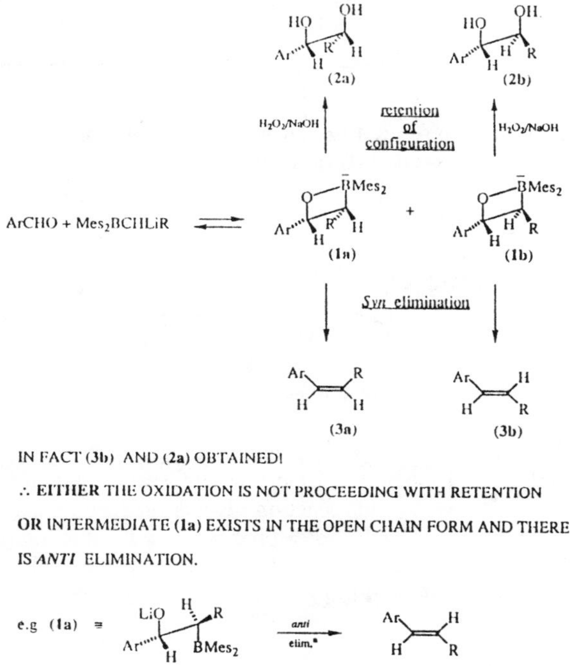

IN FACT (3b) AND (2a) OBTAINED!

∴ EITHER THE OXIDATION IS NOT PROCEEDING WITH RETENTION

OR INTERMEDIATE (1a) EXISTS IN THE OPEN CHAIN FORM AND THERE

IS *ANTI* ELIMINATION.

Figure 11

In order to check the nature of the intermediate, it was decided to trap it, and after some experimentation a procedure using low temperature addition of trimethylsilyl chloride was evolved [17]. The products proved to be stable and isolable and a study of their n.m.r. spectra showed clearly that they had the *erythro* configuration, confirming the inferences drawn from the production of diols. Of particular interest was that the silylated intermediates were produced in high yields in all cases, even for those cases that had given little or no alkenes in the direct condensation. It was therefore decided to use the silylated intermediates to attempt to produce alkenes. In practice this was readily accomplished by the addition of aqueous HF in acetonitrile to either the isolated (procedure B) or non-isolated (procedure A) silylated intermediate (Table 5). **The alkene so produced was always >95% *E*-alkene, which makes our procedure superior to any comparable condensations. Our interpretation of our results is shown in Figure 12.**

Table 5

The synthesis of alkenes by the condensation of Mes$_2$BCHLiR with ArCHO

Expt.	Ar	R	Yield (%)[a]		*E:Z* ratios[b]	
			A	B	A	B
1	Ph	Hept	95	84	84:6	97:3
2	4-MeC$_6$H$_4$	Hept	86	83	89:11	95:5
3	4-ClC$_6$H$_4$	Hept	87	84	74:26	97:3
4	4-MeOC$_6$H$_4$	Hept	84	-	100:0	-
5	4-NO$_2$C$_6$H$_4$	Hept	65	74	93:7	98:2
6	Mesityl	Hept	80	-	100:0	-
7	Ph	Me	93[c]	78[c]	86:16	98:2

[a] Yields are of isolated, characterised products, based on aldehyde.   [b] Determined by g.c. and nmr.   [c] G.c. yield using pure *trans*-2-methylstyrene for comparison.

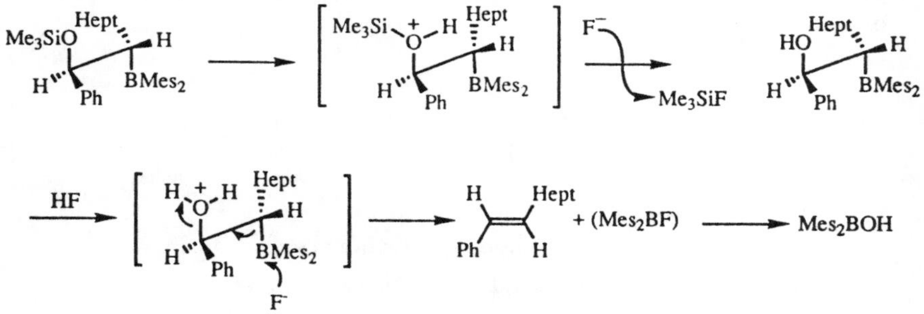

Figure 12

We then wondered whether we could induce *syn* elimination from the initial *erythro* condensation products. An acylated intermediate might favour the required conformation due to attractive forces between the carbonyl oxygen and the boron atom. A unique *syn* elimination, akin to an ester elimination but involving the elimination of an acyloxydiarylborane, was envisaged (Figure 13). In the event, addition of trifluoroacetic anhydride at low temperature yielded alkenes as shown in Table 6 [17].

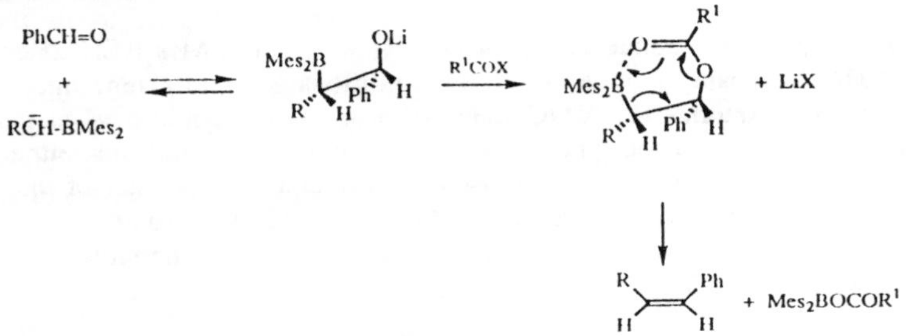

Figure 13

Table 6

The production of Z-alkenes from ArCHO and Mes$_2$BCHLiR

Exp.	Ar	R	Yield (%),[a] (h)[b]	E:Z[c]
1	4-MeC$_6$H$_4$	Hept	74(2)	4:96
2	4-ClC$_6$H$_4$	Hept	73(1.5)	20:80
3	4-MeOC$_6$H$_4$	Hept	76(6)	9:91
4	4-O$_2$NC$_6$H$_4$	Hept	72(7)	69:31
5	Mesityl	Hept	75(6)	7:93
6	Ph	Me	77[d](2)	7:93

[a] isolated, characterised products. [b] reaction time at -110°C. [c] determined by g.c. and ^{1}H nmr. [d] g.c. yield.

There is a strong shift of stereochemistry in all cases and in four experiments (1, 3, 5, 6) this leads to synthetically useful processes. When electron withdrawing groups are present (experiments 2, 4) the selectivity is less, possibly due to the lowering of the attractive forces between the carbonyl oxygen and the boron atom.

The situation as regards aromatic aldehydes is summarised in Figure 14, which shows the flexibility and usefulness of the reactions.

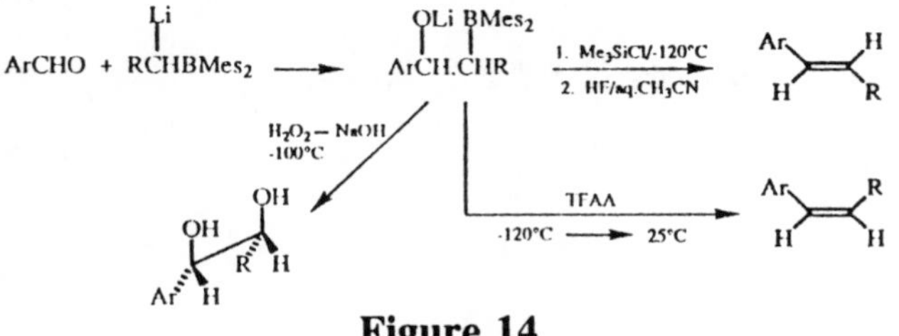

Figure 14

We then examined the reactions of aliphatic aldehydes with boron stabilised carbanions [18,19], with the expectation of producing alkenes as in equation 5.

$$Mes_2B\bar{C}HR\ Li^+ + R^1CHO \longrightarrow R^1CH=CHR + Mes_2OLi \qquad (5)$$

To our surprise, the condensation of octanal with either $Mes_2BCHLiHept$ or $Mes_2BCHLiProp$ using a wide variety of conditions gave no alkene but only low yields of ketones $HeptCOCH_2Hept(Prop)$. These could have arisen from a hydride transfer of the type shown in Figure 9. To avoid this, either TFAA or TMS chloride was added to discharge the anionic character of the intermediate. However, either one of these reagents led to a strong *enhancement* of the yields of ketones and no alkene at all!! The situation is summarised in Table 7.

Table 7

Reactions of MesBCHLiHept with HeptCHO

Expt.	Conditions	Yields (%)[a] of Products		
		HeptCOOct	HeptCH=CHHept	HeptCHOHOct
1	Me_3SiCl, -78°C	85	14	0
2	Mes_3SiCl, -115°C; aq HF/CH_3CN	20	9	55
3	TFAA, -78°C	54	4	0
4	TFAA, -115°C	90	0	0
5	TFAA, -115°C, then feflux	(70)	(17)	0
6	TFAA, -115°C, 5MHCl/5h/60	(60)	0	0
7	TFAA, -115°C, NaOH/	(100)	0	0

[a]Yields are of isolated materials, except those in parentheses, which are g.c. yields.

Our hypothesis for the unexpected course taken by the reaction is shown in Figure 15, in which TFAA acts as a hydride acceptor and so does trifluoroacetaldehyde, the product of the first reduction. This explains the more than 50% yields of ketone in the unique and useful reaction outlined in equation 6.

$$Mes_2B\bar{C}HR\ Li^+ + R^1CHO \xrightarrow{\ i,\ ii\ } R^1COCH_2R \qquad (6)$$
i , TFAA or NCS , -110°C . ii , H_2O .

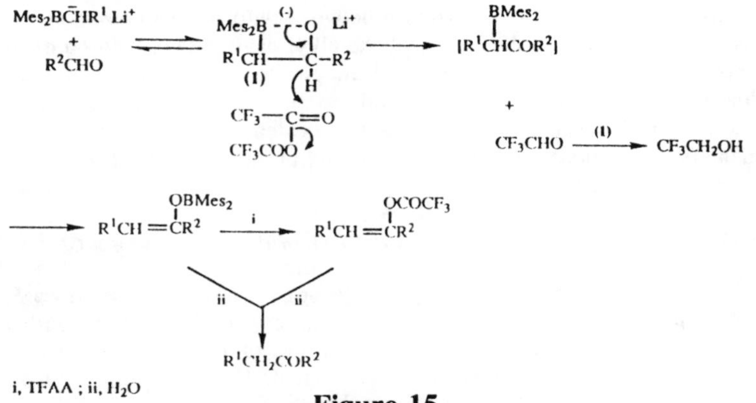

i, TFAA ; ii, H₂O

Figure 15

The process was applied to a wide variety of aldehydes with the results shown in Table 8.

Table 8

Condensation of Mes$_2$BCHLiR^1 and R^{2}CHO in the presence of TFAA

Exp.	R^1	R^2	Yield %[a] R^{1}COCH$_2$R	R^{1}CH=CHR^2
1	Hept	Me	71	0
2	Hept	Et	72	0
3	Hept	Hept	76	0
4	Hept	Bu(Et)CH	85	0
5	Hept	Bu^t	92	0
6	Hept	Cp[b]	89	0
7	Hept	Chx[c]	78	0
8	Pr	PhCH$_2$	75	0
9	Pr	Bu(Et)CH	87	0
10	Pr	Chx	82	0
11	Et	PhCH$_2$	41	0
12	Et	Bu(Et)CH	(41)	0
13	Me	Hept	d	d
14	H	Hept	0	91
15	H	Hept	0	81
16	H	Nonyl	0	74
17	H	Chx	0	(76)

[a]Yields are of isolated, purified, fully characterised compounds, except for g.c. yields given in parentheses. [b]Cp=cyclopentyl. [c]Chx=cyclohexyl. [d]No product is predominant.

The first point to note is that is when R^1 is Hept or Prop, excellent yields of ketones result from condensations with MeCHO, R^{prim}CHO, R^{sec}CHO, R^{tert}CHO and ChxCHO. We assume that anions with alkyl groups higher than propyl will all behave similarly. As the alkyl chain attached to boron becomes short (< four carbon atoms) the course of the reaction is strongly affected. When R^1=Et (exp. 11, 12) the yield of ketone is lowered and when R^1=H, then only alkene is produced. *This is an excellent method for the methylenation of aldehydes* (equation 7).

$$Mes_2BCH_2Li \ + \ RCHO \longrightarrow RCH=CH_2 \qquad (7)$$

It is clear that there is a fine balance between redox and elimination reactions that is affected by the length of the alkyl chain attached to boron. We do not know why this should be, but it may reflect the existence of the initial intermediate in either cyclic or acyclic forms. In any case, the use of a more powerful oxidant than TFAA was indicated, and after some experimentation N-chlorosuccinimide was found to be suitable. The results are given in Table 9.

Table 9

Condensation of $Mes_2BCHLiR^1$ with R^2CHO in the presence of NCS

Expt.	R^1	R^2	Product	Yield (%)[a]
1	Hept	Hept	HeptCOOct	73
2	Et	PhCH$_2$	PhCH$_2$COPr	72
3	Et	Bu(Et)CH	Bu(Et)CHCOPr	79
4	Me	Hept	HeptCOEt	72
5	Me	PhCH$_2$	PhCH$_2$COEt	57
6	Me	Bu(Et)CH	Bu(Et)CHCOEt	68
7	H	Hept	HeptCOCH$_3$	(28)[b]
8	H	Bu(Et)CH	Bu(Et)CHCOCH$_3$	21[c]
9	H	Chx	ChxCOCH$_3$	(29)[b]
10	H	PhCH$_2$	PhCH$_2$COCH$_3$	44[d]

[a]Yields of isolated, purified, fully characterised product. G.c. yields in parentheses. [b]The rest of the product was alkene by g.c. [c]Corresponding alkene (0.524g, 72%) was isolated. [d]NCS-dimethyl sulfide used.

Thus with the exception of R^1=H, the process shown in equation 6 is a unique and general method for the direct conversion of aldehydes to ketones. When R^1=H, then the general methylenation process of equation 7 results.

If the hydride transfer depends on the anionic character of the initial condensation product, then neutralisation of the charge by a reagent that was not itself reduced should result in elimination to give alkene. Therefore the aldehyde was added together with a proton source [19]. We had previously [20] used water to control the polycondensation of an anion with an aldehyde and, subsequent to this work, a brief mention appeared of the use of acetic acid to control the stereochemistry of a Wittig reaction [21]. We further argued that a strong acid might yield a protonated intermediate that would orient itself so as to give *anti*-elimination (Figure 16).

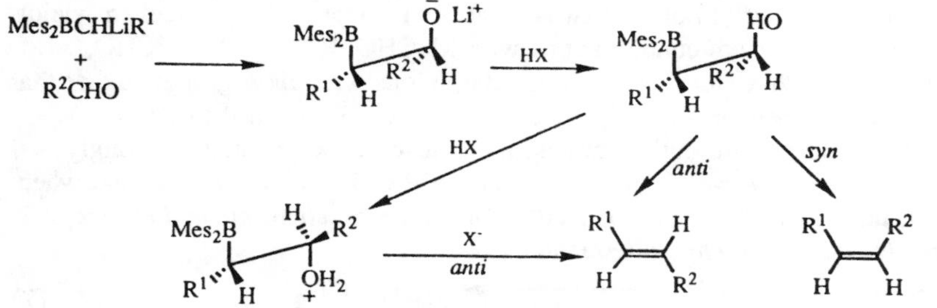

Figure 16

We were pleased to find (Table 10) that condensation did occur, *even* in the presence of strong acids, and that the products are **E-alkenes in good isolated yields in all cases.** Moreover, with $R^{sec}CHO$ and $R^{ter}CHO$, **use of weak acids leads to a high preference for Z-alkenes.** Some of the reversals are quite remarkable (see exp. 1,2; 9,10; 11,12; 13,14) and contrast greatly with both the Wittig reaction [13] and the Schlosser modification of the Wittig reaction [22]. Our results using the latter two reactions are given in Table 11.

Table 10
Condensation of $Mes_2BCHLiR^1$ with R^2CHO in the presence of HX to give $R^1CH=CHR^2$

Exp.	R^1	R^2	HX	Equiv. of HX	% Yield[a]	E:Z[b]
1	Me	$PhCH_2$	CH_3CO_2H	1.0	74	14 :86
2	Me	$PhCH_2$	CF_3SO_3H	1.2	69	93:7[c]
3	Me	Chx	CH_3CO_2H	2.0	67	63:3
4	Me	Chx	HCl	1.2	72	97:3
5	Me	Bu(Et)CH	CH_3CO_2H	1.0	52	24:76
6	Me	Bu(Et)CH	CF_3SO_3H	1.0	57	95:5
7	Me	Oct	CH_3CO_2H	1.0	77	44:56
8	Me	Oct	CF_3SO_3H	1.0	48	90:10
9	Pr	Chx	CH_3CO_2H	1.0	61	17:83
10	Pr	Chx	HCl	1.0	64	95:5
11	Pr	Bu(Et)CH	CH_3CO_2H	1.0	74	4:96[d]
12	Pr	Bu(Et)CH	CF_3SO_3H	1.0	59	92:8[d]
13	Hept	Bu^t	CH_3CO_2H	1.0	79	0:100
14	Hept	Bu^t	HCl	1.0	72	92:8
15	Hept	Et	CH_3CO_2H	1.0	75	49:51
16	Hept	Et	CF_3SO_3H	5.0	49	91:9
17	Hept	Chx	CH_3CO_2H	1.0	83	10:90

a) Yields of isolated, purified alkenes. b) Capillary g.c./i.r. c) The figures underlined are those with ≥ 90% of E or Z-alkenes. d) Established by ^{13}C nmr/ir.

Table 11
Ratios of E:Z alkenes from the condensations of $Ph_3PCHHept$ with R^2CHO to yield $HeptCH=CHR^2$

	Wittig reaction		Schlosser modification	
R^2	E:Z[a,b]	Yield %[c]	E:Z[a,d]	Yield %[c]
Me	34:66	67	77:23	42
Et	0:100[e]	73	100:0[e]	46
Hept	16:84	81	86:14	61
Me_2C	7:93	72	76:24	48
Bu^t	0:100	70	0:100	59
Chx	6:94	80	47:53	52

a) 1 equivalent of PhLi. b) Established by capillary g.c. c) Isolated, purified product. d) 2 equivalents of PhLi. e) Established by ^{13}C nmr/ir.

The trend of the Schlosser modification is to a reversal of stereo-chemistry, this becomes less marked or non-existent with increasing bulk of the aldehyde. In only one case was there a synthetically useful reversal of stereochemistry. Thus our process does not need any isolation and separation of intermediates as required by the Peterson reaction and proceeds with far better stereoselectivity than the Schlosser modification of the Wittig reaction for the production of *E*-alkenes. As such it should find general application.

Boron stabilised alkenyl carbanions [3] condense with carbonyl compounds to give allenes [23] according to equation 8. The results are given in Table 12.

$$RC{\equiv}C\text{-}SnMe_3 \longrightarrow \longrightarrow [RCH = \bar{C}BR_2^1]\,Li^+ \xrightarrow{\ R^2CHO\ } RCH = C = CR^2 \qquad (8)$$

Table 12

Expt.	R^1	R^2	R^3	Yield (%)[a] of allene (eq. 5)	
				GC	Isolated
1	Ph	Mes[b]	Ph	72(75)[c]	61
2	Hex	Chx[d]	Ph	71(76)	63
3	Hex	Mes	Ph	76	65
4	Ph	Mes	4-MeC$_6$H$_4$	67	49
5	Hex	Chx	4-MeC$_6$H$_4$	69	52
6	Ph	Mes	4-MeOC$_6$H$_4$	41(52)[e]	22
7	Hex	Chx	4-MeOC$_6$H$_4$	50(59)[e]	29
8	Ph	Mes	4-O$_2$NC$_6$H$_4$	74	58
9	Hex	Chx	4-O$_2$NC$_6$H$_4$	82	66
10	Ph	Mes	4-BrC$_6$H$_4$	7(9)	-
11	Hex	Chx	4-BrC$_6$H$_4$	4(7)	-
12	Ph	Mes	4-ClC$_6$H$_4$	11(10)	-
13	Hex	Chx	4-ClC$_6$H$_4$	10(6)	-
14	Ph	Mes	C$_9$H$_{19}$	64(72)	51
15	Hex	Mes	C$_9$H$_{19}$	74	59
16	Hex	Chx	C$_9$H$_{19}$	68(75)	61
17	Ph	Mes	Me$_2$CH	71	57
18	Hex	Chx	Me$_2$CH	73	50
19	Ph	Mes	Chx	75	56
20	Hex	Mes	Chx	76	53
21	Ph	Mes	Me$_3$C	61	46
22	Hex	Chx	Me$_3$C	57	48
23	Ph	Mes	PhCH$_2$	15(36)	4
24	Hex	Chx	PhCH$_2$	12(44)	4

[a]All yields are for the overall process, and are based on starting alkynes. [b]Mes = mesityl (2,4,6-trimethylphenyl). [c]Figures in parentheses are GC yields after addition of CdCl$_2$ to the lithio-anion. [d]Chx = cyclohexyl. [e]Yields after stirring at room temperature for 12h.

Yields are moderate in most cases (sometimes due to isolation problems), but nevertheless the reaction is a unique and useful allene synthesis.

As an offshoot of this work we had access to a variety of hindered diarylboranes from which to produce novel hydroborating and reducing agents as shown in Figure 17.

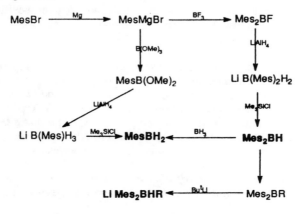

Figure 17

In particular, we have studied the following compounds.

$(Mes_2BH)_2$	$(MesBH_2)_2$	$Trip_2BH$	$(TripBH_2)_2$	$Trip_2BHEt^{(-)} Li^{(+)}$
(1)	**(2)**	**(3)**	**(4)**	**(5)**

Dimesitylborane (1), now commercially available, proved to be a most useful reagent that solved some selectivity problems that had defeated any other borane [24]. The reagent slowly hydroborated alk-1-enes but hardly touched internal alkenes at room temperature. Reactions with internal or terminal alkynes were relatively fast and in competition experiments any alkyne was hydroborated in preference to any alkene. *Thus the reagent is highly chemoselective.* Moreover, dimesitylborane proved to be the most regioselective of known hydroborating agents for the hydroboration of unsymmetrical internal alkynes (Table 13). As such it is already finding a niche in organic synthesis.

Table 13

Hydroboration of Internal Alkynes by Borane and Diorganylboranes

Reagent	Pr^n-C≡C-Me		Ph-C≡C-Me	
	↑	↑	↑	↑
$BH_3.THF$	40	60	74	26
Sia_2BH	39	61	19	81
9-BBN	27	73	63	35
Mes_2BH	10	90	2	98
$Trip_2BH$	50	50	10	90

Our hopes of finding an even more selective hydroborating agent in ditripylborane (3) foundered on the fact that this compound proved to be *the only known stable monomeric diorganylborane* [25]. Unlike (1), compound (3) was very soluble in organic solvents, even light petroleum, but unfortunately proved to be one the *least* selective of known hydroborating agents [26] (Table 13). It does however provide an interesting substrate for the study of the reactions of organoboranes. For example, hydroborations with (1) cannot proceed through an unsolvated monomer or else they would have a low regioselectivity, as evidenced by (3).

In contrast to thexylborane, which must be prepared at low temperatures and used immediately [27], monomesitylborane (2) and monotripylborane (4) [28] are readily available and stable reagents. Thus crystalline (2) had 93% of its original activity after 10 days at room temperature, and a THF solution retained 76% activity after two weeks standing. *For most purposes they can be used as replacement reagents for thexylborane but have some special properties of their own that are not paralleled by thexylborane.*

The first of these is that *both tripylborane and mesitylborane are unique amongst organoboranes in that they hydroborate terminal alkenes in a stepwise fashion* (equation 9).

$$\text{MesBH}_2 \xrightarrow{\text{R}^A\text{CH=CH}_2} \text{MesBH(CH}_2)_2\text{R}^A \xrightarrow{\text{R}^B\text{CH=CH}_2} \text{MesB[(CH}_2)_2\text{R}^A][(CH}_2)_2\text{R}^B] \qquad (9)$$

The regiospecificity in the first step is >99% and it is quantitative in the second step. Some results are given in Table 14.

Table 14

The hydroboration of alkenes with MesBH$_2$

$$\text{MesBH}_2 \rightarrow \text{MesBHR}^A \rightarrow \text{MesBR}^A\text{R}^B \rightarrow \text{MesOH} + \text{R}^A\text{OH} + \text{R}^B\text{OH}$$

Alkene A	Alkene B	Yield (%)[a]		
		MesOH	R^{A}OH	R^{B}OH
Oct-1-ene		88	91	
Hex-1-ene		94	91	
Cyclohexene		92	90	
2-Methylpent-1-ene		94	87	
1-Methylcyclopentene		91	73	
Oct-1-ene	Hex-1-ene	92	93	91
Oct-1-ene	Chx[b]	93	89	93
α-Pinene	Hex-1-ene	93	88	93
α-Pinene	Chx	91	c	43

[a] g.c. yields.   [b] Chx = cyclohexyl.   [c] 81% of α-Pinene.

Mesitylborane can add two primary alkenes *and also two secondary alkenes, without displacement of the hindered aromatic group.* However, attempts to add cyclohexene to MesBH)Ipc) led to substantial displacement of α-pinene.

In order to prove that these hydroboration products could be of synthetic use it was necessary to test the migratory aptitude of the mesityl group relative to the various alkyl groups used. To do this we used the cyanoborate (cyanidation) process [29] as shown in equation 10.

$$MesBR^AR^B \longrightarrow R^AR^BCO + MesOH \qquad (10)$$

Our results (Table 15) show that indeed the mesityl group has the low migratory aptitude required for synthetic use. In no case could we detect any products derived from the migration of a mesityl group.

Table 15
The Cyanidation of MesBR^{A}R^B

Experiment	Alkene A	Alkene B	R^{A}COR^B(%)	MesOH(%)
1	Hex-1-ene	Hex-1-ene	81	84
2	Chx	Chx	87	93
3	Chx	Oct-1-ene	77	91
4	Oct-1-ene	Hex-1-ene	86	92

Both primary and secondary alkyl groups migrate in preference to a mesityl (or tripyl) group. The unique result of the use of either mesityl or tripylboranes is illustrated by experiment 4, in which an *unsymmetrical* linear ketone is produced. This clearly greatly enlarges the scope of reactions such as the cyanidation of organoborates. Cyclic hydroborations followed by cyanidation are illustrated in equation 11. The equivalent reaction with thexylborane also goes but gives a much less pure product, perhaps due to the initial hydroboration being less selective.

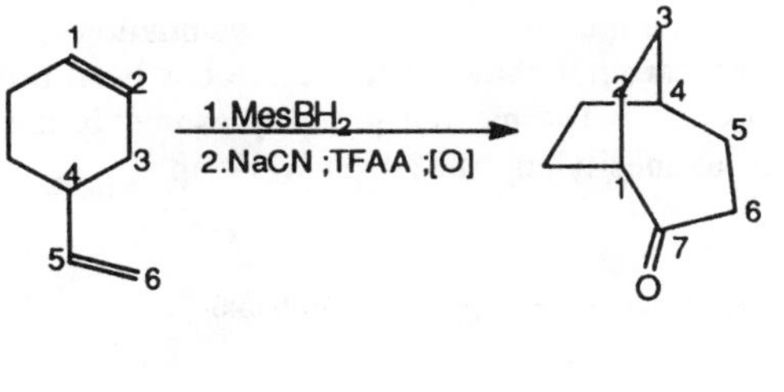

$$\qquad \qquad (11)$$

At this point the hydroboration/cyanidation of limonene (equation 12) was attempted using mesitylborane. However this sequence, which was successful with thexylborane [30], *essentially failed with mesitylborane (3% yield)*. This is due to the greater facility of thexylborane, as compared with mesitylborane, for the hydroboration of more hindered double bonds, and defines at least one area in which thexylborane will retain its use.

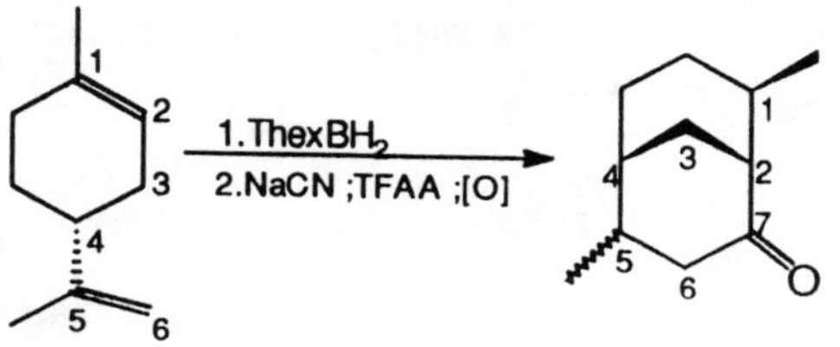

$$\text{(12)}$$

We next examined the displacement of the aromatic group from hydroboration products and found two unusual and highly selective processes that further enhance the usage of mesitylborane. The first is illustrated in equation 13.

$$\text{MesBR}^A\text{R}^B \xrightarrow{\text{Br}_2 / \text{MeOH}} \text{MesBr} + \text{MeOBR}^A\text{R}^B \xrightarrow{\text{[O]}} \text{R}^A\text{OH} + \text{R}^B\text{OH} \qquad \text{(13)}$$

Yields of alcohols are between 82-87%. More importantly, the intermediate dialkylmethoxyborane can be reacted directly with a Grignard reagent to give a fully mixed organoborane, in which all the alkyl groups are primary, and which in turn can be readily converted to the corresponding carbinol [31] (equation 14).

$$\text{MesBR}^A\text{R}^B \xrightarrow{\text{Br}_2 / \text{MeOH}} \text{MeOBR}^A\text{R}^B \xrightarrow{\text{R}^C\text{MgBr}} \text{BR}^A\text{R}^B\text{R}^C \xrightarrow[\text{ii.[O]}]{\text{i. NaCN, ii .TFAA}} \text{R}^A\text{R}^B\text{R}^C\text{C(OH)} \qquad \text{(14)}$$

In the case in which $R^A = n$-octyl, $R^B = n$-hexyl, $R^C = n$-butyl, the overall yield of isolated carbinol, *based on MesBH$_2$*, is 69%, an excellent yield given the seven steps involved.

A different method of releasing the mesityl or tripyl groups is through direct reaction with alkyl or aryllithiums (equation 15). Cyanidation converts the product to the corresponding carbinol showing that there has been little, if any, scrambling of the alkyl groups. The mesityl group, presumably displaced as mesityllithium, is recovered as mesitylene (88%) on work-up.

$$\text{MesB(Oct)(Hex)} \xrightarrow{\text{BuLi}} \text{BuB(Oct)(Hex)} \longrightarrow \longrightarrow \text{HOC(Bu)(Hex)(Oct)} \qquad \text{(15)}$$

We are currently beginning a study of the properties of compounds such as $Trip_2BHEt^{(-)} Li^{(+)}$ [31]. These are extremely stable compounds, readily made either by the action of t-butyllithium or a metal hydride on the borane. In their reductions of ketones they are highly selective (Table 16). They possess one very great advantage over, for example, lithium Selectride, in that there is no need for an oxidation step, and the *product borane is so stable that on aqueous work-up it can be recovered almost quantitatively in the light petrol fraction from a silica column. The boranes can also be used catalytically.*

Table 16

Reductions using $Trip_2\bar{B}HMe$ Li⁺ or $Trip_2\bar{B}HEt$ Li⁺

Ketone	Product Alcohol	Selectivity
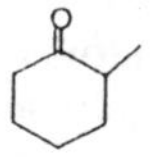	cis	>99:1
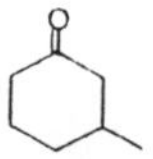	trans	99:1
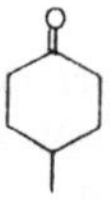	cis	97.5:2.5
	cis	>99.9:1

In the same vein we find very ready reductions of alkyl halides, the bromides and iodides in particular, to the corresponding alkanes with easy work-up and recovery of the borane. Other types of reduction will be reported in due course.

Summary. It is clear that hindered organoboranes offer a wide variety of new reagents and reactions to enlarge both the theory and practise of organic chemistry.

This work owes much to various collaborations, in particular with Dr. John Wilson and Professor Keith Smith and to many dedicated colleagues including Drs. B. Singaram, L. Warren, E. Colclough, M. Rowlands, G. Vaughan-Williams, S. Elgendy, A. Norbury and Mr. Zhao Jin. We thank the SERC for financial support.

References.

1. G. W. Spitznagel, T. Clark, J. Chandrasekhar and P. von R. Schleyer, *J. Comput. Chem.,* 1982, **3**, 363.
2. A. Pross, D. J. DeFrees, B. A. Levi, S. K. Pollack, L. Radom, W. J. Hehre, *J. Org. Chem.,* 1981, **46**, 1693.
3. A. Pelter, K. Smith, D. E. Parry and K. D. Jones, *Aust. J. Chem.,* 1992, **45**, 57.
4. A. Pelter, B. Singaram, L. Warren and J. W. Wilson, *Tetrahedron,* 1993, **49**, 2965.
5. M. M. Olstead, P. P. Power and K. J. Weese, *J. Am. Chem. Soc.,* 1987, **109**, 2541.
6. N. M. D. Brown, F. Davidson and J. W. Wilson, *J. Organomet. Chem.,* 1981, **210**, 1.
7. A. Pelter, R. Drake and M. Stewart, *Tetrahedron Lett.,* 1989, **30**, 3085.
8. D. S. Matteson and R. J. Moody, *Organometallics,* 1982, **1**, 20.
9. J. W. Wilson, A. Pelter, M. V. Garad and R. Pardasani, *Tetrahedron,* 1993, **49**, 2979.
10. A. Pelter, G. F. Vaughan-Williams and R. M. Rosser, *Tetrahedron,* 1993, **49**, 3007.
11. A. Pelter, B. Singaram and J. W. Wilson, *Tetrahedron Lett.,* 1983, **24**, 631.
12. A. Pelter, B. Singaram and J. W. Wilson, *Tetrahedron Lett.,* 1983, **24**, 635.
13. B. E. Maryanoff and A. B. Reitz, *Chem. Rev.,* 1989, **89**, 853.
14. F. F. Hudrlik, D. Peterson and R. J. Rona, *J. Org. Chem.,* 1975, **40**, 2263.
15. A. Pelter, K. Smith and H. C. Brown, 'Borane Reagents', Academic Press, London, UK, 1988.
16. A. Pelter. D. Buss and A. Pitchford, *Tetrahedron Lett.,* 1985, **26**, 5093.
17. A. Pelter, D. Buss and E. Colclough, *Tetrahedron Lett.,* 1987, **28**, 297.
18. A. Pelter, K. Smith, S. Elgendy and M. Rowlands, *Tetrahedron Lett.,* 1989, **30**, 5643.
19. A. Pelter, K. Smith, S. Elgendy and M. Rowlands, *Tetrahedron Lett.,* 1989, **30**, 5647.
20. A. Pelter, R. T. H. Al-Bayati, M. T. Ayoub, W. Lewis and P. Pardasani, *J. Chem. Soc., Perkin Trans. 1,* 1987, 717.
21. B. M. Trost, S. M. Mignani and T. N. Naninga, *J. Am. Chem. Soc.,* 1988, **110**, 1602.
22. M Schlosser and H. B. Tuong, *Angew Chem. Int. Ed. Engl.,* 1979, **18**, 633.
23. A. Pelter, K. Smith and K. D. Jones, *J. Chem. Soc., Perkin Trans. 1,* 1992, 747.
24. A. Pelter, S. Singaram and H. C. Brown, *Tetrahedron Lett.,* 1983, **24**, 1433.
25. R. A. Bartlett, H. V. R. Dias, M. M. Olmstead, P. P. Power and K. J. Weese, *Organometallics,* 1990, **9**, 146.
26. A. Pelter, K. Smith, D. Buss and A. Norbury, *Tetrahedron Lett.,* 1991, **32**, 6239.
27. H. C. Brown, 'Organic Syntheses *via* Boranes', Wiley Interscience 1975, New York.
28. K. Smith, A. Pelter and Zhao Jin, *J. Chem. Soc., Perkin Trans. 1,* 1993, 395.

29. A. Pelter, K. Smith and Zhao Jin, Unpublished results.
30. A. Pelter, K. Smith, M. G. Hutchings and K. Rowe, *J. Chem. Soc., Perkin 1*, 1975, 129.
31. A. Pelter, M. G. Hutchings, K. Rowe and K. Smith, *J. Chem. Soc., Perkin 1*, 1975, 138.
32. K. Smith, A. Pelter and A. Norbury, *Tetrahedron Lett.*, 1991, **32**, 6243.

Syntheses of Functionalized Allylic and Benzylic Boronates via Palladium-catalyzed Cross-coupling Reaction of Knochel's (Dialkoxyboryl)methylzinc Reagents with Organic Halides

N. Miyaura and A. Suzuki

DEPARTMENT OF APPLIED CHEMISTRY, FACULTY OF ENGINEERING, HOKKAIDO UNIVERSITY, SAPPORO 060, JAPAN

Although allylic boron compounds have proven to be exceptionally useful in the diastereoselective functionalization of carbonyl compounds, very little information is available for the intramolecular reaction and their stereochemical behaviors. Also, the reaction of benzylic boron compounds has been virtually unknown chemistry. We wish to describe the new synthesis of functionalized allylic or benzylic boronates by the cross-coupling reaction of the Knochel's borylmethyl zinc reagent **1** with 1-alkenyl or aryl halides (eqs 1 and 6). Our interest has focused on the potential ability of this new reagent in the synthesis of novel boron reagents which are not previously available by the conventional techniques. Although we previously reported the cross-coupling reactions of organoboron compounds with organic halides in the presence of palladium catalyst and base, the selective coupling between organic halides and the C-Zn bond is readily achieved because the C-B bond is quite inactive for the cross-coupling reaction under neutral conditions.[1]

SYNTHESIS OF OXO-2-ALKENYLBORONATES AND THEIR INTRAMOLECULAR ALLYLBORATION[2]

Allylboration of aldehydes and ketones with allylic boron reagents and the utility of the resulting homoallylic alcohols in the synthesis of complex molecules have been amply demonstrated.[3] The allylic boron compounds are most conveniently prepared by the reaction of allylmagnesium or -lithium reagents with alkoxyboranes or haloboranes. Although these approaches have been successfully applied for the synthesis of a variety of allylic boronates, they suffer from a general lack of regio- and stereoselectivity as well as inability to include functional groups sensitive to the magnesium or lithium reagents. Thus, the synthesis of oxo-2-alkenylboronates which is required for intramolecular allylboration has not been reported until very recently.[4] The palladium-catalyzed cross-coupling reaction of 1-halo-1-alkene with the Knochel's zinc reagent **1**[5] provides (*E*)- and (*Z*)- allylic boronates with high regio- and stereoselectivity (eq 1).[2] The mildness of the zinc reagents tolerates the presence of a wide variety of functional groups including aldehydes and ketones. Thus, the direct synthesis of such oxoallylboronates can be accomplished for the first successful intramolecular allylboration.

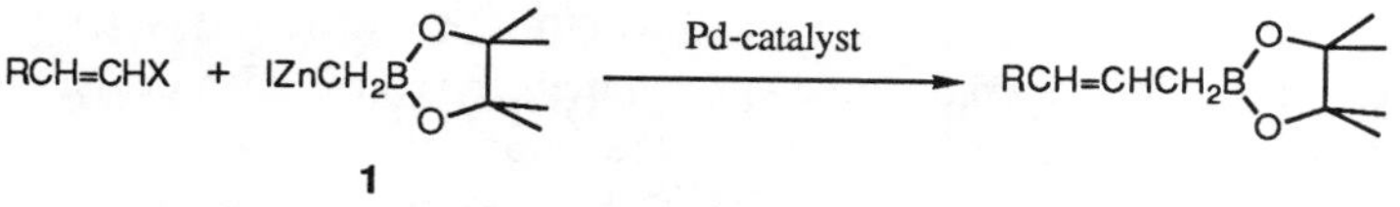

$$\text{RCH=CHX} \ + \ \text{IZnCH}_2\text{B} \underset{\mathbf{1}}{\overset{O}{\diagdown}} \xrightarrow{\text{Pd-catalyst}} \text{RCH=CHCH}_2\text{B} \overset{O}{\diagdown} \qquad (1)$$

The intramolecular allylboration reaction of oxo-2-alkenylboronates generated *in situ* by the cross-coupling reaction of **1** with halocarbonyl compounds in benzene at 70 °C in the presence of palladium catalyst (3 mol %) gives five and seven-membered exomethylene spirocyclic or bicyclic homoallyl alcohols (eqs 2-4). A rather fast coupling reaction under mild conditions is desired owing to the instability of the zinc reagent and presumably allylboronates under the coupling conditions. The use of three equivs of triphenylphosphine to palladium acetate is recognized to be more effective than Pd(PPh$_3$)$_4$. Recently, triphenylarsine was

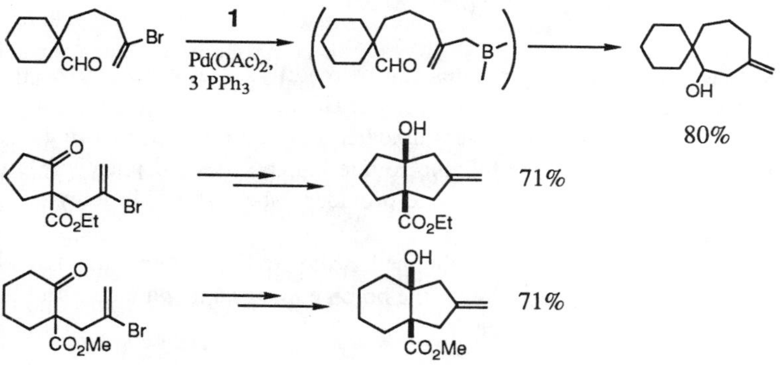

recommended as a new ligand in the palladium-catalyzed tin coupling reaction at extremely fast coupling rates. This ligand is also occasionally useful to improve the coupling yields. The intermolecular allylboration with pinacol borates is usually rather slow due to its steric bulkiness; however, the intramolecular reaction may proceed under same conditions used for the cross-coupling.

Cyclization reaction should be highly valuable if the new ring can be generated with defined stereochemistry. The cyclization of **2** at 70 °C in the presence of **1** (1.5 equiv), Pd(DBA) (3 mol %), and Ph$_3$As (12 mol %) provides directly a *trans*-vinyl alcohol **3** with high diastereoselectivity (eq. 5).

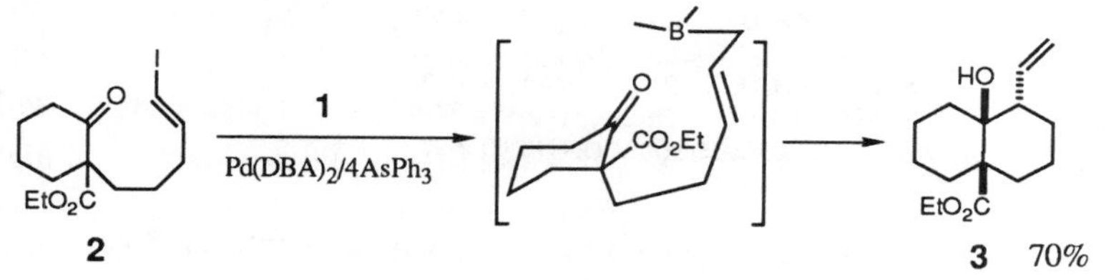

SYNTHESIS OF BENZYLIC BORONATES AND THEIR
TRANSFORMATION TO ORTHO-QUINODIMETHANES[6]

Synthetic methods for the generation of *o*-quinodimethanes have been developed by a number of research groups because of the importance of these reactive intermediates for the inter- or intramolecular Diels-Alder reaction with dienophiles.[7] Benzylic boronates having an acyl group at the *ortho*-position provide a new route to such *o*-quinodimethanes *via* their facile rearrangement to the boron enolates **6** (eqs. 6 and 7).

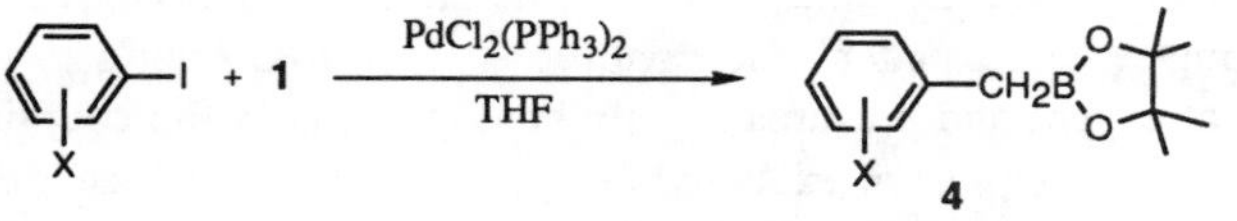

$$X = H \ (91\%); \ p\text{-MeO} \ (80\%); \ p\text{-MeO}_2C \ (76\%); \ o\text{-MeO}_2C \ (90\%);$$
$$o\text{-NC} \ (67\%); \ o\text{-Br} \ (69\%); \ o\text{-CHO} \ (78\%); \ o\text{-MeCO} \ (86\%)$$

A variety of benzylic boronates can be readily synthesized by the cross-coupling reaction of **1** with haloarenes in THF at 50 °C in the presence of PdCl$_2$(PPh$_3$)$_2$ (2-4 mol %). When less hindered boronic esters are desired, (2,2-dimethylpropanedioxyboryl)methylzinc iodide can be used in place of **1**; for example, the coupling with iodobenzene provides 2-benzyl-5,5-dimethyl-1,3,2-dioxaboronane in 90% yield.

Among benzylboronates thus obtained, *ortho*-acyl derivatives are thermally rather unstable due to the generation of the boron enolates **6** that are confirmed by trapping *in situ* with dienophiles (eq. 7).

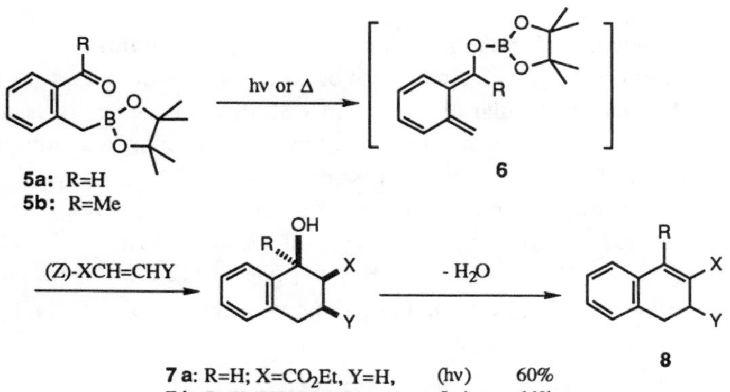

5a: R=H
5b: R=Me

7 a: R=H; X=CO$_2$Et, Y=H,   (hv)   60%
7 b: R=H; X,Y=CO$_2$Et,   (hv)   66%
7 c: R=H; X=COMe, Y=H   (hv)   65%
7 d: R=H; X,Y=CON(Ph)CO,   (110°C) 87%
7 e: R=Me; X,Y=CON(Ph)CO, (110°C) 92%

Heating a mixture of **5** and an alkyne or alkene dienophiles (3 equiv.) in toluene for 24 h at 110-180 ° affords the naphthalene or dihydronaphthalene derivatives by dehydration. In contrast, the adducts with N-phenylmaleimide are stable to permit isolation of the *cis*-alcohols **7d** and **7e** stereoselectively.

The rearrangement to boron enolate can be carried out under milder conditions by irradiation to a benzene solution of **5** in a pyrex flask with a 500W-high pressure mercury lamp for 5 h at 15 °C. Although the adducts with alkyne dienophiles still have a tendency to cause aromatization, the cycloaddition products with alkene dienophiles are less labile to dehydration. Thus, the isolation of primary adducts **7a-c** which are not available by the thermal reaction is readily achieved. The photochemical reaction again produces the cycloaddition products of all *cis*-configurations.

When the 1,5-rearrangement proceeds through coordination of boron to the carbonyl oxygen, the (Z)-boron enolate can be initially produced. However, the stereochemistry and regiochemistry of products both in the thermal and the photochemical reactions suggest the formation of (E)-quinodimethane **6** which give the *cis*-adducts through the *endo*-addition[6] of dienophiles. The photochemical rearrangement can be initiated by the intramolecular attack of an excited triplet carbonyl (n-π*) on boron, in analogy with the related photochemical S_H2 reaction of organoboranes with ketones.[8] This rearrangement is not thermally accelerated by radical initiators or is not inhibited by radical scavengers. Thus, the thermal reaction may take place in another process, presumably the intramolecular migration through the coordination of boron to the carbonyl oxygen, followed by the isomerization to (E)-*o*-quinodimethane.

In conclusion, the Knochel's zinc reagent provides a new access to functionalized organoboron compounds in combinations with the cross-coupling reaction with organic halides as well as the copper(I) induced addition reaction[5] reported by his group. The procedure appears to better tolerate various functional group variations than the conventional transmetallation methods using Grignard or lithium reagents.

REFERENCES

1. N. Miyaura, K. Yamada, H. Suginome, and A. Suzuki, *J. Am. Chem. Soc.*, **107**, 972 (1985); N. Miyaura, T. Ishiyama, H. Sasaki, M. Ishikawa, and A. Suzuki, *Ibid.*, **111**, 314 (1989).
2. T. Watanabe, N. Miyaura and A. Suzuki, *J. Organomet. Chem.*, **444**, C1 (1993).
3. R. W. Hoffmann, *Angew. Chem. Int. Ed. Engl.*, **21**, 555 (1982); D. S. Matteson, *Tetrahedron*, **45**, 1859 (1989)
4. R. W. Hoffmann and G. Niel, *Liebigs Ann. Chem.*, 1195 (1991).
5. P. Knochel, *J. Am. Chem. Soc.*, **112**, 7431 (1990).
6. G. Kanai, N. Miyaura and A. Suzuki, *Chem. Lett.*, 845 (1993).
7. C. W. Jefford, G. Bernardinelli, Y. Wang, D. C. Spellmeyer, A. Buda, and K. N. Houk, *J. Am. Chem. Soc.*, **114**, 1157 (1992); W. Carruthers "Cycloaddition Reactions in Organic Synthesis," Pergamon Press, New York (1990).
8. A. G. Davies and B. P. Roberts, "Bimolecular Homolytic Substitution at Metal Centers," in "Free Radicals," ed by J. K. Kochi, John Wiely & Sons Ltd., New York (1973), Vol. 1, Chap. 10, p. 564.

Stereoselective Synthesis of Enediynes and Enyne-allenes via γ-(Trialkylsilyl)allenylboranes

Kung K. Wang*, Yemane W. Andemichael, Yu Gui Gu, and Zhongguo Wang

DEPARTMENT OF CHEMISTRY, WEST VIRGINIA UNIVERSITY, MORGANTOWN, WV 26506, USA

Abstract: Condensation of γ-(*tert*-butyldimethylsilyl)allenylborane **2** with acetylenic aldehydes **3** followed by the Peterson olefination reaction furnished the corresponding enediynes **6** and **7** with high geometric purity. Similarly, enyne-allenes **11** and **12** were synthesized by treatment of γ-(trimethylsilyl)allenylborane **8** with allenic aldehydes **9**. This synthetic route provided an easy way for the preparation of enyne-allenes **13** and **18**, which on heating each underwent a sequence of remarkable intramolecular transformations with a cascade of energy.

The discovery of the very potent antitumor antibiotics calicheamicin, esperamicin, dynemicin A, and neocarzinostatin[1] possessing cyclic enediyne and related structures has stimulated renewed interest in the Bergman cyclization reaction.[2] Simple acyclic enyne-allenes were recently reported to undergo facile Myers cyclization reaction to produce α,3-dehydrotoluene biradicals at ambient or even subambient temperatures.[3] The palladium-catalyzed coupling reactions have been employed as one of the main synthetic tools for the construction of enediyne structure.[1,4] The Ramberg-Backlund reaction has been utilized in the synthesis of cyclic enediynes.[1] The condensation reaction between conjugated acetylenic aldehydes and ynenolates derived from α-trimethylsilyl α-allenyl carbonyl compounds is also a recent example.[5] Enyne-allenes are generally produced from acetylene to allene rearrangement of the corresponding enediyne propargylic alcohols.[3,6] Our synthetic approaches involve the use of condensation reactions between γ-(trialkylsilyl)allenylboranes and acetylenic/allenic aldehydes followed by the Peterson olefination reaction.

Allenylborane **2** was prepared by sequential treatment of 3-(*tert*-butyldimethylsilyl)-1-(trimethylsilyl)-1-propyne (**1**)[7] with *n*-butyllithium followed by *B*-methoxy-9-borabicyclo[3.3.1]nonane (*B*-MeO-9-BBN) and 4/3 $BF_3 \cdot OEt_2$ (Scheme 1).[8] Condensation of acetylenic aldehydes **3** with allenylborane **2** followed by treatment with 2-aminoethanol afforded the condensation adducts **5** with high diastereoselectivity (de > 96%) having essentially only the *SR/RS* pair. The high diastereoselectivity in forming **5** could be attributed to the preferential adoption of the *tert*-butyldimethylsilyl group and the acetylenic group of the aldehydes on the opposite sides of the six-membered pericyclic transition state **4** in order to minimize nonbonded steric interactions.[7,9] Subsequent treatment of **5** with KH to promote the Peterson olefination reaction furnished enediynes **6** (6:7 > 98:2) having only the Z isomer, whereas treatment of **5** with concentrated H_2SO_4 produced **7** (7:6 > 98:2) exclusively.

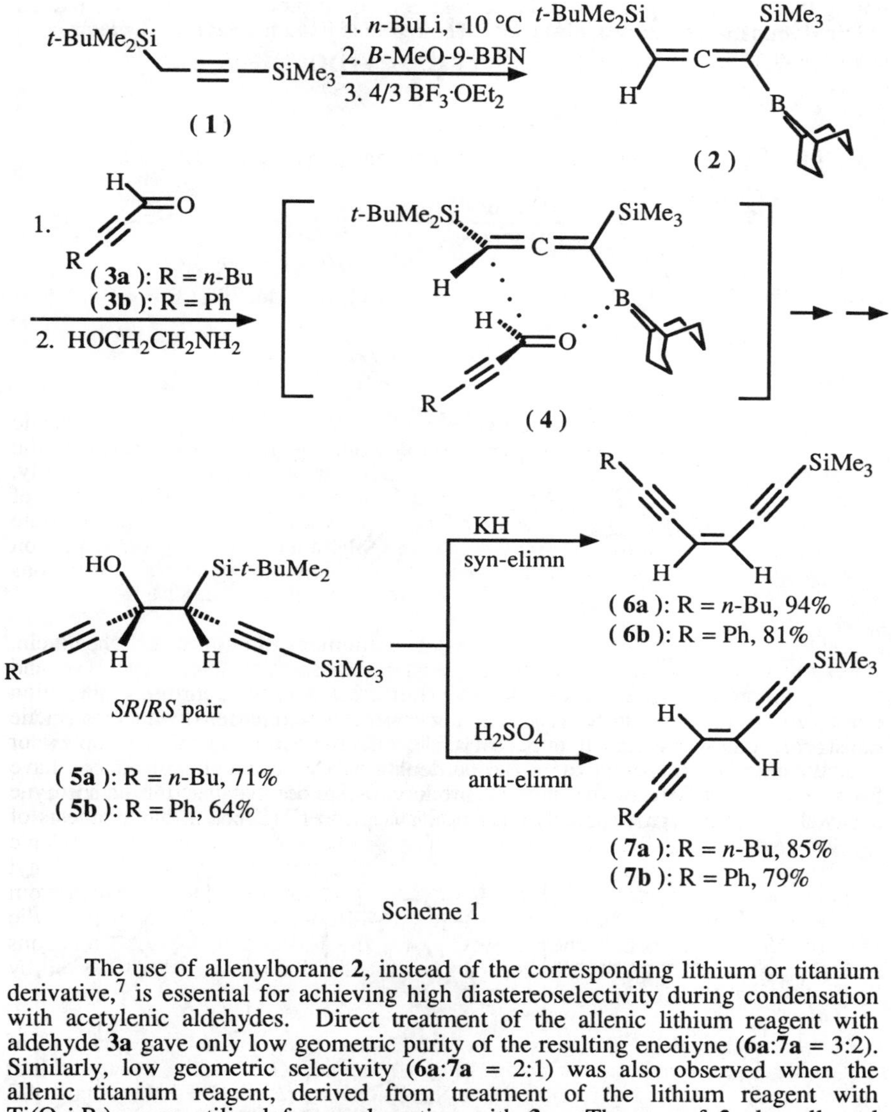

Scheme 1

The use of allenylborane **2**, instead of the corresponding lithium or titanium derivative,[7] is essential for achieving high diastereoselectivity during condensation with acetylenic aldehydes. Direct treatment of the allenic lithium reagent with aldehyde **3a** gave only low geometric purity of the resulting enediyne (**6a**:**7a** = 3:2). Similarly, low geometric selectivity (**6a**:**7a** = 2:1) was also observed when the allenic titanium reagent, derived from treatment of the lithium reagent with Ti(O-i-Pr)$_4$, was utilized for condensation with **3a**. The use of **2** also allowed isolation of the condensation adducts **5**, making it possible to prepare both the E and the Z isomers **6** and **7** with high geometric purity.

Similarly, γ-(trimethylsilyl)allenylborane **8** was synthesized from 2-(trimethylsilyl)-2,3-pentadiene.[10] Attempts to prepare the corresponding enediynes by condensation with acetylenic aldehydes resulted only in low diastereoselectivity during the condensation step, producing enediynes with low geometric purity after the Peterson olefination reaction.[11] Fortunately, condensation of allenic aldehydes **9** with allenylborane **8** proceeded smoothly and in high diastereoselectivity, leading to enyne-allenes **11** (E:Z > 96%) and **12** (Z:E > 96%) with high geometric purity and in excellent isolated yield (Scheme 2).[10] It is worth noting that a surprising reversed diastereoselectivity in producing the *RR/SS* pair as the predominant isomer was observed. The reason for such a reversal has not been determined.

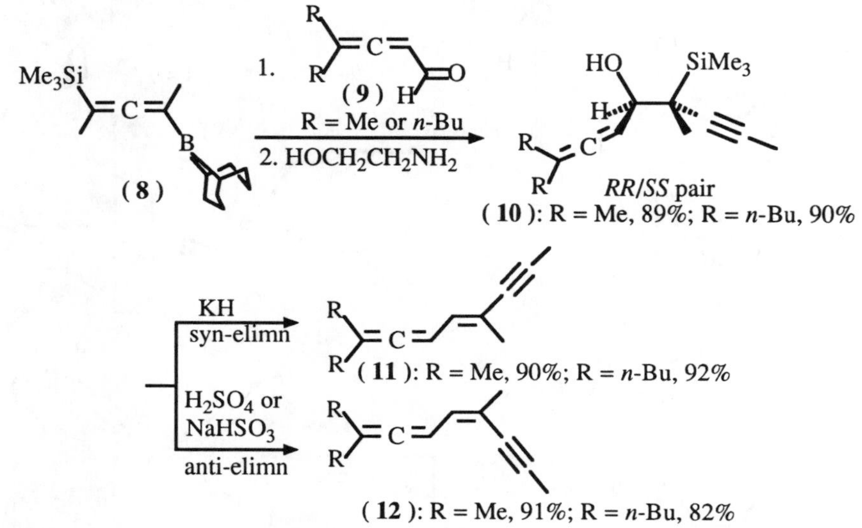

Scheme 2

The reaction sequence outlined in Scheme 2 was utilized for the synthesis of enyne-allene **13**.[10] On heating, **13** underwent a sequence of intramolecular transformations involving an initial Myers cyclization, trapping of the benzenoid radical center in **14** by the carbon-carbon double bond intramolecularly to form **15**, followed by a 1,5-hydrogen shift to produce *o*-quinodimethane **16**, which then decayed by a [1,5]-sigmatropic hydrogen shift to afford **17** (Scheme 3).

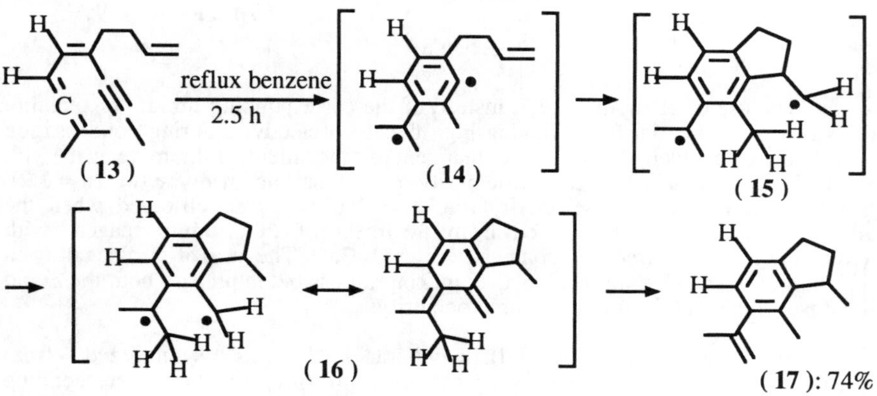

Scheme 3

By incorporating an additional carbon-carbon double bond as shown in **18**, *o*-quinodimethane **19** was captured in an intramolecular Diels-Alder reaction to furnish **20** (Scheme 4).[12] This thermally-induced chemical transformation represents a new synthetic strategy for a one-step 0 → ABCD ring construction of the tetracyclic steroidal skeleton having an aromatic C-ring.

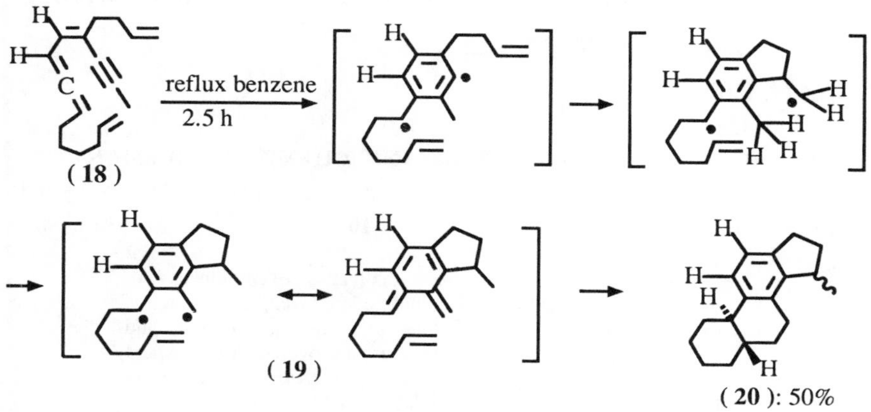

(18)

(19)

(20): 50%

Scheme 4

REFERENCES

1.   K.C. Nicolaou and W.-H. Dai, *Angew. Chem., Int. Ed. Engl.*, **1991**, *30*, 1387 and references cited therein.
2.   R.R. Jones and R.G. Bergman, *J. Am. Chem. Soc.*, **1972**, *94*, 660.
3.   (a) A.G. Myers, E.Y. Kuo, and N.S. Finney, *J. Am. Chem. Soc.*, **1989**, *111*, 8057.
      (b) A.G. Myers and P.S. Dragovich, *J. Am. Chem. Soc.*, **1989**, *111*, 9130.
      (c) R. Nagata, H. Yamanaka, E. Murahashi, and I. Saito, *Tetrahedron Lett.*, **1990**, *31*, 2907.
      (d) R. Nagata, H. Yamanaka, E. Okazaki, and I. Saito, *Tetrahedron Lett* , **1989**, *30*, 4995.
4.   P.A. Magriotis and K.D. Kim, *J. Am. Chem. Soc.*, **1993**, *115*, 2972.
5.   N.A. Petasis and K.A. Teets, *Tetrahedron Lett.*, **1993**, *34*, 805.
6.   A.G. Myers, N.S. Finney, E.Y. Kuo, *Tetrahedron Lett.*, **1989**, *30*, 5747.
7.   (a) K. Furuta, M. Ishiguro, R. Haruta, N. Ikeda, and H. Yamamoto, *Bull. Chem. Soc. Jpn.*, **1984**, *57*, 2768.
      (b) Y. Yamakado, M. Ishiguro, N. Ikeda, and H. Yamamoto, *J. Am. Chem. Soc.*, **1981**, *103*, 5568.
8.   H.C. Brown and J.A. Sinclair, *J. Organomet. Chem.*, **1977**, *131*, 163.
9.   E.J. Corey, and C. Rucker, *Tetrahedron Lett.*, **1982**, *23*, 719.
10.   Y.W. Andemichael, Y.G. Gu, and K.K. Wang, *J. Org. Chem.*, **1992**, *57*, 794.
11.   Y.G. Gu, Ph.D. Thesis, Department of Chemistry, West Virginia University, May 1991.
12.   Y.W. Andemichael, Y. Huang, and K.K. Wang, *J. Org. Chem.*, **1993**, *58*, 1651.

Reductive Mono- and *trans*-α,α'-Diallylation of Aromatic Nitrogen Heterocycles by Allylboranes

Yu. N. Bubnov

N.D. ZELINSKY INSTITUTE OF ORGANIC CHEMISTRY, RUSSIAN ACADEMY OF SCIENCES, 117913 MOSCOW, B-334, LENINSKY PROSP., 47, RUSSIA

Abstract: Reductive *trans*-α,α'-diallylation of pyridines, 4,4'-dipyridyl, pyrrole and isoquinoline as well as reductive monoallylation of pyrrole, indole, quinolines, isoquinoline and phenanthridine by allylic boranes were discovered. A convenient method for *trans*- → *cis*-isomerization of *trans*-2,6-diallyl-Δ^3-piperideines and *trans*-2,5-diallylpyrrolidine was found.

Last two decades have seen dramatic development in synthetic application of β,γ-unsaturated (allylic) boron compounds [1,2]. One of the most important types of allylborane reactions is the additions to organic compounds with multiple bonds (C=O, C=S, C=N, C≡N, C=C, C≡C). Such allylboration reactions proceed regio- and stereoselectively (2π+2π+2σ processes) and, with a proper choice of reagents, enantioselectively [1,2]. Deboration of the boron-containing adducts results in homoallylic alcohols, thiols, amines, 1,4-dienes, 1,4-enynes, etc.

As a part of our program on the use of organoboranes in synthesis [2—4], we have studied the transformations of certain aromatic nitrogen heterocycles under the action of allylic boranes and have found a series of new reactions [5,6], that unite heterocyclic and organoboron chemistries on the novel basis.

1. Reductive *trans*-diallylation of pyridines

Triallylborane reacts readily with pyridine [7a], C_5D_5N and 3-bromopyridine to form the corresponding complexes **1a—c** (Table 1). Adduct **1a** left unchanged on heating at 160°C for 20 hrs. Its IR-, Raman- [7d] and NMR spectra [7b,c] have been previously described.

Table 1. Triallylborane complexes

Complex	b.p., °C (torr)	n_D^{20}	$\delta^{11}B$
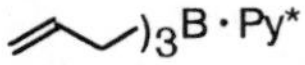)$_3$B·Py* **1a**	102 (1)	1.4535	0
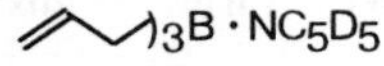)$_3$B·NC$_5$D$_5$ **1b**	103–104 (1)	1.5409	—0.60
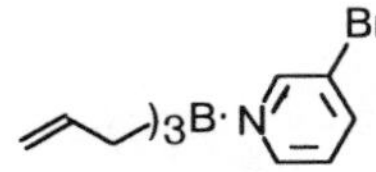)$_3$B·N (Br) **1c**	106 (1)	1.5643	—0.3

* d_4^{20} 0.932; μ=4.97D [7c]; m.p. 14–15°C.

We have found that pyridine adduct **1a** is easily transformed into *trans*-2,6-diallyl-1,2,5,6-tetrahydropyridine (*trans*-2,6-diallyl-Δ^3-piperideine) **2** in a 40—92% yield on treatment with alcohols, water or Et_2NH at 40—100°C for 2—8 hrs. Admixture of *cis*-isomer (0.5—3%) in **2** thus obtained is easily separated by chromatography on SiO_2. The yield of **2** reaches 97% if **1a** is heated (80—100°C) with *t*-BuOH or *i*-PrOH (4 equ) in the presence of pyridine (1 equ).

Scheme 1

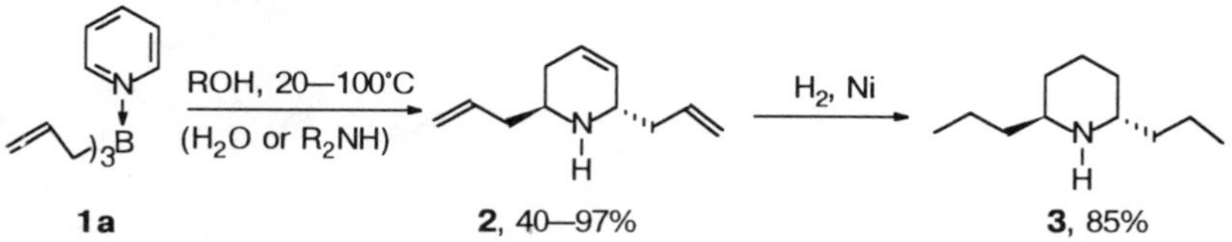

1a	**2**, 40—97%	**3**, 85%

The reaction can be carried out in ether, THF, hydrocarbons, CCl_4, etc., or without any solvent (Table 2).

Table 2. Yield of **2** (All_3B:Py:ROH = 1:1:4)

Entry	ROH (4 equ)	Solvent	T, °C	Reaction time (hs)	Yield of 2 (%
1	MeOH	THF	20	96	57
2	MeOH	ether	40—50	2—6	35—50
3	MeOH	C_6H_6	80	4	43
4	MeOH[a]	ether	45	4	63
5	EtOH	ether	60	3	50
6	EtOH	—	85	4	53
7	EtOH[a]	ether	45	4	60
8	i-PrOH	—	90	8	70
9	i-PrOH[a]	—	100	2	97
10	t-BuOH	—	95	6	85
11	t-BuOH	—	95	8	92
12	$(CH_2OH)_2$	—	90	6	36
13	Et_2NH	—	70	16	23
14	H_2O	THF	40	8	40
15	(-)-Menthol[b]	ether	45	10	66
16	MeOH[c]	ether	45	4	40

[a])Ratio triallylborane:Py:ROH = 1:2:4; [b])For **2** obtained $[\alpha]_D^{23}$ −7.30° (c = 10.00, CH_2Cl_2); [c])Allyl(dipropyl)borane was used instead of triallylborane.

More convenient preparation of **2** consists in one-pot procedure without isolation of complex **1a** as well as **1b** or **1c** (see below). A mixture of triallylborane and pyridine (1:1 or 1:2) is usually heated with 3—4 equivalents of an alcohol. After the reaction is completed, the reaction mixture is stirred with 1.2—1.3 equ. of 10% NaOH, all boron compounds (allylboronic acid and others) being transferred into aqueous layer. The product is extracted by ether or hexane. Another procedure consists in a treatment of a reaction mixture (All_3B, Py, 4 ROH) with mono- or triethanolamine followed by distilling off **2** or its extraction with a hydrocarbon solvent.

Compound **2** is also obtained in 40% yield by interaction of pyridine with allyl(dipropyl)borane in the presence of methanol (4 equ) in ether (45°C, 4 h) (entry 16).

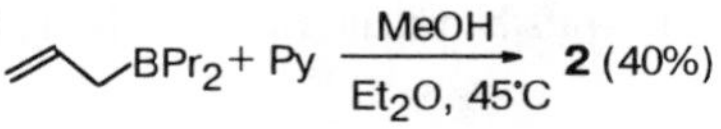

Hydrogenation of **2** in CH_3COOH over Raney nickel (100 at. H_2, 90—100°C, 6 h) leads to *trans*-2,6-dipropylpiperidine **3** (85%).

Scheme 2

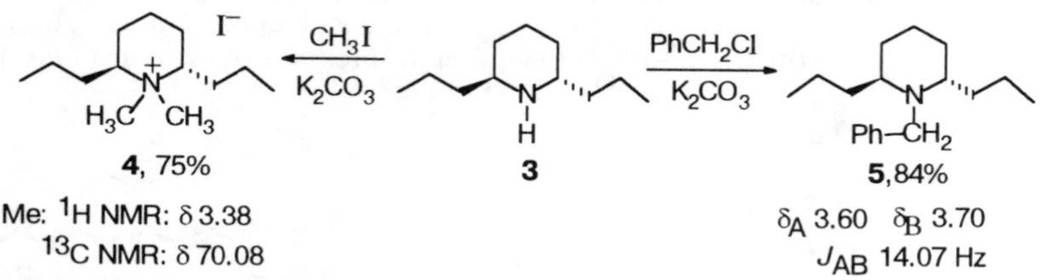

4, 75%   **3**   **5**, 84%

Me: ^1H NMR: δ 3.38 $δ_A$ 3.60 $δ_B$ 3.70
   ^{13}C NMR: δ 70.08 J_{AB} 14.07 Hz

From the latter, N,N-dimethyl (**4**, 70%) and N-benzyl derivatives (**5**, 84%) were obtained by the action of CH_3I [8a] and $PhCH_2Cl$ [8b], correspondingly.

The magnetic equivalence of both methyl groups in salt **4** was shown by NMR spectroscopy. A sharp singlet in ^1H (δ 3.38) and the only signal in ^{13}C (δ 70.08) NMR are observed. In addition, the benzyl methylene protons $(PhCH_2N)$ in ^1H NMR spectra of **5** appear as an AB quartet with $δ_A$ = 3.60 and $δ_B$ = 3.70 (J = 14.04 Hz) showing their non-equivalence. These data confirm *trans*-stereochemistry of 2,6-dipropyl compound **3** and — consequently — *trans*-configuration of 2,6-diallyl compound **2**. Similar ^1H NMR patterns have been observed in the cases of N,N-dimethylpiperidinium salt [9] and N-benzyl derivative [10] of *trans*-2,6-dimethylpiperidine.

Reaction of **1a** with CH_3OD followed by deboration with sodium hydroxide solution leads to 5-deuterio-compound **7** (78%).

Scheme 3

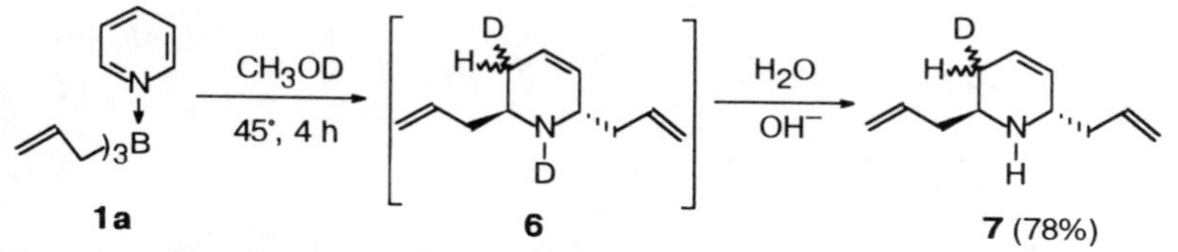

1a   **6**   **7** (78%)

There is no doubt that the 1,5-dideuterio compound **6** is the primary product of the reaction, which is converted into **7** in the course of work-up (N—D to N—H exchange).

From C_5D_5N and triallylborane, the pentadeuteriated compound **8** (74%) and *trans*-2,6-diallyl-1,2,3,4,5,5-hexadeuterio-1,2,5,6-tetrahydropyridine **9** (63%) were synthesized by heating complex **1b** with methanol (40°C, 4 h) and CH_3OD (70°C, 5 h), correspondingly.

Scheme 4

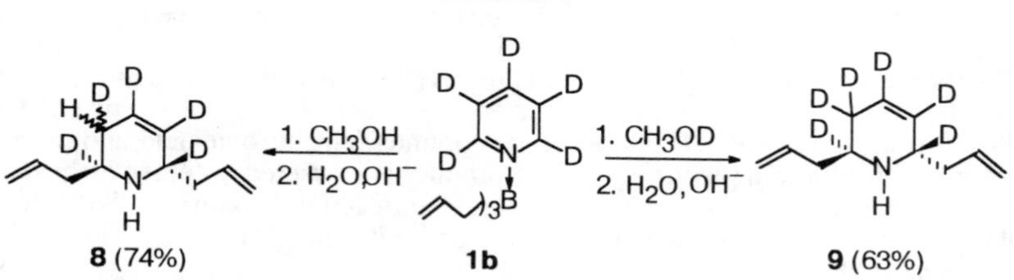

8 (74%)   **1b**   **9** (63%)

trans-Diallylation of 3-bromopyridine with triallylborane in the presence of *t*-BuOH (4 equ) also proceeds smoothly (95°C, 5 h) to give the product **10** (83%), in which bromine atom is bound to vinylic carbon atom. The structure of **10** was confirmed by X-ray analysis of its hydrochloride (**10**·HCl) [11] (Fig. 1).

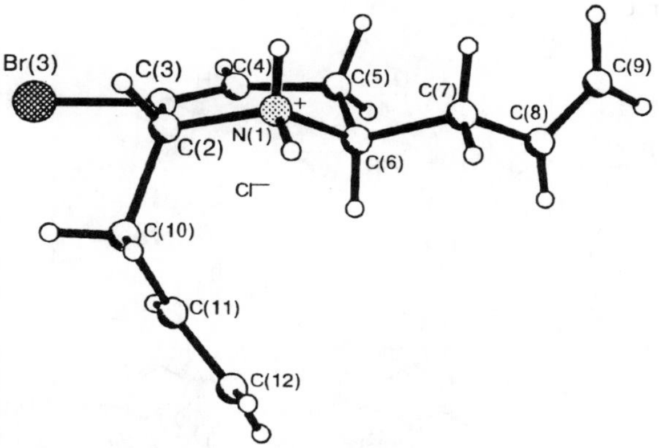

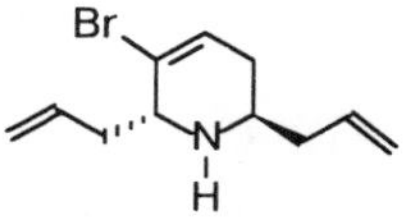

Figure 1. Crystal structure of **10**·HCl

This methodology was applied for efficient one-pot synthesis of compounds **11**, **12**, **14**, and **15** from 4-methyl-, 4-benzylpyridines and 4,4'-dipyridyl, correspondingly, as well as of **13** using trimethallylborane as the allylborating reagent.

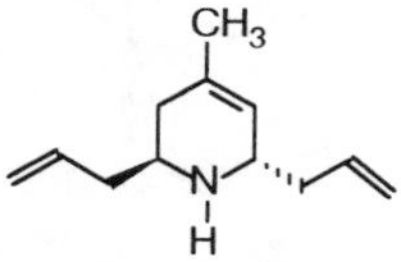

11 (75%)
~20% *cis-*

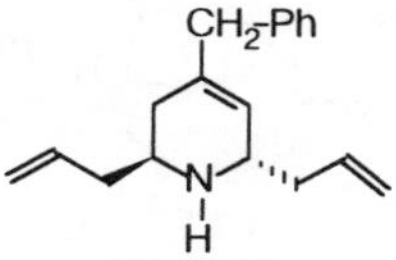

12 (75%)
~10% *cis-*

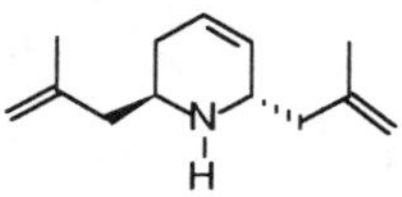

13 (87%)

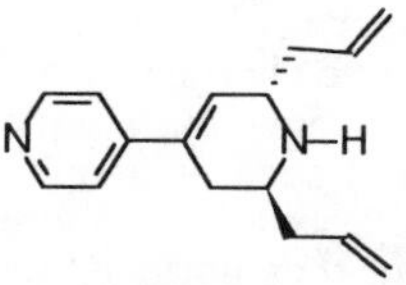

14 (76%)

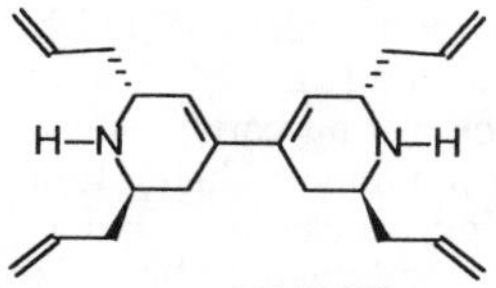

15 (20%)

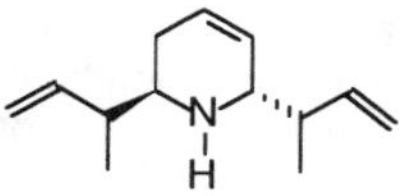

16 (63%)

Reaction of tricrotylborane with pyridine under above conditions was found to proceed with rearrangements of the both allylic moieties to give the product with terminal double bonds **16** (63%).

Actually, the presented results demonstrate that reductive *trans*-diallylation of pyridines by allylic boranes is a general reaction leading to very useful products. Two new C—C bonds are created in the process.

What is the mechanism of the reaction?

As soon as complex **1a** is not changed on prolonged heating at 160°C, it is clear that alcohol (water or R_2NH) plays a dramatic role in the course of the process. Reactions of **1a,b** with CH_3OD (Schemes 3 and 4) and of **1b** with CH_3OH (Scheme 4) show definitely that proton (deuterium) from alcohol molecule is incorporated in position 5 of nitrogen heterocycle.

Scheme 5

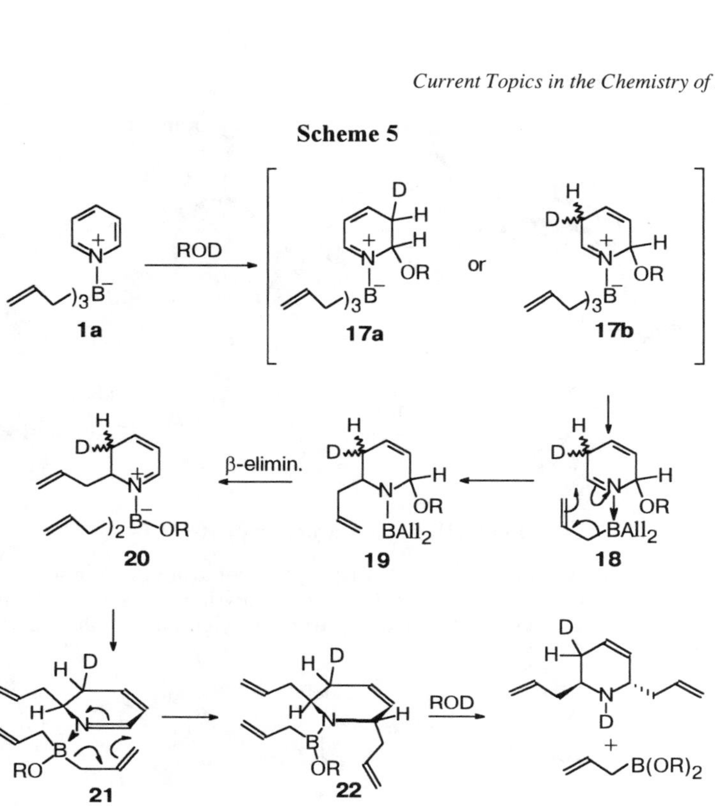

Nevertheless, the first step of the reaction is very nebulous at best.

Nitrogen atom in pyridine—triallylborane complexes of type **1** is positively charged and their behavior in some cases should be familiar to that of pyridinium salts. It has been well documented [12—14] that the positive charge in pyridinium ions favors nucleophilic attack at ring carbon atom α to nitrogen atom under mild conditions to give the corresponding adducts. Examples of such nucleophiles are hydroxide, alkoxide, sulfide, cyanide, amine, and some organometallic compounds. Sometimes these adducts can be isolated, but they normally undergo further reactions very rapidly. Typical example is the well-known oxidation of pyridinium salts with hydroxide in the presence of ferricyanide to give 2-pyridones.

A plausible mechanism of our reaction is presented in Scheme 5.

We suggest that the initial stage involves the nucleophilic alkoxide attack at ring C-2 atom to form adduct **17a** or, more likely, **17b** (1,2- or 1,4-addition of ROD to heterocycle, correspondingly). Both **17a** and **17b** contain a localized C=N bond, which immediately undergoes allylboration via six-membered transition state (**12**) to give the compound **19**. The latter is unstable and undergoes β-elimination giving rise to the complex **20**. The next stage, allylboration of the second C=N double bond, proceeds *trans*-stereoselectively with respect to the first allyl group in the ring (**21**) and this step is responsible for *trans*-stereochemistry of the final product. In aminoborane thus formed (**22**), B-N bond is cleaved at once by alcohol used in excess.

2. *Trans-* to *cis*-isomerization of *trans*-2,6-diallyl-Δ³-piperideines

We have worked out a convenient method for isomerization of *trans*-compounds **2** and **10** into *cis*-isomers **2a** and **10a** which consists in their heating with triallylborane (125—130°C, 5—6 hrs) or with allyl(dipropyl)borane

(140—150°C, 5 hrs) followed by deboration of aminoboranes formed.

2, X = H
10, X = Br

1. All$_3$B, 125—130°C, 5—6 h

2. H$_2$O, OH$^-$

2a, X = H, 80%   1—3%
10a, X = Br, 75%   6%

+ (**2** or **10**)

$)_3$B $-C_3H_6$
60—120°C

125—130°C, 6 h

H$_2$O, OH$^-$

BAll$_2$

BAll$_2$

Minor admixture of **2** and **10** in **2a** and **10a** (1—3% in **2a** and 6% in **10a**) is easily separated by chromatography on SiO$_2$ (pentane) and isomerically pure **2a** and **10a** were isolated in 80 and 75% yield, correspondingly.

2a

H$_2$, Ni

CH$_3$I
K$_2$CO$_3$

PhCH$_2$Cl
K$_2$CO$_3$

4a

3a

5a   δ 3.65 (s)

CH$_3$: ^{1}H NMR: δ 2.88; 3.40

^{13}C NMR: δ 37.24; 48.69

Hydrogenation of **2a** in acetic acid over Ra-Ni (90 atm. H$_2$, 90°C) lead to *cis*-2,6-dipropylpiperidine **3a** (90%), from which N,N-dimethylpiperidinium salt **4a** and N-benzyl derivative **5a** were obtained.

The magnetic non-equivalence of methyl groups bound to nitrogen atom in **4a** was demonstrated by NMR spectroscopy. Two sharp singlets in ^{1}H (δ 2.88 and 3.40) and two signals in ^{13}C NMR (at 37.24 and 48.69) were observed. Further evidence for *cis*-stereochemistry of **3a** and **2a** was obtained by ^{1}H NMR of **5a** in which benzyl methylene hydrogens (CH$_2$Ph) are enantiotopic and give a sharp singlet at 3.65 ppm. Similar patterns have been observed in the case of N,N-dimethyl [9] and N-benzyl derivatives [10] of *cis*-2,6-dimethylpiperidine.

In conclusion it should be stressed that the use of the only boron reagent, e.g. triallylborane, and the corresponding pyridine allows to synthesize both *trans*- and *cis*-2,6-diallyl-Δ^3-piperideines as well as *trans*- and *cis*-2,6-dipropylpiperidines in an isomerically pure state and on a large scale.

3. Reductive mono- and *trans*-1,3-diallylation of isoquinoline

Reaction of isoquinoline with triallylborane and allyl(dipropyl)borane proceeds under mild conditions (0—20°C) as a «thermal addition» of allyl-boron fragment to N=C-1 bond to give the corresponding aminoborane **23a** (δ^{11}B 47.4) and **23b** (δ^{11}B 51.6), further fate of which is determined by the conditions of work-up.

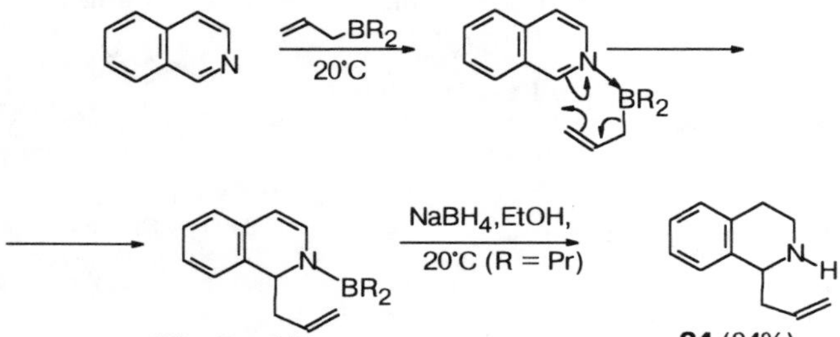

23a, R = All
b, R = Pr

24 (84%)

Reduction of **23b** by $NaBH_4$ in ethanol (20°C, 2 h) was found to give an 84% yield of 1-allyl-1,2,3,4-tetrahydroisoquinoline **24**.

On the other hand, treatment of aminoborane **23a** with methanol (3 equ, 20°C, 2 h) led to *trans*-1,3-diallyl-1,2,3,4-tetrahydroisoquinoline **25** (75%). Possible mechanism of its formation is presented below.

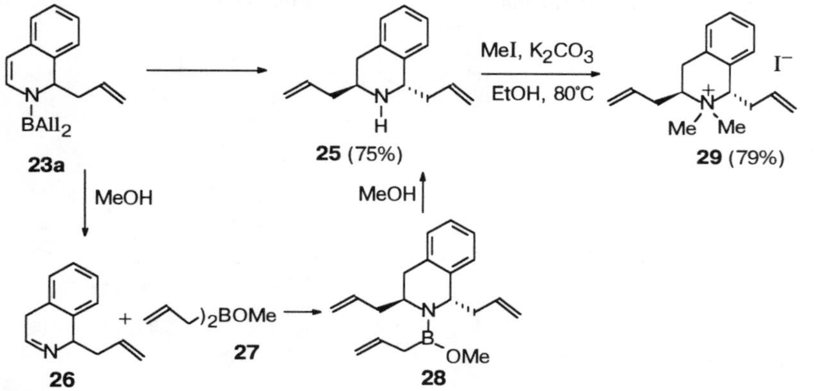

23a

25 (75%)

29 (79%)

26   **27**   **28**

Alcoholysis of **23a** proceeds with migration of double bond (proton is added to C-4 of heterocycle) leading to imine **26** and methoxy(diallyl)borane **27**. The latter allylborates rapidly **26** to give **28** and the second allylic group is added *trans*- to the first one. In aminoborane **28** formed, B—N bond is immediately cleaved by methanol which is used in excess.

Trans-stereochemistry of **3•HCl** was confirmed by X-ray analysis (Yu.T.Struchkov and M.O.Dekaprilevich).

According to [15], AllMgBr reacts with isoquinoline to give 1-allylisoquinoline.

4. Reductive monoallylation of quinolines and phenanthridine

These reactions proceed at room temperature to afford aminoboranes **30** (**30a**, $\delta^{11}B$ 46.2), deboration of which leads to the corresponding α-allylated heterocycles **31a—c** and **32**.

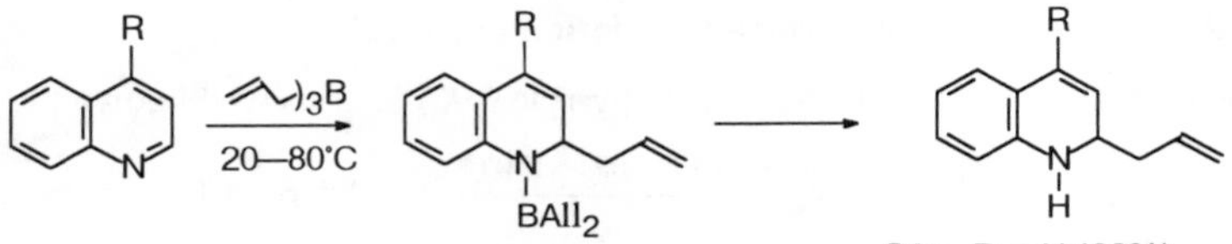

30a, R = H
b, R = Me

31a, R = H (93%)
b, R = Me (90%)

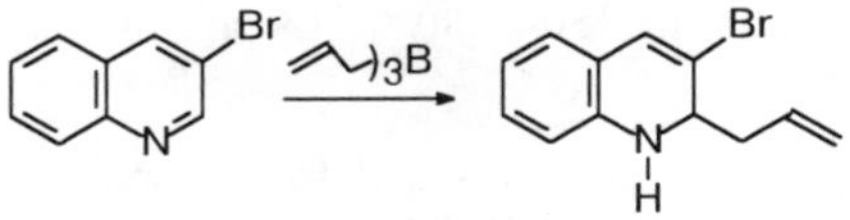

31c, (85%)

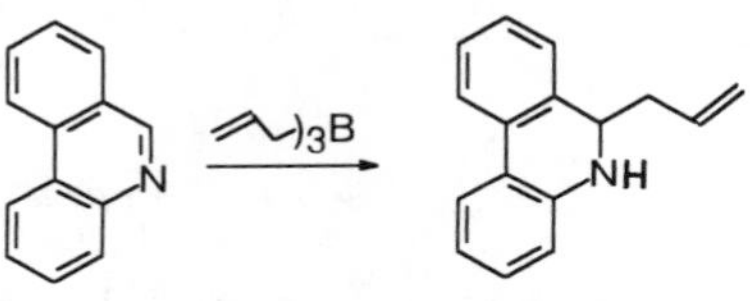

32 (98%)

Two B—C bonds of triallylborane are involved in the reaction with quinoline (ratio 1:2, 20—80°C, 1 h) to produce **31a** in 78% yield after hydrolysis of diaminoborane initially formed. Amine **31a** was also prepared with the use of allyl(dipropyl)borane as an allylborating reagent.

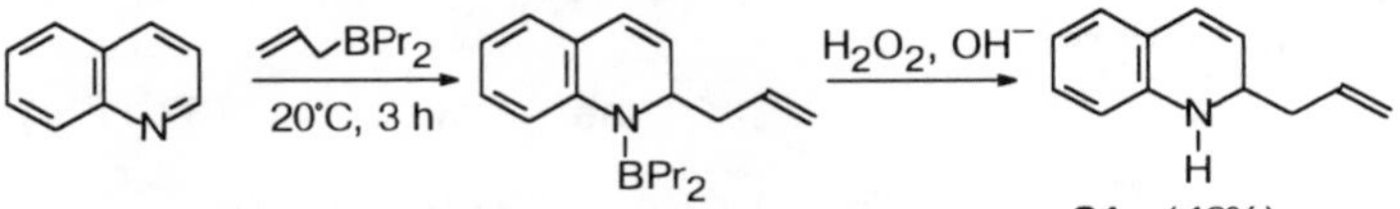

33

31a (46%)

Aminoborane **33** was isolated in a pure state (b.p. 112°C/1 torr, n_D^{20} 1.5376, δ^{11}B 56.9).

We have found that monoallylated compounds **31a—c**, **32** obtained via allylboration are stable up to 100°C in nitrogen atmosphere and compound **31a** is isomerized to 2-propylquinoline on heating at 170°C for 2 hrs.

2-Allyl-1,2-dihydroquinoline **31a** has been previously synthesized by Eisch and Comfort [16 by interaction of allylmagnesium chloride and quinoline, and **32** has been obtained similarly from phenanthridine by Gilman, Eisch and Soddy [15]. The authors have mentioned that compound **31a** is easily transformed into 2-propylquinoline through interesting hydrogen transfers even on work-up [16], and 5-allyl-5,6-dihydrophenanthridine **32** is extremely air-sensitive [15].

Carbinol **31d** was obtained from the reaction of allyl(dipropyl)borane with 4-hydroxyquinoline (2:1, 20°C) followed by treatment with methanol at −30°C.

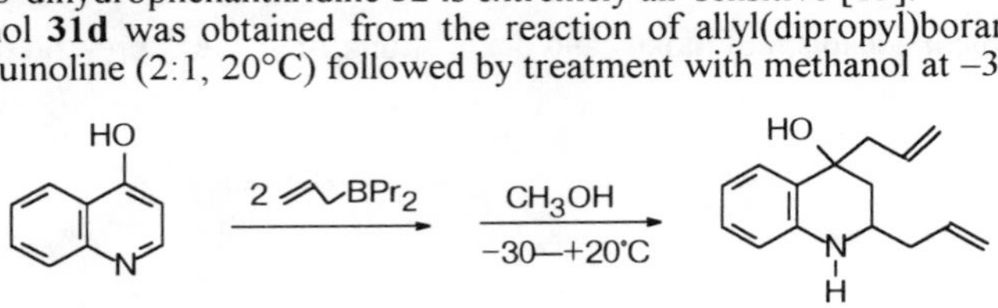

31d (40%)

5. Reductive mono- and *trans*-diallylation of pyrrole

The cleavage of RLi and RMgX with pyrrole (pK_a 17.5) is well known [17]

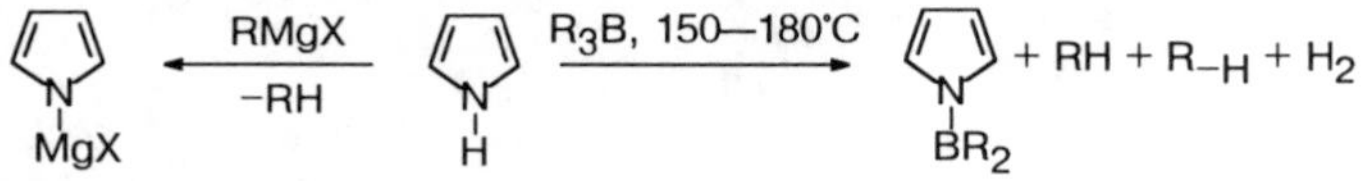

Köster and Bellut [18,19] have shown that triethyl- and tripropylborane are also cleaved by pyrrole at 150—180°C to afford the corresponding N-dialkylborylpyrrole.

Another story with allylic boranes.

We have found that allyl(dipropyl)- and triallylborane react with pyrrole at 20°C to afford a mixture of addition products **34** and **35**.

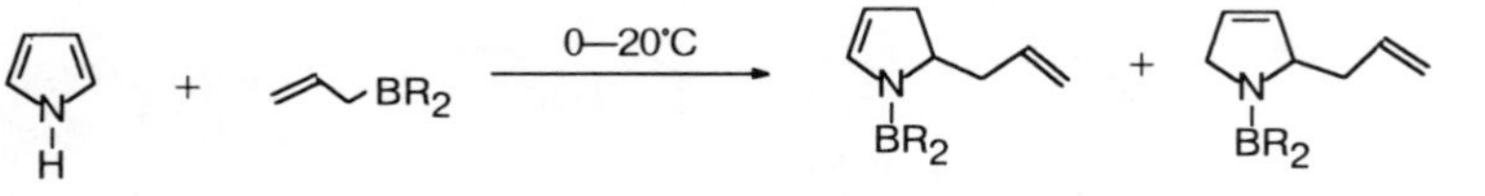

34a, R = Pr (78%)   **35a**, R = Pr (22%)
34b, R = All (70%)   **35b**, R = All (30%)

Treatment of the products formed from triallylborane (**34b** + **35b**) with methanol (3 equ, −30→20°C, 1 h) followed by NaOH (10—20%) leads to *trans*-2,5-diallylpyrrolidine **36** and 2-allyl-3-pyrroline **37** which were isolated in 61% and 15% yields, correspondingly.

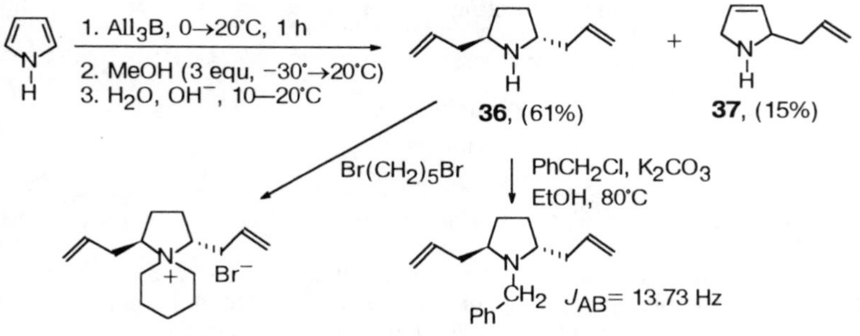

trans-Configuration of **36** was elucidated by the use of prochiral benzyl probe (**38**). As in 1-benzyl-*trans*-2,5-dimethylpyrrolidine [10], the benzyl methylene hydrogens of **38** are diastereotopic and give in the ^{1}H NMR an AB quartet, centered at 3.94 ppm (J = 13.73 Hz). Spiro compound **38a** was synthesized from **36**.

A possible mechanism of mono- and diallylation of pyrrole is presented below

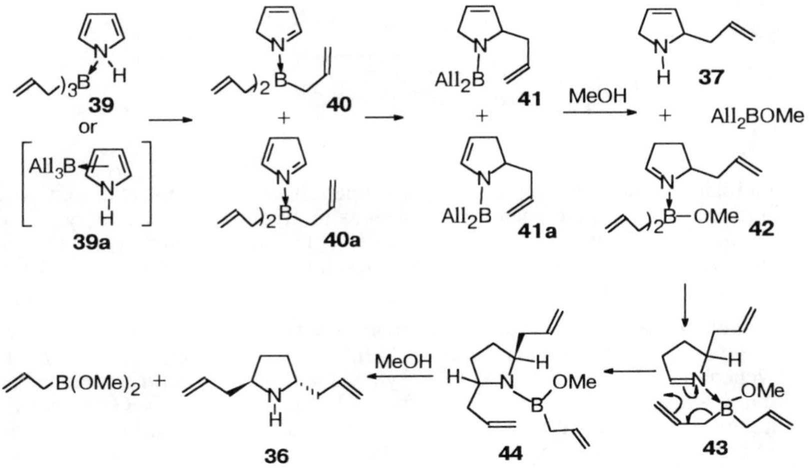

We suggest that N→B complex **39** (or π-complex **39a**) formed initially is isomerized by migration of hydrogen from nitrogen to C-3 and C-2 to give two imine adducts **40** and **40a** which undergo allylboration with the formation of monoallylated aminoboranes **41** and **41a**. Their cleavage by methanol produces the product **37** and the imine complex **42**. The latter is transformed into a new aminoborane **44** via **43** and the second allylboration reaction proceeds also *trans*-stereoselectively. Subsequent alcoholysis of **44** (the cleavage of B—N bond) affords the diallylated product **36** and dimethoxy(allyl)borane.

Both the allylborating stages of reductive *trans*-diallylation of pyrrole proceed with rearrangement. Thus, hindered amine **45** was synthesized from triprenylborane.

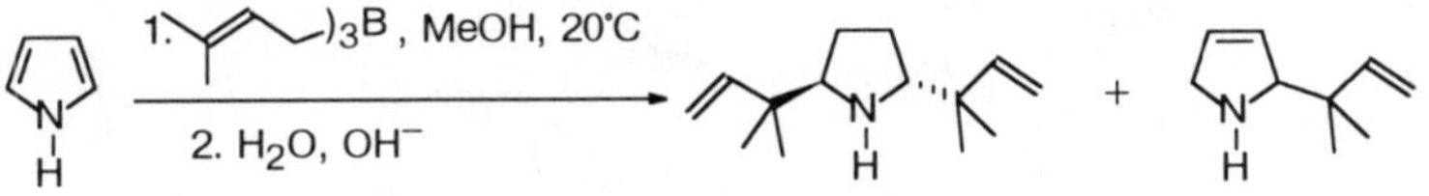

It should be stressed that reaction allows to introduce two different allylic groups into the heterocycle (**46**).

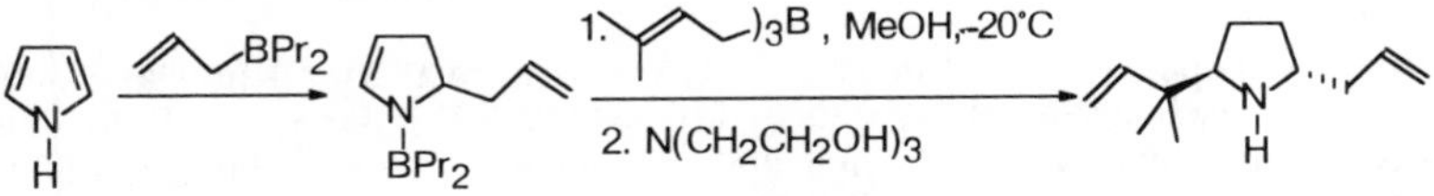

The mixture of *cis*-2,5-diallylpyrrolidine (75%) and **36** (25%) was obtained by heating of **36** with triallylborane at 160°C for 10 hs.

Appendix. Reaction of compound **47** with triallylborane in CHCl$_3$ (20°C, 1 h) followed by treatment with methanol and base leads to diallylated compound **48**, m.p. 92—94°C.

47   48 (25%)

Conclusion. The discovered reactions open new pathways and new rich possibilities in heterocyclic chemistry as well as in organoboron chemistry. There is no doubt that the reactions and compounds obtained by reductive mono- and α,α'-diallylation of nitrogen aromatic heterocycles with allylic boranes will find wide application in organic synthesis.

Acknowledgments. I wish to express my deep appreciation to Drs. E.A.Shagova, L.I.Lavrinovich and A.V.Ignatenko (NMR), and also to S.V.Evchenko, E.V.Klimkina and A.Yu.Zykov for the major contribution to this program. I am grateful to Drs. M.E.Gursky, I.D.Gridnev and A.V.Geiderikh for stimulating discussions. I also wish to acknowledge the contribution to the preparation of this manuscript by my coworker Alex Geiderikh.

This work is partly supported (1993) by the Russian Fundamental Research Foundation (grant N° 93-03-18193).

References

1.   a) B.M.Mikhailov, Yu.N.Bubnov. *Organoboron compounds in organic synthesis*, London, New York, Harwood Acad. Publishers, 1984. b) A. Pelter, K.Smith, in *Comprehensive organic chemistry*,(Eds. D.Barton, D.Ollis), Oxford, Pergamon Press, Part 14, p.688. c) A.Pelter, K.Smith, H.C.Brown, *Borane reagents*, London, Academic Press, 1988. d) R.W.Hoffmann, *Angew. Chem. Int. Ed.Engl.*, 1982, **21**,555.

2.   Yu.N.Bubnov. *Pure Appl. Chem.*, 1987, **59**, 895; 1991, **63**, 361.

3.   Yu.N.Bubnov, L.I.Lavrinovich, A.Yu.Zykov, A.V.Ignatenko, *Mendeleev Commun.*, 1992, **86**.

4.   M.E.Gurskii, T.V.Potapova, Yu.N.Bubnov, *Mendeleev Commun.*, 1993, **56**.

5.   Yu.N.Bubnov, E.A.Shagova, S.V.Evchenko, A.V.Ignatenko, I.D.Gridnev, *Izv. AN SSSR, Ser.Khim.*, 1991, 2644.

6.   Yu.N.Bubnov, S.V.Evchenko, A.V.Ignatenko, *Izv. AN. Ser. Khim.*, 1992, 2815.

7.   a) A.V.Topchiev, Ya.M.Paushkin, A.A.Prokhorova, M.V.Kurashov. *Dokl. Akad. Nauk SSSR*, 1959, **128**, 110. b) V.S.Bogdanov, V.F.Pozdnev, G.V.Lagodzinskaya, B.M.Mikhailov, *Theor. Exp. Khim.*, 1967, **3**, 488. c) V.S.Bogdanov, T.K.Baryshnikova, V.G.Kiselev, B.M.Mikhailov, *Zh. Obshch. Khim.*, 1971, **41**, 1533. d) W.Bruster, S.Schröder, K.Wittke, *Z. anorg. allgem. Chem.*, 1976, B 421(1), 89.

8.   a) E.Leetle, K.N.Juneau, *J. Am. Chem. Soc.*, 1969, **91**, 5614. b) H.Booth, J.H.Lettle, *Tetrahedron*, 1967, **23**, 291.

9.   Y.Kawazoe, M.Tsuda, M.Ohnishi, *Chem. Pharm. Bull.*, 1967, **15**, 51, 214.

10.   R.R.Hill, T.H.Chan., *Tetrahedron*, 1965, **21**, 2015.

11.   Yu.N.Bubnov,   E.A.Shagova,   S.V.Evchenko,   M.O.Dekaprilevich, Yu.T.Struchkov, *Izv. RAN, Ser. Chim.*, 1993, in press.

12.   E.F.Scriven, in: *Comprehensive Heterocyclic Chemistry*, **2**, Ch.5 (Eds.: A.J.Boulton, A.McKillop), Oxford, Pergamon, 1984, p.165.

13. A.J.Boulton, A.McKillop, see ref 12, Ch.2, p.29.

14. D.M.Smith, in: *Comprehencive organic chemistry*, **4**, Heterocyclic compounds, Ch. 16, (Ed.: P.G.Sammes), Oxford, Pergamon press, 1979.

15. H.Gilman, J.J.Eisch, T.Soddy, *J. Am. Chem. Soc.*, 1957, **79**, 1245.

16. J.J.Eisch, D.R. Comfort, *J. Org. Chem.*, 1975, **40**, 2288.

17. A.H.Jackson, see ref. 14, Ch. 17.

18. R.Köster, H.Bellut, S.Hattory, *Ann. Chem.*, 1969, **720**, 1.

19. H.Bellut, R.Köster, *Ann. Chem.*, 1970, **738**, 86.

Permanent Allylic Rearrangement (1,3-Boron Shift) – a Unique Feature of Allylic Type Organoboranes

I. D. Gridnev, M. E. Gurskii, and Yu. N. Bubnov

N.D. ZELINSKY INSTITUTE OF ORGANIC CHEMISTRY, RUSSIAN
ACADEMY OF SCIENCES, 117913 MOSCOW, B-334, LENINSKY PROSP.,
47, RUSSIA

Permanent Allylic Rearrangement (PAR) is a sigmatropic 1,3 shift of boron. It was observed for the first time in 1965 in the parent compound - triallylborane (1).[1] Later it was found that PAR is a general phenomenon for the triorganoboranes of allylic type.[2] PAR is an intramolecular reaction in the contrast to the similar metallotropic rearrangements of allylic derivatives of tin, mercury and zinc, which are intermolecular. The reason of the facile 1,3 shift of boron is a vacant 2p-atomic orbital of boron atom which decreases the energy of the cyclic four-centered transition state that normally is very high and makes it easily accessible.[3,4]

We have used modern NMR techniques and specially elaborated procedures for a computational analysis of the spectral data to study carefully and systematically the activation parameters of PAR for the vary of triorganoboranes of allylic type in order to find how the structure of allylic borane correlates with the activation parameters of PAR. Representative results are given in the Table 1.

Negative values of the activation entropies of PAR for all the compounds studied point at more regular structure of the transition state compared with the ground state of this process which corresponds to the quantum mechanics calculations.[4] Comparison of the data for the triallyl-, trimethallyl- and tricrotylboranes (1-3) with the data for the similar dipropylsubstituted compounds (6-8) indicate only slight influence of substitution of two allyl radicals to propyl ones on the free activation energy of PAR. Some decrease in the $\Delta S^{\neq}$ values in dipropylsubstituted allylboranes (6-8) presumably indicates the higher steric requirements for the propyl groups compared to allyl radicals. The values of the activation parameters of PAR for the triallylboranes (1-5) with different kind of methyl substitution indicate that β-methyl substituent does not affect significantly the rate of PAR, γ-methyl substitution slows down the PAR (the second γ-methyl group in the compound (4) makes the free activation energy still higher), while α-methyl substituent accelerates the rearrangement. As far as these effects can hardly be attributed to the electronic or steric effects, they are presumably caused by the relative thermodynamic stability of the corresponding allylboranes.

Table 1 Activation Parameters of Permanent Allylic Rearrangement for Different
Organoboranes of Allylic Type

Compound	Nucleus observed	Temperature interval, K	$\Delta G^{\neq}_{298} \dfrac{kJ}{mole}$	$\Delta S^{\neq} \dfrac{J}{mole \cdot K}$	$\Delta G^{0}_{298} \dfrac{kJ}{mole}$
$(\diagup\!\!\!\diagdown)_3 B$ (1)	13C	210-335	61.65±0.09	-42±3	—
$(\diagdown\!\!\!\diagup)_3 B$ (2)	^{13}C	233-353	62.43±0.09	-49±5	—
$(\sim\!\!\diagup\!\!\!\diagdown)_3 B$ (3)	^{13}C	306-412	73.0±0.1	-120±2	1.79±0.05
$(\diagup\!\!\!\diagdown)_3 B$ (4)	^1H	330-430	77.5±0.2	-120±1	—
$(\diagdown\!\!\!\diagup\!\!\!\diagdown)_3 B$ (5)	^1H	232-364	54.60±0.05	-45±3	—
$\diagup\!\!\!\diagdown\,BPr_2$ (6)	^{13}C	277-343	66.2±0.2	-54±10	—
$\diagdown\!\!\!\diagup\,BPr_2$ (7)	^1H	233-344	63.8±0.1	-68±2	—
$\sim\!\!\diagup\!\!\!\diagdown\,BPr_2$ (8)	^{13}C	305-408	70.1±0.1	-157±2	1.01±0.06
$=\!\!\diagup^{BPr_2}_{BPr_2}$ (9)	^1H	213-392	54.5±0.1	-8±3	—
$\diagdown\!\!\!<\!\!\diagup^{BPr_2}_{BPr_2}$ (10)	^1H	294-406	74.5±0.2	-24±3	—

Interesting correlation of the structure and the rate of PAR is observed for the two representative allylic diboranes (9) and (10). The activation barrier of PAR in 1,1-bis(dipropylborylmethyl)ethylene (9) is about 10 kJ/mole lower than in dipropylmethallylborane (7). This accelerating effect of the neighbouring second boron atom is probably stipulated for the possibility of homoallylic interaction of the π–MO of the double bond and the 2p-AO of two boron atoms instead of one in monoallylboranes.[5] In accordance with this assumption is a relatively high activation barrier of PAR for the allylic diborane (10) where two boron atoms are separated. The slowing down of PAR in this case is evidently caused by the conjugation of the allylic borane system with the second double bond. The similar effect was observed for other polyunsaturated triorganoboranes of allylic type.[6,7]

In unsymmetrically substituted triorganoboranes of allylic type PAR causes fast establishment of thermodynamic equilibrium. In the simplest case of tricrotylborane (3) only Z- and E- isomers are observed in the NMR spectra, their ratio 1:3 being only slightly dependent on the temperature. The α-methylallylic form was not detected under any conditions, though it is undoubtedly an intermediate in the interconversion of the Z- and E-isomers detectable at high temperature by NMR spectroscopy.

For more complex unsymmetrically substituted triorganoboranes of allylic type usually only the most stable isomer can be detected in the spectra. For example in the various cyclic boranes below only for the compound with four-membered ring (12) can a presence of two coexisting forms be detected. Compound containing three-membered ring (11) exists exclusively in a methylenecyclopropane form. This is probably the single strictly established example of non-fluxional triorganoborane of allylic type,[8] otherwise it should decompose rapidly since allylic boranes are known to react with cyclopropenes.[2] For compounds with five- and six-membered rings (13) and (14) only exo-cyclic form can be detected in the spectra due to its relative thermodynamic stability. These results correlate perfectly with the relative stability of corresponding hydrocarbons.[9]

(11)

(12), $\Delta G^{\neq}(298)=53.5\pm0.2$ kJ/mol,
$\Delta S^{\neq}=-3\pm3$ J/mol K, $\Delta G^{\bullet}_{298}=0.98\pm0.01$ kJ/mole

(13)

(14)

2,4-Pentadienyldipropylborane (15) is the simplest compound of the type where not only 1,3-B, but also 1,5-B shift could be expected. For example, intramolecular 1,3-Sn shift usually is not observed in allyltin derivatives, but in Z-triphenylpentadienyltin 1,5-Sn shift was found.[10] Our results showed, that E- and Z-isomers of 2,4-pentadienyldipropylborane (15a) and (15b) interconvert reversibly which indicated clearly the 1,3-B shift.[6] Generally speaking, it is impossible to guess whether there is only 1,3-B shift in the Z-isomer of 2,4-pentadienylborane (15b) or a combination of both 1,3-B and 1,5-B shifts. However, accurate line shape analysis of the temperature dependent ^{13}C NMR spectra of the compound (15) allowed us to obtain the values of activation parameters for all three kinds of interconversion which occur due to PAR:

(15a)   [1,3]-B   BPr₂   [1,3]-B   (15b) BR₂

$\Delta G^{\bullet}_{ab}=3.5$ kJ/mol, $\Delta G^{\neq}_{aa}=82.3$ kJ/mol, $\Delta G^{\neq}_{bb}=91.5$ kJ/mol, $\Delta G^{\neq}_{ab}=85.5$ kJ/mol
(Values at 360 K)

As far as the interconversion is slower in the Z-isomer, which is geometrically favored for the 1,5-B shift, only 1,3-B shift occurs in the compound (15) in the contrast with the pentadienyl derivative of tin.

Dynamic behaviour of 2,4,6-heptatrienyldipropylborane (16) proved to be most fascinating. As could be expected, just after preparation exists as a mixture of four geometric isomers in a ratio reflecting their relative thermodynamic stability. However, after heating for several minutes at 100°C compound (16) is almost completely stereoselectively transformed into E,Z,Z-1,3,5-heptatrienyldipropylborane (17)[7, 11]. This unusual reaction is a result of the 1,7-H shift occurring in the minor Z,Z-isomer (16a) of compound (16). The reaction is reversible, but due to the thermodynamic stability of the vinyl-boron fragment, the resulting equilibrium mixture contains 92% of (17) and 8% of (16).

The last reaction is a good illustration of the synthetic potential of fluxional organoboron compounds. Due to the PAR only one or few most stable isomers will always exist in no matter how complex triorganoborane of allylic type. This fact is itself useful for synthesis. Moreover, if some reaction for structural reasons can proceed in only one of the possible isomers, even if this isomer is a minor one, reaction will be complete. This approach allowed us to synthesize for the first time Z,Z-1,3,5-heptatriene (18) by the treatment of the vinylic borane (17) with acetic acid:[7, 11]

Synthesis of (18) can be considered as the conversion of unconjugated E-heptatriene (19) into the conjugated Z,Z-heptatriene (18) which becomes possible due to the PAR in 2,4,6-heptatrienyldipropylborane (16).

REFERENCES

1. B.M. Mikhailov, V.S. Bogdanov, G.V. Lagodzinskaya and V.F. Pozdnev, Izv. Acad. Nauk SSSR, Ser. Khim., 1966, 386.
2. B.M. Mikhailov and Yu.N. Bubnov, 'Organoboron Compounds in Organic Synthesis', Harwood Acad. Sci. Publ.; Chur, Utrecht, London, Paris, New York, 1984.
3. S. Iganaki, H. Fujimoto and K. Fukui, J. Am. Chem. Soc., 1976, 98, 4054.
4. M. Bühl, P.v.R. Schleyer, M.A. Ibrahim and T. Clark, J. Am. Chem. Soc., 1991, 113, 2466.
5. Yu.N. Bubnov, M.E. Gurskii, I.D. Gridnev, A.V. Ignatenko, Yu.A. Ustynyuk and V.I. Mstislavsky, J. Organomet. Chem., 1992, 424, 127.
6. M.E. Gurskii, I.D. Gridnev, A.V. Geiderich, A.V. Ignatenko, Yu.N. Bubnov, V.I. Mstislavsky and Yu.A. Ustynyuk, Organometallics, 1992, 11, 4056.
7. I.D. Gridnev, M.E. Gurskii, A.V. Ignatenko, Yu.N. Bubnov and Yu.V. Il'ichev, Organometallics, 1993, 12, in press.
8. Yu.N. Bubnov, M.E. Gurskii and A.V. Ignatenko, Izv. Akad. Nauk SSSR, Ser. Khim., 1988, 2184.
9. K.B. Wiberg, R.A. Fenoglio, J. Am. Chem. Soc., 1968, 90, 3395.
10. M.J. Hails, B.E. Mann, C.M. Spencer, J. Chem. Soc. Dalton Trans., 1983, 729.
11. M.E. Gurskii, I.D. Gridnev, Yu.V. Il'ichev, A.V. Ignatenko and Yu.N. Bubnov, Angew. Chem., Int. Ed. Engl., 1992, 31, 781.

Two Metals Are Better than One: Advantages of Transmetallations between Metals and Boron

Thomas E. Cole and Ramona Quintanilla

DEPARTMENT OF CHEMISTRY, SAN DIEGO STATE UNIVERSITY, SAN DIEGO, CA 92182, USA

We are exploring the migration of organic groups between metals and a metalloid, boron. This is an area that we feel will provide significant utility to both synthetic as well as organometallic chemists. Organoboranes have attained a reputation of being valuable synthetic intermediates for organic transformations. A variety of important and structurally diverse products have been prepared via hydroboration and organoborane reactions. General characteristics of these reactions are high selectivity and yield, using mild reaction conditions. Despite the many advances in borane chemistry, there are limitations as to what can be accomplished using boron alone. Recent investigation of other organometallic complexes has demonstrated similar high selectivities, under mild conditions, that frequently complement organoborane reactions. Unfortunately, in the vast majority of cases, these metal systems do not undergo further transformations as cleanly or have not been as fully developed as the organoborane reactions. A potential solution to this problem is to migrate the formed groups from one metal center to other metals or metalloids, who's chemistry is better established and developed for a particular type of transformation. Hopefully, these transmetallations would proceed with complete retention of stereochemistry of the organic fragment. The following is a brief summary of our efforts in combining the utility of two metal systems toward synthetic applications.

Transmetallations from Zirconium to Boron

The hydrozirconation reaction is similar to the hydroboration reaction in its regioselectivity adding in a *cis* fashion across carbon-carbon multiple bonds. There are a number of organozirconium reactions which complement hydroboration reactions. However, the organozirconiums have limited ability in their subsequent transformations, thus limiting their synthetic applications.[1] There has been a renewed interest in zirconium's ability to form carbon-carbon and carbon-hetero atom bonds. Again, the limited ability of zirconium to undergo other reactions limit the value of this metal in synthetic applications.[2]

We began our investigations of the migration of organic groups from zirconium to boron with the 1-alkenyl group.[3] These alkenylzirconium complexes are readily prepared from the hydrozirconation of a terminal alkyne. The transmetallation is rapid and essentially quantitative for a wide variety of haloboranes (X= Cl, Br), forming products free of other boron containing side-products (eq 1).

$$+ \quad \text{X}-\text{B} \quad \xrightarrow[\ 0\,^\circ\text{C}\ ]{\text{CH}_2\text{Cl}_2} \quad + \quad \text{Cp}_2\text{ZrClX} \downarrow \quad (1)$$

Alkoxy, alkyl and aryl groups may be present on the haloborane forming a variety of vinylboranes with minimal steric influence. Transmetallations proceed stepwise, allowing one to form mixed di- and trivinylboranes. The migration occurs with complete retuntion of the stereochemistry. The hydrozirconation of internal alkynes with a 3a5 mole % excess of $\text{Cp}_2\text{Zr(H)Cl}$ generally gives superior regioselectivities to the most selective hydroborating agents.[4] These disubstituted alkenyl groups transfer to boron like the 1-alkenyl groups,[5] although steric effects may be important in some cases (eq 2).

$$+ \quad \text{B}-\text{Br} \quad \xrightarrow[\ 0\,^\circ\text{C}\ ]{\text{CH}_2\text{Cl}_2} \quad \underset{84\%}{\quad} \quad + \quad \text{Cp}_2\text{ZrClBr}\downarrow \quad (2)$$

One of the notable aspects of the hydrozirconation of internal alkenes is the facile isomerization forming the 1-alkyl zirconium in extremely high regioselectivities, provided no tertiary or quaternary carbons are in the migration pathway. The transmetallation of these alkyl groups proceeds quickly, with good to excellent yields, to a variety of trihalo- and organohaloboranes (X= F, Cl, Br).[6] This gives a general route to primary alkylfluoroboranes. Migration of these groups can occur stepwise to form a variety of mixed organoboranes. This allows one to prepare a variety of organoboranes with very high regiopurity to the terminal alkyl position.

Applications of Transmetallations from Zirconium to Boron

The hydrozirconation of 2-methyl-2-pentene followed by migration to boron trichloride resulted in the formation of the alkyldichloroborane. Oxidation under basic conditions of alkyldichloroborane resulted in a 68% GC yield of the 4-methyl-1-pentanol as the only alcohol present (eq 3).[6]

$$\xrightarrow[\ 2)\ [\text{O}]\ ]{\ 1)\ \text{BCl}_3\ } \quad \underset{68\%}{\text{HO}} \quad (3)$$

with $\text{Cp}_2\text{Zr(H)Cl}$ above the arrow.

Unsymmetrical *trans-trans* dienes can be prepared by transmetallation of a terminal alkenyl group to an α-bromovinylborane using a base to induce migration followed by protonation.[3] The high regioselectivity of the hydrozirconation of alkenes and unsymmetrical alkynes allows for successive migrations to boron trichloride. Using classical boron chemistry, unsymmetrical trisubstituted olefins can be prepared in good yields with high regiopurities (eq 4).[7]

$$(4)$$

Transmetallations from Boron to Magnesium

The Grignard reagents and organolithium reagents historically have been used to prepare organoboranes. The types of boranes that can be formed by this reaction are limited only by availability of the corresponding organometallic reagent. With the discovery of the hydroboration reaction and subsequent development of many organoborane reactions, the types and diversity of organoboranes that can be formed have dramatically increased. These reactions give essentially pure organic fragments with well defined regio- and stereochemistry frequently with chiral groups. It would be of considerable use if these valuable fragments could be transferred to other metals, who's intrinsic reactivity complements organoborane reactions. Most organoborane reactions involve the electrophilic nature of the boron atom. The reactions of Grignard reagents with boron generally form "ate" complexes. There is comparatively little known about the nature, stability and reactivity of these tetracoordinate species.

Organic groups have been found to exchange between boron and magnesium. This exchange reaction is an equilibrium.[8] To be useful, the equilibrium must be shifted, driving the reaction to completion. Previous researchers have used different methods to shift the reaction.[8,9,10] The utility of this reaction has not been fully realized. Our approach to shift the reaction is based on the fact that the phenyl-boron bond is stronger than an alkyl-boron bond by approximately 20 kcal/mol.[11] Exchange of an alkyl for aryl groups on the boron allows one to readily follow the extent of the reaction by ^{11}B NMR. Our initial studies have shown a large solvent dependency on the exchange of the alkyl groups.[12] The reaction of trihexylborane and PhMgBr, in ether, yields a single hexyl group migration to magnesium.

$$(C_6H_{13})_3B \ + \ 2 \ PhMgBr \ \xrightarrow{\text{Et}_2O} \ C_6H_{13}MgBr \ + \ [MgBr]^+ \ [(C_6H_{13})_2BPh_2] \quad (5)$$

In the same reaction, additional equivalents of PhMgBr give two or essentially all three equivalents of the corresponding hexylmagnesium bromide. Tricyclohexylborane undergoes migration of the cyclohexyl group, although this reaction is sluggish and the yields are lower. In the migration of a single alkyl group, to minimize the loss of valuable organic fragments, mixed organoborane would be desirable. We have found that there are different migratory aptitudes in these exchange reactions. For example, *B*-decyl-9-BBN transmetallates the alkyl group to magnesium in an approximate 60% yield.

$$\text{B}-\text{C}_{10}\text{H}_{21} + 2\,\text{PhMgBr} \longrightarrow \xrightarrow{\text{D}_2\text{O}} \text{C}_{10}\text{H}_{21}\text{D} \quad (6)$$

The reaction of boronic esters with four equivalents of PhMgBr gives modest yields of the formed Grignard reagent. The presence of the borate salts in these reaction mixtures does not appear to adversely affect the classical Grignard reagent type reactions. Examples of these reactions are shown below.

$$(\text{C}_{10}\text{H}_{21})_3\text{B} + 2\,\text{PhMgBr} \longrightarrow \xrightarrow[\text{Li}_2\text{CuCl}_4]{} \quad \diagup\!\!\diagup\!\!\diagdown\text{C}_{10}\text{H}_{21} \quad (7)$$

$$(\text{C}_6\text{H}_{13})\text{B} + 2\,\text{PhMgBr} \longrightarrow \xrightarrow[\text{2 MeOH/BF}_3]{\text{1 CO}_2} \quad \text{MeO}\overset{\text{O}}{\underset{}{\diagdown\!\!\diagup}}\text{C}_6\text{H}_{13} \quad (8)$$

Conclusion

The preliminary studies presented here begin to demonstrate the potential importance of the migration of alkyl groups between metals and boron. The transmetallation of groups from zirconium to boron gives a superior route to fragments which would be difficult to prepare otherwise. These formed organoboranes can then be used in established organoborane reactions. The exchange of organic fragments from a large variety of organoboranes to magnesium may allow new types of Grignard reagents for organic synthesis.

REFERENCES

1. D. B. and J. Schwartz <u>J. Am. Chem. Soc</u> ,1979, <u>101</u>, 3521.
2. S. L. Buchwald and R. B. Nielsen <u>Chem. Rev.</u>, 198, <u>7</u>, 1044 and references therein.
3. T. E. Cole, R. Quintanilla and S. Rodewald <u>Organometallics</u> ,1991, <u>10</u>, 3777.
4. D. Hart, T. F. Blackburn and J. Schwartz <u>J. Am. Chem. Soc</u> ,1979, <u>97</u>, 679.
5. T. E. Cole and R. Quintanilla <u>J. Org. Chem</u> ,1992, <u>57</u>, 7366.
6. T. E. Cole, S. Rodewald and C. L. Watson <u>Tetrahedron Lett.</u> ,1992, <u>33</u>, 5295.
7. T. E. Cole and R. Quintanilla, manuscript in preparation.
8. K. Kondo and S. I. Murahashi, <u>Tetrahedron Lett.</u>, 1979, <u>14</u>, 1237.
9. K. Ziegler, R. Köster and W. Grimes, U. S. Patent 3,217,020, Nov. 9, 1965.
10. J. D. Buhler, PhD Thesis, Purdue University, 1973.
11. A. Finch, P. J. Gardner, E. J. Pearn and G. B. Watts, <u>Trans. Faraday Soc.</u>, 1967, <u>63</u>, 1880.
12. Manuscripts in preparation.

Selective Cleavage of the Carbon–Zirconium Bond in 1,1-Bimetalloalkanes of Boron and Zirconium. A Convenient Method for Preparation of α-Bromoboranes

Bin Zheng and Morris Srebnik*

DEPARTMENT OF CHEMISTRY, THE UNIVERSITY OF TOLEDO, TOLEDO, OH 43606, USA

There is much current interest in development of organozirconium compounds which have a number of unique properties enabling them to be used as highly reactive reagents in organic synthesis[1], and also as effective catalysts for organic reactions and polymerizations of unsaturated hydrocarbons[2].

Hydrozirconation has been developed as a procedure for functionalizing alkenes, alkynes, etc. by Schwartz's reagent, H(Cl)ZrCp$_2$, to give organozirconium compounds[3]. 1,1-Bimetallics of transition metals are undergoing active study as intermediates for olefinations with carbonyl derivatives[4] and selective functionalization with two different electrophiles[5]. 1,1-Bimetallic compounds containing zirconium as well as aluminum[6], and zinc[6b,7] have been investigated for olefination with a ketone. As part of our investigations on the preparation and use of new substituted and highly functionalized zirconium and boron 1,1-bimetallic reagents led us to approach a convenient way to achieve α-bromoboranes[8] which represent a useful class of boron intermediates for organic synthesis[9].

In this study, we have now found that various B-alkenyl borabicyclo[3.3.1]nones 1 (B-alkenyl-9-BBN), which are the products of hydroboration of alkynes with 9-BBN[10], smoothly hydrozirconated by Schwartz's reagent, H(Cl)ZrCp$_2$[11], furnishing 1,1-bimetallics of boron and zirconium depicted as 2 (eq 1).

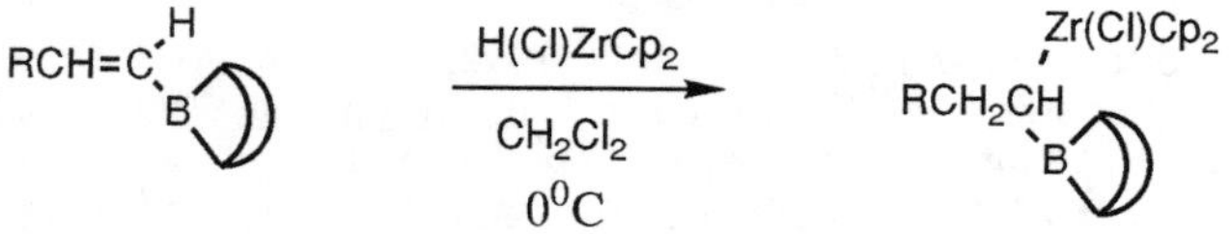

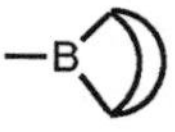

$$ (1) $$

We investigated the effect of solvents on the hydrozirconation step. The solvent study showed that there was no hydrozirconation of alkenylboranes with

Schwartz's reagent in both diethyl ether and hexanes. In benzene, the reaction was extremely slow. However hydrozirconation can be completed in dichloromethane and THF, and took 1 hr and 1.5 hrs respectively at $0^{\circ}C$ for B-hexenyl-9-BBN. This result suggested that factors other than the electronic donor ability of a solvent were involved in the hydrozirconation process. This is different from the solvent effect observed in hydroboration where both THF and diethyl ether are suitable solvents.

Cyclic alkenylboronic esters have also been investigated. It was found that hydrozirconation of alkenylboronic esters such as pinacol hexenyl boronate and ethylene glycol hexenylboronate were much slower than that of 9-BBN counterparts, and 3 equiv. of Schwartz's reagent was required to complete the reaction. However, the products, 1,1-bimetallics based on boronic esters and zirconium, were relatively stable, and were characterized by NMR and X-ray diffraction[12], etc.

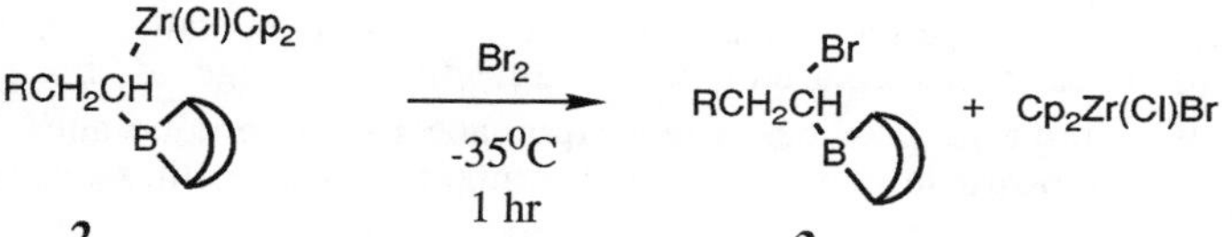

$$ (2) $$

Table I. Preparation of α-Bromoboranes **3** by Bromination of Bimetallics **2**

R in alkenyl-9-BBN, **1**	Hydrozirconation Time, hr	Bromination Product, **3**	^{1}H NMR δ of α-H(dd)	Yield[a] %
n-butyl	1	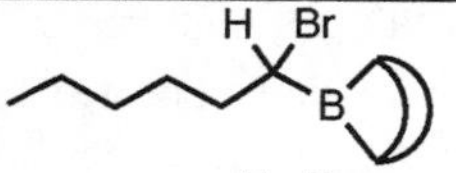	4.13	97
3-chloropropyl	1	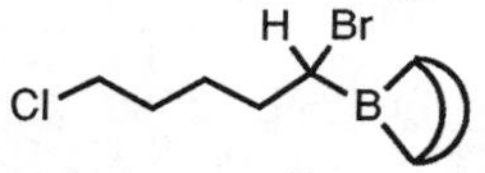	4.08	99
1-methylpropyl	1.5	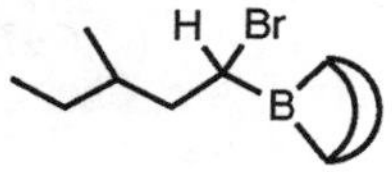	4.10	95
3-phenylpropyl	1.5	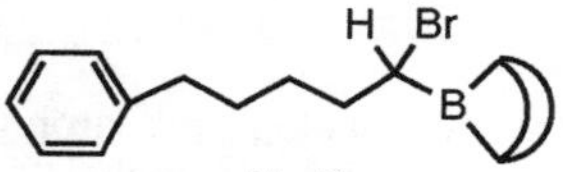	4.26	99
cyclopentyl	2	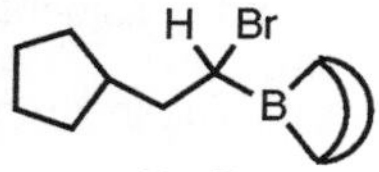	4.19	91
t-butyl	6	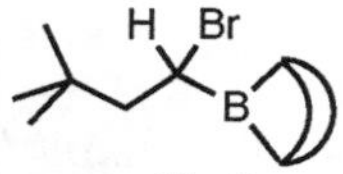	4.24	87
phenyl	6	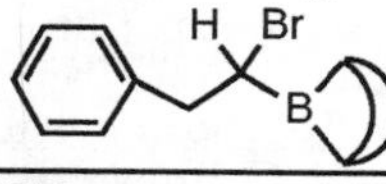	4.27	83

[a]crude yields, percent based on alkenyl-9-BBN

Substituted groups on boron larger than 9-BBN also decrease the rate of hydrozirconation. Thus diisopinocampheyl hexenylborane and diisocaranyl hexenylborane were incompletely hydrozirconated at reduced rate. Obviously both electronic and steric factors greatly influence the course of hydrozirconation of alkenylboranes.

α-Bromoboranes were achieved by addition of bromine in situ (eq 2). The reaction mixture changed color from yellow to colorless, and a white precipitate slowly formed. The proton NMR of this isolated solid is consistent with a zirconocene dihalide. Through appropriate work-up, the crude oily α-bromoboranes[13] were obtained in high yields. The results are summarized in Table I. This reaction serves two purposes. It confirms the regioselectivity of the hydrozirconation step[14], and provides a convenient approach to α-bromoboranes which can be converted into a multitude of desired organic products.

In summary, we have shown that boron and zirconium bimetallic reagents can be readily prepared by hydrozirconation. Selective cleavage of this type of bimetallics with bromine provides a new approach to α-bromoboranes. The determination of the structure and the further synthetic scope of these reagents is underway in our laboratories.

ACKNOWLEDGMENT. We thank to The University of Toledo for financial support which made this project possible, and also to the State of Ohio Academic Challenges Program for providing funds for a high field NMR spectrometer. We also thank Dr. D. Giolando for initial productive conversations.

REFERENCES AND NOTES

1.   a) J. Schwartz and J. A. Labinger, <u>Angew. Chem., Int. Ed., Engl.</u>, 1976, <u>15</u>, 333
      b) J. Schwartz, <u>Pure Appl. Chem.</u>, 1980, <u>52</u>, 733.
      c) E. Nigishi and T. Takahashi, <u>Synthesis</u>, 1988, 1.
2.   G. Erker, <u>Pure Appl. Chem.</u>, 1991, <u>62</u>, 797.
3.   D. W. Hart and J. Schwartz, <u>J. Am. Chem. Soc.</u>, 1974, <u>96</u>, 8115.
4.   a) S. M. Clift and J. Schwartz, <u>J. Am. Chem. Soc.</u>, 1984, <u>106</u>, 8300.
      b) A. Aguero, J. A. Kress and J. Osbom, <u>J. Chem. Soc., Chem. Commun.</u>, 1986, 531.
      c) P. Knochel and J. F. Normant, <u>Tetrahedron Lett.</u>, 1986, <u>27</u>, 1039.
      d) J. R. Wass, A. R. Sidduri and P. Knochel, <u>Tetrahedron Lett.</u>, 1992, <u>33</u>, 3717, and references therein.
5.   P. Knochel and J. F. Normant, <u>Tetrahedron Lett.</u>, 1986, <u>27</u>, 1043, 4427, 4431, and references therein.
6.   a) F. W. Hartner, J. Schwartz and S. M. Clift, <u>J. Am. Chem. Soc.</u>,1983, <u>105</u>, 640.
      b) F. W. Hartner and J. Schwartz, <u>J. Am. Chem. Soc.</u>, 1981, <u>103</u>, 4979.
      c) T. Yashida and E. Nigishi, J. Am. Chem. Soc., 1981, 107, 1276.
7.   C. E. Tucker and P. Knochel, <u>J. Am. Chem. Soc.</u>, 1991, <u>113</u>, 9888.
8.   B. Zheng and M. Srebnik, <u>Tetrahedron Lett.</u>, 1993, in press.
9.   a) A. Pelter, K. Smith and H. C. Brown, 'Borane Reagents', Academic Press, Lodon, 1988.
      b) D. S. Matteson, <u>Tetrahedron</u>, 1989, <u>45</u>, 1859.

c) H. C. Brown, G. W. Kramer, A. B. Levy, and M. M. Midland, 'Organic Synthesis via Boranes', Wily-Interscience, New York, 1975.

10. H. C. Brown, C. G. Scouten and R. Liotta, <u>J. Am. Chem. Soc.</u>, 1979, <u>101</u>, 96.

11. S. L. Buchwald, S. J. LaMaire, K. B. Nielson, B. T. Watson and S. M. King, <u>Tetrahedron Lett.</u>, 1987, <u>28</u>, 3895.

12. E. Skrzypczak-Jankun, B. Cheesman, B. Zheng and M. Srebnik, manuscript in preparation.

13. Typical procedure.   Preparation of B-(-1-bromohexyl)-9-borabicyclo[3.3.1]-nonane: All reactions and operations were under argon.   To a stirred ice-cooled suspension of Schwartz's reagent (0.26g, 1mmol) in dry CH_2Cl_2 (1ml) was added a solution of B-hexenyl-9-BBN (0.20g, 1 mmol) in dry CH_2Cl_2 (1 ml).The cloudy suspension of the mixture became clear yellow solution in 1 hr at 0 °C (or 10 min at ambient temperature).   After cooling to -35 °C, bromine (0.16g, 1mmol) in 1ml of CH_2Cl_2 was added dropwise.   As the reaction mixture became colorless, a white precipitate slowly formed.   The resulting mixture was stirred 1 hr, and warmed to ambient temperature.   After pumping off CH_2Cl_2, dry hexane (2x2ml) was added, and the reaction mixture extracted.   Filtration of this hexane solution and evaporation of the solvent from the filtrate afforded the crude product, α-bromohexyl-9-BBN, as a clear colorless oil (0.27g, 97%).   Organic products were identified by ^{1}H NMR.

14. B. Zheng, M. Srebnik, unpublished result.   Schwartz and co-workers demonstrated that hydrozirconation of olefins proceeded to place the zirconium moiety at the sterically least hindered position of the olefin chain. We found that hydrozirconation of disubstituted B-alkenyl-9-BBN, such as 2-butenyl-9-BBN **4,** did not necessarily place zirconium moiety on the least

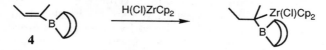

hindered terminal carbon.   Therefore this regioselectivity of hydrozirconation step was further proved.

New Developments in the Transition Metal-Catalyzed Hydroboration of Alkenes

Stephen A. Westcott[1], Paul Nguyen[1], Henk P. Blom[1],
Nicholas J. Taylor[1], Todd B. Marder*[,1], R. Tom Baker*[,2] and
Joseph C. Calabrese[2]

[1] DEPARTMENT OF CHEMISTRY, UNIVERSITY OF WATERLOO,
WATERLOO, ONTARIO, CANADA N2L 3GI
[2] CENTRAL RESEARCH AND DEVELOPMENT, DUPONT SCIENCE AND
ENGINEERING LABORATORIES, EXPERIMENTAL STATION,
WILMINGTON, DE 19880-0328, USA

The discovery[1] that certain transition metal complexes promote the addition of B-H bonds to unsaturated organic moieties has received considerable attention recently[2] for applications in organic synthesis.[3] The advantage of employing a catalyst has been best exemplified in hydroborations utilizing catecholborane (HBcat, cat =1,2-$O_2C_6H_4$). While addition of HBcat to alkenes generally requires harsh reaction conditions (T > 100°C), analogous metal-catalyzed hydroborations proceed readily at much lower temperatures. More important, however, is that under these milder reaction conditions, hydroboration products can be prepared that have chemo-, regio- and/or stereoselectivities that are complementary to products obtained from the uncatalyzed variants.[4] For example, Nöth originally reported that the alkene fragment in 5-hexen-2-one can be hydroborated preferentially in the presence of the much more reactive carbonyl functionality by employing HBcat and a catalytic amount of $RhCl(PPh_3)_3$ (1).[1]

To understand the role the metal centre plays in such transformations, our initial investigations focussed on examining the interaction of HBcat with 1. While early reports[1,5] showed that HBcat oxidatively adds to 1, with concomitant loss of PPh_3, to afford the unsaturated boryl (BR_2) species $RhHCl(Bcat)(PPh_3)_2$ (2), we have demonstrated[6] that this reaction is much more complex and gives complicated product distributions that are dependent upon solvent, temperature and rate and order of addition of HBcat. The presence of these extraneous rhodium complexes in solution has a negligible effect on product distributions in hydroborations of simple alkenes such as oct-1-ene and unsubstituted styrenes, giving the desired alcohols, upon oxidative workup, in excellent yields.[2a] Conversely, we found that hydroborations of sterically hindered 2-phenylpropene with 1 gave products derived from a number of competing reaction pathways (Scheme 1) including hydrogenation and dehydrogenative borylation.[7]

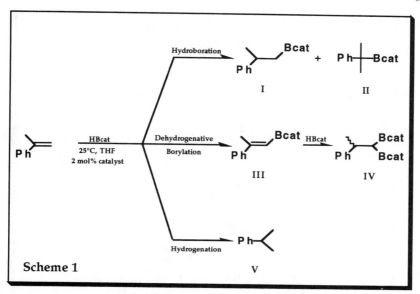

Scheme 1

That vinylboronate esters are formed in these reactions[8] suggests that metal-catalyzed hydroborations may proceed, at least in part, via insertion of alkene into the M-B and not the M-H bond. In order to test the feasibility of this hypothesis, we examined the reactivity of the novel bis(boryl) compound $RhCl(Bcat)_2(PPh_3)_2$ (3) with a variety of unsaturated organic substrates and recently reported the first discrete examples of insertion of alkenes into Rh-B bonds.[9] Formation of 1,2-bis(boronate esters), however, is accompanied by significant amounts of (E)-vinylboronate esters and hydroboration products due to a competing β-H elimination pathway (Scheme 2). Indeed, while addition of B_2cat_2 to alkenes is catalyzed by 1, selectivity for the diborated products is poor.

Table 1. Effect of Catalyst Precursor on Product Distribution

Catalyst	I	II	III	IV	V
$[RhCl(COE)_2]_2/2\ PPh_3$	98	—	—	—	2
$[Rh(COD)(DPPB)]BF_4$	30	70	—	—	—
$Rh(\eta^6\text{-catBcat})(DPPB)$	5	95	—	—	—
$RhCl(PPh_3)_3$	14	3	53	27	3
$RhCl(PPh_3)_3/10\ PPh_3$	10	1	70	18	1
$[RhCl\{PPh_2(o\text{-tol})\}_2]_2$	15	—	76	8	1

DPPB = 1,4-bis(diphenylphosphino)butane; COD = 1,5-cyclooctadiene; COE = cyclooctene; acac = acetylacetonate; tol = $-C_6H_4Me$

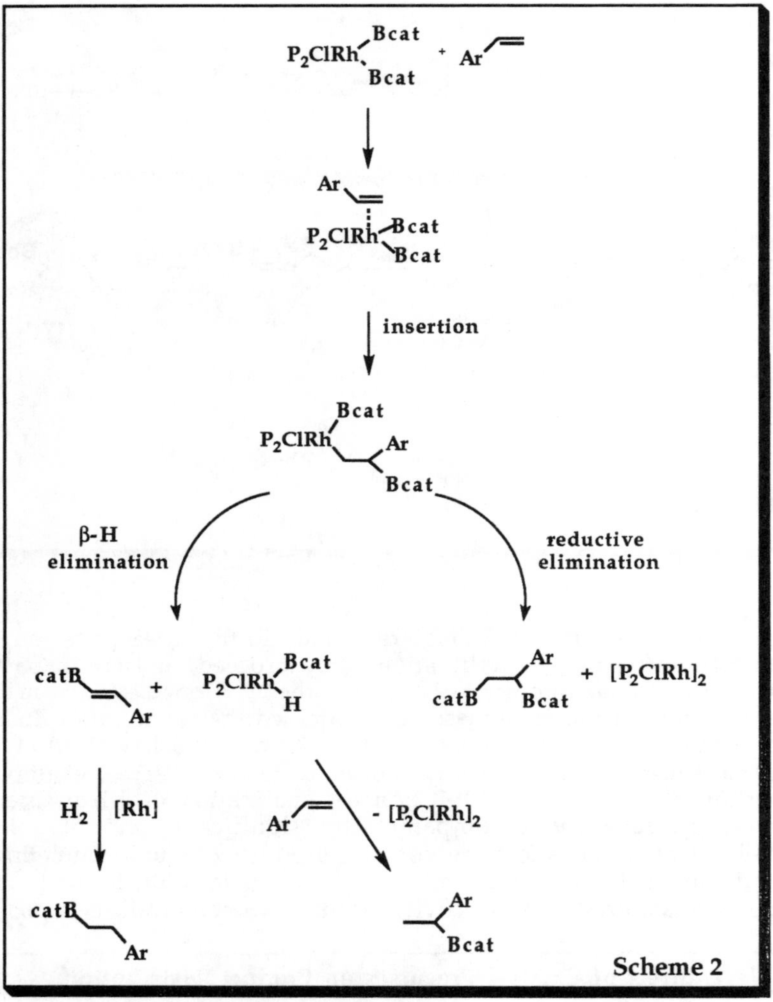

$P_2ClRh\overset{Bcat}{\underset{Bcat}{\diagdown}}$ + $Ar\diagup\!\!=$

$\downarrow$

$Ar\diagdown\!\!=$

$P_2ClRh\overset{Bcat}{\underset{Bcat}{\diagdown}}$

$\downarrow$ insertion

$P_2ClRh\diagdown\overset{Bcat}{\underset{Ar}{\diagup}}\overset{}{\underset{Bcat}{\diagdown}}$

β-H elimination   reductive elimination

catB$\diagdown$Ar   + $P_2ClRh\overset{Bcat}{\underset{H}{\diagdown}}$ $\overset{Ar}{\underset{catB\diagdown Bcat}{}}$ + $[P_2ClRh]_2$

$H_2 | [Rh]$ $Ar\diagup\!\!=$ $- [P_2ClRh]_2$

catB$\diagdown$Ar $\overset{Ar}{\underset{Bcat}{}}$

Scheme 2

In order to overcome this limitation, a number of Rh and Ir complexes were used for the catalyzed hydroboration of 2-phenylpropene. As shown in the Table,[9] selectivities varied from 98% terminal to 95% internal hydroboration product. The latter was achieved using the new zwitterionic catalyst, Rh(DPPB)(η^6-catBcat) (4), which is formed by treatment of commercial Rh(acac)(η^2-C$_2$H$_4$)$_2$ with the chelating phosphine ligand, followed by excess HBcat.

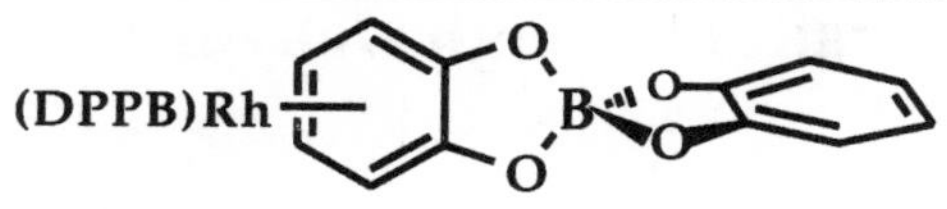

(DPPB)Rh

Figure 1 **4**

Unlike the more commonly employed cationic Rh and Ir catalysts, **4** is the sole metal-containing complex in solution during and upon completion of catalysis and is believed to be the catalyst resting state. We have shown recently that **4** is an extremely active and selective catalyst prescursor for a wide range of alkenes.[2g] As hydroborations employing **4** are usually not complicated by competing pathways, we are currently investigating the diboration of alkenes using this novel catalyst precursor.

References

1.   D. Männig, H. Nöth, Angew. Chem. Intl. Ed. Eng. 1985, 24, 878.
2.   For leading references see: (a) K. Burgess, M. J. Ohlmeyer, Chem. Rev. 1991, 91, 1179. (b) D. A. Evans, G. C. Fu., A. H. Hoveyda, J. Am. Chem. Soc. 1992, 114, 6671. (c) M. Satoh, N. Miyaura, A. Suzuki, Tetrahedron Lett. 1990, 31, 231. (d) J. M. Brown, G. C. Lloyd-Jones, Tetrahedron Asymm. 1990, 1, 869. (e) T. Hayashi, Y. Matsumoto, Y. Ito, Tetrahedron Asymm. 1991, 2, 601. (f) J. Zhang, B. Lou, G. Guo, L. Dai, J. Org. Chem. 1991, 56, 1670. (g) S. A. Westcott, H. P. Blom, T. B. Marder, R. T. Baker, J. Am. Chem. Soc. 1992, 114, 8863. (h) K. N. Harrison, T. J. Marks, J. Am. Chem. Soc. 1992, 114, 9220.
3.   (a) D. A. Evans, G. S. Sheppard, J. Org. Chem. 1990, 55, 5192. (b) P. Kocienski, K. Jarowicki, S. Marczak, Synthesis 1991, 1191.
4.   A. Pelter, K. Smith, H. C. Brown, 'Borane Reagents' Academic Press, New York, NY 1988.
5.   H. Kono, K. Ito, Y. Nagai, Chem. Lett. 1975, 1095.
6.   K. Burgess, W. A. van der Donk, S. A. Westcott, T. B. Marder, R. T. Baker, J. C. Calabrese, J. Am. Chem. Soc. 1992, 114, 9350.
7.   S. A. Westcott, T. B. Marder, R. T. Baker, Organometallics 1993, 12, 975.
8.   J. M. Brown, G. C. Lloyd-Jones, J. Chem. Soc., Chem. Commun. 1992, 710.
9.   R. T. Baker, J. C. Calabrese, S. A. Westcott, P. Nguyen, T. B. Marder, J. Am. Chem. Soc. 1993, 115, 4367.

trans-Vinylboranes through Dehydroborylation

J. A. Soderquist, J. C. Colberg, A. Rane, and J. Vaquer

DEPARTMENT OF CHEMISTRY, UNIVERSITY OF PUERTO RICO, RIO
PIEDRAS, PUERTO RICO 00931

1 INTRODUCTION

In contrast to the convenient preparation of *trans*-vinylboranes
(2),[1,2] through the monohydroboration of terminal alkynes (1) with
most of the common monofunctional hydroborating agents, 9-
borabicyclo[3.3.1]nonane (9-BBN-H) dimer[1,3] also produces significant quantities of 1,1-diboryl adducts (3) from (1) with a 1:1
stoichiometry, a problem only partially solved by employing a
100% excess of (1)[3] or by using silylated derivatives.[5a,b] The
convenience of employing a stable, highly regioselective reagent
which produces thermally stable adducts,[1] provided the impetus
for the discovery of a new synthetic route to (2), a process
which we term, dehydroborylation (eqn 1).[4]

$$R \!=\!\equiv\!-H \xrightarrow{\text{9-BBN-H}} (3) \xrightarrow[\text{- (4)}]{\text{ArCHO}} (2) \tag{1}$$

(1)   (3)   (2)

2 RESULTS AND DISCUSSION

Hydroboration of (1) with 9-BBN-H

The examination by 96 MHz ^{11}B NMR of the hydroboration
of (1) with 9-BBN-H under optimal conditions (THF, 18 h, 0 °C)[3]
revealed resolved signals for (2) and (3) allowing the product distribution to be directly assessed (Table 1). These results were
corroborated (±2%) for (1b) by GC[3b] and ^{13}C NMR.[5b] This recon-
firms that a 100% excess of (1) significantly enhances monohydroboration.[3] We isolated (2b-d) in pure form by their distillation
from these reaction mixtures (Table 1), noting that, with the
exception of SiMe$_3$,[5b,6] larger R groups favor (2) *vs* (3).

Table 1 Hydroboration of 1-Alkynes (1) with 9-BBN-H.[a]

Series	R	Ratio[b]	(2)[c]	(3)[d]
a	Me	1:1	20	40
		2:1	56	22
b	n-Bu	1:1	56	22
		2:1	80 (70)	10
c	Ph	1:1	92	4
		2:1	96 (79)	2
d	i-Pr	1:1	88	6
		2:1	94 (79)	3
e	SiMe$_3$	1:1	38	31
		2:1	74	13

[a] Reactions were carried out at 0 °C [0.5 M 9-BBN-H in THF]. [b] Mol ratio of (1):9-BBN-H employed. [c] Yields are calculated based upon 9-BBN-H and are rounded to the next highest even percentage. (Isolated yields of (2) by distillation). [d] ^{11}B NMR (THF/C$_6$D$_6$) δ 87.0 (3a), 86.9 (3b), 87.6 (3c), 87.0 (3d), 85.1 (3e) ppm. Calculated from the peak areas/2.

Dehydroborylation of (3)

The facile dihydroboration of (1) with 9-BBN-H makes the quantitative preparation of (3) from (1) with 2 equiv of 9-BBN-H a trivial process.[1,3] It occurred to us that if (3) could be quantitatively converted to (2), only 9-BBN-H, but not (1), would be sacrificed. The reduction of carbonyl compounds with trialkylboranes (Midland reductions),[7] viewed from the opposite perspective, offered a potential solution through the dehydroborylation of (3). The limited information on the fate of the alkene by-product in the Midland reduction was not encouraging with the reported sluggish reaction of s-Bu-9-BBN (5) with aldehydes producing 2-butene as a c/t mixture (35/65).[7a] However, we felt that the behavior of (3) with two proximate 9-BBN groups may prove to be exceptional.

Indeed, with the addition of 1 equiv of PhCHO to (3a), in 2 h at 25 °C, B-PhCH$_2$O-9-BBN (4a) and (2a) were quantitatively formed as the only detectable products, the latter with exclusively the *trans* configuration. Moreover, the (3)→(2) conversion is quite general with aromatic aldehydes, occurring smoothly for all of our representative systems to produce (2) quantitatively as the *trans* isomer exclusively together with an equal quantity of (4) (Table 2). With the exception of (2c), the pure vinylborane was isolated in >80% yield by distillative separation from (4). To rule out a dehydroboration mechanism, we added styrene to (3a), observing no detectable reaction after 1.5 h at 25 °C.[7a,8] However, once PhCHO is added to this mixture, the smooth production of (4a) and (2a), but not B-(PhCH$_2$CH$_2$)-9-BBN, is observed, a result which confirms the absence of free 9-BBN-H in the process.

Current Topics in the Chemistry of Boron

<u>Table 2.</u> *trans*-Vinylboranes (2) via the Dehydroborylation of 3

Series	R	Ar	Yield of (2)[a,b]
a	Me	Ph	89[b]
		1-Naph	81[b]
b	*n*-Bu	Ph	100[c]
		1-Naph	86[b]
c	Ph	Ph	100[c]
		1-Naph	100[c]
		o-C$_6$H$_5$CHO	68[c,d]
d	*i*-Pr	1-Naph	85[b]
e	SiMe$_3$	1-Naph	82[b]

[a] Reactions were carried out in THF, 25 °C, 4 h. [b] Isolated yield of pure (2).
[c] ^{11}B NMR yield. [d] *o*-C$_6$H$_4$(CHO)$_2$/(3) = 0.5.

Kinetic and Mechanistic Features

MMX-minimized structures were determined for (3a) which revealed a significantly greater energetic preference (2.6 kcal/mol) for the *gauche* conformation represented in Figure 1 over the alternative bisecting *gauche* conformation compared to normal *B*-alkyl 9-BBN derivatives such as (5) (0.7 kcal/mol). This suggests that larger 1,3 repulsions in (3) (*i.e.* 9-BBN/R) compared to those in (5) (*i.e.* Me/Me) may account, in part, for the unusually high *trans* selectivity in the dehydroborylation reaction.

The second-order rate constants for the reaction of (3a) with five *para*-substituted benzaldehydes were determined resulting in the Hammett plot illustrated in Figure 2 from which a ρ value of +0.42 was calculated, very close to the +0.49 value obtained for Alpine-borane® (7).[7a] Midland also observed that both *beta* alkyl substitution and a synperiplanar arrangement of the H-C-C-B array in *B*-alkyl-9-BBN derivatives enhance the reduction rate.[7a] We carried out competitive experiments with 1 equiv of PhCHO, (3a) and either (5), *B*-siamyl-9-BBN (6) or (7) at 23 °C, which revealed an ordering of k$_{(7)}$ (4.5) > k$_{(3a)}$ (1.0) > k$_{(6)}$ (0.34) >> k$_{(5)}$. Thus, the reactivity of (3a), with only a single *beta* substituent and no fixed synperiplanar relationship of the H-C-C-B array as is present in (7), is exceptionally high.

Borane reductions are thought to occur through reversible *O*-complexation.[7a,c,9] With a low equilibrium concentration, it may not be observable by ^{11}B NMR, but its *anti* geometry (*i.e.* R-C=O-B)[10] may well be responsible for the favored configuration in Midland's proposed boat-like electrocyclic hydride transfer transition state.[7a] These considerations suggest the model depicted in

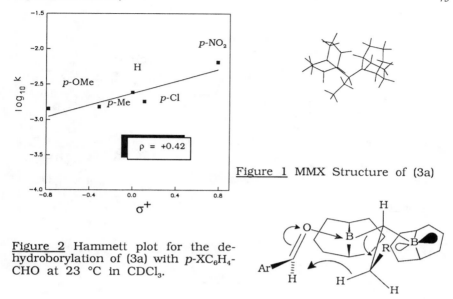

Figure 1 MMX Structure of (3a)

Figure 2 Hammett plot for the dehydroborylation of (3a) with p-XC$_6$H$_4$-CHO at 23 °C in CDCl$_3$.

Figure 3 Proposed TS for the Dehydroborylation of (3)

Figure 3 for the dehydroborylation process. The stabilizing influence of the empty p orbital on the transition state energy[6,11] which allows reaction to occur at lower temperatures than for (5), together with the steric considerations previously discussed, probably accounts for the high *trans* selectivity observed in the dehydroborylation process.

Vinylborane Conversions

Selected reactions of (2) were carried out to demonstrate the versatility of B-vinyl-9-BBN derivatives (Scheme 1). Thus, deuterated alkenes (8)[1] and air-stable 9-oxa-10-borabicyclo[3.3.2]decanes, (9),[5a,f,g,h] are efficiently prepared. The "Grignard-like" addition of (2) to aldehydes, unique to vinyl-9-BBN derivatives,[5e,12] was used to prepare (10) (89%). The Suzuki coupling[3b,5l,13,14] of (2) with PhBr produced pure *trans*-1-phenyl-1-propene (11) in 87% yield. This coupling can also be conducted with (2c) generated *in situ* (cf. (12)). The clean formation of (15), while not directly accessible from hydroboration, was easily accomplished by the dehydroborylation of (14) and further converted to (16) (64%) (Scheme 2). Finally, the remarkable "tandem walk"[5b] is utilized to isomerize the initially formed 1,2-diboryl adduct to its 3,3-isomer which is subsequently converted through (19) to the *trans*-3-deuterioallylsilane, (20) (62%) in a one-pot sequence from (17) (Scheme 3).

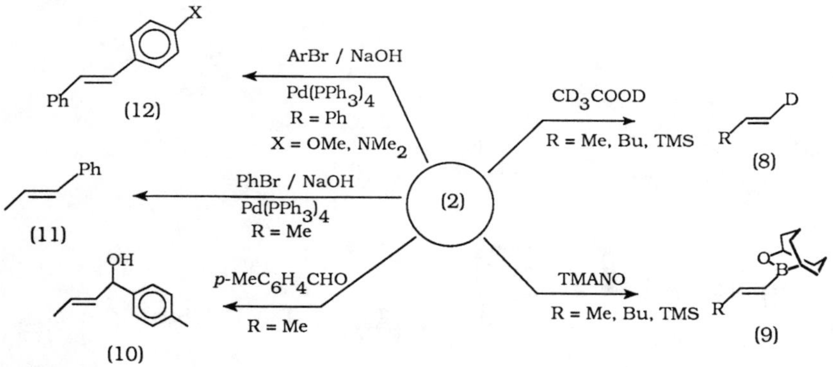

Scheme 1

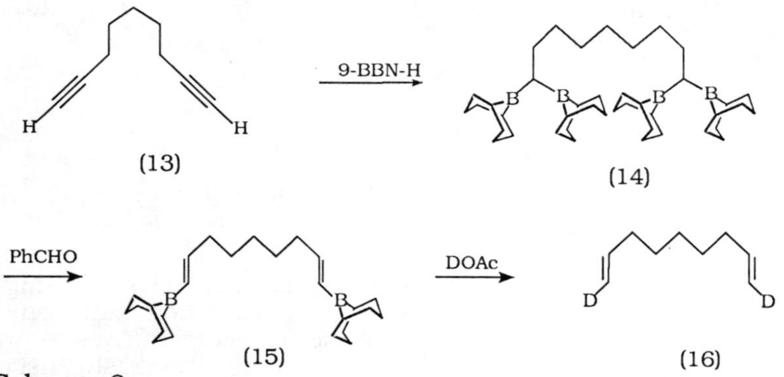

Scheme 2

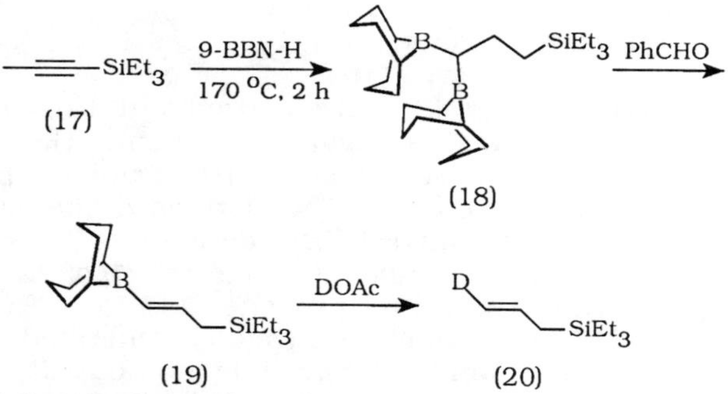

Scheme 3

Acknowledgment

The support of the National Science Foundation (EPSCoR) and the National Institutes of Health (SO6-GM08102) is gratefully acknowledged.

REFERENCES

1. (a) A. Pelter, K. Smith, H. C. Brown, "Borane Reagents", Academic Press, London, 1988. (b) J. A. Soderquist, H. C. Brown, J. Org. Chem., 1981, 46, 4559. (c) J. A. Soderquist, A. Negron, Org. Synth., 1991, 70, 169. (d) R. Köster, M. Yalpani, Pure Appl. Chem., 1991, 63, 387.

2. (a) H. C. Brown, J. B. Campbell, Jr., Aldrichimica Acta, 1981, 14, 3. (b) N. Miyaura, K. Yamada, H. Suginome, A. Suzuki, J. Am. Chem. Soc., 1985, 107, 972. (c) P. Martinez-Fresneda, M. Vaultier, Tetrahedron Lett., 1989, 30, 2929. (d) W. R. Roush, K. J. Moriarty, B. B. Brown, ibid., 1990, 31, 6509. (e) T. E. Cole, R. Quintanilla, S. Rodewald, Organometallics, 1991, 10, 3777. (f) W. Oppolzer, R. N. Radinov, J. Am. Chem. Soc., 1993, 115, 1593.

3. (a) H. C. Brown, C. G. Scouten, R. Liotta, J. Am. Chem. Soc., 1979, 101, 96. (b) K. K. Wang, C. G. Scouten, H. C. Brown, ibid., 1982, 104, 531.

4. J. C. Colberg, A. Rane, J. Vaquer, J. A. Soderquist, J. Am. Chem. Soc., 1993, 115, 6065.

5. (a) J. A. Soderquist, M. R. Najafi, J. Org. Chem., 1986, 51, 1330. (b) J. A. Soderquist, J. C. Colberg, L. Del Valle, J. Am. Chem. Soc., 1989, 111, 4873. (c) J. A. Soderquist, I. Rivera, Tetrahedron Lett., 1989, 30, 3919. (d) J. A. Soderquist, B. Santiago, ibid., 1990, 31, 5113. (e) J. A. Soderquist, J. Vaquer, ibid., 1990, 31, 4545. (f) J. A. Soderquist, B. Santiago, I. Rivera, ibid., 1990, 31, 4981. (g) J. A. Soderquist, B. Santiago, ibid., 1990, 31, 5541. (h) B. Santiago, J. A. Soderquist, J. Org. Chem., 1992, 57, 5844. (i) I. Rivera, J. C. Colberg, J. A. Soderquist, Tetrahedron Lett., 1992, 33, 6915.

6. D. A. Singleton, J. P. Martinez, Tetrahedron Lett., 1991, 32, 7365.

7. (a) M. M. Midland, Chem. Rev., 1989, 89, 1553 and references cited therein. (b) H. C. Brown, P. V. Ramachandran, S. A. Weissman, S. Swaminathan, J. Org. Chem., 1990, 55, 6328. (c) J. A. Soderquist, C. L. Anderson, E. I. Miranda, I. Rivera, G. W. Kabalka, Tetrahedron Lett., 1990, 31, 4677. (d) H. C. Brown, P. V. Ramachandran, Accts. Chem. Res., 1992, 25, 16.

8. (a) M. M. Midland, J. E. Petre, S. A. Zderic, J. Organomet. Chem., 1979, 182, C53. (b) M. M. Midland, J. E. Petre, S. A. Zderic, A. Kazubski, J. Am. Chem. Soc., 1982, 104, 528.

9. R. Bolton, Aust. J. Chem., 1990, 43, 493.

10. (a) M. T. Reetz, M. Hullmann, W. Massa, S. Berger, P. Rademacher, P. Heymanns, J. Am. Chem. Soc., 1986, 108, 2405. (b) S. E. Denmark, B. R. Henke, E. Webber ibid., 1987, 109, 2512.

11. See, for example: (a) P. G. Gassman, D. A. Singleton, Tetrahedron Lett., 1987, 28, 5969. (b) P. G. Gassman, D. B. Gorman, J. Am. Chem. Soc., 1990, 112, 8623.

12. P. Jacob, III, H. C. Brown, J. Org. Chem., 1977, 42, 579.

13. (a) A. Suzuki, Pure Appl. Chem., 1991, 63, 419. (b) N. Miyaura, T. Ishiyama, H. Sasaki, M. Ishikawa, M. Satoh, A. Suzuki, J. Am. Chem. Soc., 1989, 111, 314. (c) J. A. Soderquist, J. C. Colberg, Synlett, 1989, 25. (d) T. Ishiyama, N. Miyaura, A. Suzuki, ibid., 1991, 687. (e) T. Oh-e, N. Miyaura, A. Suzuki, J. Org. Chem., 1993, 58, 2201.

14. J. A. Soderquist, G. León-Colón, Tetrahedron Lett., 1991, 32, 43.

The 1,3-Dipolar Cycloaddition of Nitrile Oxides to Vinylboronic Esters

Richard H. Wallace*, K. K. Zong, and Melissa P. Schoene

DEPARTMENT OF CHEMISTRY, UNIVERSITY OF ALABAMA,
TUSCALOOSA, AL 35487, USA

1 INTRODUCTION

The Δ^2-isoxazolines produced by the 1,3-dipolar cycloaddition of a nitrile oxide to an alkene have proven to be extremely useful compounds in organic chemistry. [1,2] Among the various classes of compounds which have been prepared from these cycloadducts are enones, 1,3-amino alcohols, and β-hydroxy ketones. Boronic esters have also enjoyed considerable use in organic synthesis. [3,4] Procedures exist for the direct conversion of boronic esters into alcohols, aldehydes, carboxylic acids, and amines. [5] It has also been shown that it is possible to homologate boronic esters by inserting a methylene into the carbon-boron bond. [3,4] The ability to carry out the homologation of a boronic ester coupled with the large number of functional groups which it represents has resulted in the boronic ester becoming an extremely versatile functional group. One area in which we have been interested involves combining the synthetic utility of the Δ^2-isoxazoline with that of the boronic ester by carrying out nitrile oxide cycloadditions with vinylboronic esters to afford boronic ester substituted Δ^2-isoxazolines. [6]

The first 1,3-dipolar cycloaddition of nitrile oxides to vinylboronic esters was reported in 1966 by Grünager and coworkers. [7,8] In this work they detailed the reaction of three aromatic nitrile oxides with dibutyl vinylboronate. We disclose in this report our recent results in the area of nitrile oxide cycloadditions to vinylboronic esters to afford boronic ester substituted Δ^2-isoxazolines. [9]

2 RESULTS AND DISCUSSION

Since our initial report was published we have explored the use of several nitrile oxides, both functionalized and unfunctionalized, in the 1,3-dipolar cycloaddition of nitrile oxides to pinacol vinylboronate (1). In our studies we have found that the vinylboronic ester is compatible with both the

Mukaiyama and Huisgen methods of nitrile oxide generation. [1,2] The compounds prepared in these reactions and the isolated yields are shown in Scheme 1.

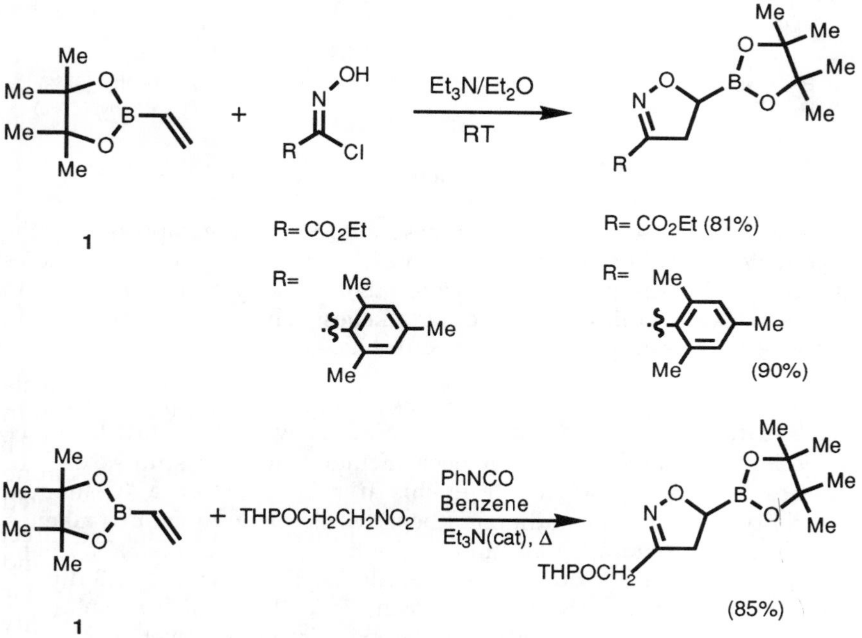

R= CO$_2$Et

R=

R= CO$_2$Et (81%)

R=

(90%)

(85%)

1

Scheme 1

Encouraged by the fact that the dipolar cycloaddition of such a wide variety of nitrile oxides to pinacol vinylboronate had been carried out with good success and the problem of one-carbon homologation of the resulting cycloadducts had been solved,[6] we wanted to explore how variations in the vinylboronic ester structure would impact on the reaction. In particular, we wanted to investigate the use of 1,1-di-substituted vinylboronic esters as dipolarophiles in these cycloaddition reactions.

The procedure used for the preparation of the di-substituted vinylboronic esters (**2 & 3**) chosen for use in these reactions is analogous to that employed for the preparation of pinacol vinylboronate (**1**). [6] The 1,3-dipolar cycloaddition of benzonitrile oxide to the vinylboronic esters **2 & 3** was carried out by treatment of an ether solution of the vinylboronic ester and phenyl hydroximic acid chloride with an ether solution of triethylamine. The reactions were stirred at room temperature overnight, filtered and evaporated to yield the cycloadducts. The Δ^2-isoxazolines were purified by Kugelrohr distillation and the isolated yields are shown in Scheme 2.

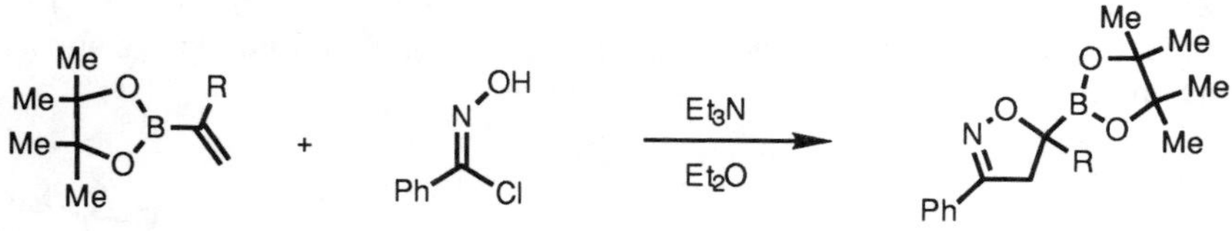

2 R = Methyl
3 R = Phenyl

R = Methyl (97%)
R = Phenyl (84%)

Scheme 2

In the course of carrying out these studies we were impressed by the reactivity of vinylboronic esters with nitrile oxides. This prompted studies directed at determining the relative reactivity of vinylboronic esters with nitrile oxides in comparison to other alkenes. These studies resulted in some very interesting and unexpected results.

Although a number of studies have addressed the question of the relative rates of reaction of various alkenes with benzonitrile oxide, vinylboronic esters have never been included in these studies. [11] The studies carried out previously in this area have shown a "U-shaped" reactivity curve for the addition of benzonitrile oxide to various alkenes. This is evidence for a significant contribution from both the HOMO and LUMO of the nitrile oxide in these reactions. These types of 1,3-dipoles have been classified by Sustmann as being Type II, and their reactions are faster with alkenes which bear electron withdrawing as well as electron donating substituents, when compared to simple alkenes. [11] Previous studies have determined the relative reactivity of various alkenes with benzonitrile oxide relative to ethylene. The results of these studies have shown: methyl acrylate is 8.3 times more reactive than ethylene, butyl vinyl ether is 2.1 times more reactive than ethylene, styrene is 1.15 times more reactive than ethylene, and norbornene is 15.3 times more reactive than ethylene. [11]

Our studies were carried out by stirring one equivalent of pinacol vinylboronate (**1**), one equivalent of the other alkene (butyl vinyl ether, styrene, methyl acrylate, or norbornene) and one equivalent of phenylhydroxamic acid chloride in ether at room temperature. Triethylamine in ether (one equivalent) was added dropwise and the reaction was stirred overnight at room temperature. The reaction was filtered, evaporated, and analyzed by ^{1}H NMR (360MHz). The ratio of the two cycloadducts was obtained by integration of the proton on the 5-position of the Δ^2-isoxazolines. The ratios of the cycloadducts obtained from these studies are shown in Scheme 3. As can be seen from the data vinylboronic esters are very reactive dipolarophiles in nitrile oxide cycloaddition reactions. Pinacol vinylboronate (**1**) proved to even be more reactive than norbornene, which is one of the most reactive dipolarophiles known in nitrile oxide cycloadditions.

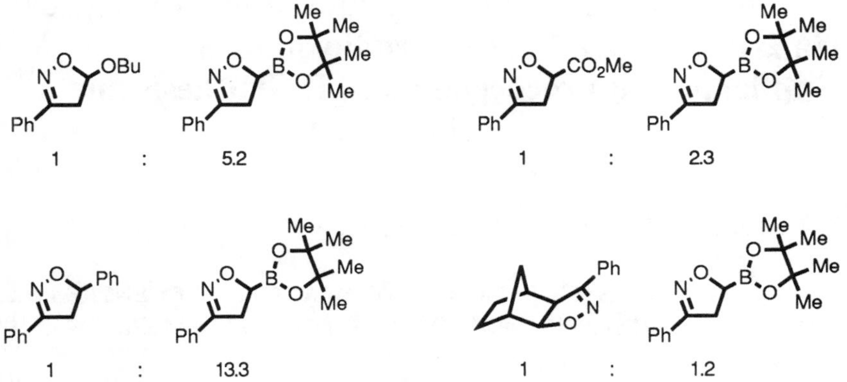

1 : 5.2   1 : 2.3

1 : 13.3   1 : 1.2

Scheme 3

Further studies in this area including the exploration of the possibility of employing optically active diols in an effort to obtain asymmetric induction in the resulting cycloadducts, and the application of this methodology for the synthesis of natural products are currently underway.

3 REFERENCES

1. P. Grünanger and P. Vita-Finzi, "Isoxazoles, Part I" ; John Wiley and Sons; E. C. Taylor, ed.; 1991, 417-621.
2. P. Caramella and P. Grünanger, "1,3-Dipolar Cycloaddition Chemistry", Wiley-Interscience, Vol. I.; A. Padwa, ed.; 1984, 291.
3. D. S. Matteson, Tetrahedron, 1989, 45, 1859.
4. A. Pelter, K. Smith, and H. C. Brown, "Borane Reagents", Academic Press, New York, 1988.
5. For a recent paper which summarizes some of the interconversions possible with boronic esters, see: M. V. Rangaishenvi, B. Singaram, and H. C. Brown, J. Org. Chem. 1991, 56, 3286.
6. R. H. Wallace and K. K. Zong, Tetrahedron Lett., 1992, 33, 6941.
7. G. Bianchi, A. Cogoli, and P. Grünanger, J. Organometal. Chem. 1966, 6, 598.
8. G. Bianchi, A. Cogoli, and P. Grünanger, Ric. Sci. 1966, 132.
9. For a paper dealing with the use of dibutyl vinylboronate as a dipolarophile with ethyl diazoacetate and diphenyldiazomethane, see: D. S. Matteson J. Org. Chem. 1962, 27, 4293. For another report dealing with the use of ethyl diazoacetate as a 1,3-dipole in a cycloaddition with a vinylboronic ester, see: W. G. Woods and I. S. Bengelsdorf J. Org. Chem. 1966, 31, 2769.
10. All the nitrile oxides employed in this work are known compounds, and were prepared by the literature procedures.[1]
11. P. Grünanger and P. Vita-Finzi, "Isoxazoles, Part I" ; John Wiley and Sons; E. C. Taylor, ed.; 1991, 489-497, and references therein.

Synthesis of Polycyclic Chromans and Chromenes via 2-Phenyl-4*H*-1,3,2-Benzodioxaborins. Application in Precocene I and II Synthesis and a Formal Synthesis of Robustadial

S. Bissada, C. K. Lau, M. A. Bernstein, and C. Dufresne*

MEDICINAL CHEMISTRY DEPARTMENT, MERCK FROSST CENTRE FOR
THERAPEUTIC RESEARCH, PO BOX 1005, POINTE CLAIRE-DORVAL,
QUEBEC, CANADA H9R 4P8

1   INTRODUCTION

The widespread occurrence of phenolic compounds as precursors to polycyclic aromatic natural products has stimulated efforts directed towards the synthetic exploitation of orthoquinone-methides.[1]   There are several elegant synthesis already reported that are based on Diels-Alder trapping or Michael addition reaction of orthoquinone-methide intermediates.[2]   We have recently reported the use of 2-phenyl-4H-1,3,2-benzodioxaborins as a stable orthoquinone-methide precursor and its reaction with various nucleophile and dienophile.[3]   We are now extending these methodologies to a new synthesis of polycyclic chromans via tricyclic-2-phenyl-4H-1,3,2-benzodioxaborins.

Synthesis of Tricyclic-2-Phenyl-4H-1,3,2-Benzodioxaborins Starting Materials

An annulation reaction was developed in the formation of tricyclic benzodioxaborins. Thus, an intramolecular reaction of a phenol possessing a meta substituted aliphatic aldehyde of various chain length (2), prepared via a palladium coupling of 3-iodophenol (1) and a terminal olefin alcohol[4], with phenylboronic acid and a catalytic amount of propionic acid gave the corresponding tricyclic benzodioxaborin of various ring size (3), shown in Scheme 1.

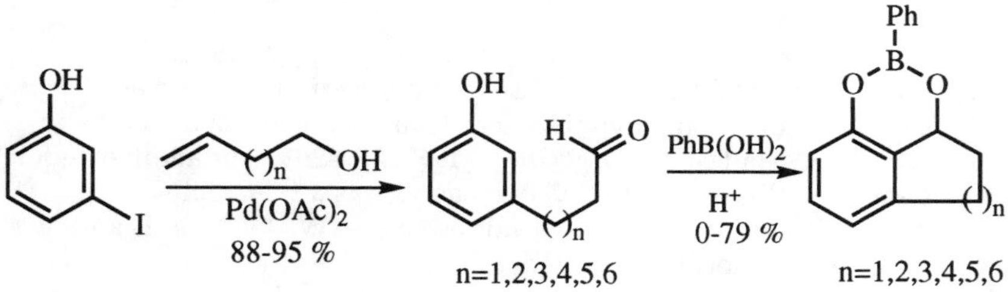

Scheme 1   Annulation reaction via the formation of tricyclic-2-phenyl-4H-1,3,2 - benzodioxaborins

2 POLYCYLIC CHROMAN SYNTHESIS

The tricyclic benzodioxaborins (3) thus obtained were then treated with various allyl trimethylsilane derivatives in the presence of BF_3-etherate. Acting as nucleophile, in a Michael addition reaction, allyl trimethylsilane derivatives react with the bicyclic orthoquinone-methide intermediate generated from the dioxaborin to give a bicyclic benzylic substituted homoallylic phenol (4). Intramolecular cyclization of the phenol to the olefin under trifluoroacetic acid condition gave the corresponding polycyclic chroman ring system (5) outlined in Table 1.

Table 1   Polycyclic Chromans via Tricyclic-2-phenyl-4H-1,3,2-benzodioxaborins

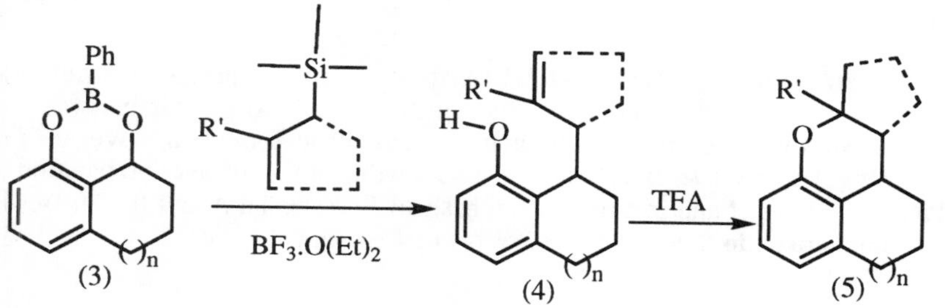

Entry n	R' Nu	Products R'	%	Chromans R'	%
3a,1	H, allylTMS	4a H	60	5a H	49
3b,1	CH$_2$OAc, "	4b CH$_2$OAc	41	5b CH$_2$OAc	58
3c,2	H, allylTMS	4c H	70	5c H	87
3d,2	CH$_2$OAc, "	4d CH$_2$OAc	50	5d CH$_2$OAc	84
3e,2	H, cyclopentene TMS	4e H,cyclo-pentene	74	5e H,cyclo-pentane	50

As we mentioned earlier, the tricyclic dioxaborin serving as a precursor of the corresponding bicyclic orthoquinone-methide, should undergo thermal [4+2] cycloaddition reaction with electron rich dienophiles. Indeed this is illustrated in Table 2 by the reaction of (3) with ethyl vinyl ether which gave the corresponding 2-ethoxy tricyclic chromans (6).

Table 2   [4+2] Cycloaddition Reactions

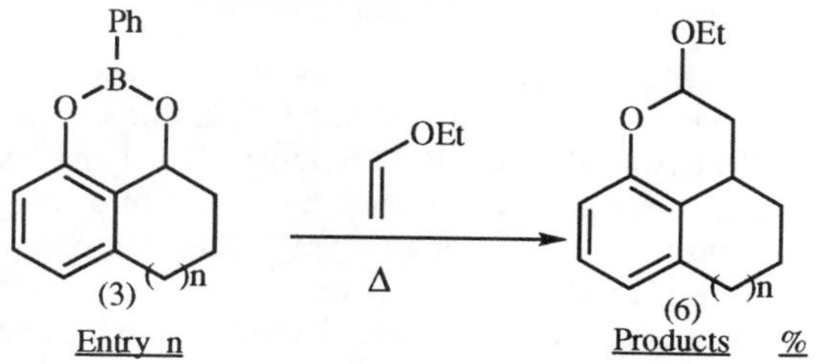

Entry n	Products	%
3a 1	6a	60
3b 2	6b	78

3   APPLICATION: PRECOCENE I AND II, ROBUSTADIAL

The title compounds are natural products which are part of the benzopyran family. Precocene I and II, exert strong antihormonal activity in sensitive insects.[5] On the other hand Robustadial A and B are used in the treatment of malaria.[6]

The classical method for chromene ring system synthesis requires three steps: (a) an acid catalyzed cyclization to chromanone from a phenol and a α,β-unsaturated acid, (b) a reduction and (c) a dehydration.[7] Another approach involves heating pyridine, a bisphenol and an α,β-unsaturated aldehyde to give a chromenylation reaction.[8] We have recently reported the reaction of α,β-unsaturated aldehydes, phenols and phenylboronic acid results in the formation of chromene ring system.[3]

Substituted-2-phenyl-4H-1,3,2-benzodioxaborins are produced in situ which decompose to an orthoquinone-methide intermediate. The latter undergoes an electrocyclization reaction resulting in the chromene formation. We are now reporting on the successful use of this methodology in the preparation of Precocene I and II and in a formal synthesis of Robustadial A and B. The results are summarized in Scheme 2, for Precocene I (9) and II (11).

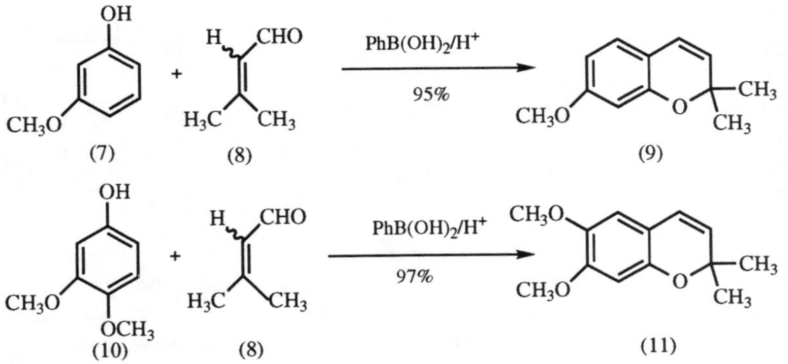

Scheme 2   One pot synthesis of precocene **I** and precocene **II**

These two one pot and efficient preparations of chromene ring systems serve to demonstrate the viability of the methodology. This process was applied to the synthesis of the Robustadial A and B. The most challenging task in synthesizing Robustadial is the stereoselective construction of the spiro ring junction at the 2-position of a benzopyran ring, while preserving the integrity of the 6,6-dimethylbicyclo[3.1.1]heptane ring systems, known to be sensitive to rearrangement under acid catalysis. First, the Robustadial skeleton was prepared via our standard reaction condition using 3,5-dimethoxy phenol (12), an α,β-unsaturated homologated ß-pinene aldehyde (13) and phenylboronic acid to give a mixture (14a) and (14b) in 3:2 ratio of isomers. Secondly, elaboration of C_3-C_4 double bond was done via hydrogenation (15a) and (15b) and oxidation with

CAN, leading to the desired chomanone (16a) and (16b). The two isomers were separated by flash chromatography and correlated with the literature ^{1}H NMR nOe and 2D-NMR spectral data for these precursors of Robustadial A and B, therefore constituting a formal of Robustadial A and B,[9] Scheme 3.

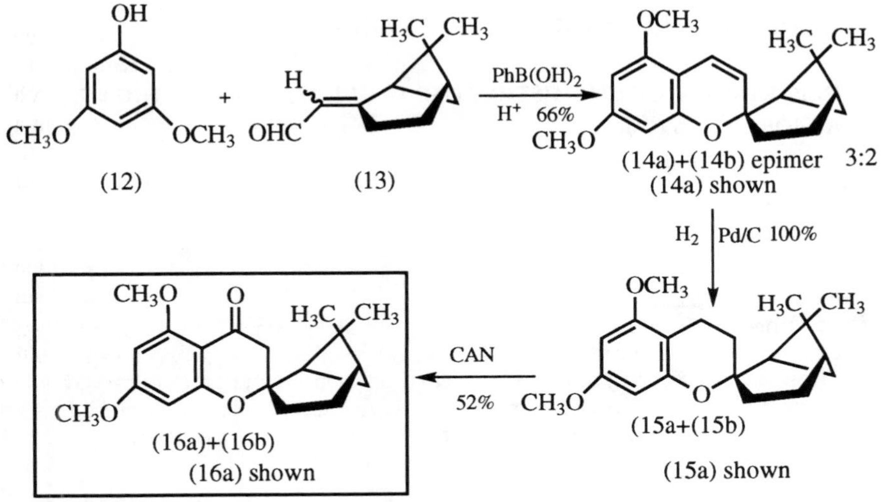

<u>Scheme 3</u>   Formal synthesis of robustadial A and B

In summary, we have shown a novel and efficient route to polycyclic chromans, Precocene I and II via 2-phenyl-4H-1,3,2-benzodioxaborin intermediates. This method has also been applied to the synthesis of Robustadial skeleton which has been elaborated into a key chromanone ring system.

REFERENCES

1.   D.L. Boger and S.N. Weinreb, 'Hetero Diels-Alder Methodology in Organic Synthesis', Academic Press, New York, 1987.
2.   a) A. Arduini, A. Bosi, A. Pochini and R. Ungaro, <u>Tetrahedron</u>, 1985, <u>41</u>, 3095. (b) M. Moreau, R. Quagliaro, R. Longeray and J. Dreux, <u>Bull. Soc. Chim. Fr.</u>, 1968, 4251. (c) D.A. Bolon, <u>J. Org. Chem.</u>, 1970, <u>35</u>, 366. (d) O.L. Chapman and C.L. McIntosh, <u>Chem. Commun.</u>, 1971, 383.
3.   J.D. Chambers, J. Crawford, H.W.R. Williams, C. Dufresne, J. Scheigetz, M.A. Bernstein and C.K. Lau, <u>Can. J. Chem.</u>, 1992, <u>70</u>, 1717.
4.   J.B. Melpolder, R. Heck, <u>J. Org. Chem.</u>, 1976, <u>41</u>, 265.
5.   a) S.Y. Dike, J.R. Merchant, N.Y. Supré, <u>Tetrahedron</u>, 1991, <u>47</u>, 4775. (b) T. Timar, S. Hosztafi, J.C. Jászberényi, <u>Acta Chim. Hung.</u>, 1988, <u>125</u>, 617.
6.   G.W. Qin, Z.X. Chen, H.C. Wang, M.K. Qian, <u>Acta Chim. Sin.</u>, 1981, <u>39</u>, 83.
7.   T. Timár, J. Csaba Jászberényi, <u>J. Heterocyclic Chem.</u>, 1988, <u>25</u>, 871.
8.   M. Tiabi, H. Zamarlik, <u>Tetrahedron Lett.</u>, 1991, <u>32</u>, 7251.
9.   R.G. Salomon, S.M. Mazza, K. Lal, <u>J. Org. Chem.</u>, 1989, <u>54</u>, 1562.

Borepins Are Aromatic. The Synthesis, Structure, and Chemistry of 1-Substituted Borepins

A. J. Ashe, III, J. W. Kampf, and W. Klein

DEPARTMENT OF CHEMISTRY, THE UNIVERSITY OF MICHIGAN, ANN ARBOR, MI 48109, USA

Borepins (1) have long interested chemists because of their isoelectronic relationship with the aromatic tropylium cation (2). Prior work has involved heavily substituted borepins.[1-7] The recent availability of 1-substituted borepins[8-11] has allowed a critical examination of the aromatic character of borepins which are uncomplicated by C-ring substitution.

Borepins (1) are prepared by the exchange reaction of 1,1-dialkylstannepin 3 with boron halides. 1-Chloroborepin (1f) is particularly useful since it can be converted to a large number of 1-substituted borepins (Scheme 1).

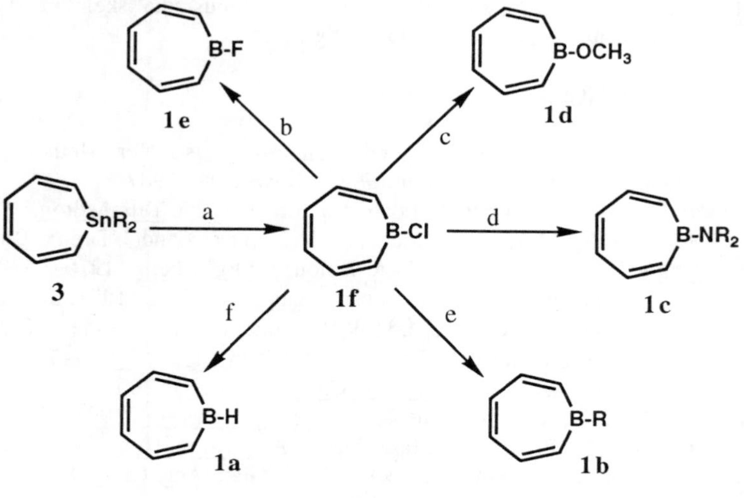

Scheme 1. Preparation of 1-substituted borepins.
(a) BCl_3 (b) SbF_3 (c) CH_3OH (d) HNR_2 (e) RLi (f) Bu_3SnH

Borepins 1b-1f are liquids at room temperature. However 1f forms X-ray quality crystals at its freezing point (-37° C). The ring is completely planar (see Figure 1).[10] The C-C bond distances range from 1.37 to 1.42 Å, which is similar to range of distances in napthalene. The B-C distance of 1.51 Å is significantly shorter than that of Me_3B (1.58 Å)[12], indicating multiple bonding between boron and carbon. Thus, the structure of 1-chloroborepin is that of a heteroaromatic compound.

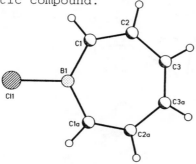

Figure 1. The molecular structure of 1-chloroborepin (**1f**).

The H-NMR spectra of all borepins (**1**) consist of an [AA'BB'CC'] pattern due to the signals of the six ring protons. In all cases the signals due to the γ-proton occur as a symmetrical multiple at highest field. Since the γ-protons are remote from the boron substituents, it is most appropriate to use their chemical shift values to evaluate ring current effects. The 1H, alkyl, aryl and halo-borepins (1a,b,fe) show low field signals (δ7.38-7.18) indicating an appreciable ring current. However borepins with donor substituents (**1c,d**) show higher field signals (δ6.58-6.89) suggesting a more modest ring current. Apparently the π-donors saturate the boron and thereby attenuate the C-B π-bonding.

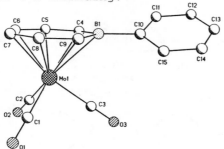

Figure 2. The molecular structure of tricarbonyl(1-phenylborepin)molybdenum (**4a**). The hydrogen atoms have been omitted for clarity.

Aromatic rings are the most common ligands in organometallic chemistry. We find that borepins are excellent ligands with strong acceptor properties. [2c,6] Borepins react with $Mo(CO)_3(C_5H_5N)_3/BF_3 \bullet OEt_2$ to give adducts **4**.[9,10] The crystal structures of the 1-chloro[4f] and the 1-phenyl[4g] complexes show that the $Mo(CO)_3$ group is η^7-coordinated to the borepin rings (Figures 2 and 3). It is interesting that the borepin rather than the phenyl ring of **4h** is coordinated.

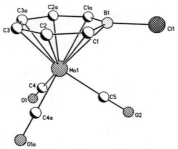

Figure 3. The molecular structure of tricarbonyl (1-chloroborepin) molybdenium (**4f**).

The comparison of the structures of **1f** and **4f** clearly shows the changes effected by complexation of the borepin ring (Figure 3). In contrast to the planar **1f**, the boron atom of **4f** is displaced out of the ring plane away from Mo by 0.05 Å. Similar displacements are found[13] for most other complexed boron heterocycles. Relative to **1f** complexation in **4f** causes bonds (C1-C2, C3-C3a) to lengthen, while bonds (B-C1, C2-C3) are unchanged. Therefore the Mo seems to remove electron density primarily from the electron rich regions of the borepin ring which are trans to the CO ligands. The result is to reduce the C-C bond alternation from ±.058 Å in **1f** to ±.023 Å in **4f**. Thus the delocalization of π-bonding is enhanced on complexation.

Like **1f** the complexed chloride **4f** can serve as a precursor to a variety of 1-substituted borepin complexes(**4**). The reduction of **4f** with lithium triethylborate gave the complex of 1H-borepin (**4a**) as a stable crystalline compound. See Scheme 2.

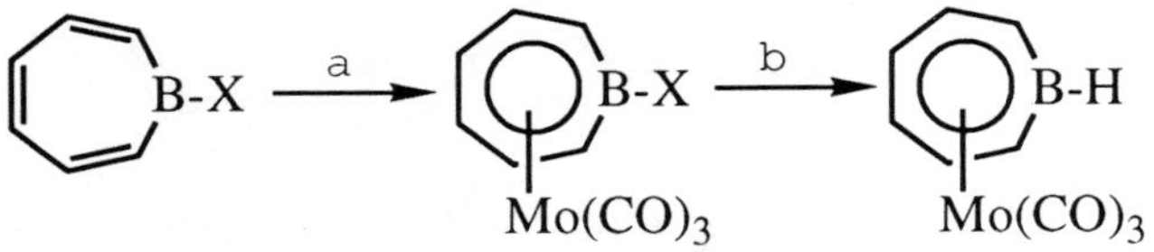

B-X $\xrightarrow{\text{a}}$ B-X $\xrightarrow{\text{b}}$ B-H

$Mo(CO)_3$ $Mo(CO)_3$

1   4   4 a

Scheme 2. a) $Mo(CO)_3(C_5H_5N)_3/BF_3 \bullet OEt_2$ b) $LiBHEt_3$

REFERENCES

1. E. E. van Tamelen, G. Brieger and K. G. Untch, Tetrahedron Lett., 1960, No. 8, 14.
2. (a) A. J. Leusink, W. Drenth, J. G. Noltes and G. J. M. van der Kerk, Tetrahedron Lett., 1967, 1263; (b) G. Axelrad and D. Halpern, Chem. Commun., 1971, 291; (c) A. J. Ashe, III, J. W. Kampf, C. M. Kausch, H. Konishi, M. O. Kristen and J. Kroker, Organometallics, 1990, 9, 2944.
3. A. T. Jefferies, III and S. Gronowitz, Chem. Scr., 1973, 4, 183.
4. J. J. Eisch and J. E. Galle, J. Am. Chem. Soc., 1975, 97, 4436.
5. W. Schacht and D. Kaufmann, Angew. Chem. Int. Ed. Engl., 1987, 26, 665.
6. (a) A. J. Ashe, III and F. J. Drone, J. Am. Chem.Soc., 1987, 109, 1879; 1988, 110, 6599. (b) A. J. Ashe, III, F. J. Drone, C.M. Kausch, J. Kroker and S. M. Al-Taweel, Pure Appl. Chem., 1990, 62, 513.
7. Y. Sugihara, T. Yagi, I. Murata and A. Imamura, J. Am. Chem. Soc., 1992, 114, 1479.
8. Y. Nakadaira, R. Sato and H. Sakurai, Chem. Lett., 1987, 1451.
9. A. J. Ashe, III, J. W. Kampf, Y. Nakadaira and J. M. Pace, Angew. Chem. Int. Ed. Engl., 1992, 31, 1255.
10. A. J. Ashe, III, J. W. Kampf, W. Klein, and R. Rousseau, Angew. Chem. Int. Ed. Engl., 1993, 32, in press.
11. A. J. Ashe, III, W. Klein and R. Rousseau, Organometallics, 1993, 12, in press.
12. L. S. Bartell and B. L. Caroll, J. Chem. Phys., 1965, 42, 3076.
13. (a) G. E. Herberich, W. Boveleth, B. Hessner, D. P. J. Köffer, M. Negele and R. Saive, J. Organomet. Chem., 1986, 308, 153. (b) W. Siebert, M. Bochmann, J. Edwin, C. Krüger and Y.-H. H. Tsay, Z. Naturforsch, B, 1978, 33, 1410. (c) G. Huttner and W. Gartzke, Chem. Ber., 1974, 107, 3786. (d) G. E. Herberich, B. Hessner, M. Negele and J. A. K. Howard, J. Organomet. Chem., 1987, 336, 29.

Stereochemistry of 1,3-Imidazolidine-N-Boranes

Armando Ariza-Castolo and Rosalinda Contreras*

DEPARTAMENTO DE QUIMICA, CENTRO DE INVESTIGACION Y DE
ESTUDIOS AVANZADOS DEL INSTITUTO POLITÉCNICO NACIONAL,
AP 14-740, MÉXICO, DF, 07000, MÉXICO

We have found that formation of a N-BH₃ adduct in
six or five membered ring nitrogen heterocycles freezes
the ring and the nitrogen inversion affording anchored
conformers[1-3]. We have studied the conformation and ring
inversion of several 1,3-diazolidines[4], and hence we were
interested in investigating the effect of N-BH₃ addition
on these systems. We report herein the NMR analysis of
the electronic and steric effects of the structure of
N-BH₃ mono-adducts (**1a-8a**) and di-adducts (**1b-8b**) of
1,3-diazolidines (**1-8**) and the determination of the N and
C atoms configuration and the ring conformation, Fig. 1.
We have previously reported a ^{11}B NMR study of these
compounds[1]. Now, we present a more detailed and precise
study of their configuration and conformation.

a N₁ = N→BH₃
b N₁→BH₃   N₃→BH₃

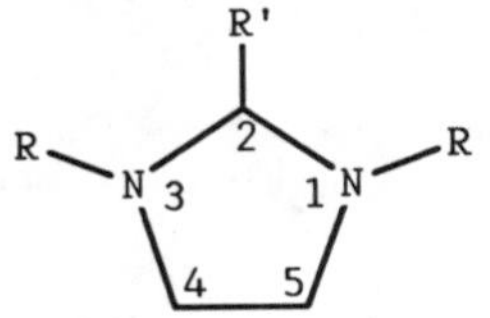

Compd	R	R'	Compd	R	R'
1	CH₃	H	5	CH₂C₆H₅	CH₃
2	CH₂C₆H₅	H	6	CH₃	C₆H₅
3	CH(CH₃)₂	H	7	CH₂C₆H₅	C₆H₅
4	CH₃	CH₃	8	CH(CH₃)₂	C₆H₅

Figure 1.

1,3-Diazolidines are symmetric heterocycles; in
compounds **1-3** addition of one molecule of BH₃ makes N-1 a
chiral center and a *dl* pair is formed. If N-inversion of
N-3 is stopped the latter becomes also a second center of
chirality. When the heterocycle has a prochiral center
(**4,8**) the presence of BH₃ produces at one time two (N-1
and C-2) or three chiral centers (N-1, C-2 and N-3) and
four or eight isomers, respectively, are in principle
expected. When two molecules of BH₃ are introduced to the
1,3-diazolidines two *dl* pairs and a meso are possible for
1-3, and one *dl* pair and two meso isomers for **4-9**.

In order to check the regio- and stereoselectivity in the formation of BH₃ adducts of 1,3-diazolidines, we have made to react the heterocycles with one equivalent of BH₃-THF at low temperature, evaporated the solvent and studied the structure by NMR methods. The addition of one equivalent of BH₃ gave in all cases selectively and quantitatively the monoborane adduct. Formation of diadducts is also a clean reaction that affords pure compounds. It occurs only when more than one equivalent of BH₃ is added to the free diazolidines indicating a different energy for the second bond formation when compared to the first addition.

The free heterocycles (**1-8**) have a symmetry plane perpendicular to the ring. Their spectra are simple owing to symmetry, ring pseudorotation and inversion of the N atoms. For the BH₃ adducts the ^{1}H, ^{11}B, and ^{13}C NMR give valuable information about the stereochemistry. The ^{1}H spectra were first order in almost all BH₃ products and were analyzed by comparison with the calculated spectra. The assignment was made considering, as it was previously reported, that the α-H *syn* to the BH₃ were shifted to high frequency[2]. The ^{13}C NMR spectra were assigned based on the δ at higher frequency for the C-α to the BH₃ group[3]. Also from the higher coupling constant values ^{1}J(^{13}C/^{1}H) for the C-α to the N-B bond[6]. The ^{1}H and ^{13}C assignments were confirmed by the 2D ^{13}C/^{1}H HETCOR. The ^{11}B[5,7] and ^{13}C[7] NMR δ give useful information because they are sensitive to the steric hindrance.

<u>MONOADDUCTS.</u> We will discuss compound **1a** as an example of monoadducts. The different NMR signals for N-1-CH₃ and N-3-CH₃ show that the N-B bond is stable in these conditions and that the BH₃ is not in a fast equilibrium between the two N atoms. The preferred conformation was deduced from ^{3}J (^{1}H-^{1}H) *trans*, Fig. 2.

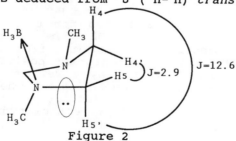

Figure 2

Compound **1a** was found in a preferred envelope conformation with the C-4 out of the plane. The N-3-CH₃ is in pseudoequatorial position and *trans* to N-1-CH₃. The spatial effects of the different substituents at the N atoms are reflected by the δ differences found at C4-CH₂ and C5-CH₂ by effect of the lone pair and the BH₃ group. The Hb *syn* to BH₃ is shifted to high frequency, followed by the Hc syn to the lone pair, Fig. 3. Compounds **2a** and **3a** have the same conformational behaviour as **1a**, as it can be found from the similarity of spectral data. The slow inversion of N-3 was evident in **2a** and **3a** when N-3 becomes a chiral center and in consequence the benzylic or isopropylic-methyl groups becomes diasterotopic.

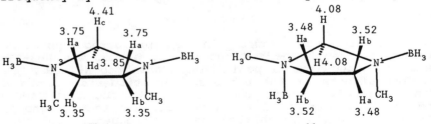

Figure 3, ^{13}C (right) and ^{1}H NMR data (left) of **1a**

Monoadducts, in 1,3-diazolidines with a prochiral center. Compounds **4-8** give in all cases only the *dl* pair of the *trans* isomers (BH3 in opposite face to the C-2 substituent). The adducts were in a preferred envelope conformation with the N-3 out of the plane and the C-2 substituent in *pseudo*equatorial position, Figure 4. This conformation was confirmed in **4a** from the $^{3}J(^{13}C-^{1}H)$ values of the N-CH3 groups coupled with the neighboring methylenes and from the $^{3}J(^{1}H/^{1}H)$. The same conformation was found for **5a-8a**.

DIADDUCTS. The introduction of the second BH3 is stereoselective to give the *trans* isomers. Each BH3 group is found in opposite faces of the heterocycle. Only in one case (**1b**) were both isomers *cis* and *trans* obtained. This was a fortunate fact that has helped to understand the conformational behaviour, the symmetry and the spectra of these systems. Compound **1b-cis** is a *meso* isomer; the symmetry plane makes equivalent Ha, Hb, the CH3 and BH3 groups. The BH3 and CH3 positions make different Ha from Hb and Hc from Hd. The ^{1}H NMR spectrum of **1b-cis** shows an AB system for C2-H owing to faces differentiation, and an AA'BB' for the other methylene groups. The molecule was found in a frozen conformer with the BH3 groups in *pseudo*equatorial position. Its ^{1}H NMR spectrum was not modified by lowering the temperature indicating an anchored molecule at rt. The protons in the same face as the BH3 are clearly shifted to high frequency by the electronic effect of hydrides, Fig. 4.

Figure 4.- ^{1}H NMR chemical shifts of compounds **1b**.

In compound **1b-trans** a C2 axis makes equivalent the CH3, the BH3 groups and the protons of C-2. The position of BH3 and CH3 makes different Ha from Hb, Surprisingly, the *trans* isomer has a very simple spectrum: the CH2 show only two singlets in a ratio 1:2, one for the methylene of C-2 and another for those of C-4 and C-5. This simple pattern can not be attributed to a fast exchange between the BH3 groups by analogy with **1a** or **1b-cis** that are not in equilibrium. The equivalence of the chemical shifts in **1b** *trans* could neither be attributed to a magnetic equivalence of groups N-BH3 with N-CH3 that could also be evident in **1a** and **1b-cis**. Our explanation is that the molecule is in a ring fast pseudorotation. It has two N-BH3, and two N-CH3 groups; each one of these in pseudoequatorial and in pseudoaxial positions which induce a conformational equilibrium. By lowering the temperature the singlets are transformed in a very close AB system indicating a very low energy of pseudorotation $\Delta G^{\ast} \cong 68.2$ KJ/mol.

The *trans* geometry can be deduced from the ^{13}C and ^{11}B chemical shifts. In compounds **4b-8b** there are different signals for each carbon atom indicating that the molecules are asymmetric. In NMR the carbon and the boron *syn* to the C-2 substituent are shifted to low frequency owing to the steric effect, an example of this behaviour is compound **4b**, shown in Fig. 5.

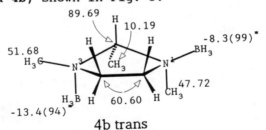

4b trans

Figure 5. $^{11}B^{\ast}$(J, Hz) and ^{13}C NMR δ, ppm, of **4b-trans**.

References.

1.-A. Flores-Parra, N. Farfán, A. I. Hernandez Bautista, L. Fernández Sanchez and R. Contreras, *Tetrahedron,* 1991, **47**, 6903.
2.-R. Contreras, F. Santiesteban, M. A. Paz-Sandoval and B. Wrackmeyer, *Tetrahedron,* 1984, **40**, 3829.
3-M. A. Paz-Sandoval, F. Santiesteban and R. Contreras, *Magn. Reson. Chem.,* 1985, **23**, 428.
4.- A. Ariza-Castolo and R. Contreras, Submitted.
5.- R. Contreras, H. R. Morales, M. L. Mendoza and C. Domínguez, *Spectrochim. Acta,* 1987, **43A**, 43 .
6.-I. I. Padilla-Martínez, A. Ariza-Castolo and R. Contreras, *Magn. Reson. Chem.,* 1993, **31**, 189.
7.-F. Santiesteban, T. Mancilla, A. Klaébé and R. Contreras, *Tetrahedron Lett.,* 1983, **24**, 739.

Kinetics and Mechanism of the Acid Catalyzed Amine–Borane Reduction of Nitrite

K. E. Bell and H. C. Kelly

DEPARTMENT OF CHEMISTRY, TEXAS CHRISTIAN UNIVERSITY, FORT WORTH, TX 76129, USA

The high degree of kinetic stability of selected alkyl and heterocyclic amine-boranes renders them useful as reducing agents in aqueous acid solution, and a manifestation of this activity is found in the acid catalyzed reduction of nitrite by trimethylamine-borane. This reaction yields a mixture of H_2 and N_2O, the latter being identified by the mass spectrum of the product gas which shows a parent ion (N_2O^+) peak at m/e = 44, and by the gas phase infrared spectrum. Stoichiometric studies, carried out by iodometric titration of the hydride remaining after exposure of excess amine-borane to acidified nitrite solutions, indicate a molar equivalence $[HNO_2]_o/[(CH_3)_3NBH_3] = 2/1$ consistent with equation 1.

$$(CH_3)_3NBH_3 + 2HNO_2 + H_3O^+ \longrightarrow$$
$$(CH_3)_3NH^+ + B(OH)_3 + N_2O + H_2 + H_2O \qquad (1)$$

Kinetic studies involved measurement of the decrease with time of total nitrite absorbance at 210-220 nm by stopped-flow or conventional spectrophotometry. The solvent usually consisted of 1% dioxane/99% H_2O; the dioxane serving to enhance the solubility of trimethylamine-borane.

In excess amine-borane and at constant hydronium ion concentration, the rate is first-order in stoichiometric nitrite concentration $[HNO_2]_o$.

$$-d[HNO_2]_o/dt = k_{obs}[HNO_2]_o \qquad (2)$$

A first-order dependence on amine-borane is evident from the linearity of k_{obs} with amine-borane concentration at constant $[H_3O^+]$. The resulting second-order rate constant, k_2 is also linear in $[H_3O^+]$ in the region pH = 0.5-3.0, but increases markedly with increasing acidity at $[H_3O^+]>0.2M$. A linear correlation between k_2 and the Hammett Acidity Function, h_o, is apparent (unit slope) from h_o 0.1 to 0.6.

A small normal substrate isotope effect is observed between 0.1 and 0.2M H_3O^+ ($k_{(CH_3)_3NBH_3}/k_{(CH_3)_3NBD_3} = 1.3$). However, at comparable acidity, a significant inverse

Table 1   Solvent Isotope Effect in the reaction of tri-
          methylamine-borane with nitrite in 2% dioxane-
          98% $H_2O(D_2O)$; t = 25°C.

$[H_2SO_4]$ (M)	$[D_2SO_4]$ (M)	pH(pD)[a]	k_{obs}[b] (s^{-1})	k_2[c] ($M^{-1}s^{-1}$)	k_{D_2O}/k_{H_2O}
0.0995		0.96	1.31	161[d]	2.83
	0.0970	(1.05)	3.03	456	
0.0535		1.21	0.668	89.6[e]	2.63
	0.0530	(1.25)	1.57	235	

[a]pD = pH + 0.41;   [b]k_{obs} = $-dln[HNO_2]_0/dt$;   [c]k_2 =
$k_{obs}/[(CH_3)_3NBH_3]$;   [d]value on extrapolation to pH 1.05
[e]value on extrapolation to pH 1.25.

solvent isotope effect is apparent with $k_{D_2O}/k_{H_2O} \approx 2.7$
(Table 1).

     The combined solvent isotope effect and correlation
of rate with h_0 suggest a stoichiometric mechanism for
nitrite reduction involving rapid pre-equilibrium protonation of substrate (HNO_2) followed by a rate-determining
attack by amine-borane on the resulting conjugate acid
which may also be conceptually regarded as hydrated nitrosonium ion, NO^+ (Scheme I).

Scheme I

$$HONO + H^+ \rightleftharpoons H_2ONO^+ \qquad \text{(rapid)} \qquad (3)$$

$$(CH_3)_3NBH_3 + H_2ONO^+ \xrightarrow{k_r} \ddagger \longrightarrow \qquad \text{(slow)} \qquad (4)$$
$$HNO + H_2O + \text{borane residue}$$

Subsequent rapid reactions presumably include formation of
a second nitrosyl hydride (HNO) via hydride transfer from
the borane biproduct of (4) to H_2ONO^+; formation of N_2O
via bimolecular association of HNO species; hydrolysis of
a third (borane derived) hydridic hydrogen; and equilibra-
tion of amine with its conjugate acid.

$$H^- + H_2ONO^+ \longrightarrow HNO + H_2O \qquad (5)$$

$$2HNO \longrightarrow N_2O + H_2O \qquad (6)$$

$$H^- + H_3O^+ \longrightarrow H_2 + H_2O \qquad (7)$$

$$(CH_3)_3N + H_3O^+ \rightleftharpoons (CH_3)_3NH^+ + H_2O \qquad (8)$$

     The kinetic consequence of Scheme I is given in equa-
tion 9 where $[HNO_2]_0 = [H_2ONO^+] + [HONO] + [NO_2^-]$ and K_a
and K_a' represent the acid dissociation constants of HONO
and H_2ONO^+ respectively.

$$\frac{-d[HNO_2]_0}{dt} = \frac{k_r[(CH_3)_3NBH_3][HNO_2]_0[H_3O^+]^2}{K_a'[K_a + [H_3O^+]]} \qquad (9)$$

A first-order dependence of rate on $[H_3O^+]$ is found at high acidity where $[H_3O^+] \gg K_a$. Consistent with (and demanded by) this scheme is the transition from a first-order to second-order dependence on $[H_3O^+]$ at pH values greater than pK_a for HONO (3.4). Such a transition is shown in Figure 1.

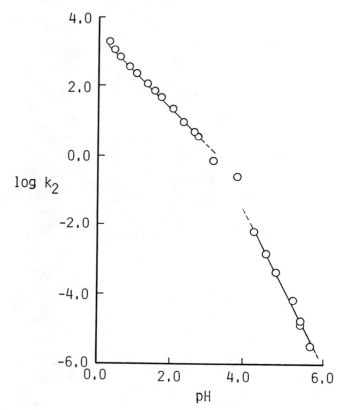

<u>Figure 1</u>   Log k_2 vs pH.   Solvent: 1% dioxane/99% H_2O.
          From pH 0.5 -3, slope = -1.08; from pH 4-
          5.8, slope = -1.98; t = 25°C.

Although the substrate isotope effect is small, it is not unlike those observed in other reactions of boranes in which some degree of hydride transfer from boron is proposed to occur in the corresponding activated complex.[1,2] The formation of nitrosyl hydride (HNO) has been proposed in numerous studies of nitrosation mechanisms[3] and the dynamics of HNO dimerization and subsequent decomposition to N_2O and H_2O has been extensively investigated.[4,5] Whether nitrosonium ion exists as free NO^+ or as nitrous acidium ion, H_2ONO^+ is a matter of conjecture. Williams has suggested this to be dependent on the acidity of the medium employed in nitrosation reactions.[3] In either event, the formal pattern of the kinetics of the reduction of nitrite by amine-borane is not affected.

Equation 10 relates the experimental second-order rate constant, k_2, to parameters derived from Scheme I.

$$k_2 = k_r[H_3O^+]^2/[K_a'[K_a + [H_3O^+]]] \qquad (10)$$

Data taken in the region of pH 1-3 in Figure 1 lead to a value for the third-order rate constant, $k_r/K_a \simeq 2500M^{-2}s^{-1}$. Interestingly, this is the approximate value of rate constants obtained in a series of nitrosation reactions thought to involve near diffusion controlled attack of substrate by NO^+.[6] In addition, Bayless and co-workers have described nitrous acid protonation in terms of equation 11 for which K_{eq} is calculated to be 3×10^{-7}.[7] If

$$HONO + H^+ \rightleftharpoons NO^+ + H_2O \qquad (11)$$

this is now taken as the pre-equilibrium step in Scheme I with subsequent attack by amine-borane occurring on NO^+ rather than H_2ONO^+, K_a^{-1} becomes equated to 3×10^{-7} and $k_r \simeq 8\times10^9 M^{-1}s^{-1}$ which indeed implies a near diffusion controlled reaction of nitrosonium ion with amine-borane.

This reaction of trimethylamine-borane with nitrite represents but one of many examples of the enhanced reducing capacity of selected amine-boranes in acidic media. It suggests the potential use of amine-boranes in the trapping of nitrosating agents with possible extension to biological systems. Also suggested is the possible application of hydrolytically stable hydrides as scavengers of other nitrogen/oxygen species and related environmental pollutants in solution.

REFERENCES

1. M.F. Hawthorne and E.S. Lewis, J. Am. Chem. Soc., 1958, 80, 4269.
2. S.S. White, Jr. and H.C. Kelly, J. Am. Chem. Soc., 1970, 92, 4203.
3. D.L.H. Williams, Adv. Phys. Org. Chem., 1983, 19, 381.
4. F.C. Kohout and F.W. Lampe, J. Chem. Phys., 1967, 46, 4075.
5. D.A. Bazylinski and T.C. Hollocher, Inorg. Chem., 1985, 24, 4285.
6. J.H.Ridd, Adv. Phys. Org. Chem., 1978, 16, 1.
7. N.S. Bayless , R. Dingle, D.W. Watts and R. Wilkie, Aust. J. Chem., 1963, 16, 933.

CHIRAL ORGANOBORANES

Recent Advances in the Boron Route to Asymmetric Synthesis

Herbert C. Brown* and P. V. Ramachandran

H. C. BROWN AND R. B. WETHERILL LABORATORIES OF CHEMISTRY,
PURDUE UNIVERSITY, WEST LAFAYETTE, IN 47907, USA

INTRODUCTION

It was serendipity that we discovered the hydroboration reaction, the remarkably facile addition of diborane in ether solvents to alkenes and alkynes (eq 1).[1]

$$\text{C=C} \quad + \quad H-B \quad \longrightarrow \quad H-C-C-B \tag{1}$$

This reaction was initially received without enthusiasm. Little was known about the chemistry of organoboranes. But, we studied the characteristics of hydroboration systematically.[2] These studies gave birth to several new reagents. Systematic study of the organoborane products made available by hydroboration revealed their remarkable versatility (Fig. 1).[3]

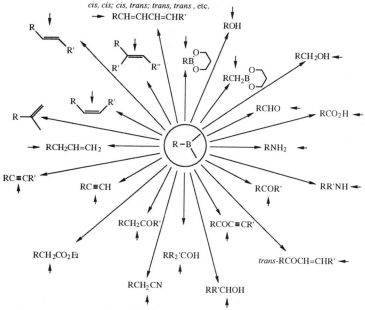

Fig. 1. The versatile organoboranes. Substitution with retention.

An unexpected feature revealed by these studies was the fact that the great majority of the substitution reactions of organoboranes proceed with complete retention of configuration in the organic group that is transferred from boron to some other element or group (Fig 2).

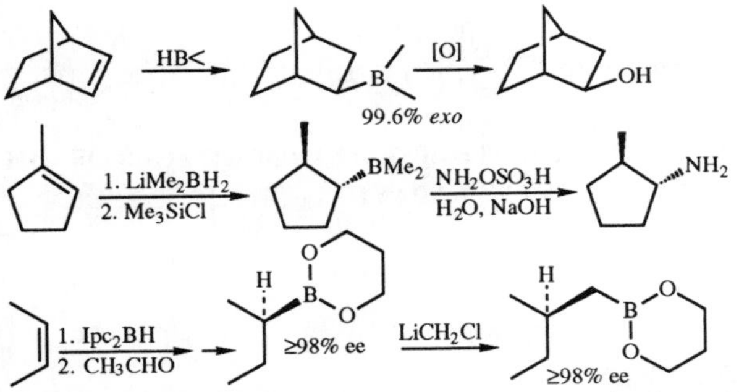

Fig. 2. Representative examples of reactions involving substitution with retention.

ASYMMETRIC HYDROBORATION

The discovery of asymmetric hydroboration marked the beginning of practical asymmetric synthesis and a new era in organic chemistry. It began with a study of the hydroboration of α-pinene, a reaction carried out to test for possible rearrangements during the hydroboration of sensitive olefins. This experiment led to diisopinocampheylborane, Ipc$_2$BH, which hydroborated *cis*-2-butene to provide 2-butanol (after oxidation) in 87% ee, the highest asymmetric yield ever achieved at the time.[4a] This result was even better than it appeared since the α-pinene used to prepare the reagent was only 92% ee. Later we developed procedures for preparing Ipc$_2$BH of ≥99% ee,[4b] which provided 2-butanol of 98% ee (Fig. 3).[4c] The reaction is general for most types of *cis*-olefins, including heterocyclic olefins .[4d]

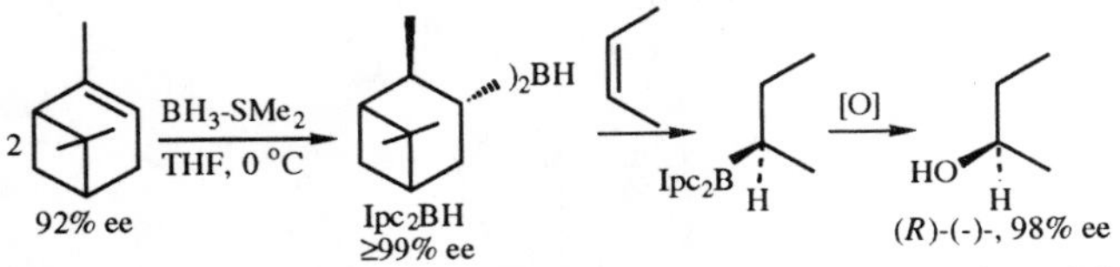

Fig. 3. Synthesis of Ipc$_2$BH and asymmetric hydroboration of *cis*-2-butene.

However, Ipc$_2$BH failed for the asymmetric hydroboration of other more hindered classes of olefins. This led to monoisopinocampheylborane, IpcBH$_2$, as a useful reagent for the asymmetric hydroboration of the more hindered, *trans-* and trisubstituted olefins (Fig. 4).[5]

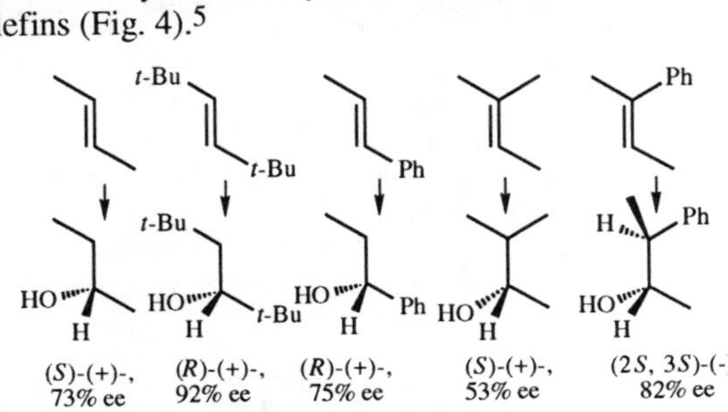

Fig. 4. Asymmetric hydroboration of *trans-* and trisubstituted olefins with IpcBH$_2$.

A GENERAL ASYMMETRIC SYNTHESIS

Initially, the application of chiral organoboranes was limited to the synthesis of alcohols. It was desirable to recycle the chiral auxiliary (α-pinene) and to use the R*B<, free of the chiral auxiliary for further modification. We successfully achieved both of these aims and developed procedures to convert the trialkylboranes Ipc_2BR^* and $IpcBR^*R'$ to $R^*B(OR)_2$ and $R^*R'BOR$, optically active boronates and borinates, respectively, by treatment with an aldehyde.[6] A bonus in this reaction was the easy recovery of the chiral auxiliary, α-pinene, without loss of optical activity. These boronates and borinates could be easily converted into optically active mono- or dialkylboron intermediates readily utilized in our program on general asymmetric synthesis via chiral organoboranes (Fig. 5). The small arrows in fig. 5 indicate the syntheses that have already been demonstrated.

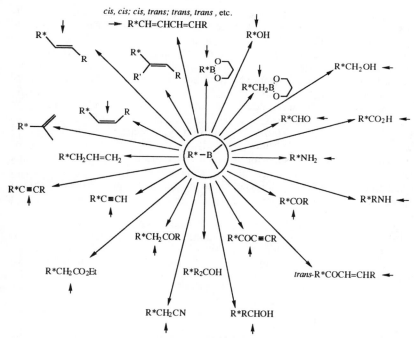

Fig. 5. A general asymmetric synthesis *via* asymmetric hydroboration.

Recent Applications in Asymmetric Synthesis: Around Figure 5.

A decade ago, we began our systematic program of transforming the chiral boron intermediates into the desired optically active molecules. Several representative reactions from Fig. 5 have been discussed earlier.[7] In this chapter, we report the more recent developments in asymmetric synthesis as part of the program to confirm all of the reactions shown on the chart in Fig. 5.

Since the alkyl group on boron is most efficiently utilized in transformations with boronic esters, these esters have emerged as important organoborane intermediates for asymmetric synthesis. We have achieved continued success in demonstrating the utility of these esters for the synthesis of optically active compounds. Unfortunately, all of the hydroboration reactions do not yield optically pure organoboranes. We have overcome this difficulty by developing simple upgradation procedures whereby the initial hydroboration product of lower ee is upgraded by one of the following three methods.

Optically pure borinates can be synthesized by successive hydroboration of two appropriate alkenes with IpcBH$_2$, with crystallization of the initially formed IpcR*BH.[8a] Treatment with aldehyde eliminates α-pinene to provide a single enantiomer of the product. An alternate approach for obtaining optically pure boronic esters *via* asymmetric hydroboration consists of treatment of the initial boronates with a chelating agent to furnish a crystalline material. This is then recrystallized to yield the boronic ester derivatives in essentially optically pure form.[8b] This method of optical upgrading can also be applied to borinic esters. Optical enrichment of the initial hydroboration product can also be achieved by a kinetic resolution with less than one equivalent of benzaldehyde. This controlled treatment liberates α-pinene selectively from one isomer thus providing the boronate ester in essentially 100% ee.[8c]

Synthesis of α-chiral *sec*-amines, alkenones, olefins, and acetylenes. The synthesis of α-chiral primary amines was reported earlier.[9a] We have now achieved an excellent route for the synthesis of α-chiral *sec*-amines.[9b] Our established procedures for the synthesis of α-chiral ketones was applied for the synthesis of α-chiral-α'-*trans*-alkenyl ketones.[10] The synthesis of both [E] and [Z]-olefins were achieved in high yields and ee from chiral boronates using procedures established during our organoborane program.[11] Chiral [Z]-alkenylborinates when treated with iodine in the presence of sodium methoxide provide [Z]-alkenes in high yields and ee.[11] α-Chiral [E]-alkenyl boronates can be converted into optically pure [E]-olefins by treating IpcR*BH with 1-bromoalkyne followed by, in steps, (1) acetaldehyde, (2) aqueous sodium hydroxide, (3) trimethylene glycol, and (4) protonolysis with acetic acid.[11] This procedure is superior to our earlier multistep procedure for the preparation of α-chiral [E]-alkenes starting from R*B(Thx)H.[12] Earlier we had established procedures for the synthesis of chiral internal acetylenes. We extended this for the synthesis of terminal acetylenes *via* the silyl acetylenes.[13]

Asymmetric cyclic hydroboration. Asymmetric hydroboration of appropriate dienes, followed by high pressure carbonylation, provides a general method for the synthesis of optically active *trans*-fused bicyclic ketones. IpcBH$_2$ provides the ketones in low ee. Utilization of IpcBHCl for the initial hydroboration, followed by the before-mentioned upgradation by kinetic resolution during the preparation of the borinate ester,[8c] increases the ee to ≥99% (Fig. 6).[14]

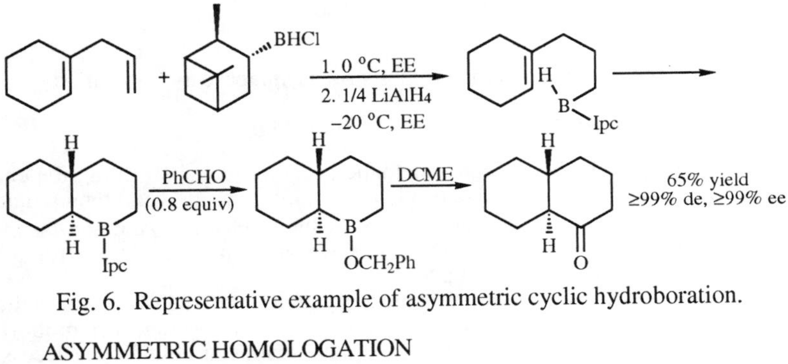

Fig. 6. Representative example of asymmetric cyclic hydroboration.

ASYMMETRIC HOMOLOGATION

Matteson and his co-workers developed an ingenious and elegant procedure for the preparation of α-chiral boronate esters *via* an asymmetric homologation.[15] Thus the reaction of cyclic boronate esters derived from pinanediol with pre-formed dichloromethyllithium, LiCHCl$_2$, at –100 °C, followed by transfer of the organic group from boron to carbon, induced by anhydrous ZnCl$_2$, provides the asymmetric boronate ester with ≥99% enantioselectivity.

While the Matteson procedure provides a method for the synthesis of many optically pure compounds that cannot be achieved *via* direct hydroboration, it has certain drawbacks. The rigorous requirement for inconvenient reaction temperatures, such as –100 °C, for the preformation of LiCHCl$_2$ and the difficulty in recycling the chiral auxiliary, pinanediol, limited the scaling-up of this procedure. We overcame these limitations by utilizing LiCHCl$_2$ prepared *in situ* and carrying out the reaction at –78 °C. Procedures were also developed for recycling the pinanediol chiral auxiliary without loss of optical activity. We applied the chiral boronates derived by the modified procedure to the synthesis of aldehydes, acids, amines, ketones etc. using our established procedures (Fig. 7).[16]

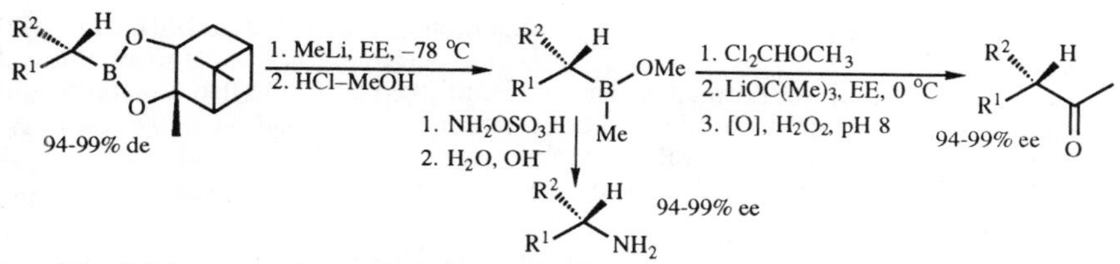

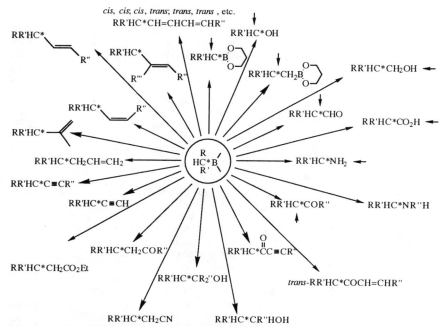

Fig. 7. Asymmetric synthesis of α-chiral ketones and amines inaccessible *via* asymmetric hydroboration.

Thus, our established synthetic schemes can also be applied to the boronates prepared *via* Matteson's homologation procedure, the synthesis of which are difficult to realize *via* asymmetric hydroboration (Fig. 8). (The small arrows in the chart indicate completed work).

Fig. 8. A general asymmetric synthesis *via* asymmetric homologation.

Alternatives for asymmetric hydroboration of 2-substituted-1-alkenes. A major deficiency in our asymmetric hydroboration procedure was circumvented by Matteson's homologation methodology. Using a one-carbon homologation of α-chiral boronates we are now in a position to synthesize the boronates that cannot be obtained in high ee *via* the direct hydroboration of 2-substituted-1-alkenes. We applied this homologation procedure to our α-chiral boronates prepared from asymmetric hydroboration and succeeded in the synthesis of β-chiral boronate esters (Fig. 9).[17]

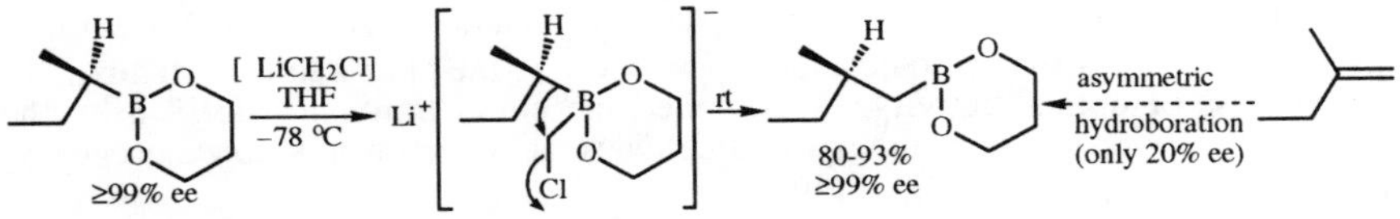

Fig. 9. Alternative for asymmetric hydroboration of 2-substituted-1-alkenes.

The above β-chiral boronates can now be utilized in all the usual reactions of organoboranes to prepare β-chiral molecules. Another homologation of the above β-chiral boronates provides γ-chiral boronates; and a third homologation, δ-chiral boronates. These β-, γ- and δ-chiral boronates can be used for the general asymmetric synthesis shown in Fig. 8.

Three-carbon homologation

The utility of the Matteson homologation procedure for the synthesis of medium ring boracyclanes has been established. We synthesized up to the 12-membered boracyclanes starting from borinane, increasing the ring size one carbon at a time.[18a] Recently we have developed a three-carbon homologation process utilizing (α-chloro)allyllithium generated *in situ* from allyl chloride and LDA at –78 °C (Fig. 10).[18b,c]

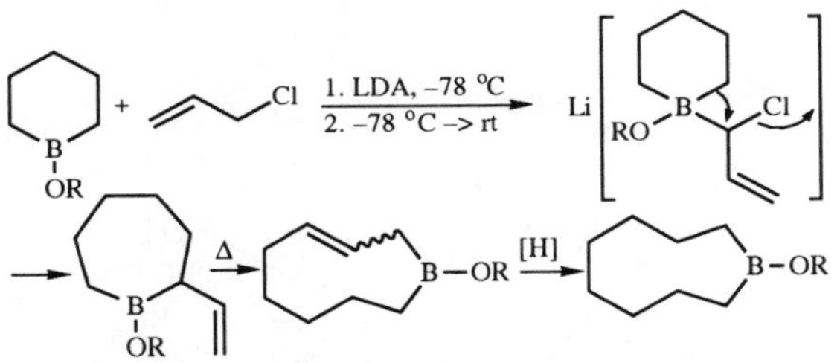

Fig. 10. Three-carbon homologation of boracyclanes.

Broad scope of organoborane chemistry

The easy synthesis of chiral boronate esters by hydroboration with Ipc_2BH or $IpcBH_2$, or by Matteson's homologation procedure, has expanded the scope of this asymmetric synthesis to an unimaginable extent. We have now achieved the synthesis of 34 optically pure boronates *via* hydroboration. Since both enantiomers of α-pinene are readily available, we can synthesize 68 pure enantiomers. A comparable number should be easily synthesized *via* asymmetric homologation. This doubles the number of optically active boron intermediates to 136. A simple one-carbon homologation doubles the number of compounds to 272. A second homologation triples the original number (408). A third sequence makes a total of 544 pure enantiomers.

We have shown 24 major reactions in Figs. 6 and 18 (other, less important reactions are also known). Each of the above boronates can undergo the 24 major reactions in the chart. This makes a total of 13,056 optically pure compounds. Many of the functional groups contained in some of these 13,056 compounds can be transformed to new functional groups. Thus we are now capable of synthesizing more than 100,000 pure enantiomers using simple organoborane chemistry! Based on our successes in a relatively short period of time, we believe that greater success awaits those willing to undertake new applications of this chemistry. Needless to say, this chemistry is still very young.

ASYMMETRIC REDUCTION

The capability of synthesizing optically active organoboranes by hydroborating suitable optically active terpenes led to the possibility of achieving asymmetric reduction of prochiral ketones, another boron based synthesis of pure enantiomers.

M. M. Midland and coworkers developed *B*-isopinocampheyl-9-borabicyclo-[3.3.1]nonane (Aldrich: Alpine-Borane®) as the first successful chiral organoborane reducing agent. He utilized it to prepare a number of optically pure primary 1-deuteroalcohols by reduction of the deuteroaldehydes, RCDO (Fig. 11).[19a] However, Alpine-Borane fails to reduce simple prochiral ketones, such as acetophenone and 3-methyl-2-butanone. Yet certain reactive carbonyls, such as α,β-acetylenic ketones, α-keto esters, and α-halo ketones, can be converted to the corresponding alcohols in very high ee with Alpine-Borane (Fig. 11).[19b] The poor selectivity in the reduction of simple ketones with Alpine-Borane is presumed to be due to a concurrent dehydroboration of the reagent in slow reductions, followed by an achiral reduction of the carbonyl group by the 9-BBN produced in this stage.[20] This problem can be overcome by minimizing the dissociation either by conducting the reductions in high concentrations at room temperature,[21a] or at greatly elevated pressures.[21b] However these modifications are still incapable of achieving the chiral reduction of unactivated ketones in high ee.

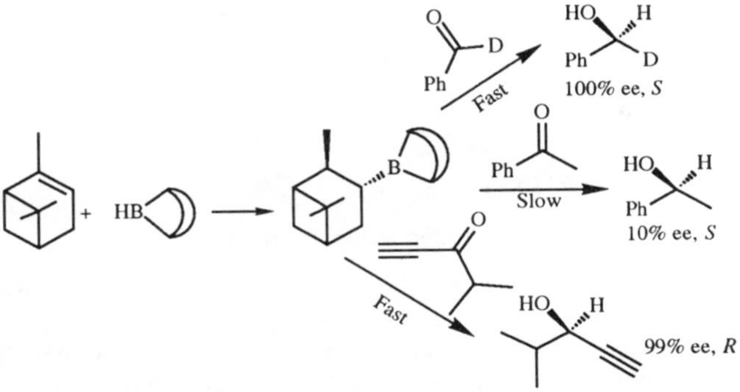

Fig. 11. Preparation and asymmetric reduction with Alpine-Borane.

B-Chlorodiisopinocampheylborane (Ipc$_2$BCl, DIP-Chloride™)

Another method examined for increasing the rate of reduction was a change in the electronic environment of the boron atom. Our investigations had indicated that sterically hindered R$_2$BCl derivatives are more stable toward dissociation than R$_3$B. Accordingly, we synthesized *B*-chlorodiisopinocampheylborane (Aldrich: DIP-Chloride™) which consistently reduced aralkyl ketones extremely efficiently with predictable stereochemistry.[22] The original workup procedure for DIP-Chloride reductions involved a non-oxidative removal of the boron by-product as the diethanolamine complex. We have since developed a considerably improved workup procedure for the isolation of product alcohols after reduction (Fig. 12). This achieves the complete recovery of α-pinene from the reagent for recycle.[23]

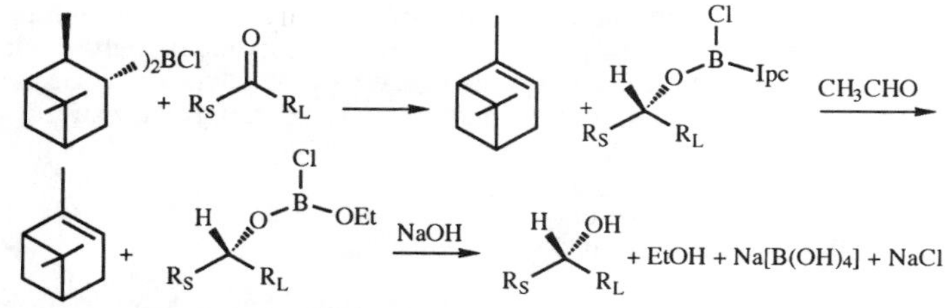

Fig. 12. Improved work-up procedure for DIP-Chloride reductions.

Perfluoroalkyl ketones. Fluorinated compounds are gaining importance in organic, medicinal, biological, and agricultural chemistry. Application of DIP-Chloride for the reduction of perfluoroalkyl ketones provide the products in very high ee.[24] Both aromatic and aliphatic ketones are reduced with equal efficiency. One significant feature in the reduction of fluoroalkyl ketones is the fact that the reduction products possess the opposite configuration, compared to the hydrogen analogs. Apparently, the electronic and steric effects of the fluorine atoms alter the course of the reduction.

Modified reagents

The effectiveness of DIP-Chloride for the reduction of hindered ketones persuaded us to consider carefully the proposed transition state for the reaction. It appeared that an increase in the steric requirement of the group at the 2-position of the apopinene moiety might increase the ee achieved in the chiral reduction (Fig. 13).[25]

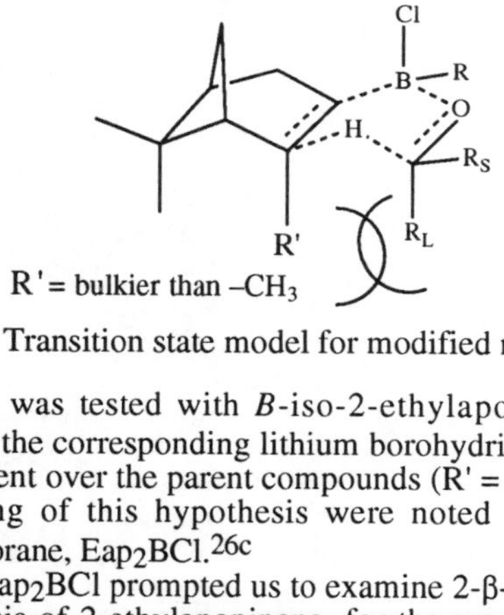

Fig. 13. Transition state model for modified reagents.

This hypothesis was tested with B-iso-2-ethylapopinocampheyl-9-BBN (Eapine-Borane)[26a] and the corresponding lithium borohydride (Eapine-Hydride).[26b] Considerable improvement over the parent compounds (R' = Me) was realized. Even better results supporting of this hypothesis were noted with B-chlorodiiso-2-ethylapopinocampheylborane, Eap_2BCl.[26c]

The success of Eap_2BCl prompted us to examine 2-β-chloroethylapopinene, a precursor in the synthesis of 2-ethylapopinene, for the preparation of a new chiral reducing agent. Indeed, diiso-2-β-chloroethylapopinocampheylborane ($Cleap_2BCl$), synthesized from 2-β-chloroethylapopinene and chloroborane-methyl sulfide complex proved highly favorable for the chiral reduction of eight of the ten classes of ketones in high ee and a ninth class in moderate ee.[27] We do not foresee any major difficulty in synthesizing reagents with increased steric requirement at the 2-position of apopinene that can handle all classes of ketones.

ASYMMETRIC ALLYL- AND CROTYLBORATION

The art of asymmetric synthesis has become highly sophisticated in conformationally non-rigid systems such as macrolide and ionophore antibiotics with a plethora of stereodefined *vic*-diols or β-methyl alcohols.[28] Here, not only the enantioselectivity, but the diastereoselectivity of the reaction is also highly important. Accordingly, numerous searches for the most efficient reagent that can achieve both these selectivities in a single step have been made. Chiral organoboranes have also revealed their uniqueness and advantages for these desired transformations.

B-Allyldiisopinocampheylborane. Based on our successes in asymmetric hydroboration with Ipc_2BH, we envisaged that *B*-allyldiisopinocampheylborane, Ipc_2BAll, might be a successful reagent for asymmetric allylboration. The synthesis of the reagent from Ipc_2BH was simple and the reaction with aldehydes at −78 °C, followed by either alkaline hydrogen peroxide or ethanolamine work-up, provided very good yields of the homoallylic alcohols in very high ee (Fig. 14).[29]

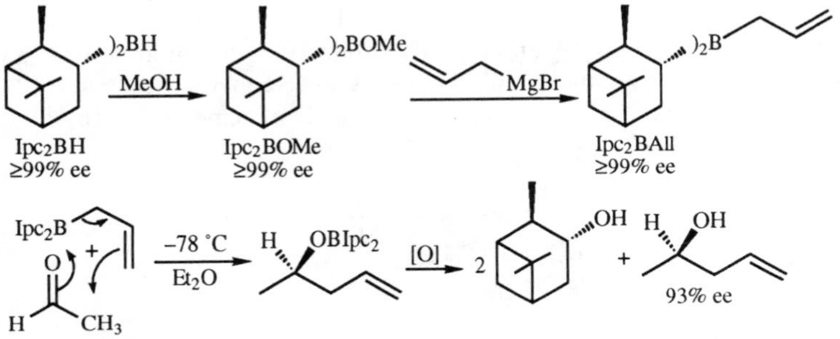

Fig. 14. Ipc_2BAll achieves excellent asymmetric alylborations.

The success of Ipc_2BAll led to several other derivatives, such as *B*-methallyldiisopinocampheylborane, 3,3-dimethylallyldiisopinocampheylborane, [Z]-3-methoxyallyldiisopinocampheylborane, etc. and all of them proved highly successful.[7a]

B-[*E*]- and [*Z*]-Crotyldiisopinocampheylborane

Based on our successes with various allylborating agents, there was no reason for us not to believe that crotylborations will also be highly successful with our reliable chiral auxiliary, α-pinene. Practical procedures were developed for the synthesis of pure Ipc_2BCrt^E and Ipc_2BCrt^Z starting with isomerically pure crotylpotassium. Asymmetric crotylboration of aldehydes with these derivatives proceeded with remarkable optical and geometric efficiencies. Consequently, it is now possible to synthesize, at will, each of the four possible isomers of β-methylhomoallylic alcohols (Fig. 15).[30]

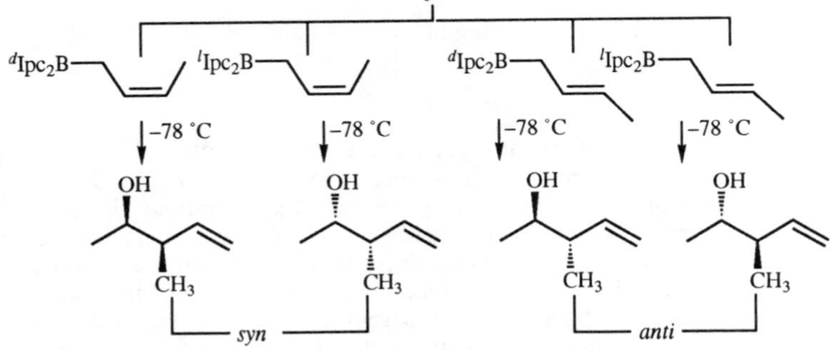

Fig. 15. Synthesis of four possible isomers of β-methylhomoallylic alcohols.

B-2'-Isoprenyldiisopinocampheylborane. Our success with crotylpotassium persuaded us to prepare the *B*-2'-isoprenyldiisopinocampheylborane from isoprenylpotassium and *B*-methoxydiisopinocampheylborane. Condensation of this reagent with aldehydes provided isoprenylated chiral alcohols. This methodology was applied for an efficient one-pot synthesis of both enantiomers of the bark beetle *Ips paraconfusus* Lanier, ipsenol and ipsdienol.[31] This simple synthesis is the sharp contrast to multistep syntheses (13 and 17 steps) by Mori.[32]

Improved reagents. Though we did not have much success with chiral hydroborating agents derived from other terpenes, such as 2-carene, 3-carene, limonene, and longifolene, the allylboration reagents synthesized from 2-carene and 3-carene are proving to be even more efficient than that from α-pinene.[33]

Synthesis of γ-butyrolactones. The optically active homoallylic alcohols readily available *via* asymmetric allylborations were converted to γ-substituted-γ-butyrolactones without loss of optical activity using the scheme in Fig. 16.[34]

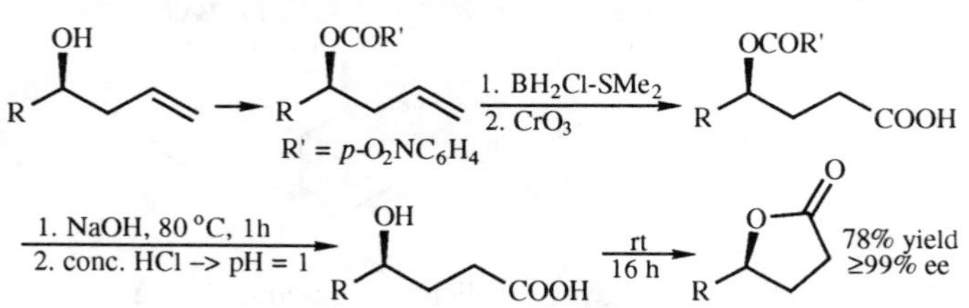

Fig. 16. Synthesis of chiral γ-butyrolactones.

ASYMMETRIC CLEAVAGE OF MESO-EPOXIDES

We carried out a systematic study of the asymmetric ring cleavage of *meso*-epoxides using mono- and diisopinocampheylhaloboranes and succesfully accomplished the preparation of 1,2-halohydrins in good to excellent ee.[35] We found that diisopinocampheyliodoborane is the best suited reagent (95-≥99% ee) to achieve the cleavage in an anti-periplanar manner, with an S_N2 type reaction pathway (Fig. 17). This reaction sequence provides highly valuable optically active difunctionalized compounds for asymmetric synthesis.

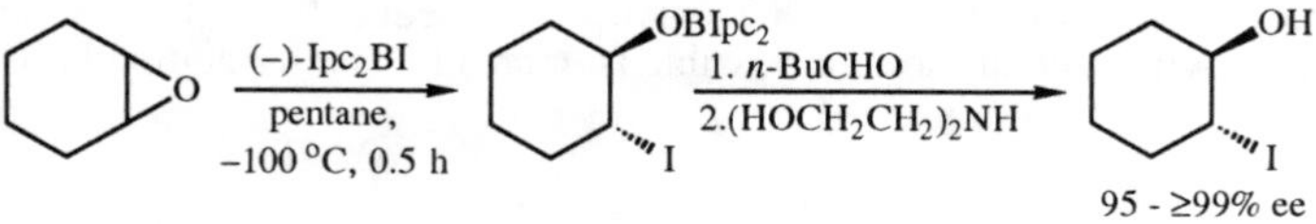

Fig. 17. Asymmetric ring opening of meso-cyclohexene oxide with Ipc₂BI.

CONCLUSIONS

The reaction of α-pinene with borane, just another routine reaction during the study of the characteristics of hydroboration, turned out to be one of those grandiose reactions that is a researcher's dream. α-Pinene proved to be a very fortuitous choice as a chiral auxiliary for asymmetric synthesis. It satisfies most of the conditions that are tests for an excellent chiral auxiliary, such as (1) both isomers of α-pinene are readily available in high ee. (2) Optical upgradation of the commercial material is easily attained during hydroboration. (3) The preparation of the reagents and the reaction conditions in most of the reactions are very simple and convenient. (4) The workup is easy. (5) The chiral auxiliary is readily recovered in all of the reactions without loss of any optical activity in an easily recyclable form. (6) A tentative mechanism is known for all of the reactions which helps in modification, wherever necessary. (7) The configuration of the products can be predicted based on the mechansim, with rare exceptions. (8) The scaling up of the reactions is easy, and, most important of all, (9) the enantiomeric excesses achieved in most reactions are very high.

Enantiomeric excesses in the range of >95% are obtained for the following established procedures; (1) the asymmetric hydroboration of three of the four classes of olefins (the product from the fourth (Class I olefin) can be obtained indirectly *via* asymmetric homologation); (2) in the general asymmetric synthesis *via* boronates and borinates obtained from hydroboration and homologation; (3) in asymmetric homologation; (4) in asymmetric allyl- and crotylboration; (5) in asymmetric reductions; (6) in asymmetric enolboration-aldol reactions, and (7) in the cleavage of epoxides. To our knowledge, there is no other chiral auxiliary and reaction comparable to α-pinene and its hydroboration leading to chiral organoboranes that are capable of achieving so many different types of asymmetric reactions in such high efficiency (Fig. 18). This asymmetric synthesis *via* chiral organoboranes is truly general.

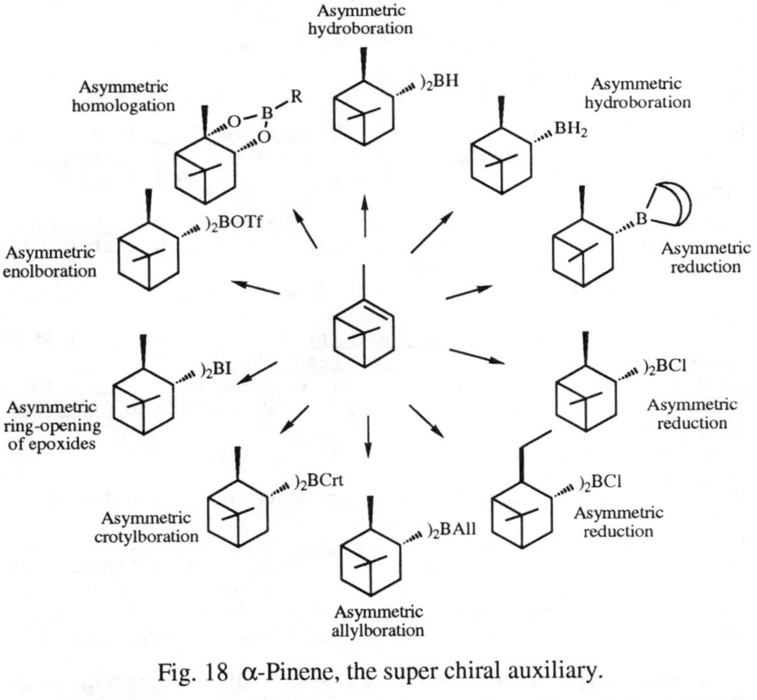

Fig. 18 α-Pinene, the super chiral auxiliary.

Acknowledgements. We wish to express our deep appreciation to the National Institutes of Health (GM 10937-30), the National Science Foundation (CHE 9012236) and the U. S. Army Research Office (DAAL 03-91-G-0024) for financial support which made possible the various studies herein described.

REFERENCES

1. H. C. Brown and B. C. Subba Rao, *J. Am. Chem. Soc.*, 1956, 78, 5694.
2. H. C. Brown, 'Hydroboration', Benjamin, New York, 1962.
3. (a) H. C. Brown, 'Boranes in Organic Chemistry', Cornell University Press, Ithaca, New York, 1972. (b) H. C. Brown, G. W. Kramer, A. B. Levy and M. M. Midland, 'Organic Synthesis via Boranes', Wiley-Interscience, New York, 1975.

4.   (a) H. C. Brown and G. Zweifel, J. Am. Chem. Soc., 1961, 83, 486. (b) H. C. Brown and N. M. Yoon, Isr. J. Chem., 1977, 15, 12. (c) H. C. Brown, P. K. Jadhav and M. C. Desai, J. Org. Chem., 1982, 47, 5065. (d) H. C. Brown and J. V. N. V. Prasad, Heterocycles, 1987, 25, 641.

5.   (a) H. C. Brown, P. K. Jadhav, J. Org. Chem., 1981, 46, 5047. (b) H. C. Brown, P. K. Jadhav and A. K. Mandal, ibid. 1982, 47, 5074.

6.   H. C. Brown, P. K. Jadhav and M. C. Desai, J. Am. Chem. Soc., 1982, 104, 4303.

7.   (a) H. C. Brown and P. V. Ramachandran, Pure & Appl. Chem., 1991, 63, 307. (b) H. C. Brown and B. Singaram, ibid., 1987, 59, 879.

8.   (a) H. C. Brown and B. Singaram, J. Am. Chem. Soc., 1984, 106, 1797. (b) H. C. Brown and J. V. N. V. Prasad, J. Org. Chem., 1986, 51, 4526. (c) N. N. Joshi, C. Pyun, V. K. Mahindroo, B. Singaram and H. C. Brown, ibid., 1992, 57, 504.

9.   (a) H. C. Brown, K. W. Kim, T. E. Cole and B. Singaram, J. Am. Chem. Soc., 1986, 108, 6761. (b) H. C. Brown, A. M. Salunkhe and B. Singaram, J. Org. Chem., 1991, 56, 3286.

10.   H. C. Brown and V. K. Mahindroo, Synlett, 1992, 626.

11.   H. C. Brown, R. R. Iyer, V. K. Mahindroo and N. G. Bhat, Tetrahedron: Asym., 1991, 2, 277.

12.   H. C. Brown, R. K. Bakshi and B. Singaram J. Am. Chem. Soc., 1988, 110, 1529.

13.   H. C. Brown, V. K. Mahindroo, N. G. Bhat, B. Singaram J. Org. Chem., 1991, 56, 1500.

14.   H. C. Brown, Pathare, P. M.; V. K. Mahindroo, Unpublished results.

15.   D. S. Matteson, Acc. Chem. Res. 1988, 21, 294.

16.   M. V. Rangaishenvi, B. Singaram and H. C. Brown, J. Org. Chem., 1991, 56, 3286 (1991).

17.   H. C. Brown, S. M. Singh, M. V. Rangaishenvi, J. Org. Chem. 1986, 51, 3150.

18.   (a) H. C. Brown, A. S. Phadke, and M. V. Rangaishenvi, J. Am. Chem. Soc., 1988, 110, 6263. (b) M. V. Rangaishenvi and H. C. Brown, Tetrahedron Lett., 1990, 31, 7115. (c) H. C. Brown, S. Jayaraman ibid. 1993, 34, 3997.

19.   (a) M. M. Midland, S. Greer, A. Tramontano, S. A. Zderic, J. Am. Chem. Soc., 1979, 101, 2352.   (b) M. M. Midland, D. C. McDowell, R. L Hatch and A. Tramontano, ibid., 1980, 102, 867.

20.   M. M. Midland, J. E. Petre, S. A. Zderic, A. Kazubski, J. Am. Chem. Soc., 1982, 104, 528.

21   (a).H. C. Brown and G. G. Pai, J. Org. Chem., 1985, 50, 1384. (b) M. M. Midland, J. I. McLoughlin and J. Gabriel, J. Org. Chem., 1989, 54, 159.

22.   (a) H. C. Brown, J. Chandrasekharan and P. V. Ramachandran, J. Am. Chem. Soc., 1988, 110, 1539. (b) H. C. Brown and P. V. Ramachandran, Acc. Chem. Res., 1992, 25, 16.

23.   P. V. Ramachandran, A. V. Teodorovic, M. V. Rangaishenvi and H. C. Brown, J. Org. Chem., 1992, 57, 2379.

24.   P. V. Ramachandran, A. V. Teodorovic and H. C. Brown, Tetrahedron, 1993, 49, 1725.

25.   H. C. Brown and P. V. Ramachandran, J. Org. Chem. 1989, 54, 4504.

26.   (a) H. C. Brown, P. V. Ramachandran, S. A. Weissman and S. Swaminathan J. Org. Chem., 1990, 55, 6328.   (b) P. V. Ramachandran, H. C. Brown, and S. Swaminathan Tetrahedron: Asym. 1990, 1, 433.   (c) H. C. Brown, P. V. Ramachandran, A. V. Teodorovic and S. Swaminathan Tetrahedron Lett. 1991, 32, 6691.

27.   For a preliminary communication see a chapter by P. V. Ramachandran and H. C. Brown in this volume.

28.   R. W. Hoffmann, Angew. Chem. Int. Ed. Engl.,1992, 31, 1124.

29.   H. C. Brown and P. K. Jadhav, J. Am. Chem. Soc., 1983, 105, 2092.

30.   H. C. Brown and K. S. Bhat, J. Am. Chem. Soc., 1986, 108, 5919.

31.   H. C. Brown and R. S. Randad, Tetrahedron, 1990, 46, 4463.

32.   K. Mori, Tetrahedron, 1976, 32, 1101.; Tetrahedron Lett., 1976, 1609.

33.   H. C. Brown, R. S. Randad, K. S. Bhat, M. Zaidlewicz and U. S. Racherla, J. Am. Chem. Soc., 1990, 112, 2389.

34.   H. C. Brown, S. V. Kulkarni and U. S. Racherla Unpublished results.

35.   H. C. Brown, N. N. Joshi, M. Srebnik, Isr. J. Chem., 1989, 29, 229.

Allylboronates in Organic Synthesis: Design of an Improved Chiral Auxiliary and Synthesis of the Trioxadecalin Nucleus of Mycalamides A and B

William R. Roush*, Paul T. Grover, and Thomas G. Marron

DEPARTMENT OF CHEMISTRY, INDIANA UNIVERSITY BLOOMINGTON, IN 47405, USA

AN IMPROVED CHIRAL AUXILIARY FOR THE ASYMMETRIC ALLYLBORATION REACTION

The allylboration reaction is an important method for the synthesis of stereochemically rich acyclic molecules.[1,2] While the tartrate ester modified allylboronates developed in our laboratory are highly practical reagents and have been applied in a number of complex problems,[2] they give only good to moderate levels of enantioselectivity in reactions with achiral aldehydes (typically 60-87% e.e.,[3] except for reactions with metal carbonyl complexed unsaturated aldehydes which give 83-98% e.e.),[4] and diminished enantioselectivity frequently results with β-alkoxy aldehydes.[5] Moreover, double asymmetric reactions with chiral aldehydes are usually most selective when performed in the matched stereochemical series,[2] especially if the substrate contains a conformationally unconstrained β-alkoxy substituent.[5] These considerations have encouraged us to continue work on the development of improved, second generation reagents that exhibit high enantioselectivity with a wide range of substrates.

Several years ago we reported the synthesis of allylboronate **4** containing a conformationally rigid tartramide auxiliary.[6] This reagent, designed on the basis of our rationale of the mechanism of asymmetric induction in allylborations of the tartrate ester based reagents **1-3**,[7] was found to be substantially more enantio-selective than **1-3** (94-97% e.e. for most substrates except PhCHO).[6] The Banfi reagent **4** still ranks among the most highly enantioselective allylboration agents in the literature.[8] However, the very poor solubility characteristics of **4** results in impractically long reaction times and poor conversions under conditions which give maximum asymmetric induction (toluene, -78°C).[6] Thus, reagent **4** has not yet found any applications in total synthesis.

(R,R)-**1**, $R_E = R_Z = H$
(R,R)-**2**, $R_E = Me$, $R_Z = H$
(R,R)-**3**, $R_E = H$, $R_Z = Me$

(R,R)-**4**, R = CH_2Ph
(R,R)-**5**, R= $CH_2C_6H_{11}$
(R,R)-**6**, R= CH_2CF_3

(R,R)-**7**, R = CH_2Ph
(R,R)-**8**, R = $CH_2C_6H_{11}$
(R,R)-**9**, R = CH_2CF_3

Searching for a reagent with improved solubility characteristics, we were encouraged initially by observations that 0.5 M toluene solutions of **5** (prepared from auxiliary **8**, in turn available by hydrogenation of **7** over Rh/Al$_2$O$_3$) are homogeneous at -78°C, that the reaction of **5** and C$_6$H$_{11}$CHO is complete within 5 h under these conditions, and that the corresponding homoallylic alcohol was produced in 95% e.e. However, reaction times with other aldehydes were still unacceptably long, and so our search for an improved diol auxiliary continued.

Previous studies have established that tartrate allylboronate **1** is substantially more reactive than allylboronates containing tartramides or any other chiral or achiral diols.[9] This has been rationalized by the inductive effect of the carbonyl carbons that increases the Lewis acidity of the boron center.[9a] Reasoning that increased reactivity of tartramide-based reagents might be achieved by using strongly electron withdrawing substituents on nitrogen, we targeted N,N'-bis-trifluoroethyl-N,N'-ethylene tartramide **9** as a new auxiliary for the allylboration reaction.

Diol **9** was synthesized by a simple three step sequence starting from the bis trifluoroacetamide **10**. Thus, reduction of **10** with LiAlH$_4$ and DCC coupling of the resulting diamine with benzylidene tartaric acid provided **11** in 47% yield. Removal of the benzylidene acetal by catalytic hydrogenation in 10% HOAc-EtOH provided diol **9** in 97% yield, which was then converted into allylboronate **6** by an exchange reaction with the diethanolamine complex of allylboronic acid.[3c]

Results of reactions of **6** with several achiral and chiral aldehydes are summarized in the following figure. Comparative reactions of allylboronates **5** and **6** with C$_6$H$_{11}$CHO (entries 2, 3), TBDPSO(CH$_2$)$_3$CHO (entries 6, 7), and pentylidene D-glyceraldehyde (entries 11, 12) clearly showed that **6** is considerably more reactive than **5**. However, **6** is only sparingly soluble in toluene at room temperature (ca. 0.1 M), and considerably less so at -78°C. We found it

RCHO	Reagent	Conditions	% Yield	Selectivity
C$_6$H$_{11}$CHO	(R,R)-**6**	THF, 0.5 M, 5 h	91%	94% e.e.
C$_6$H$_{11}$CHO	(R,R)-**6**	toluene, 0.1 M, 1 h	71% (a)	94% e.e.
C$_6$H$_{11}$CHO	(R,R)-**5**	toluene, 0.5 M, 5 h	82 %	95% e.e.
n-C$_5$H$_{11}$CHO	(R,R)-**6**	THF, 0.5 M, 5 h	95%	96% e.e.
n-C$_5$H$_{11}$CHO	(R,R)-**5**	toluene, 0.5 M, 7 h	93%	97% e.e.
TBDPSO(CH$_2$)$_3$CHO	(R,R)-**6**	THF, 0.5 M, 5 h	80%	95% e.e.
TBDPSO(CH$_2$)$_3$CHO	(R,R)-**5**	toluene, 0.5 M, 36 h	72 %(b)	94% e.e.
(**13**)	(R,R)-**6**	THF, 0.5 M, 8 h	83%	97 : 3
(**13**)	(S,S)-**6**	THF, 0.5 M, 8 h	82%	8 : 92
(**14**)	(R,R)-**6**	THF, 0.5 M, 8 h	73%	99.4 : 0.6
(**14**)	(R,R)-**6**	toluene, 0.1 M, 6 h	43% (c)	99.4 : 0.6 (e)
(**14**)	(R,R)-**5**	toluene, 0.5 M, 36 h	53% (d)	99.4 : 0.6 (e)
(**14**)	(S,S)-**6**	THF, 0.5 M, 8 h	86%	5 : 95

(a) 87% conversion by GC; (b) 85% conversion; (c) 54% conversion by GC; (d) 70% conversion; (e) pentylidene D-glyceraldehyde was used as substrate.

15, R$_1$ = OH, R$_2$ = H
16, R$_1$ = H, R$_2$ = OH

17, R$_1$ = OH, R$_2$ = H
18, R$_1$ = H, R$_2$ = OH

most convenient, therefore, to use THF as the solvent for allylborations of **6** since 0.5 M solutions remained homogeneous at -78°C. Fortunately, the allylborations of **6** were comparably enantioselective in either solvent, unlike the tartrate allylboronates **1-3** for which the enantioselectivity is highly solvent dependent.[3]

The data summarized in the figure show that the new allylboronates **5** and **6** are highly enantioselective (94-97% e.e.) and are also highly diastereoselective in reactions with chiral aldehydes **13** and **14**. The matched double asymmetric reactions of (R,R)-**6** and **13** and **14** provided the anti homoallylic alcohols **15** and **17** as the major products of 97 : 3 and 99.4 : 0.6 mixtures, respectively, while the mismatched double asymmetric reactions performed with (S,S)-**6** provided **16** and **18** with 92 : 8 and 95 : 5 selectivity. These data compare favorably with those previously obtained by using the Banfi reagent **4**, and are significantly improved compared to results obtained with the parent tartrate allylboronate **1**.[6]

Because allylboronate **6** is easy to prepare, is highly enantioselective and has favorable solubility and reactivity characteristics, it appears ideally suited for application to problems in total synthesis. Studies with the corresponding (E)-crotylboronate also have been initiated, and results obtained thus far indicate that this reagent is also considerably improved compared to the parent tartrate (E)-crotylboronate **2**. Thus, a practical and highly enantioselective chiral auxiliary has been developed for the asymmetric allylboration reaction, which should considerably extend the scope and applicability of this technology in total synthesis.

A HIGHLY STEREOSELECTIVE SYNTHESIS OF THE TRIOXADECALIN NUCLEUS OF MYCALAMIDES A AND B

Mycalamides A (**19**) and B (**20**) are potent anti-tumor and anti-viral antibiotics isolated from marine sponges.[10,11] They are structurally related to pederin,[12] an insect toxin, and to the onnamides which are another group of marine sponge metabolites.[13] In planning to undertake total syntheses of these compounds,[14] we wished to devise a strategy that was highly stereoselective and which exercised complete stereochemical control over the interesting hemi-aminal center at C(10). Kishi has suggested that this unit plays an important role in the biological properties of these compounds. Toward this end, we have developed a highly stereoselective synthesis of the model trioxadecalin nucleus **21** by a route that features two new chiral allylboronates in highly stereoselective allylboration reactions.

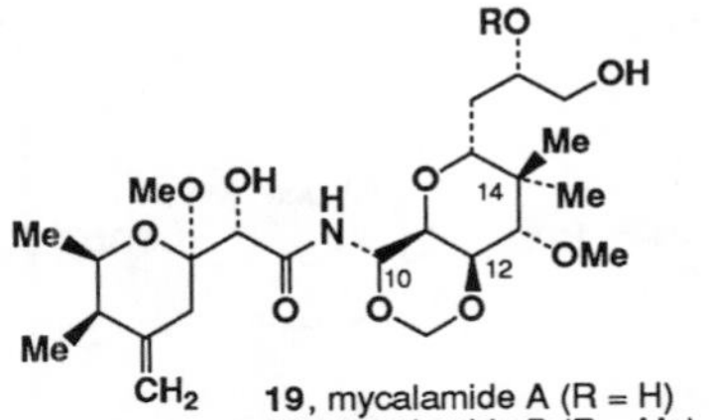

19, mycalamide A (R = H)
20, mycalamide B (R = Me)

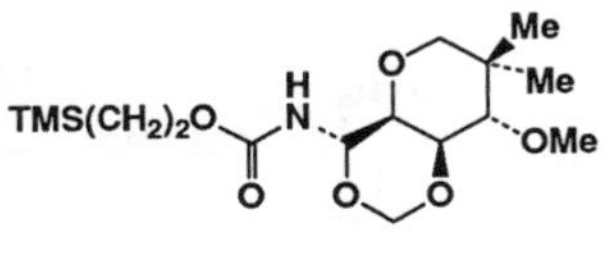

21

Our synthesis of **21** begins with the reaction of pentylidene D-glyceraldehyde **22**[15] and (R,R)-diisopropyl tartrate modified prenylboronate **23**.[16] Prenylboronate **23** was prepared from 3-methyl-1-butene by using slight modifications of our standard crotylboronate synthesis.[3c] The reaction of **22** and (R,R)-**23** provided **24** in 79% yield and with ≥99 : 1 selectivity as determined by GC analysis. Use of other achiral prenylating reagents such as Me$_2$C=CHCH$_2$Br-CrCl$_2$ gave mixtures of the two stereoisomers that were difficult to separate. Thus, the enantioselectivity **23** is critical to the success of this transformation.

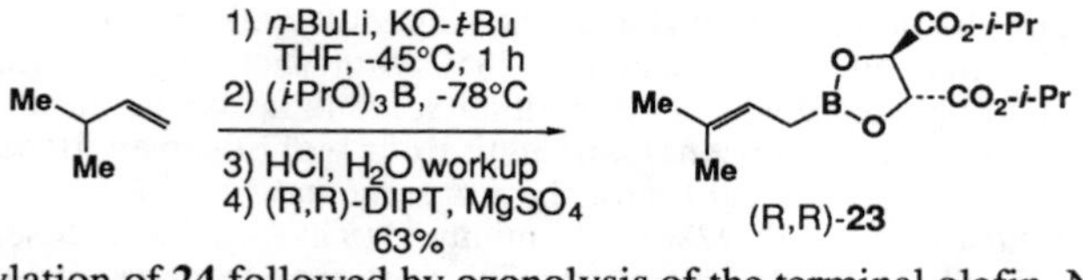

Methylation of **24** followed by ozonolysis of the terminal olefin, NaBH$_4$ reduction of the ozonide, and protection of the resulting primary alcohol gave TBDPS ether **25** in 69% overall yield. Hydrolysis of the pentylidene ketal and periodate cleavage of the diol afforded aldehyde **26**, the substrate for the next C-C bond forming reaction. Addition of **26** to a -78°C solution of (R, R)-(E)-γ-[dimethylphenylsilyl]allylboronate **27**[2d] in toluene provided **28** in 86% yield and with ≥99 : 1 diastereoselectivity. The outstanding selectivity of this reaction is presumably the result of matched double diastereoselection. Oxidation of the terminal olefin with dimethyldioxirane followed by acid catalyzed Petersen elimination gave allylic alcohol **29** in 86% yield. Asymmetric epoxidation of **29** by using standard Sharpless conditions provided epoxy diol **30** in 92% yield as a single isomer (>99:1).[17] Simultaneous deprotection of the TBDPS ether and cyclization of the epoxyalcohol was carried out by treating **30** with a large excess of HF (50 equiv.) in acetonitrile at ambient temperature, thereby providing **31** in 86% yield. After selective protection of the primary hydroxy group of **31** as a trityl ether, the methylene acetal was introduced via the reaction of the internal 1,3-diol with bromochloromethane and NaH in DMF.[18] Removal of the trityl ether by treatment with acetic acid at 60°C then gave **32** in 49% yield from **30**.

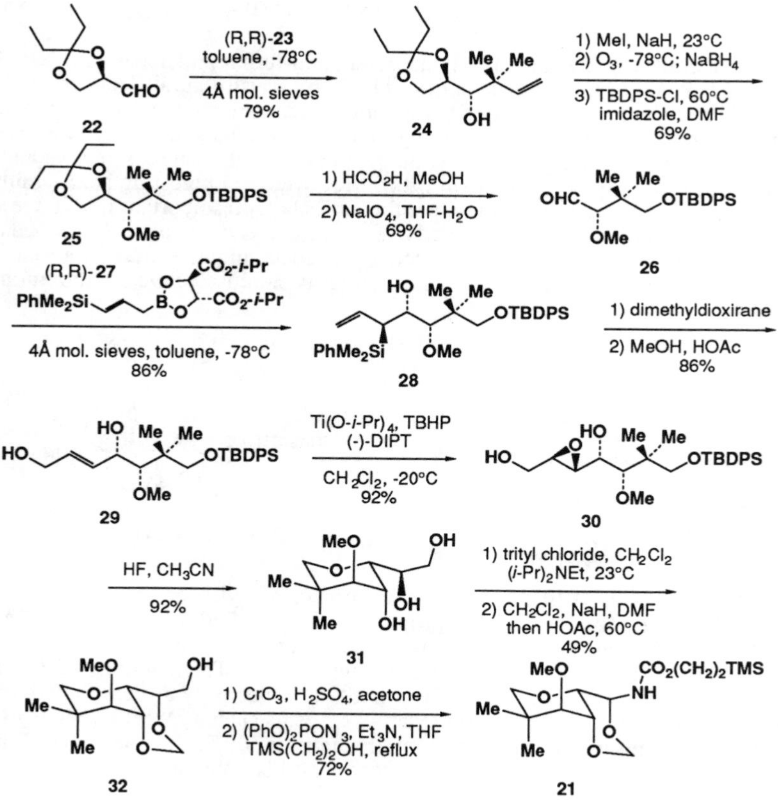

Finally, Jones' oxidation of the primary alcohol (91%) followed by Curtius rearrangement[19] in the presence of 2-trimethylsilylethanol provided the desired carbamate **21** in 79% yield.

The stereochemistry of **21** was confirmed by a NOE study performed on the carboxylic acid precursor **33**. The NOE and coupling constant data summarized below for **33** and **21** indicate that they preferentially adopt conformations with the C(13)-methoxyl and C(10) substituents in axial positions. Primary alcohol **32** also adopts a similar conformation. Molecular mechanics calculations (MMX) confirmed that the indicated conformation of **33** is ca. 5.5 kcal/mol more stable than the alternative cis-decalin conformer. However, conformers **21a** and **21b** were calculated to be within ca. 0.5 kcal/mol of each other, which is consistent with greater $J_{10,11}$ and $J_{12,13}$ values for **21** compared to **33**.

$J_{10,11} = 1.2$ Hz
$J_{11,12} = 1.4$ Hz
$J_{12,13} = 2$ Hz
$J_{13,15eq} = 1.2$ Hz

33

$J_{10,11} = 4.2$ Hz
$J_{11,12} = 4.0$ Hz
$J_{12,13} = 5.6$ Hz
(at 80°C)

21a

21b

The preferred conformations deduced for **32**, **33** and **21** corresponds to the minor conformation of the trioxadecalin unit of the natural products.[10] One of the factors that presumably contributes to the unusual conformational preferences of **21**, **32**, and **33** is that the equatorial C(13) methoxyl group in conformer **21b** experiences destabilizing gauche pentane interactions either with the gem dimethyl groups at C(14) or with the methylene acetal CH_2 in any of the three staggered conformations about the C(13)-O bond. In contrast, the axial methoxyl groups in **33** and **21a** can adopt a conformation in which it does not experience destabilizing 1,5 interactions. It is also interesting to note that the preferred conformation of the trioxadecalin unit of the mycalamides and onnamides corresponds to **21b** presumably since the C(15) side chain is equatorial in **21b** but is axial in the alternative cis-decalin conformer analogous to **21a**.

In summary, we have developed an efficient and highly diastereoselective synthesis of the mycalamide trioxadecalin nucleus **21**. Current efforts are focusing on introduction of the C(15) side chain, and ultimately the completion of the total syntheses of these interesting, biologically active natural products.

ACKNOWLEDGEMENT

Support provided by the National Institute of General Medical Sciences (GM 38536 and 38907) is gratefully acknowledged.

REFERENCES

1. For a recent review: W. R. Roush, in "Comprehensive Organic Synthesis" 1991, Vol. 2, p. 1; C. Heathcock, ed.; Pergamon Press, Oxford.

2. For several recent examples from our laboratory: (a) W. R. Roush and A. D. Palkowitz, *J. Org. Chem.*, 1989, <u>54</u>, 3009. (b) W. R. Roush, A. D. Palkowitz, and K. A. Ando, *J. Am. Chem. Soc.*, 1990, <u>112</u>, 6348. (c) W. R.

Roush, J. A. Straub, and M. S. VanNieuwenhze, J. Org. Chem., 1991, 56, 1636. (d) W. R. Roush and P. T. Grover, Tetrahedron, 1992, 48, 1981. (e) W. R. Roush and B. B. Brown, J. Am. Chem. Soc., 1993, 115, 2268.

3.   (a) W. R. Roush, L. K. Hoong, M. A. J. Palmer and J. C. Park, J. Org. Chem. 1990, 55, 4109. (b) W. R. Roush, L. K. Hoong, M. A. J. Palmer, J. A. Straub, and A. D. Palkowitz, Ibid. 1990, 55, 4117. (c) W. R. Roush, K. Ando, D. B. Powers, R. L. Halterman, and A. D. Palkowitz, J. Am. Chem. Soc. 1990, 112, 6339.

4.   (a) W. R. Roush and J. C. Park, J. Org. Chem., 1990, 55, 1143. (b) W. R. Roush and J. C. Park, Tetrahedron Lett., 1990, 31, 4707. (c) W. R. Roush and J. C. Park, Ibid., 1991, 32, 6285.

5.   W. R. Roush, L. K. Hoong, M. A. J. Palmer, J. A. Straub, and A. D. Palkowitz, J. Org. Chem. 1990, 55, 4117.

6.   W. R. Roush and L. Banfi, J. Am. Chem. Soc., 1988, 110, 3979.

7.   (a) W. R. Roush, A. E. Walts, and L. K. Hoong, J. Am. Chem. Soc., 1985, 107, 8186. (b) For supporting evidence based on solid-state and solution conformational analyses of tartrate-derived 1,3-dioxolanes and 1,3,2-dioxaborolanes: W. R. Roush, A. M. Ratz, and J. A. Jablonowski, J. Org. Chem., 1992, 57, 2047.

8.   For leading references to the most highly enantioselective classes of chiral allyl- and crotylboron reagents: (a) P. K. Jadhav, K. S. Bhat, P. T. Perumal, and H. C. Brown, J. Org. Chem., 1986, 51, 432. (b) H. C. Brown, and K. S. Bhat, J. Am. Chem. Soc., 1986, 108, 5919. (c) J. Garcia, B-M. Kim, and S. Masamune, J. Org. Chem., 1987, 52, 4831. (d) R. W. Hoffmann, S. Dresely, and J. W. Lanz, Chem. Ber., 1988, 121, 1501, and previous papers in this series. (e) M. T. Reetz, and T. Zierke, Chem. and Ind. (London), 1988, 663. (f) R. W. Hoffmann, and S. Dresely, Chem. Ber. 1989, 122, 903. (g) R. W. Hoffmann, K. Ditrich, G. Köster, and R. Stürmer, Ibid., 1989, 122, 1783. (h) R. P. Short, and S. Masamune, J. Am. Chem. Soc., 1989, 111, 1892. (i) E. J. Corey, C.-M. Yu, and S. S. Kim, J. Am. Chem. Soc., 1989, 111, 5495. (j) H. C. Brown, R. S. Randad, K. S. Bhat, M. Zaidlewicz, and U. S. Racherla, Ibid., 1990, 112, 2389.

9.   (a) W. R. Roush, L. Banfi, J. C. Park, and L. K. Hoong, Tetrahedron Lett., 1989, 30, 6457. (b) H. C. Brown, U. S. Racherla, and P. J. Pellechia, J. Org. Chem., 1990, 55, 1868.

10.   (a) N. Perry, J. Blunt, and M. Munro, J. Am. Chem. Soc., 1988, 110, 4850; J. Org. Chem., 1990, 55, 223.

11.   (a) N. S. Burres, and J. J. Clement, Cancer Res., 1989, 49, 2935. (b) H. Ogawara, K. Higashi, K. Uchino, and N. B. Perry, Chem. Pharm. Bull. 1991, 39, 2152.

12.   For a leading reference on the synthesis and biological properties of pederin: R. Hoffmann, and A. Schlapbach, Tetrahedron, 1992, 48, 1959.

13.   (a) S. Sakemi, T. Ichiba, S. Kohmoto, G. Saucey, and T. Higa, J. Am. Chem. Soc., 1988, 110, 4851. (b) S. Matsunaga, N. Fusetani, and Y. Nakao, Tetrahedron, 1992, 48, 8369.

14.   For the first total syntheses of the mycalamides and onnamide A: (a) C. Y. Hong and Y. Kishi, J. Org. Chem., 1990, 55, 4242. (b) C. Y. Hong and Y. Kishi, J. Am. Chem. Soc., 1991, 113, 9693.

15.   C. R. Schmidt and D. A. Bradley, Synthesis, 1992, 587.

16.   For other chiral prenylating reagents, see ref. 12 and H. C. Brown and P. K. Jadhav, Tetrahedron Lett., 1984, 25, 1215.

17.   Y. Gao, R. M. Hanson, J. M. Klunder, S. Y. Ko, H. Masamune, and K. B. Sharpless, J. Am. Chem. Soc., 1987, 109, 5765.

18.   R. E. Zelle and W. J. McClellan Tetrahedron Lett., 1991, 32, 2461.

19.   T. Shiori, K. Ninomiya, K., and S. Yamada, J. Am. Chem. Soc., 1972, 94, 6203.

Ultraselective Asymmetric Synthesis with Boronic Esters

Donald S. Matteson

DEPARTMENT OF CHEMISTRY, WASHINGTON STATE UNIVERSITY, PULLMAN WA 99166, USA

1   SEQUENTIAL DOUBLE DIASTEREOSELECTION

Chain extension of an alkylboronic ester of a diol of C_2 symmetry with (dichloromethyl)lithium followed by alkylation with a Grignard reagent can yield diastereomeric ratios as high as 1000:1 for each new chiral center introduced.[1] This ultrahigh ratio is the result of sequential double diastereoselection. For example, (S,S)-1,2-diisopropylethanediol ["(S,S)-DIPED"] propylboronate (1) with LiCHCl$_2$ yields a borate complex (2), which rearranges in the presence of zinc chloride with displacement of chloride by alkyl to form (S,S)-DIPED (1R)-(1-chlorobutyl)boronate (3) in ~99% diastereomeric purity. Alkylation of 3 with methylmagnesium bromide leads to a second borate complex (4), which rearranges on warming with displacement of the second chloride to form (S,S)-DIPED (1R)-(1-methylbutyl)boronate (5), which has turned out to be ~99.9% diastereomerically pure.[1]

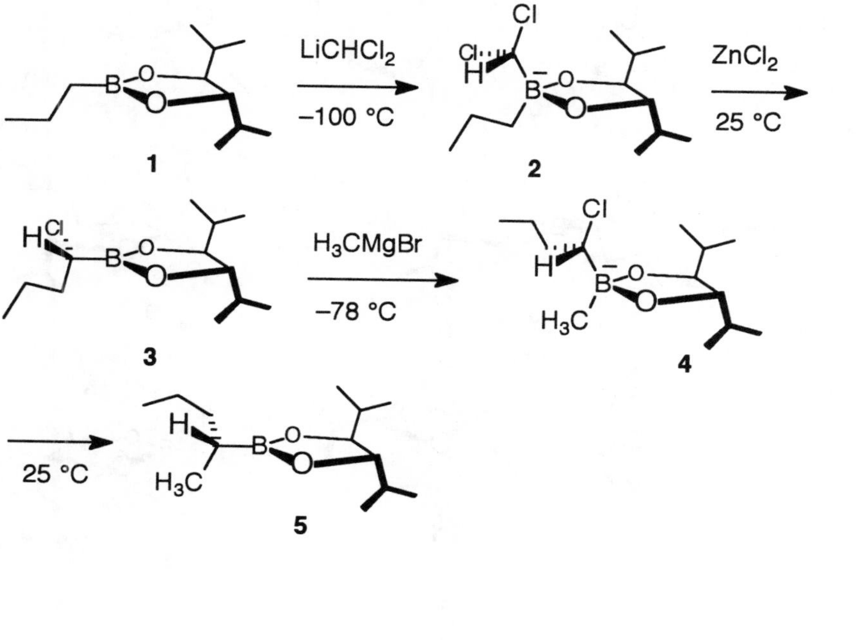

Why is there a second diastereoselection in the final step, the conversion of **4** to **5**? This result was in no way anticipated, though in hindsight it may be noted that assuming similar conformational placement for the larger groups in borate intermediates **2** and **4** sets both up for migration of an alkyl group with intramolecular displacement of chloride with inversion. Thus, inversion of the carbon atom during displacement of the first chlorine puts the remaining chlorine in the right configuration for the subsequent displacement.

The observations that led to discovery of the double stereo-differentiation were failed attempts to convert (*S,S*)-DIPED (1*S*)-(1-chloro-butyl)boronate (**8**), the diastereomer of **3** formed from **1** and LiCHCl$_2$ to the extent of perhaps 1%, to (*S,S*)-DIPED (1*S*)-(1-methylbutyl)boronate (**1 0**), a diastereomer of **5**. Preparation of **8** was accomplished by trans-esterification of (*R,R*)-2,3-butanediol (1*S*)-(1-chlorobutyl)boronate (**6**) to the diethanolamine ester (**7**), which on treatment with dilute acid followed by (*S,S*)-DIPED yielded **8**. Treatment of **8** with methylmagnesium bromide presumably yielded borate complex **9**, in which the favored conformation illustrated places the chlorine atom in the wrong position to allow displacement by a migrating methyl group. Instead, an oxygen atom migrates, expanding the five-membered dioxaborolane ring to a six-membered ring borinic ester (**1 1**), which proved extremely oxygen-sensitive and could be characterized by NMR only if suitable precautions were taken to prevent exposure of the sample to air. The product isolated in high yield was (*S,S*)-DIPED methylboronate (**1 2**), and butyraldehyde was detected by ^{1}H-NMR analysis.

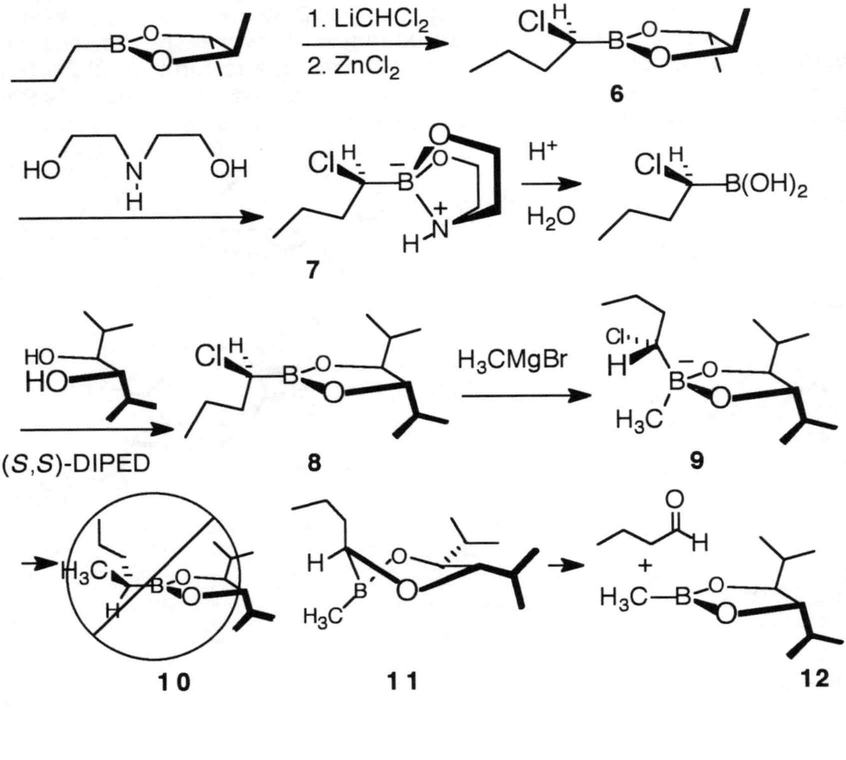

The outcome of the reaction of chloro boronic ester **8** with methylmagnesium bromide strongly implies that the normal reaction sequence exemplified by the chain extension of **1** via ~99% diastereomerically pure **3** should yield **5** of ultrahigh diastereomeric purity. Our failure to find **1 0** among the products of reaction of **8** is not an indication of total absence, but implies that the yield of **8** is at most a few per cent.

For another test of the selectivity of this type of process, we made a 50:50 mixture of diastereomers **1 3** and **1 6** from racemic ethylene glycol 1-chloro-2-phenylethylboronate and (*S,S*)-DIPED and treated it with LiCHCl$_2$ followed by CH$_3$MgBr.[1] The (1*R*) boronic ester **1 3** yielded **1 4**, which was oxidized to (*R*)-1-phenyl-2-propanol (**1 5**), enantiomeric excess 95 to ≥99%. The (1*S*) boronic ester **1 6** evidently yielded **1 7**, which in air formed phenylacetaldehyde and (*S,S*)-DIPED methylboronate (**1 2**).

1 3   **1 4**   **1 5**

1 6   **1 7**   **+ 12**

If each of two successive diastereoselective processes, for example conversion of **1** to **3** and **3** to **5**, produces a 100:1 diastereomer ratio, then the overall selection is 10,000:1. While it is conceivable that some sequences may be inherently that selective, practical considerations intervene. For example, a liquid chiral starting material such as **1** is difficult to purify to the 99.9% level, and the presence of 0.1% of an achiral organoborane impurity in **1** would be difficult to detect.

A measurement of the high diastereomeric ratios attainable with this chemistry is provided by the synthesis of (3*S*,4*S*)-4-methyl-3-heptanol (**2 1**), a component of the aggregation pheromone of the elm bark beetle, *Scolytus multistriatus*, in which a 700:1 diastereomer ratio was achieved.[1] In order for the mole fractions of the two enantiomers of the minor diastereomer to sum to 1/700, the configurational ratio must be ≥1400:1 at one stereogenic center and between 700:1 and 1400:1 at the other. The fraction of enantiomer of **2 1** from paired random errors is ≤1/1,960,000.

A 500:1 diastereomer ratio, or an average 1000:1 ratio at each chiral center, was obtained in an analogous synthesis of (3*S*,4*R*)-4-methyl-3-heptanol (**2 3**), the trail pheromone of a southeast Asian ponerine ant, *Leptogenys diminuta*.[1] For each of these syntheses, the strategy involved choosing the correct chiral director, (*R,R*)- or (*S,S*)-DIPED, to produce the second chiral center introduced in boronic ester intermediates **2 0** and **2 2**. In order to obtain the correct configuration at the first chiral center, the route to **2 1** began with **1 8** (the enantiomer of **1**) with the propyl group in place and introduced the methyl substituent second. The route to **23** began from **12** with the methyl group in place and connected the propyl group second to produce **10**. The absolute configurations of the α-carbons of diastereomers **19** and **10** are the same.

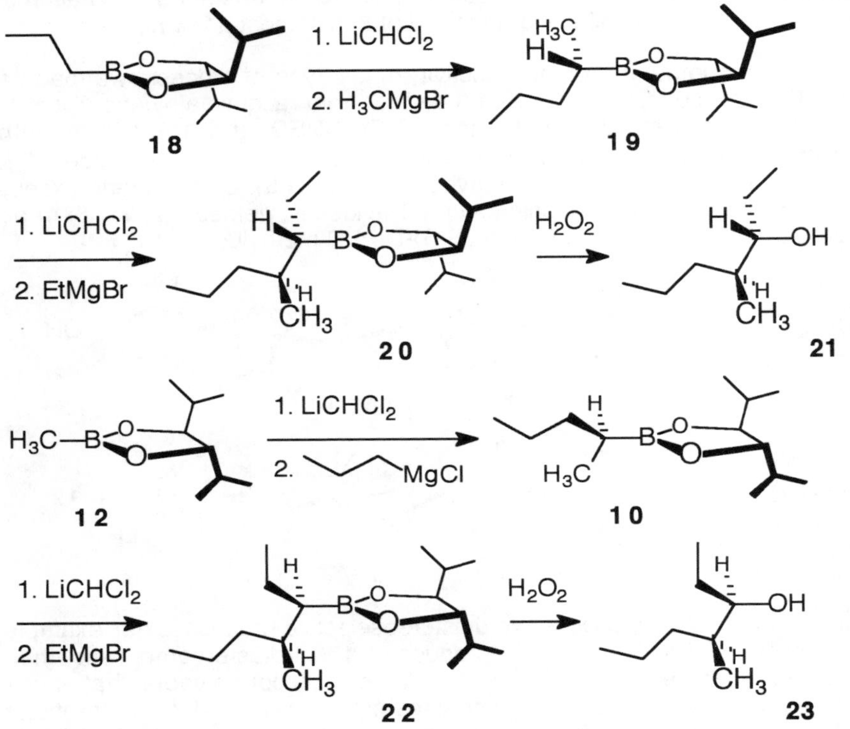

It was found that *Leptogenys diminuta* would readily follow an artificial trail of **23**, but not of **21** or of the enantiomer of **21** or **23**.[2] One of the inactive isomers was found to be about 1/500 as attractive as **23** to the ants, an order of magnitude of activity consistent with the amount of **23** present as an impurity.

2   ENOLATES WITH α-HALO BORONIC ESTERS

Although the foregoing chemistry provides ultrahigh diastereomeric and enantiomeric purities, it only introduces one chiral center at a time. For some synthetic purposes, it is advantageous to introduce two chiral centers at a time, even at the cost of some degree of stereoselectivity. An approach which has yielded useful diastereomer ratios is the reaction of the lithium *E*(O)-enolate of *tert*-butyl propionate with (*S*,*S*)-DIPED (alkyl)-(bromo)methyl boronic esters (**24**).[3] For R = isopropyl the *threo/erythro* ratio in the product **25** was 15:1 before any purification, based on NMR analysis of the hydroxy ester **26** derived by treatment with H_2O_2. For R = *n*-butyl, the *threo/erythro* ratio of **25** was 60:1 after chromatography with a broad cut taken. These are useful ratios, especially when it is considered that the enantiomeric excesses are likely to be 99.5-99.8% if the same factors operate that were found in the alkylation of α-chloro boronic esters.

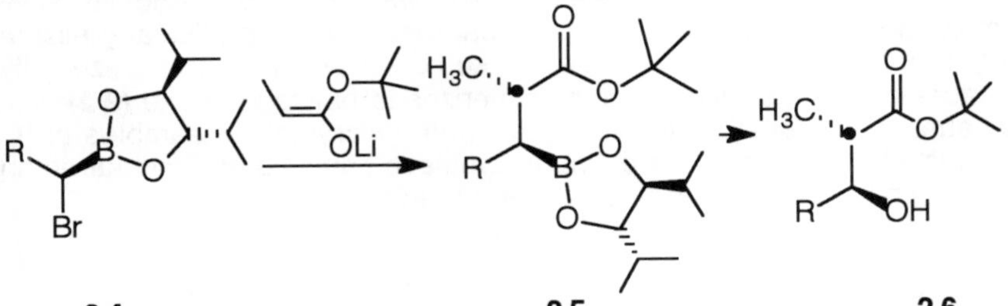

24   **25**   **26**

This chemistry has been found to have several limitations. Chloro boronic esters react sluggishly, and enolates other than carboxylic ester enolates failed to give good yields. Recently, we made an attempt to turn the stereochemistry around to produce *erythro* product by using an amide Z(O)-enolate, but the *threo* product predominated.[4]

3   LABILE TARGETS

Ultraselective asymmetric synthesis has now been found useful for making stereolabile structures, where the mild conditions used for boronic ester chemistry can leave epimerizable chiral centers intact. A demonstration is the synthesis of (2S,3S)-2-methyl-3-benzyloxypentanal (**29**) from (R,R)-dicyclohexylethanediol[5,6] ethylboronate (**27**) in 65% overall isolated yield after purification by simple distillation.[4] No intermediate was chromatographed except for analytical samples.

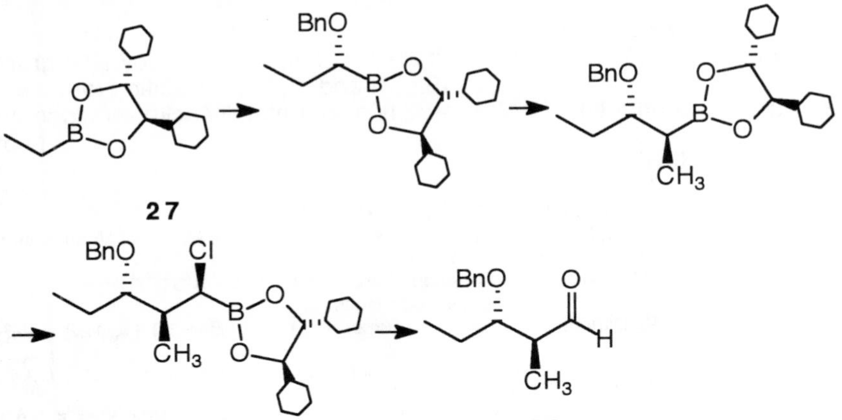

27

28   **29**

Although hydrogen peroxide oxidations of α-chloro boronic esters have given poor yields,[7] control of the pH to ~9 resulted in efficient conversion of **28** to **29**, diastereomeric purity 99-99.5% after distillation (by 300-MHz ^{1}H-NMR). Chromatography of labile **29** on alumina resulted in formation of ~50% (2R,3R)-epimer, on silica ~10%.[4]

We have explored the utility of **2 5** and **2 9** for the synthesis of stegobiol (**3 0**) and stegobinone (**3 1**),[4] the constituents of the aggregation pheromone of the drugstore beetle, *Stegobium paniceum* (Anobiidae),[8,9] a major pest of stored grain. Stegobinone has been a particularly elusive target, not previously synthesized in sufficient purity to crystallize fully, because of the ease with which it epimerizes to epistegobinone (**3 2**).[10,11] Because **3 2** is repellent to the insects and noncrystalline samples of **3 1** are unstable to storage, the previous synthetic samples had low attractant activity when fresh, and they deteriorated rapidly.

3 0 stegobiol   **31** stegobinone   **32** epistegobinone

The ester intermediate used by Mori and Ebata[10,12] was prepared from **2 5** (R = ethyl), the other required portion[10] having been constructed via boronic ester chemistry as well. Attempts to improve the reported inefficient ring closure[10,12] failed.[4] In recent work which is not quite ready for formal publication yet, a different approach based on aldehyde **2 9** as the source of the side chain and related boron chemistry for assembling the chiral centers of the ring has led to samples of stegobiol (**3 0**) that were easily purified. Oxidation by a standard method has converted pure **3 0** to crystalline stegobinone (**3 1**). Crystalline **3 1** stored in a freezer at −15 °C for 9 months remained ≥99.5% pure by 500-MHz ^{1}H-NMR analysis.[4]

Acknowledgment. We thank the National Science Foundation for grant CHE-8922672. Support of the WSU NMR Center by NIH grant RR 0631401, NSF grant CHE-9115282, and Battelle Pacific Northwest Laboratories Contract 12-097718-A-L2 is also gratefully acknowledged.

4   REFERENCES

1.   Tripathy, P. B.; Matteson, D. S. *Synthesis* **1 9 9 0**, 200-206.
2.   Steghaus-Kovâc, S.; Maschwitz, U.; Attygalle, A. B.; Frighetto, R. T. S.; Vostrowsky, O.; Bestmann, H. J. *Experientia* **1 9 9 2**, *48*, 690-694.
3.   Matteson, D. S.; Michnick, T. J. *Organometallics* **1 9 9 0**, *9*, 3171-3177.
4.   Matteson, D. S.; Man, H.-W. Unpublished results.
5.   Hoffmann, R. W.; Ditrich,K.; Köster, G.; Stürmer, R. *Chem. Ber.* **1 9 8 9**, *122*, 1783-1789.
6.   Jacobsen, E. N.; Markó, I.; Mungall, W. S.; Schröder, G.; Sharpless, K. B. *J. Am. Chem. Soc.* **1 9 8 8**, *110*, 1968-1970.
7.   Brown, H. C.; Imai, T.; Desai, M. C.; Singaram, B. *J. Am. Chem. Soc.* **1 9 8 5**, *107*, 4980.
8.   Kodama, H.; Ono, M.; Kohno, M.; Ohnishi, A. *J. Chem. Ecol.* **1 9 8 7**, *13*, 1871-1879.
9.   Kuwahara, Y.; Fukami, H.; Howard, R.; Ishii, S.; Matsumura, F.; Burkholder, W. E. *Tetrahedron* **1 9 7 8**, *34*, 1769-1774.
10. Mori, K.; Ebata, T. *Tetrahedron* **1 9 8 6**, *42*, 4413-4420.
11. Hoffmann, R. W.; Ladner, W.; Steinbach, K.; Massa, W.; Schmidt, R.; Snatzke, G. *Chem. Ber.* **1 9 8 1**, *114*, 2786-2801.
12. Mori, K.; Ebata, T. *Tetrahedron* **1 9 8 6**, *42*, 4685-4689.

An Approach to a General Reagent for Asymmetric Reduction

P. V. Ramachandran and Herbert C. Brown*

H. C. BROWN AND R. B. WETHERILL LABORATORIES OF CHEMISTRY, PURDUE UNIVERSITY, WEST LAFAYETTE, IN 47907, USA

Asymmetric reduction of prochiral ketones is known for about half a century. However, good chiral reducing agents became available only towards the beginning of the last decade. Prof. Mark Midland introduced B-isopinocampheyl-9-borabicyclo[3.3.1]nonane (Aldrich: Alpine-Borane®, 1) as the first successful organo-borane chiral reducing agent (1). Until then there was only one report of a moderately successful chiral reducing agent prepared by modifying lithium aluminum hydride with (2S,3R)-(+)-4-(dimethylamino)-1,2-diphenyl-3-methyl-2-butanol (Darvon alcohol) (2). During the same period when Alpine-Borane was introduced, Prof. Noyori introduced a chiral reducing agent by modifying lithium aluminum hydride with binaphthol and ethanol (Binal-H) (3). With Alpine-Borane and Binal-H began a new era in asymmetric reductions.

Alpine-Borane is very efficient in reducing highly reactive carbonyl groups, such as α-deuteroaldehydes, α,β-acetylenic ketones, α-keto esters, α-halo ketones and acyl cyanides (4). However, chiral reduction of slower reacting ketones, such as aralkyl and dialkyl ketones, was not successful probably because the dehydroboration of 1 in slow reductions is followed by an achiral reduction of the carbonyl group by 9-BBN produced in the dehydroboration stage (5). By conducting the reductions under neat conditions it proved possible to suppress the dehydroboration and achieve high ee (6). Midland suppressed the dehydroboration by conducting the reductions under very high hydrostatic pressures (7).

Alpine-Borane, 1   DIP-Chloride, 2   Ipc(t-Bu)BCl, 3

With a different approach, increasing the Lewis acidity of the boron atom in the reagent to provide for a stronger coordination of the carbonyl oxygen with the boron atom, we tested B-chlorodiisopinocampheylborane (Aldrich: DIP-Chloride®, 2) (8). This proved to be an excellent reagent for the reduction of aralkyl ketones, achieving maximum chiral induction for most types of aralkyl ketones. Testing the reagent for a series of aralkyl ketones substituted with representative functional groups showed that most substituents do not affect the chiral outcome (9). α-Quaternary alkyl ketones are another class of ketones for which 2 is effective (10). The success achieved in the reduction of such ketones could probably be accounted for by a transition state in which the bulky α-quaternary alkyl group interacts sterically with the methyl group at the 2-position of the isopinocampheyl moiety (Fig. 1) (8). When the steric interaction between the reagent and ketone is less, as in the reduction of 3-methyl-2-butanone (32% ee) and 2-butanone (4% ee), the chiral induction is poor.

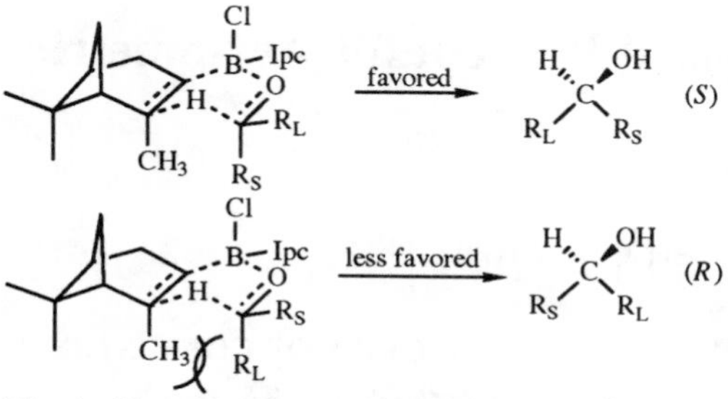

Fig. 1. Transition State Model For Asymmetric
Reduction With (–)-DIP-Chloride.

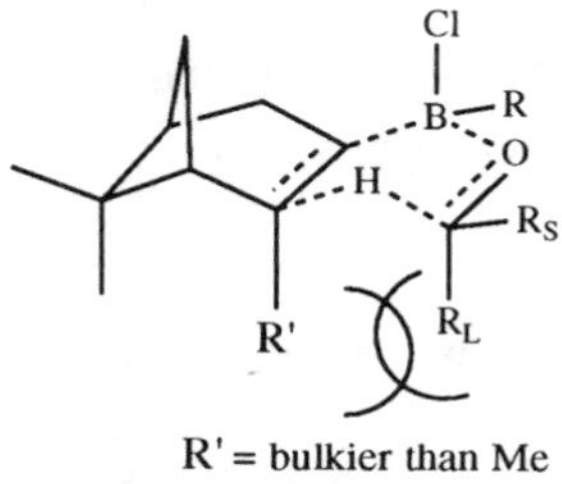

Fig. 2. Transition State Model
For Improved Reagents.

DIP-Chloride has since been shown to be effective for the chiral reduction of acyl silanes (11), hindered acetylenic ketones (12), and trifluoromethyl ketones (13). This reagent has found numerous applications in the synthesis of important pharmaceuticals and chiral synthons (14). We simplified the work-up procedure for the isolation of the product alcohol, and can now recycle the chiral auxiliary, α-pinene, completely (12).

We then turned our attention to synthesizing improved reagents which will reduce those classes of ketones which cannot be handled by DIP-Chloride. We studied the effect of substituting one of the isopinocampheyl moieties in **2** with alkyl groups of increasing steric requirements such as Me, Et, i-Pr, t-Bu, etc. and found a correlation between steric requirements of the alkyl group R in the reagent IpcBRCl and the chiral induction realized in the reductions (15). The reagent isopinocampheyl-t-butylchloroborane (**3**) gave results comparable to **2** though the rates of reductions were slightly slower.

The effectiveness of DIP-Chloride for the reduction of hindered ketones persuaded us to consider carefully the proposed transition state for the reaction. It appeared that an increase in the steric requirements of the group at the 2-position of the apopinene moiety might increase the ee achieved in the chiral reduction (Fig. 2) (16).

This hypothesis was tested with *B*-iso-2-ethylapopinocampheyl-9-BBN (Eapine-Borane, **4**) (17) and the corresponding lithium borohydride (Eapine-Hydride, **5**) (18). Considerable improvement over the parent compounds (R'=Me) was realized.

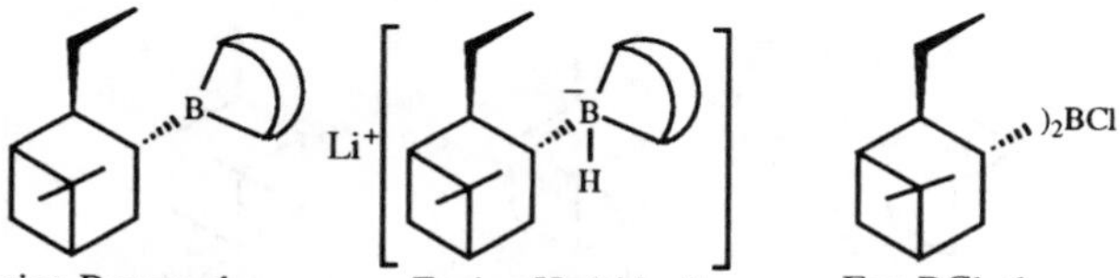

Eapine-Borane, **4**   Eapine-Hydride, **5**   Eap$_2$BCl, **6**

Even better results supporting this hypothesis were noted with *B*-chlorodiiso-2-ethylapopinocampheylborane (Eap$_2$BCl, **6**) (19). The reagent was synthesized by hydroborating 2-ethylapopinene with H$_2$BCl.EE or H$_2$BCl.SMe$_2$ and asymmetric reduction of our ten standard ketones (20) showed the reagent to be excellent for six classes of ketones and good for other two classes (Fig. 3).

The success of Eap$_2$BCl prompted us to examine 2-β-chloroethylapopinene, a precursor in the synthesis of 2-ethylapopinene, for the preparation of a new chiral reducing agent. We were interested in observing the influence that the chlorine atom in the R' group might have in the reduction of ketones, particularly those with strong electronic environments, such as α,β-acetylenic ketones and α-keto esters. Indeed, *B*-chlorodiiso-2-β-chloroethylapopinocampheylborane (Cleap$_2$BCl, **7**), synthesized from

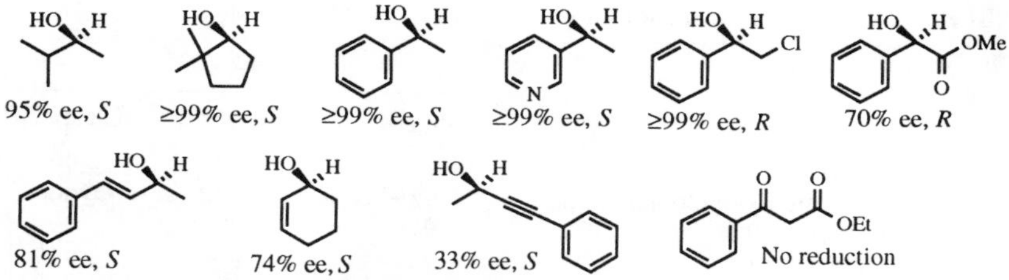
95% ee, S
≥99% ee, S
≥99% ee, S
≥99% ee, S
≥99% ee, R
70% ee, R
81% ee, S
74% ee, S
33% ee, S
No reduction

Fig. 3. Asymmetric reduction of ten standard ketones with Eap_2BCl.

2-β-chloroethylapopinene and chloroborane-methyl sulfide complex proved highly favorable for the chiral reduction of the above two classes of ketones (Fig. 4).

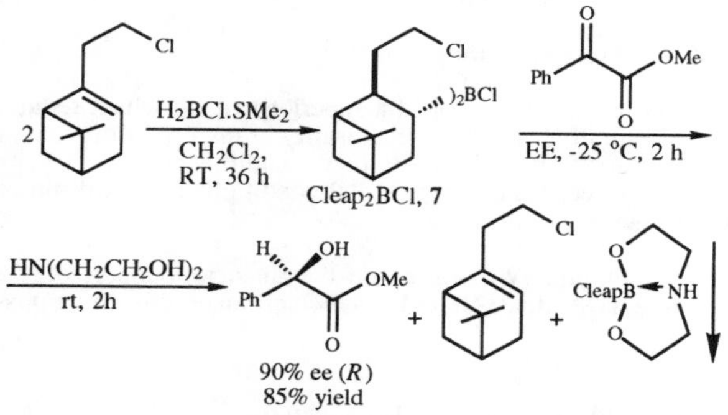
$H_2BCl.SMe_2$
CH_2Cl_2, RT, 36 h
$Cleap_2BCl$, 7
EE, -25 °C, 2 h
$HN(CH_2CH_2OH)_2$
rt, 2h
90% ee (R)
85% yield
CleapB

Fig. 4. Synthesis and Reaction of $Cleap_2BCl$.

Chiral reduction of methyl benzoylformate in EE at –25 °C, with reagent **7** prepared from 2- β-chloroethylapopinene and chloroborane, was complete in less than 2 h (^{11}B NMR of a methanolyzed aliquot at δ 32 ppm). The steric bulk at the 2-position of apopinene, while it increases the enantiomeric excess of the product alcohols from reductions, decreases the rate of reduction (16). However reactive carbonyls such as α-keto esters and acetylenic ketones were reduced with an appreciable rate. When the reduction was complete, the reaction mixture was warmed to rt and 2.2 equiv of diethanolamine was added. The precipitated diethanolamine complex was filtered, washed with pentane, and the filtrate concentrated. The liberated alcohol and 2-β-chloroethylapopinene were separated by distillation. The product alcohol was analyzed for the enantiomeric excess as its α-methoxy-α-(trifluoromethyl)phenylacetate (MTPA) derivative on a capillary gas chromatograph which showed it to be of 90% ee. This result is an improvement over the one provided by **2** (50% ee), and by **6** (70% ee). A similar effect was observed in the reduction of 4-phenyl-3-butyn-2-one when product alcohol of 66% ee was obtained with **7**, as compared to the 21% ee realized with **2** and 33% ee with **6**. 2-Cyclohexen-1-one was reduced to the corresponding alcohol in 80% ee as against the 36% ee provided by **2**. The results of reduction of the ten standard ketones with **7** are summarized in Table 1. As can be seen from the table, reagent **7** reduces 7 out of the 10 classes of ketones to the alcohols in excellent ee. Two classes of ketones are reduced in the 66-80% ee range and β-keto ester protonolyzed $Cleap_2BCl$.

To our knowledge this is the best result yet achieved by a single reagent for the reduction of several classes of ketones. Though various reagents have been developed in the past, most of them handle aralkyl ketones, acetylenic ketones and/ or α-keto esters. DIP-Chloride handles nicely relatively hindered ketones, with the *t*-Bu moiety. Eap_2BCl extended asymmetric reduction to ketones of the 3-methyl-2-butanone type. Now we have a reagent that can handle α-keto esters effectively and α,β-acetylenic ketones moderately. We are continuing our search for a new reagent which will handle the $COCH_3$ and $-COCH_2CH_3$ moieties.

TABLE 1. Asymmetric Reduction of Prochiral Ketones with R*$_2$BCl at –25 $^{\circ}$C

class of ketone	ketone	%ee		
		Ipc$_2$BCl	Eap$_2$BCl	Cleap$_2$BCl
1	acetylcyclohexane	26	97	≥99
2	2,2-dimethylcyclopentanone	98[a]	≥99[a]	≥99[a]
3	acetophenone	98	≥99	≥99
4	acetylpyridine	92	≥99	≥99[b]
5	2-chloroacetophenone	95	≥99	95[b]
6	methyl benzoylformate	50	70	90
7	ethyl benzoylacetate	no reduction		
8	*trans*-4-phenyl-3-butene-2-one	8 1	82	
9	2-cyclohexen-1-one	36	74	80[b]
10	4-phenyl-3-butyn-2-one	2 1	33	66

[a]For a reaction at rt. [b]For a reaction at 0->10 $^{\circ}$C.

In conclusion, we have developed a superior reagent which reduces many classes of ketones with essentially quantitative chirality transfer. Both enantiomers of the reagent can be synthesized at will from either enantiomer of α- or β-pinene which is readily available in high ee. The experimental conditions and work-up are easy. These factors make the reagent very attractive.

Acknowledgements. We appreciate the financial support of the United States Army Research Office (DAAL 03-91-G-0024) which made this study possible.

REFERENCES

1.   (a) M. M. Midland, S. Greer, A. Tramontano, S. A. Zderic, J. Am. Chem. Soc., 1979, 101, 2352. (b) M. M. Midland, D. C. McDowell, R. L Hatch and A. Tramontano, ibid., 1980, 102, 867.
2.   S. Yamaguchi and H. S. Mosher J. Org. Chem., 1973, 38, 1870.
3.   R. Noyori, I. Tomino, M. Yamada and M. Nishizawa, J. Am. Chem. Soc., 1984, 106, 6717.
4.   M. M. Midland and P. E. Lee, J. Org. Chem., 1985, 50, 3237.
5.   M. M. Midland, J. E. Petre, S. A. Zderic and A. Kazubski, J. Am. Chem. Soc., 1982, 104, 528.
6.   H. C. Brown and G. G. Pai, J. Org. Chem., 1985, 50, 1384.
7.   M. M. Midland, J. I. McLoughlin and J. Gabriel, J. Org. Chem., 1989, 54, 159.
8.   H. C. Brown, J. Chandrasekharan and P. V. Ramachandran, J. Am. Chem. Soc., 1988, 110, 1539.
9.   H. C. Brown, B. Q. Gong and P. V. Ramachandran, Unpublished work.
10.   H. C. Brown, J. Chandrasekharan and P. V. Ramachandran, J. Org. Chem., 1986, 51, 3394.
11.   J. A. Soderquist, G. W. Kabalka et al. Tetrahedron Lett., 1990, 31, 4677.
12.   P. V. Ramachandran, A. V. Teodorovic, M. V. Rangaishenvi and H. C. Brown, J. Org. Chem., 1992, 57, 2379.
13.   P. V. Ramachandran, A. V. Teodorovic and H. C. Brown, Tetrahedron, 1993, 49, 1725.
14.   H. C. Brown and P. V. Ramachandran, Acc. Chem. Res., 1992, 25, 16.
15.   H. C. Brown, M. Srebnik and P. V. Ramachandran, J. Org. Chem., 1989, 54, 1577.
16.   H. C. Brown and P. V. Ramachandran, J. Org. Chem. 1989, 54, 4504.
17.   H. C. Brown, P. V. Ramachandran, S. A. Weissman and S. Swaminathan J. Org. Chem., 1990, 55, 6328.
18.   P. V. Ramachandran, H. C. Brown, and S. Swaminathan Tetrahedron: Asym. 1990, 1, 433.
19.   H. C. Brown, P. V. Ramachandran, A. V. Teodorovic and S. Swaminathan Tetrahedron Lett. 1991, 32, 6691.
20.   H. C. Brown, W. S. Park, B. T. Cho and P. V. Ramachandran, J. Org. Chem., 1987, 52, 5406.

Synthesis with Allylic and Chiral Organoboranes

M. Zaidlewicz, Z. Walasek, and M. Krzemiński

INSTITUTE OF CHEMISTRY, NICOLAUS COPERNICUS UNIVERSITY, 87-100 TORUŃ, POLAND

1. HYDROXYMETHYLATION, FORMYLATION AND ACYLATION OF CYCLOALKENES

Allylic hydroxymethyl derivatives of cycloalkenes and the corresponding aldehydes are important synthetic intermediates.[1] A direct approach to these compounds by one carbon atom homologation of cycloalkenes is used in the Prins reaction. However, the synthetic utility of this reaction is limited due to the formation of mixtures, although in some cases single products are obtained. Similarly, Friedel-Crafts acylation of olefins, providing a simple access to unsaturated ketones has limitations. For example, acylation of 5- and 6-membered ring 1-methylcycloalkenes can be controlled to give either 1-acyl-2-methyl-2-cycloalkenes or 1-acyl-2-methyl-1-cycloalkenes depending on the reaction conditions. The third possible product, 2-methylene-1-acylcycloalkane, is formed in low yield or not at all.[2]

Over the past few years we have developed the synthesis of allylic diethylboranes via metalation of olefins with trimethylsilylmethylpotassium followed by the reaction with chlorodiethylborane. These organoboranes and their organopotassium precursors are useful intermediates for the transformation of cycloalkenes into homoallylic hydoxyalkyl and acyl derivatives, not readily available by the Prins and Friedel-Crafts reactions[1-4]. Thus, the allylic organopotassium compounds derived from unsubstituted cycloalkenes react readily with aldehydes to give the corresponding homoallylic alcohols which can be oxidized to 1-acyl-2- and 1-acyl-1-cycloalkenes. Mixtures of products are obtained from substituted cycloalkenes, e.g., 1-methylcycloalkenes, by the reaction of their organopotassium derivatives with aldehydes. Fortunately, allylic diethylboranes obtained from these olefins add to aldehydes stereoselectively with allylic rearrangement providing 1-(2-methylenecycloalkyl)alkanols which can be oxidized to 1-acyl-2-methylenecycloalkanes or 1-acyl-2-methyl-1-cycloalkenes. Hydroxymethylation and formylation of cycloalkenes is readily achieved using formaldehyde in these reactions.

The addition of allylic diethylboranes to ketene provides the corresponding enolates which are hydrolyzed to 1-acetyl-1-cycloalkenes. Consequently, the oxidation step can be omitted by using ketene instead of acetaldehyde (Scheme 1).

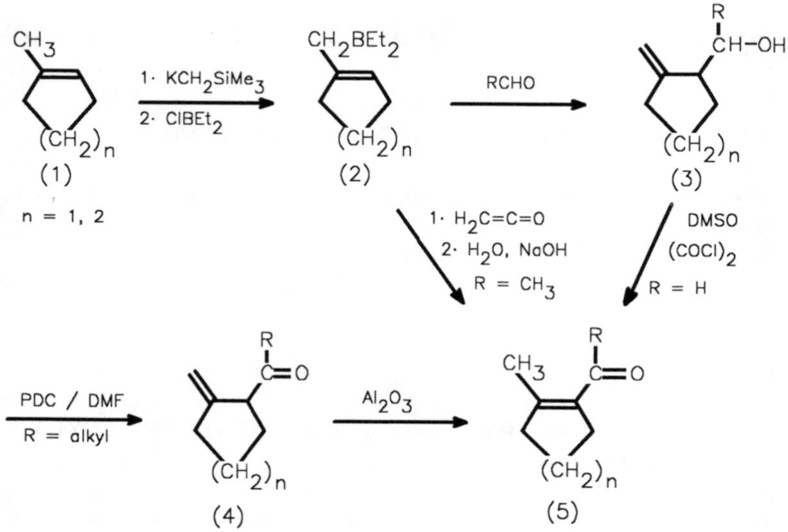

Scheme 1

2. THE ADDITION OF ALLYLDIETHYLBORANE TO VINYLIC EPOXIDES.

Vinylic epoxides react with allyldiethylborane to give the 1,2- and 1,4-addition products. The reactions with 1,3-butadiene and isoprene monoepoxides are slightly exothermic and can be carried out without a solvent.

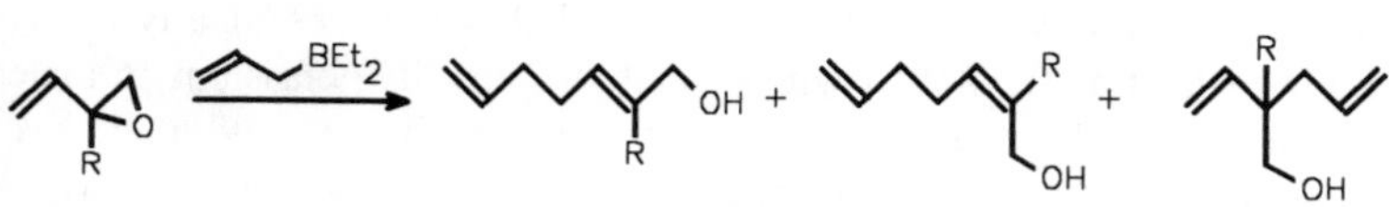

| R = H | 38 % | 11 % | 51 % |
| R = CH$_3$ | 61 % | 11 % | 28 % |

Cyclopentadiene monoepoxide reacts with allyldiethylborane with ring opening and 1,Z5,7-octatrien-4-ol is obtained. Its configuration suggests the rearrangement of the epoxide to Z2,4-pentadienal followed by the addition of allyldiethylborane.

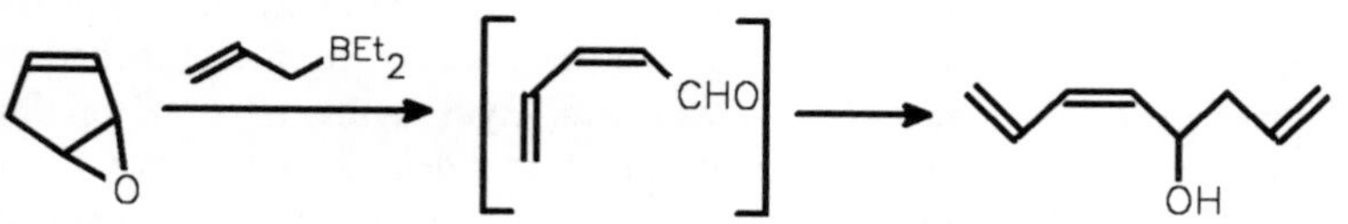

1,3-Cyclohexadiene monoepoxide does not undergo the rearrangement and the main addition product is cis- and trans 2-allyl-3-cyclohexenol.

3. ENANTIOSELECTIVE SYNTHESIS OF CYCLOALKENOLS.

In contrast to the enantioselective synthesis of alcohols by asymmetric hydroboration of olefins,[5] the corresponding transformation of 1,3-dienes into allylic
alcohols has been much less studied.[6] Although both reactions are similar, there
are also differences. First, acyclic 1,3-dienes are less reactive than their monohydroboration products and dihydroboration prevails.[7] Second, conjugated
cycloalkadienes undergoing monohydroboration may give mixtures of allylic and
homoallylic alcohols rendering single product isolation difficult, Third, allylic
organoboranes are prone to racemization by allylic isomerization. To avoid racemization both the hydroboration and oxidation steps should be carried at low
temperature. Recently, it has been shown that diisopinocampheylborane hydrobo-
rates unsubstituted 1,3-cycloalkadienes with high enantioselectivity. The intermediate allylic organoboranes have been used for the allylboration of aldehydes.[6]

We applied di-2- and di-4-isocaranylboranes (2-Icr$_2$BH and 4-Icr$_2$BH) of high
optical purity obtained from (+)-2- and (+)-3-carene,[8] respectively, for the asymmetric hydroboration of 1,3-cycloalkadienes and representative methyl-substituted
1,3-cyclohexadienes. The regioselectivity of hydroboration of these dienes is
shown in Figure 1. 1,3-Cyclooctadiene is not included in the Figure since some
migration of the boron atom around the ring occurs.

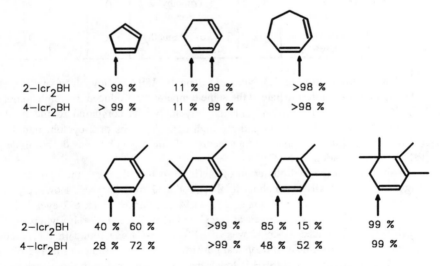

| 2-Icr$_2$BH | > 99 % | 11 % 89 % | >98 % |
| 4-Icr$_2$BH | > 99 % | 11 % 89 % | >98 % |

| 2-Icr$_2$BH | 40 % 60 % | >99 % | 85 % 15 % | 99 % |
| 4-Icr$_2$BH | 28 % 72 % | >99 % | 48 % 52 % | 99 % |

Figure 1.   Regioselectivity of the monohydroboration of 1,3-cycloalkadienes
with diisocaranylboranes.

In most cases studied, both reagents show similar regioselectivity. As follows
from Fig.1, it is possible to direct the boron atom either to the allylic or
homoallylic position by proper location of the substituents.

The monohydroboration of cycloalkadienes with diisocaranylboranes was
carried out in tetrahydrofuran at -5° C using a 50 % excess of the diene. The
organoborane intermediates were oxidized at -10° C by the addition of one molar

equivalent of triethylamine followed with 30 % hydrogen peroxide. Under these conditions the oxidation is very fast, practically complete after the addition of hydrogen peroxide. All three organic groups attached to the boron atom are oxidized. The oxidation is effective even at -30⁰ C. The product alcohols obtained with di-2-isocaranylborane, isolated by distillation and purified by preparative GLC, are shown in Table 1.

Table 1.   Synthesis of alcohols by monohydroboration of conjugated cyclo-
           alkadienes with di-2-isocaranylborane at -5⁰ C.

Diene	Time	Product alcohol			
	h	Alcohol	Conf.	ee %	Yield %
1,3-Cyclohexadiene	16	2-Cyclohexenol	R	68	84
1,3-Cycloheptadiene	18	2-Cycloheptenol	R	93	76
1,3-Cyclooctadiene	16	2-Cyclooctenol	R	98	77
1-Methyl-1,3-cyclo-hexadiene	25	3-Methyl-2-cyclo-hexenol	R	79	50
1,5,5-Trimethyl-1,3-cyclohexadiene	22	3,5,5-Trimethyl-2-cyclohexenol	R	84	94
1,2,6,6-Tetramethyl-1,3-cyclohexadiene	32	3,4,5,5-Tetramethyl-3-cyclohexenol	R	60	91

The hydroboration of 1,3-cyclohexadiene at -15⁰ C followed by oxidation of the organoborane intermediate at the same temperature produced 2-cyclohexenol of 72 % ee. Its optical purity is only slightly higher as compared to the product obtained at -5⁰ C. This result and the high optical purites of 2-cycloheptenol and 2-cyclooctenol indicate that the intermediate allylic organoboranes do not undergo allylic isomerization at -5⁰ C.

Similarly to the hydroboration of olefins,[8] 4-Icr$_2$BH gives the opposite con-figuration of the allylic alcohols to that produced by 2-Icr$_2$BH. However, their optical purites vary over a wide range (2-81 % ee). For example, S-2-cyclohexenol of 13 and 16 % ee was obtained at -5⁰ C and -15⁰ C, respectively, whereas S-2-cyclooctenol produced at -5⁰ C was of 81 % ee. 3-Carene found in the reaction products when alcohols of low optical purities are obtained, indicates dehydrobora-tion. The reactivity of 4-Icr$_2$BH is lower than 2-Icr$_2$BH and hydroborations below 0⁰ C with 4-Icr$_2$BH are sluggish. Clearly, only 2-Icr$_2$BH is a convenient reagent for the asymmetric hydroboration of conjugated cycloalkadienes producing constantly alcohols of the same configuration for similar structures in moderate to high optical purity.

4.   SYNTHESIS OF ISOLIMONYLBOROLANE

The isolation of alcohols with boiling points close to the oxidation products of the hydroborating agent is often a problem. In such cases a hydroborating agent, e.g., a boraheterocycle, producing a high boiling diol upon oxidation would be advantageous. Unfortunately, only very few boraheterocyclic asymmetric hydroborating agents are known. Limonylborane having a borinane ring is moderately enantioselective.[9] Trans-2,5-dimethylborolane shows excellent enantioselectivity but the reagent is not readily available.[10]

Consequently, we decided to prepare the substituted borolanes (12) and (13) by cyclic hydroboration of (+)-3,8(9)-p-menthadiene (9) synthesized from (+)-isoterpinolene (6),[11] (Scheme 2). Its monoepoxidation proceeds at the more substituted double bond to give a mixture of stereoisomeric epoxides (7). The reaction of the mixture with borane in tetrahydrofuran is accompained by the evolution of hydrogen and does not lead to (+)-3-p-menthen-8-ol (8) as might be expected from our earlier studies on the reduction of vinylic epoxides.[12] The alcohol (8) is obtained by the reduction of (7) with calcium in liquid ammonia and dehydrated with potassium hydrogen sulfate to give (+)-3,8(9)-p-menthadiene in 52% overall yield from (+)-isoterpinolene. Cyclic hydroboration of the diene with monochloroborane, methanolysis and distillation yields a mixture of diastereomeric B-methoxyisolimonylborolanes (10) and (11) which is reduced to (12) and (13) by lithium aluminium hydride. Separation of (10) and (11) is under current investigation.

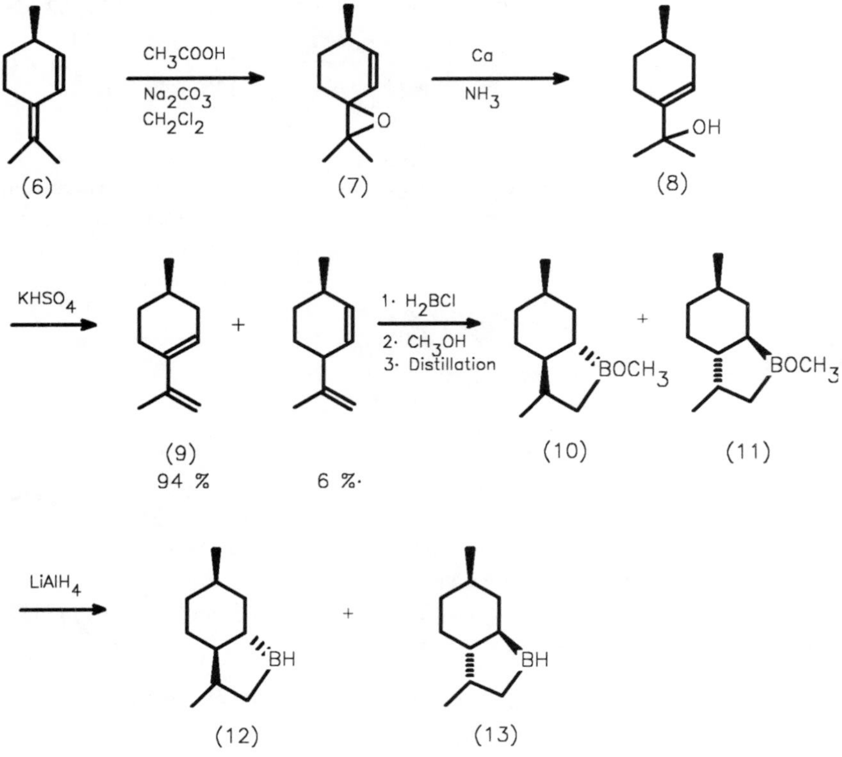

Scheme 2.

REFERENCES

1. M. Zaidlewicz, J. Organomet. Chem., 1991, 409, 103, and references therein.
2. M. Zaidlewicz, Synthesis, 1988, 701, and references therein.
3. M. Zaidlewicz, J. Organomet. Chem., 1985, 293, 139.
4. M. Zaidlewicz, Tetrahedron Lett., 1986, 27, 5135.
5. H.C. Brown and P.K. Jadhav "Asymmetric Synthesis", Academic Press, Orlando, 1983, Vol. 2, Part A, Chapter 1, p.1.
6. H.C. Brown, K.S. Bhat and P.K. Jadhav, J. Chem. Soc. Perkin Trans. 1, 1991, 2663.
7. M. Zaidlewicz, "Comprehensive Organometallic Chemistry", Pergamon Press, Oxford, 1982, Vol. 7, Chapter 45, p.199.
8. H.C. Brown, J.V.N. Vara Prasad and M. Zaidlewicz, J. Org. Chem., 1988, 53, 2911.
9. P.K. Jadhav and S.U. Kulkarni, Heterocycles, 1982, 18, 169.
10. S. Masamune, B.M. Kim, J.S. Petersen, T. Sato, S.J. Veenstra and T. Imai, J. Am. Chem. Soc., 1985, 107, 4549.
11. A free sample generously supplied by Camphor and Allied Products, 133 Mahatma Ghandi Road, 400001 Bombay, India.
12. M. Zaidlewicz, A. Uzarewicz and A.Sarnowski, Synthesis, 1979, 62.

Enantiospecific Synthesis of 2-Alkylpyrrolidine and Piperidine Derivatives Mediated by Organoborane Reagents

B. Singaram, T. Nguyen, D. Sherman, D. Ball, and M. Solow

DEPARTMENT OF CHEMISTRY, THE UNIVERSITY OF CALIFORNIA,
SANTA CRUZ, CA 95064, USA

Introduction

Various alkyl substituted pyrrolidines and piperidines have been reported as venomous constituents of ant venom alkaloids and have necrotoxicity, hemolytic and antibiotic activities.[1] In recent years, a number of groups have reported the synthesis of 2-substituted pyrrolidines and piperidines.[2] Many of these syntheses involve either a resolution or the use of chiral synthons from "chiral pool".[3] During the course of our program on asymmetric hydroboration of enamines,[4] we became interested in the synthesis of enantiomerically pure 2-alkylpyrrolidines and 2-alkylpiperidines. We speculated that the 2-substituted pyrrolidines and piperidines can be readily obtained from enantiomerically pure homoallylic alcohols.[5] Retrosynthetic analyses of both 2-alkylpyrrolidines and 2-alkyl piperidines indicates the potential of homoallylic alcohols (Scheme).

Scheme:

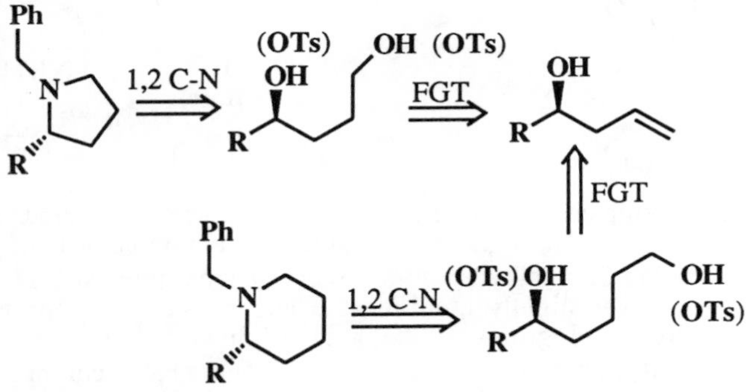

We have found that the hydroboration-methanolysis reaction converts homoallylic alcohols[5] predominantly into 6-alkyl-2-methoxy-1,2-oxaborinanes **1**. We were interested in developing a method for converting **1** into 2-alkylpyrrolidines and 2-alkylpiperidines (eq. 1). We now describe our preliminary results on the conversion of **1** into 2-alkylpyrrolidine and 2-alkylpiperidine derivatives.

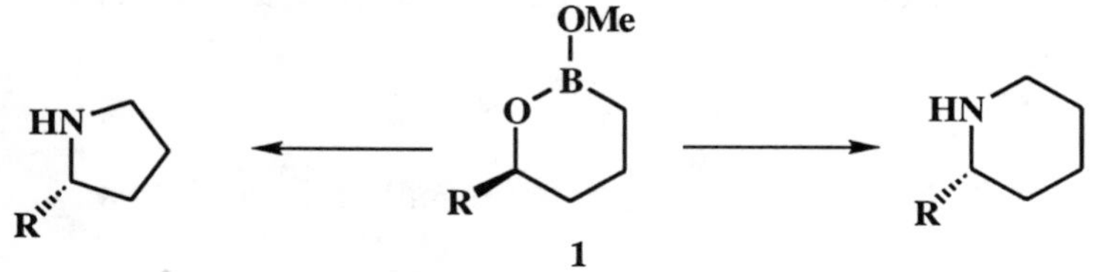

$$(1)$$

Results and Discussion

This procedure is experimentally convenient and provides final products in >98% enantiomeric purity. Starting with commercially available aldehydes and either isomer of diisopinocampheylallylborane (Ipc$_2$BALL), enantiomerically pure homoallylic alcohols were readily obtained.[5] Subsequent hydroboration and treatment with an equivalent of methanol produced cyclic boronic esters (**1**) of known configuration.

Thus, the hydroboration-methanolysis of (*R*)-1-octen-4-ol with borane-dimethylsulfide (BMS) provided (*R*)-6-butyl-2-methoxy-1,2-oxaborinane in essentially quantitative yield. This boronic ester was oxidized to the corresponding 1,4-diol which was converted into a dimesylate and cyclized with benzylamine[6] to give (*S*)-*N*-benzyl-2-butylpyrrolidine. Similarly, (*S*)-1-octen-4-ol was converted into (*R*)-*N*-benzyl-2-butylpyrrolidine. Following this general procedure several *N*-benzyl-2-alkylpyrrolidines of known configuration were synthesized (eqs. 2).

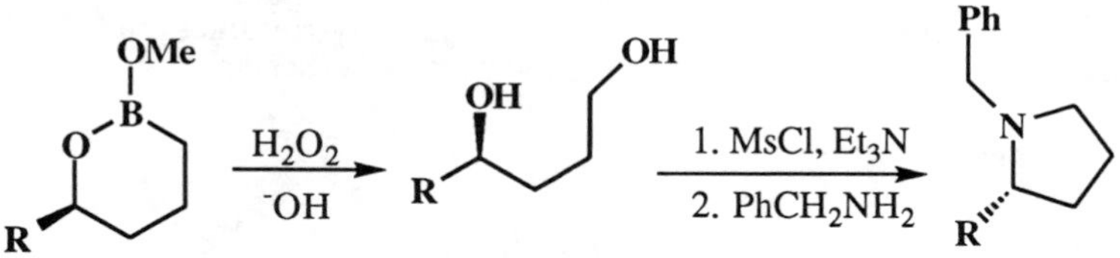

$$(2)$$

R	Yield, %
n-Butyl	76
n-Heptyl	62
3-Pentyl	70

We then attempted to extend this methodology to the synthesis of 2,5-dialkylpyrrolidines. In the light of biological and entomological importance, pathways producing 2,5-dialkylpyrrolidines are of great interest.[1] In order to achieve this goal, we slightly modified the above synthetic approach. We protected the hydroxyl group of the homoallylic alcohol by acetylation. Hydroboration followed by oxidation with pyridinium chloro chromate (PCC)[7] afforded the corresponding acetoxy aldehyde. Allylboration followed by deprotection gave the desired 1,4-diol which on mesylation and subsequent cyclization with benzylamine produced *N*-benzyl-2-alkyl-5-allyl-pyrrolidines. Stereochemistry at the 5-position can be controlled by the choice of an appropriate enantiomer of Ipc$_2$BALL. Thus, (*R*)-4-acetoxy-1-octene (>98% ee) was converted into (2*S*,5*R*)-2-butyl-5-allyl-*N*-benzylpyrrolidine of very high enantiomeric purity in moderate yield (eq. 3).

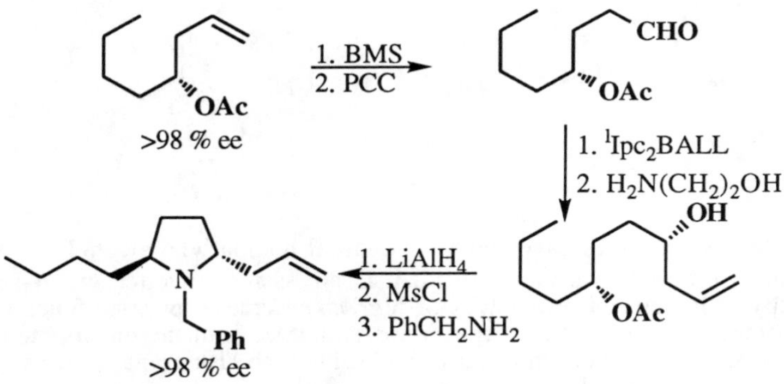

The common boronic ester intermediate (**1**) was also converted into the corresponding *N*-benzyl-2-alkylpiperidines. Homologation of the common boronic ester gave the corresponding one-carbon homologated cyclic boronic ester.[8] The homologation is envisioned to proceed through an "ate" complex formed by the addition of chloromethyl anion (eq. 4).[8]

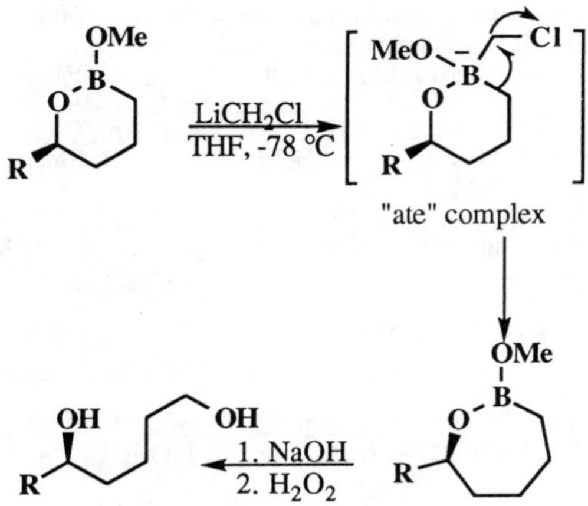

Oxidation of the homologated intermediate using alkaline hydrogen peroxide gave the desired 1,5-diol in essentially quantitative yield. Thus, (*R*)-1, 5-octane diol upon mesylation and subsequent cyclization with benzylamine gave (*S*)-*N*-benzyl-2-propylpiperidine (*N*-benzylconiine)(eq. 5).

For analytical purposes, portions of these *N*-benzyl-2-alkylpyrrolidines and piperidines were debenzylated using Pearlman's catalyst, Pd(OH)$_2$-C/H$_2$, and converted into the corresponding MTPA-amides.[9] Capillary GC analysis on a 30 m Methylsilicone column showed that these 2-alkylpyrrolidines and piperidines were >98% enantiomerically pure.

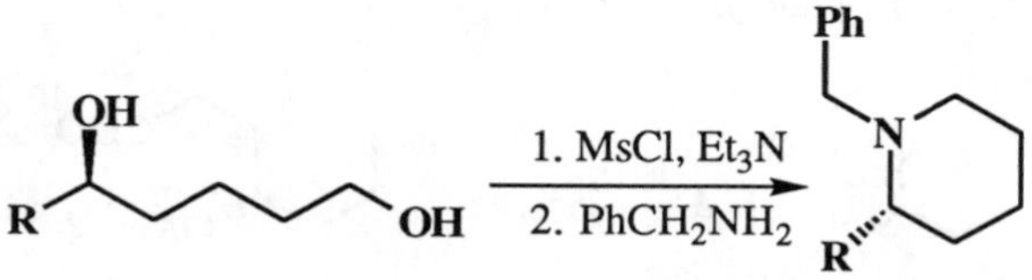

$$\underset{R}{\overset{OH}{\wedge}}\!\!\!\!\!\wedge\!\!\!\!\wedge\!\!\!\!OH \quad \xrightarrow[\text{2. PhCH}_2\text{NH}_2]{\text{1. MsCl, Et}_3\text{N}} \quad \text{(piperidine structure)} \tag{5}$$

R	Yield, %
Me	71
n-Pr	64

In summary, we have shown that chiral homoallylic alcohols (>98% ee), obtained from aldehydes and ^{d}Ipc$_2$BALL, are easily converted into *N*-benzyl-2-alkylpyrrolidines and piperidines of high enantiomeric purity. Since both the enantiomers of Ipc$_2$BALL reagents are available, both the enantiomers of the heterocyclic compounds are readily obtained in high enantiomeric purity. We are currently evaluating the generality of this method for the synthesis of optically active *cis*- and *trans*-2,5-dialkylpyrrolidines and 2,6-dialkylpiperidines.

References

1. (a) T. H. Jones, M. S. Blum and H. M. Fales, Tetrahedron, 1982, 38, 1949.
    (b) G. W. J. Fleet and J. C. Son, Tetrahedron, 1988, 44, 2637.
2. R. R. Frazer and S. Passannanti, Synthesis, 1976, 540. (b) T. Nagasaka, H. Hayashi and F. Hamaguchi, Hetreocycles, 1988, 27, 1685.
3. (a) K. Shiosaki and H. Rapaport, J. Org. Chem, 1985, 50, 1229. (b) R. P. Polniazek, S. E. Belmont and R. Alvarez, J. Org. Chem, 1990, 55, 215. (c) L. Guerrier, J. Royer, D. S. Grieerson and H.-P. Husson, J. Am. Chem. Soc, 1983, 105, 7754. (d) A. I. Meyers, D. A. Dickman and T. R. Baily, J. Am Chem. Soc 1985, 107, 7974.
4. B. Singaram, C. T. Goralski and G. B. Fisher, J. Org. Chem, 1991, 56, 5691.
5. (a) H. C. Brown, P. K. Jadhav and B. Singaram, Modern Synthetic Methods, 1986, 4, 307. (b) U. S. Racherlla and H. C. Brown, J. Org. Chem, 1991, 56, 401.
6. R. P. Sharp, R. M. Kennedy and S. Masamune, J. Org. Chem, 1989, 54, 1755.
7. C. G. Rao, S. U. Kulkarni and H. C. Brown, J. Organometal. Chem., 1979, 172, C20.
8. (a) D. S. Matteson and D. Mujumdar, J. Am. Chem. Soc, 1980, 102, 7588. (b) H. C. Brown, S. M. Singh and M. V. Rangaishenvi, J. Org. Chem, 1986, 51, 3150.
9. J. A. Dale, D. L. Dull and H. S. Mosher, J. Org. Chem, 1969, 34, 2543.
10. For experimental details see: T. Nguyen, D. Sherman, D. Ball, M. Solow and B. Singaram, Tetrahedron: Asymmetry, 1993, 4, 189.

(1R,2S,3S,5R)-2-Dialkylamino-6,6-dimethylbicyclo[3.1.1]heptan-3-ols. Synthesis from R-(+)-Nopinone Enamines via Hydroboration/Oxidation and Utility as Chiral Auxiliaries for the Addition of Diethylzinc to Benzaldehyde

C. T. Goralski[*,1], D. L. Hasha[2], L. W. Nicholson[2], P. R. Rudolf[2], and B. Singaram[3]

[1]PHARMACEUTICALS PROCESS RESEARCH AND [2]ANALYTICAL SCIENCES, CORE R&D, THE DOW CHEMICAL COMPANY, MIDLAND MI 48674, USA
[3]THE DEPARTMENT OF CHEMISTRY AND BIOCHEMISTRY, UNIVERSITY OF CALIFORNIA, SANTA CRUZ, SANTA CRUZ, CA 95064, USA

β-Amino alcohols, especially those containing one or more stereogenic centers, are important in medicinal chemistry[1] and as chiral auxiliaries for a wide variety of asymmetric organic reactions.[2] We recently described the synthesis of several series of β-amino alcohols via the hydroboration/methanolysis/oxidation of ketone and aldehyde enamines.[3] This paper describes the extension of this procedure to the preparation of (1R, 2S, 3S, 5R)-2-dialkylamino-6,6-dimethylbicyclo[3.1.1]heptan-3-ols (3a-d) from the corresponding enamines 2a-d of the chiral ketone (1R,5S)-(+)-nopinone (1).[4]

The reaction of 1 with a secondary amine in refluxing cyclohexane afforded excellent yields of the corresponding enamines 2a-d (eq 1). Hydroboration of 2a-d with borane methyl sulfide (BMS) in tetrahydrofuran (THF) followed by methanolysis and oxidation with basic hydrogen peroxide afforded moderate to good yields of the β-amino alcohols 3a-d (eq 2).

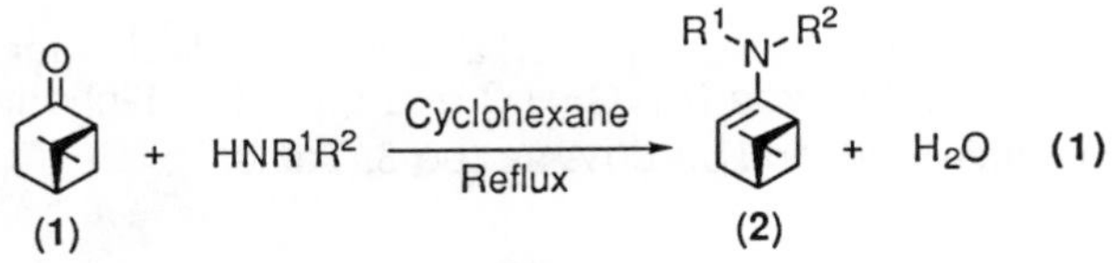

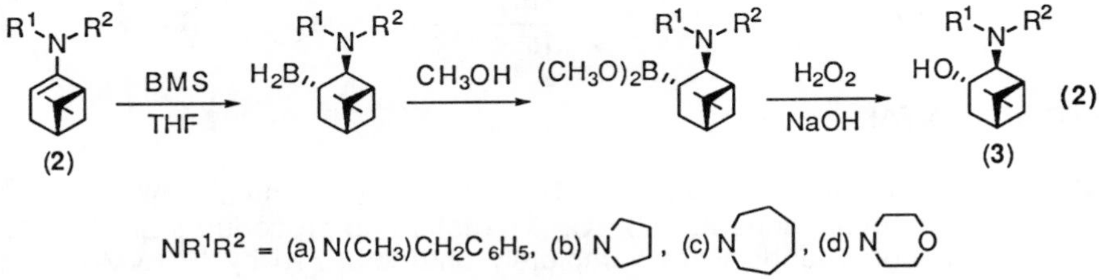

$$NR^1R^2 = \text{(a) } N(CH_3)CH_2C_6H_5, \text{ (b) } N\bigcirc, \text{ (c) } N\bigcirc, \text{ (d) } N\bigcirc O$$

The structures of the β-amino alcohols 3a-d were confirmed by detailed analyses of their ^1H and ^{13}C NMR spectra, and data for compound 3a are given in Table 1.

<u>Table 1</u> NMR Parameters of (1R,2S,3S,5R)-2-Benzylmethylamino-6,6-di-methylbicyclo[3.1.1]heptan-3-ol (CDCl$_3$)

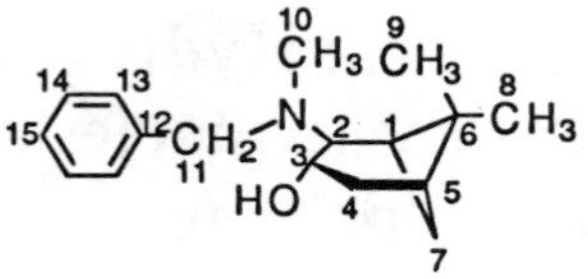

Position	^{13}C (δ ppm)	^{1}H (δ ppm)	$|J_{HH}|^a$ (Hz)
1	42.7	2.43	$J_{1,2}$=2.9; $J_{1,7}\cong$0
2	77.1	2.64	$J_{2,3}$=3.1
3	70.0	4.29	$J_{3,4}$=3.2; $J_{3,4^*}$=8.5
4,4* b	38.6	1.77, 2.66	$J_{4,4^*}$=14.0; $J_{4,5}$=3.2
5	41.1	1.99	$J_{4^*,5}$=2.4; $J_{5,7}\cong$0
6	36.6	-----	
7,7*	29.6	1.19, 2.39	$J_{7,7^*}$=9.2
8	27.0	1.26	
9	22.8	1.12	
10	39.7	2.10	
11,11*	59.8	3.46, 3.77	$J_{11,11^*}$=13.6
12	140.5	-----	
13	128.4 or 128.2	~7.45	
14	128.2 or 128.4	~7.45	
15	126.6	7.24	

^{a}Absolute value of the coupling constant is presented; ±0.2 Hz. All other couplings were not determined. ^{b}The * designates the downfield proton of a nonequivalent methylene group.

The structure of **3a** was further confirmed by single crystal X-ray analysis (Figure 1).

The amino alcohols 3a-d were employed as chiral auxiliaries for the addition of diethylzinc to benzaldehyde (Ar=C$_6$H$_5$) and gave (R)-1-phenyl-1-propanol in nearly quantitative yield and 63-80% ee (eq 3, Table 2).

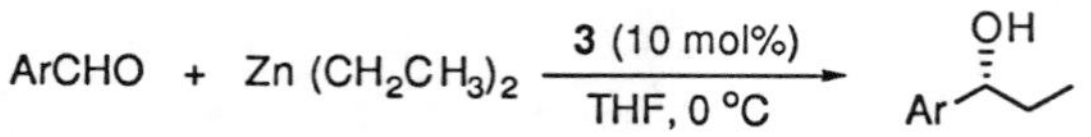

$$\text{ArCHO} + \text{Zn (CH}_2\text{CH}_3)_2 \xrightarrow[\text{THF, 0 °C}]{\textbf{3} \text{ (10 mol%)}} \underset{\text{Ar}}{\overset{\text{OH}}{\diagup\!\!\diagdown}} \quad (3)$$

EXPERIMENTAL

General. All reactions were performed under nitrogen with glassware which had been previously dried at 140 °C. The following procedures are typical, and all new compounds were completely characterized by IR, NMR, and elemental analyses.

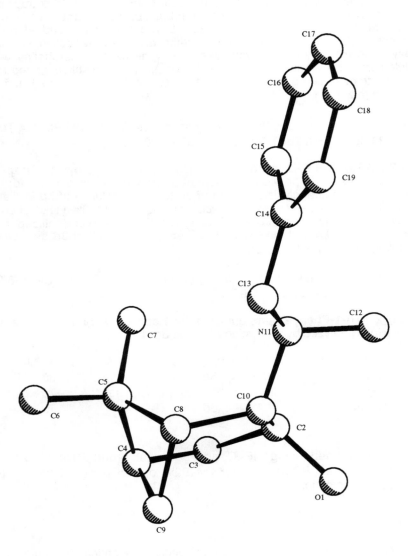

<u>Figure 1</u> Single Crystal X-Ray Structure of (1*R*,2*S*,3*S*,5*R*)-2-Benzylmethyl-amino-6,6-dimethylbicyclo[3.1.1]heptan-3-ol

Morpholine Enamine of (1*R*,5*S*)-(+)-Nopinone (2d).   A 50-mL, single-neck flask equipped with a magnetic stirrer and a Dean-Stark trap fitted with a reflux condenser bearing a nitrogen bubbler was charged with 4.84 g (35 mmol) of (1*R*,5*S*)-(+)-nopinone, 25 mL of cyclohexane, and 11 mL of morpholine. The mixture was heated at reflux and the water collected. After 7 days, the majority of the cyclohexane and the excess morpholine were distilled into the Dean-Stark trap. The residue was distilled at reduced pressure to give 6.64 g (91.5% yield) of 2d as a colorless, viscous liquid, bp 76-78 °C (0.4 Torr).

The same procedure gave the following enamines: **2a**, bp 130 °C (0.4 Torr); **2b**, bp 83-85 °C (0.6 Torr); **2c**, bp 100-103 °C (0.9 Torr).

(1*R*,2*S*,3*S*,5*R*)-6,6-Dimethyl-2-(4-morpholino)bicyclo[3.1.1]heptan-3-ol (3d).   A previously described procedure[3] was used for the hydroboration/methanolysis/oxidation of **2d** to give a solution of **3d** in diethyl ether. The ether was removed *in vacuo* leaving 3.68 g (98% yield) of crude **3d** which rapidly crystallized. The crude **3d** was recrystallized from approximately 30 mL of hexane to give 2.65 g (71% yield) of **3d** as a white, crystalline solid, mp 84.5-86.5 °C.

The same procedure gave the following amino alcohols: **2a**, mp 134-136 °C; **2b**, mp 67-69 °C; **2c**, mp 96.5-97.5 °C.

<u>Table 2</u> (*R*)-1-Phenyl-1-propanol *via* the Reaction of Benzaldehyde with Diethylzinc Catalyzed by the Amino Alcohols **3a-d**

Ar	Amino Alcohol	ee, %[a,b]
C_6H_5	3a	80
C_6H_5	3b	76
C_6H_5	3c	70
C_6H_5	3d	63

[a]Determined using HPLC and a Daicel brand CHIRALPAK AD stationary phase. [b]Corrected for the 92% optical purity of the starting (1*R*,5*S*)-(+)-nopinone.

REFERENCES

1. M. Grayson, Ed. , 'Kirk-Othmer Encyclopedia of Chemical Technology, 3rd ed.', Wiley-Interscience, New York, 1982, Vol. 17, pp. 311-345.
2. (a) K. Tomioka, *Synthesis*, 1990, 541. (b) R. Noyori and M. Kitamura, *Angew. Chem. Int. Ed. Engl.*, 1991, *1*, 49.
3. C.T. Goralski, B. Singaram, and H.C. Brown, *J.Org. Chem.*, 1987, *52*, 4014. (b) B. Singaram, C.T. Goralski, and G.B. Fisher, *J. Org. Chem.*, 1991, *56*, 5691.
4. P. Lavallee and G. Bouthillier, *J. Org. Chem.*, 1986, *51*, 1362.

MEDICAL APPLICATIONS OF BORON

The Rationale for the Development of Boron Containing Nucleosides for Boron Neutron Capture Therapy

A. H. Soloway[1], W. Tjarks[1], A. K. M. Anisuzzaman[1], F.-G. Rong[1], A. J. Lunato[1], I. M. Wyzlic[1], and R. F. Barth[2]

[1]COLLEGE OF PHARMACY AND [2]DEPARTMENT OF PATHOLOGY, THE OHIO STATE UNIVERSITY, 500 W. 12TH AVENUE, COLUMBUS OH 43210, USA

In order to enhance the utility of Boron Neutron Capture Therapy (BNCT) in the treatment of malignant tumors, the development of more effective agents which possess greater specificity for tumor versus contiguous normal tissues and blood remains one of the crucial needs (1). Compounds must not only demonstrate selectivity for and cellular persistence in the tumor but attain tissue concentrations in the range of 10^9 boron atoms per cancer cell which translates into approximately 30 micrograms of boron per gram of tumor. Thus, both specificity and adequate concentration levels are the two key requirements which must be fulfilled for any suitable agent.

To maximize the effects of the high linear energy transfer (LET) particles produced in the capture reaction, the compounds should localize intracellularly and ideally become incorporated or attached to the tumor cell nucleus. This locus will increase the relative biological effectiveness (RBE) by at least two fold and possibly an order of magnitude greater than if the compounds were localized in the cytoplasm (2,3). A key question has been how can compounds be designed which will achieve such a locus?

A number of researchers have explored the possibility of using chemical building blocks of cellular constituents with the view that malignant cells, many of which have significantly shorter doubling times than do their normal counterparts, would require larger numbers of such biochemical substrates. It is, of course, essential that any boron-containing analogues have the capability of mimicking normal substrates by the action of cellular enzymes. One of the approaches which we have undertaken is to design boron-containing nucleosides which may ultimately become incorporated in tumor nucleic acids.

One of the initial approaches in the development of such structures was the research by C.C. Cheng and his collaborators in the synthesis of 5-dihydroxyboryluracil (4). Its synthesis encouraged Schinazi and Prusoff to prepare the first boron-containing pyrimidine nucleoside, 5-dihydroxyboryl-2'-deoxyuridine (DBDU) (5). If such a boron-containing nucleoside is to be incorporated into DNA, it must be effectively transported into the cell, phosphorylated successively to the mono-, di- and triphosphate and the latter subsequently incorporated into nucleic acids by the action of DNA polymerases. Biological research with DBDU indicated that this thymidine analogue was relatively nontoxic and from radiobiological studies that approximately 5-15% replaced thymidine. However, there was no chemical evidence that DNA incorporation had been achieved.

All of these initial compounds contained a single boron atom. Incorporation of a carboranyl moiety into a nucleoside would: (1) increase the boron content tenfold compared with the corresponding boronic acid analogues and; (2) greatly enhance the compound's lipophilicity. The first carboranyl nucleoside prepared was 2'-O-(o-carboran-1-ylmethyl)uridine(2'-CBU) (6). More recently, Yamamoto (7) and Schinazi (8) have attached the carborane nucleus directly to the 5-position on the pyrimidine nucleosides. However, the critical question is not the number of boron atoms but whether these structures can be converted intracellularly into the corresponding nucleotide and subsequently into nucleic acids? 2'-CBU though tightly bound to cells, was not transformed to its nucleotide (9). However, 5-carboranyldeoxyuridine was metabolized to its nucleotide (8) with little, if any conversion to the di- and triphosphates nor was there evidence of incorporation into DNA. Bulky groups attached directly at the 5-position have the potential for inhibiting enzyme activity. For example, the activity of deoxycytidine kinase was affected by the size of the substituent in the 5 position. The enzyme affinity for 5-fluorodeoxycytidine was comparable to deoxycytidine but lower for the 5-methyl analogue and considerably reduced for the 5-bromo and 5-iodo analogues (10). It has been shown with the interaction of kinases with immobilized nucleotides that the accessibility and enzyme binding was enhanced by the use of polymethylene spacers which in effect insulate and project the nucleotide from the matrix surface. The binding was weak when the extension arms contained four or fewer methylene groups. When the tether was increased to 10 Å by interposing additional methylene groups, 6 to 8, there was a substantial increase in enzyme affinity (11). This same principle has now been applied in the design and synthesis of boron-containing nucleosides in which the boron moiety is attached at the end of a flexible chain to the 5-position of the pyrimidine nucleus.

Since much of our research has focused on the use of BNCT for the treatment of malignant brain tumors, an additional important requirement of the designed compounds is that they possess the capability of penetrating the blood-brain barrier (BBB) and target disseminated islands of tumor cells in the normal brain. For certain nucleosides, there appears to be a carrier-mediated transport operating that is independent of the compound's lipophilicity (12). The compounds handled in this way include uridines and thymidines in contrast with the cytidines. This observation has provided the basis for the pyrimidines selected for design and synthesis.

Two general types of pyrimidine nucleosides substituted in the 5 position have been selected: (1) dihydroxyboryl compounds; and (2) carboranes. Among the structures being synthesized are the following target compounds:

$(HO)_2B(CH_2)_n$ — **A**

$HC\overset{\diagdown O\diagup}{\equiv}C(CH_2)_n$ / $B_{10}H_{10}$ — **B**

$HC\overset{\diagdown O\diagup}{\equiv}C(CH_2)_2CONH(CH_2)_m$ / $B_{10}H_{10}$ — **C**

R_1 = 2-deoxy-D-ribofuranose
n = 6-8
m = 2-4

It is unclear at this juncture what are the structural features which will permit a boron-containing nucleoside to become incorporated into tumor DNA? For this reason, compounds containing both a single, nonbulky boron atom, as in a boronic acid, to those containing a carborane moiety with ten boron atoms are being synthesized. The tether chain should be sufficiently flexible so as not to interfere with the action of the enzymes in their phosphorylation and ultimate incorporation into DNA. When the three dimensional structures of all the enzymes and their active sites are fully elaborated, then computer drug modeling may provide the answer for the optimal structure of such a boron-containing nucleoside.

All of the foregoing structures have a hydrocarbon chain attached directly at the 5-position and may be considered as either 5-substituted uridines or as extensions of thymidines. In addition, it has been reported (13) that 5-methylmercapto-2'-deoxyuridine (MMDU) is equal or superior to thymidine in its incorporation into cellular and viral DNA. Among the compounds being synthesized are target compounds D and E.

$(HO)_2B(CH_2)_nS$ — **D**

$HC\overset{\diagdown O\diagup}{\equiv}C(CH_2)_nS$ / $B_{10}H_{10}$ — **E**

R_1 = 2-deoxy-D-ribofuranose
n = 5-7

These may be viewed as higher homologues of MMDU. In these structures as well, the observations noted in affinity chromatographic studies (10) are being applied.

In the past, boron compounds have been evaluated by *in vitro* cellular uptake and retention studies and by *in vivo* pharmacokinetic studies in tumor-bearing animals. However, for these nucleosides, it will also be important to determine whether these structures are phosphorylated and enter the biosynthetic pathways leading to incorporation into DNA. For this purpose nucleotides will be synthesized (14) as reference compounds for use in determining whether they are formed in cell-free systems, and *in vitro* and *in vivo* from their corresponding nucleosides. Such biochemical and subcellular distribution studies are important in determining whether these compounds are indeed nucleic acid precursors in tumor cells, as well as the mechanism by which incorporation is achieved.

REFERENCES

1.   R.F. Barth, A.H. Soloway, R.G. Fairchild and R.M. Brugger, Cancer, 1992, 70, 2995.
2.   T. Kobayashi and K. Kanda, Radiat. Res., 1982, 91, 77.
3.   D. Gabel, S. Foster and R.G. Fairchild, Radiat. Res., 1987, 111, 14.
4.   T.K. Liao, E.G. Pondrebarac and C.C. Cheng, J. Am. Chem. Soc., 1964, 86, 1869.
5.   R.F. Schinazi and W.H. Prusoff, Tetrahedron Lett., 1985, 50, 841.
6.   A.K.M. Anisuzzaman, F. Alam and A.H. Soloway, Polyhedron, 1990, 9, 891.
7.   Y. Yamamoto, T. Seko, H. Nakamura, H. Nemoto, H. Hojo, N. Mukai and Y. Hashimoto, J. Chem. Soc., Chem. Commun., 1992, 157.
8.   R.F. Schinazi, N. Goudgaon, J. Soria and D.C. Liotta, Fifth International Symposium on Neutron Capture Therapy for Cancer, September 13-17, 1992, 11.
9.   W. Tjarks, A.H. Soloway, L. Liu and R.F. Barth, unpublished.
10.   G.M. Cooper and S. Greer, Molec. Pharmacol., 1973, 9, 704.
11.   C.R. Lowe and P.D.G. Dean, 'Affinity Chromatography', Wiley & Sons, New York, 1973, pp 25-26.
12.   J.M. Collins, R.W. Klecker, J.A. Kelley, J.S. Roth, C.L. McCully, F.M. Balis and D.G. Poplack, J. Pharmacol. Exp. Therap., 1988, 245, 466.
13.   R. Hardi, R.G. Hughes, Y.K. Ho, K.C. Chadha and T.J. Bardos, Antimicrob. Agents and Chemotherap., 1976, 10, 62.
14.   W. Tjarks, A.K.M. Anisuzzaman and A.H. Soloway, unpublished.

ACKNOWLEDGEMENTS

We wish to thank the U.S. Department of Energy and the National Cancer Institute for their financial support of our research and the Callery Chemical Company for kindly providing us with certain boron compounds.

Synthesis of New ^{10}B Carriers and Their Selective Uptake by Cancer Cells

Yoshinori Yamamoto*, Hisao Nemoto, Hirofumi Nakamura, and Satoshi Iwamoto

DEPARTMENT OF CHEMISTRY, FACULTY OF SCIENCE, TOHOKU UNIVERSITY, SENDAI 980, JAPAN

1   INTRODUCTION

Much attention has been paid to the design and synthesis of boron-10 carriers that deliver adequate concentration of ^{10}B atoms to tumors.[1] Thus biologically active and tumor-seeking organic moieties such as nucleosides, porphyrins, sugars, and lipids, have been bonded to carboranes. We chose uridine and aziridine derivatives as a functional group which may be prone to be accumulated in certain tumor cells. We report the synthesis of carboranes containing such functional groups and their biological properties. Further it occurred to us that there must be a relationship among water solubility of ^{10}B carriers, their cytotoxicity and their uptake by cancer cells. Accordingly, we prepared a water soluble BPA derivative and investigated its biological property.

2   Synthetic Chemistry

Cellular and molecular studies indicated that under aerobic conditions RSU-1069 **2** and RSU-1131 **3** alkylate DNA at the phosphate and purine bases via the aziridine group, a process that leads to DNA strand breakage.[2] Accordingly, we thought that a carborane-containing aziridine group may become a potentially useful boron carrier.[3] The synthesis of MACB **1** was accomplished by the reaction of epoxy-1-allyl-carborane with the higher order cyano cuprate type reagent[4] of aziridine (eq 1).[3]

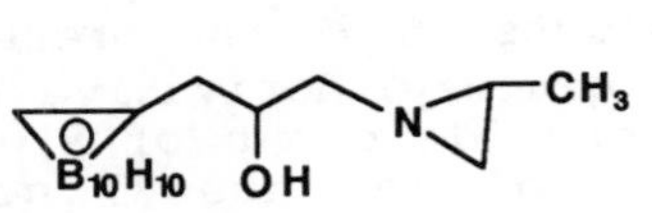

1 MACB

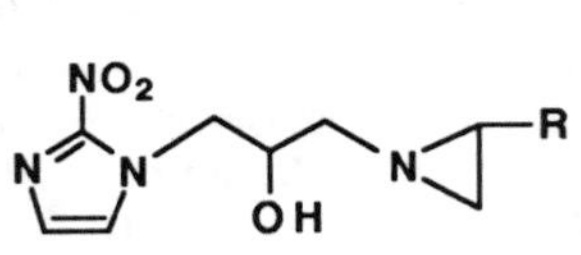

2: R=H
3: R=CH$_3$

The thermal reaction of the epoxy carborane with methyl-aziridine did not provide the desired product. The use of metal amides at lower temperature was essential to obtain 1 in good yield. Among metal amides examined, the copper amide reagent afforded the best result.

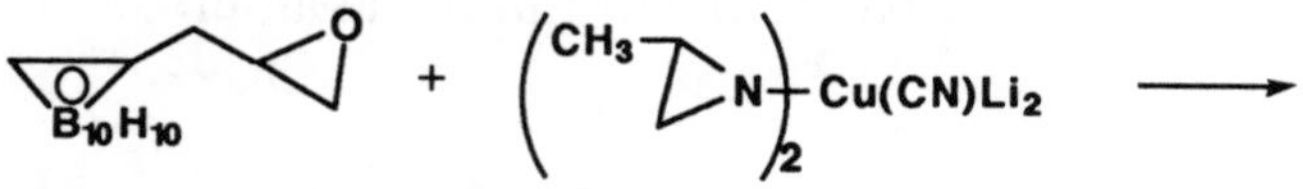

$$\text{(eq 1)}$$

5-Carboranyluridine 4 (5-$B_{10}U$), 5-carboranyldeoxy-uridine 5 (5-$B_{10}UD$) and 5-hydroxymethylcarboranyluridine 6 (5-$HB_{10}U$) were prepared from the corresponding acetylenes and decaborane (eq 2).[5] The choice of Lewis base in the decaboration step was important to obtain high chemical yields; the use of propionitrile gave the best result for the preparation of 4 and 5, whereas the use of Et_2S produced a higher yield in the case of 6. The benzoyl groups of the resulting decaboration products were removed with NaOMe-MeOH, giving 4 in 67% yield, 5 in 55% yield, and 6 in 40% yield.

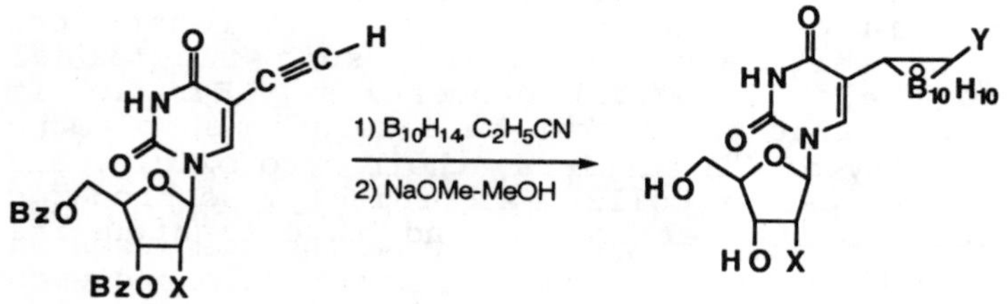

$$\text{(eq 2)}$$

4; X = OH, Y = H (5-$B_{10}U$)
5; X = H, Y = H (5-$B_{10}U$)
6; X = OH, Y = CH_2OH (5-$HB_{10}U$)

The water solubility of the ^{10}B carriers 1,4-6 is not so high. In order to elucidate a relationship among water solubility, cytotoxicity, and accumulation in cancer cells, we next prepared a more water-soluble carboranyl uridine. As a water-solubilizing moiety, phosphoric acid (-P(O)(OH)$_2$), sugars such as D-glucose,[6] or cascade type polyols[6] which have no asymmetric center may be conceivable. The acetonide protected derivative 7 was treated with α-bromo-D-glucose 8 in the presence of mercuric bromide to give the glucosyl derivative 9 in 85% yield. Removal of the acetonide group of 9 with HOAc-MeOH followed by removal of the acetyl group afforded 10 (5'Gly5$B_{10}U$) in good yield (eq 3).

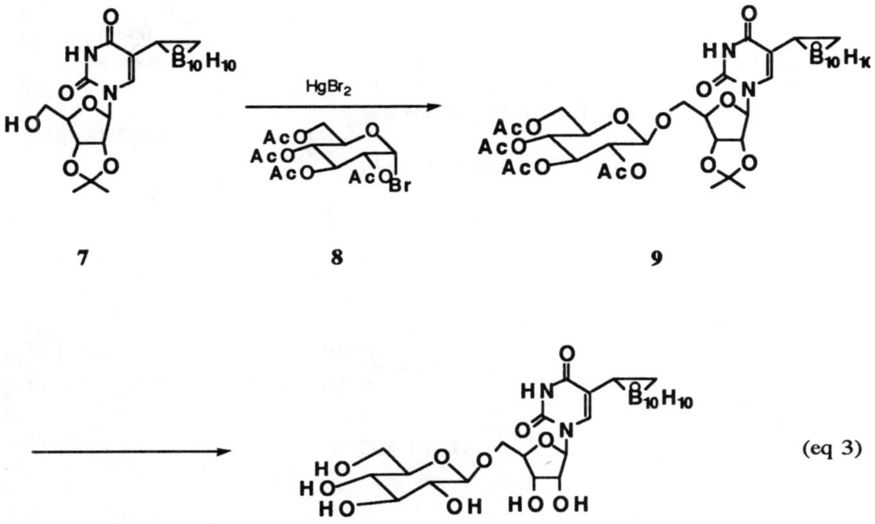

7   8   9

(eq 3)

10

(±)-BPA (p-boronophenylalanine) has been used clinically for neutron capture therapy of human-melanoma cancer. Because of its low solubility in water, BPA has been used as hydrochloric acid or alkali metal salt. More recently, the monosaccharide complexes of BPA have been used to enhance water solubility. However, BPA itself seems to be easily released from the complexes in vivo because of the labile chemical interaction between the monosaccharide and BPA. We have synthesized BPA derivatives **11** and **12** having cascade type polyols as a water solubilizing moiety.[7] It was expected that the bonding between BPA unit and the polyols would not be cleaved easily in vivo because of its tight covalent bond. Protection of the amino group of BPA with Cbz-Cl/aq NaOH followed by treatment with 2 equiv of MeN(CH₂CH₂OH)₂ afforded **13** which was treated with **14** in the presence of N-hydroxybenzotriazole and C₂H₅N=C=NCH₂CH₂CH₂NMe₂·HCl giving the benzyl and Cbz-protected form of **11** in 95% yield. Removal of the protecting groups was carried out by hydrogenation with Pd(OH)₂/C in ethanol/HCl to give the hydrochloric acid salt of **11** in 77% yield. Purification of the salt by cation exchange resin and HPLC gave BPA(OH)₂ **11** (eq 4). Synthesis of BPA(OH)₄ **12** was performed similarly (eq 5).

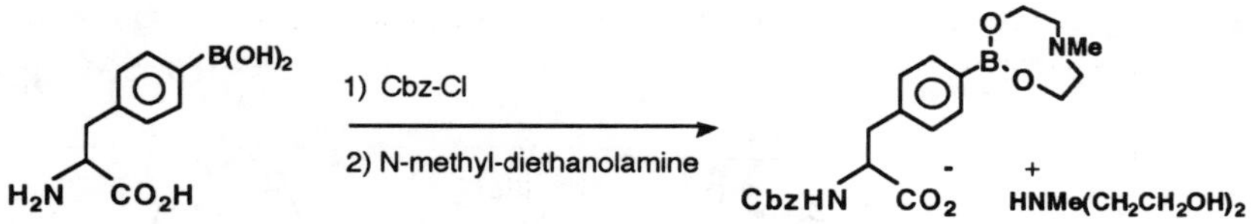

BPA   13

(eq 4)

11 BPA(OH)$_2$

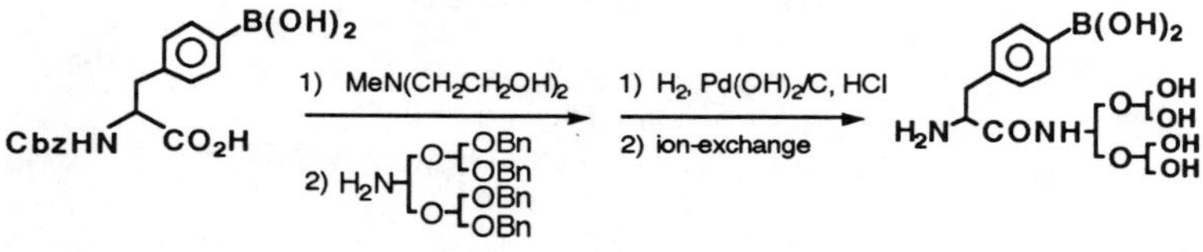

(eq 5)

15   12

The Cbz protected form of BPA was treated with **15**, and then the resulting tetraol derivative was hydrogenated with Pd(OH)$_2$/C. Purification with ion-exchange resin gave **12**.

3   Biological Property

The growth inhibition of B-16 melanoma cell lines and normal cells (TIG-1-20) with the [10]B carriers **1, 4- 6, 10-11**, BPA and BSH is summarized in Table 1. The IC$_{50}$ value of BPA·HCl toward the cancer and normal cells was ca. 8×10^{-3}M, indicating BPA exhibits similar cytotoxicity to both cells. BSH (sulfhydryl boron hydride monomer Na$_2$B$_{12}$H$_{11}$SH), clinically used for treatment of brain tumor, also showed relatively low cytotoxicity (ca. 1.6×10^{-3}M) whereas the cytotoxicity toward TIG-1-20 increased about one order of magnitude over that of B-16. MACB **1** exhibited relatively high growth inhibition toward B-16 melanoma, with IC$_{50}$ value of the order of 10^{-6}M. Very interestingly, this value was significantly lower than that of TIG-1-20, indicating that MACB possesses selective cytotoxicity toward certain cancer cells. The

Table 1. The growth inhibition of B-16 melanoma and normal cells (TIG-1-20)

compound	IC50 (M)	
	B-16 (tumor)	TIG-1-20 (normal)
1 (MACB)	5.44×10^{-6}	1.83×10^{-5}
4 (5B$_{10}$U)	3.78×10^{-5}	2.46×10^{-5}
5 (5B$_{10}$DU)	4.86×10^{-5}	1.67×10^{-5}
6 (5HB$_{10}$U)	1.15×10^{-4}	3.61×10^{-4}
10 (5'Gly5B$_{10}$U)	7.29×10^{-4}	1.39×10^{-3}
11 (BPA(OH)$_2$)	2.04×10^{-2}	
BPA•HCl	8.55×10^{-3}	8.35×10^{-3}
BSH	1.59×10^{-3}	1.48×10^{-4}

IC$_{50}$ values of carboranyluridines **4** and **5** were the order of 10^{-5}M, but the hydroxymethyl analogue **6** was less toxic; IC$_{50}$ values were the order of 10^{-4}M. More water soluble **10** exhibited lower toxicity toward B-16 and TIG-1-20 than **4-6**, and very interestingly selective cytotoxicity was observed: **10** was more toxic toward B-16 than TIG-1-20. BPA(OH)$_2$ **11** was the least toxic among the ^{10}B carriers in Table 1.

Boron accumulation in B-16 and TIG-1-20 cells was measured by using ICP (induced coupled plasma) method. The cells [$(4.5-5.0)\times10^{6}$] were incubated for 1-24h with Eagle-MEM medium containing the ^{10}B carriers. The concentrations of the carriers were adjusted to those of their IC$_{50}$ values. At 1, 2, 3, 6, and 24h, the cells were washed three times with PBS-(-) (Ca-Mg free phosphate-buffered saline) and processed for boron measurement by ICP. The relationship among water solubility, cytotoxicity, and boron incorporation is shown in Table 2. The water solubility increased as the carriers change from **1** to **11**. On the other hand, the cytotoxicity decreased with the change of carriers from **1** to **11**. The ratio of boron incorporation to the amount of boron incubated, namely the efficiency of boron uptake by the cancer cells, also decreased with this order. Accordingly, the relationship among the three biological properties may be shown as follows; (+), positive relationship means that a magnitude of one of the two properties is enhanced as that of the other property increases; (-), negative relationship stands for that a magnitude of one of the two properties is enhanced as that of the other property decreases. In conclusion, it is important to control a subtle balance among the three properties in order to create a better carrier for NCT.

Table 2. Relationship among water solubility, cytotoxicity toward B-16 melanoma, and boron incorporation

compound	water solubility (M)	cytotoxicity $IC_{50}(M)$	boron incorporation[a] ($\mu gB/10^6$cells)
1		5.44×10^{-6}	0.32 (13%)
4	4.91×10^{-4}	4.86×10^{-5}	0.17 (1.3%)
5	1.91×10^{-4}	3.78×10^{-5}	0.13 (1.1%)
6	4.60×10^{-3}	1.15×10^{-4}	0.11 (0.8%)
10	$>8.84 \times 10^{-2}$	7.29×10^{-4}	0.37 (0.9%)
11	6.67×10^{-1}	$>1.77 \times 10^{-2}$	0.42 (0.7%)
BPA	7.66×10^{-3}	8.55×10^{-3}	0.31 (0.7%)

[a]The concentrations of carriers were adjusted to those of their IC_{50} values. The boron incorporation into the cells at 24h is shown in $\mu g/10^6$ cells. The percentage in a parenthesis means the ratio of boron incorporation to the amount of boron incubated.

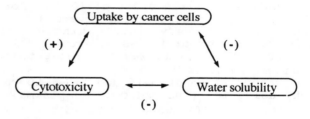

REFERENCES

1. "Progress in Neutron Capture Therapy for Cancer" Ed by B.J. Allen, D.E. Moore, B.V. Harrington, Plenum Press, New York, 1992. Y. Yamamoto, Pure Appl. Chem., 1991, 63, 423.
2. A. R. J. Silver and P. O'Neill, Biochem. Pharmacol., 1986, 35, 1107.
3. Y. Yamamoto and H. Nakamura, J. Med. Chem., 1993, in press.
4. Y. Yamamoto, N. Asao, M. Meguro, N. Tsukada, H. Nemoto, N. Sadayori, J. G. Wilson and H. Nakamura, J. Chem. Soc. Chem. Commun., 1993, in press. Y. Yamamoto, N. Asao and T. Uyehara, J. Am. Chem. Soc., 1992, 114, 5427.
5. (a) Y. Yamamoto, T. Seko, H. Nakamura, H. Nemoto, H. Hojo, N. Mukai and Y. Hashimoto, J. Chem. Soc. Chem. Commun., 1992, 157. (b) Y. Yamamoto, T. Seko, H. Nakamura and H. Nemoto, Heteroatom Chem., 1992, 3, 239.
6. H. Nemoto, J. G. Wilson, H. Nakamura and Y. Yamamoto, J. Org. Chem., 1992, 57, 435.
7. H. Nemoto, S. Iwamoto, H. Nakamura and Y. Yamamoto, Chem. Lett., 1993, 465.

Recent Developments in Tumor-selective Targeting of Boron for Neutron Capture Therapy: Carboranylalanine

S. B. Kahl and P. A. Radel

DEPARTMENT OF PHARMACEUTICAL CHEMISTRY, UNIVERSITY OF CALIFORNIA, SAN FRANCISCO, CA 94143, USA

Boron neutron capture therapy (BNCT) holds great promise as a selective chemoradiotherapeutic treatment for a wide variety of cancers and other diseases. Its appeal lies in its potential to deliver a lethal radiation dose selectively to target cells *via* cell-specific delivery of a radiation sensitizer, ^{10}B, which can be activated independently of the administration of a boron-containing drug. A noncytotoxic drug containing ^{10}B is first introduced to the patient by conventional means which may be either systemic administration of the drug-boron complex or by perilesional administration of a suitable formulation containing a source of the sensitizer. After allowing adequate time for the drug to be selectively taken up by the target cell population and cleared by the plasma and extravascular compartments, the target cell area is exposed to a thermal or epithermal beam of neutrons. Currently, such a beam can be obtained from the cascade of neutrons produced in a fission reactor, but in the near future such beams will likely be available from linear accelerators. Regardless of the neutron source, the absorption of low energy neutrons by ^{10}B results in a prompt fission reaction, producing an alpha particle, a lithium nucleus and a gamma ray with a total energy of about 2.8 MeV. These fission fragments deposit the vast majority of their energy within the originating cell, resulting in reproductive death of the cell through irreparable double stranded DNA breakage. None of the normal elemental constituents of human tissue (C,H,N,O, etc.) absorb low energy neutrons to the same degree as ^{10}B, and the selectivity of BNCT thus relies on the ability of boron chemists to develop site-selective drugs containing the ^{10}B sensitizer.

Indeed, despite the passage of nearly sixty years since Locher first suggested that this nuclear reaction could be used in the treatment of human disease[1] and despite the breathtaking advances in boron chemistry over the past three decades, no truly tumor-selective boron compound has reached the clinical application stage. Hatanaka has used $B_{12}H_{11}SH^{2-}$ (BSH) for the apparently successful BNCT treatment of brain tumors, but this agent is widely regarded as not being truly selective for the tumor nor as being retained by tumor cells over normal cells.[2]

In 1958, Snyder noted the potential use of metabolic mechanisms for localizing boron in tumors and synthesized p-boronophenylalanine (BPA) with the specific intention of using the accentuated aromatic amino acid uptake system found in many melanomas to load boron for BNCT.[3] Phenylalanine and tyrosine are the natural substrates for tyrosinase, an obligatory enzyme in the biosynthesis of melanin. Very little is known regarding the biochemical specificity of this active transport system, but it seems likely that a phenyl ring is a minimal structural requirement. The system is stereospecific, however, and will only transport the (L)-enantiomer of aromatic amino acids. Mishima and coworkers have actively exploited this rationale using BPA for the successful suppression of malignant melanoma in hamsters, pigs and recently several human patients with BNCT[4]. However, their success does not appear to be due to BPA incorporation into melanin. Based on an excellent series of studies by Coderre et al.[5], it appears that BPA, either as the free amino acid or as the fructose complex, is transiently accumulated in a variety of tumors including the 9L gliosarcoma and Harding-Passey melanoma. It thus appears that BPA and perhaps other boronated amino acid mimics of phenylalanine could be used to treat a variety of human tumors using BNCT.

Several lines of reasoning strongly support the concept of developing an amino acid delivery system in which the aromatic phenyl ring of phenylalanine has been replaced by a carborane cage. Such a structure is shown below and is known as carboranylalanine (nicknamed Car in keeping with the standard, three letter amino acid abbreviation system). Car is 46.8% boron by weight compared to 5.2% for BPA. The icosahedral carborane cage is extremely resistant to chemical degradation and is essentially non-toxic in the absence of thermal neutrons. Most simple carborane derivatives are characterized by LD_{50}'s in the 2000-3000 mg/kg range in rodents. Perhaps most significantly, the carborane cage and the phenyl ring are nearly bioisosteric. In a series of computer generated molecular models, we have demonstrated that the van der Waals surface of a phenyl ring freely rotating around its C(1)-C(4) axis is only slightly smaller than the volume occupied by a carborane cage. This computer modeling suggests that Car might exhibit strong biochemical mimicry to Phe.

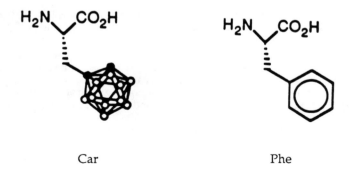

Car   Phe

Figure 1. Structures of carboranylalanine and phenylalanine

Carboranylalanine was first synthesized by Schwyzer's group using a laborious process which produced the L-enantiomer.[6] A higher yielding sequence is desirable since the active isotope, ^{10}B, is only 20% naturally abundant and is relatively expensive in the form of >95% enriched borane precursors. It is also of interest to develop an enantioselective sequence that would permit the synthesis of either enantiomer as desired and which would also afford greater access to homologs. Sjöberg and coworkers have recently reported the successful enantioselective preparation of Car but the details of the synthetic procedure have not yet appeared.[7]

In considering potential synthetic routes, many of those ordinarily available for amino acids are unsuited for Car. For example, bromomethyl carborane is totally unreactive towards nucleophilic displacement of the bromine so reactions such as O'Donnell's methods are impossible.[8] A route utilizing Vederas' strategy of stereoselective ring opening of di-N-protected serine lactones by copper-catalyzed carboranyl Gignards was investigated in our laboratory, but was unsuccessful. Little is known about the behaviour of Cu-carborane reagents and identifying the proper combination of temperature sequence, solvent mixture and metallic salt proved insurmountable. The 1,4 addition of the carboranyl-copper species to the lactone must be carried out under fairly polar conditions which results in extensive disproportionation of the carborane species. We conclude that neither ortho nor meta carborane carbanions are sufficiently nucleophilic to carry out 1,4-additions at the low temperature required for the stereoselective ring opening.

A longer but potentially high-yielding alternative route was then examined, as shown in Figure 2. Diastereoselective introduction of an azido group to imide systems has been reported by Evans and coworkers,[9] and this approach, while longer, has the advantage of ease of removal of possible diastereomeric side products.

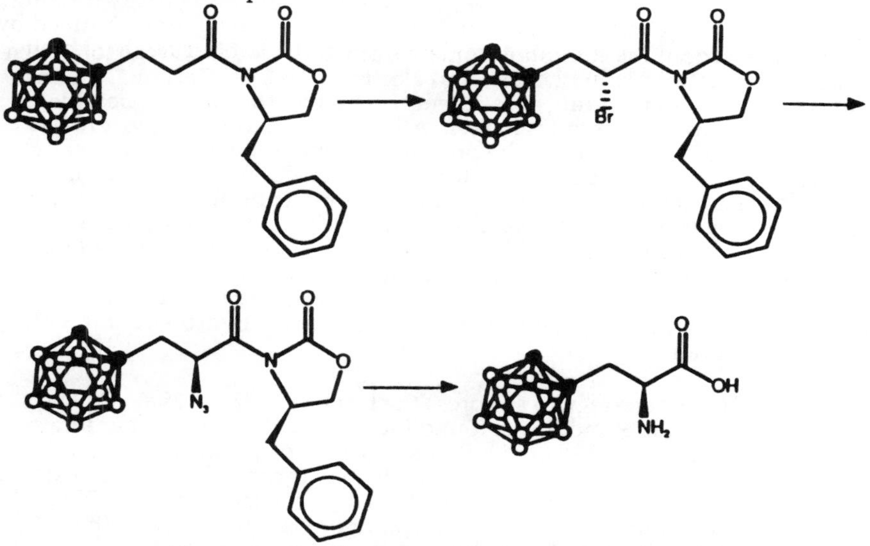

Figure 2. Synthetic strategy for preparation of Car

This strategy required the synthesis 3-carboranyl propionic acid. A limited variety of synthetic methods are available for the preparation of carborane carboxylic acids, none of which is compatible with the requirements of high yield and ease of purification. The most practical route appeared to be hydroboration of the readily available allyl carborane followed by oxidation of the resultant 1° alcohol to the acid as shown in Figure 3. Despite the enormous power and flexibility of hydroborations for alkene functionalization, there are no literature reports of such reactions on carboranyl alkenes. Hydroboration of allyl carborane with borane: THF and workup with buffered alkaline peroxide completely destroyed the carborane cage, leading to the conclusion that the cage is not stable to alkaline hydroperoxide. Use of the very mild and elegant sodium perborate oxidation of Kabalka[10] resulted in 90% yield of alcoholic products, but 35-40% was the undesired 2° alcohol.

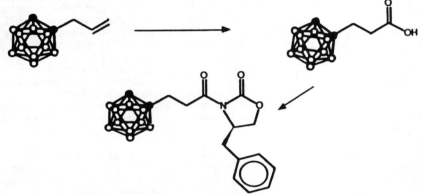

Figure 3. Synthetic strategy for carboranyl oxazolidone

The distribution of alcohol isomers from hydroboration reactions such as these is influenced by both steric and electronic factors. Terminal olefins bearing sterically demanding substituents internal to the olefin generally give high regioselectivity for the 1° alcohol. Allyl benzene, for example, gives a 90:10 mixture of 1°:2° alcohol. The product distribution observed with allyl carborne suggests that electronic factors in the transition state play a major role. In a series of styrenes, Brown and coworkers found that the 2° alcohol was favored by increasing the electron withdrawing ability of a para substituent: styrene (20:80), p-chlorostyrene (27:73), p-CF$_3$ styrene (34:66).[11] The latter case and that of 1-chloro-2-propene (40:66) suggest that the powerful electron withdrawing ability of the carborane cage plays a significant role in the regiochemistry of allyl carborane hydroboration with unhindered boranes.

The obvious strategy to suppress the formation of the 2° alcohol was to use a hindered borane, and it was found that 9-BBN followed by perborate gave >90% yield of the 1° alcohol with no evidence of 2° alcohol. Oxidation with Jones reagent gave the desired acid which was converted to the acid chloride with SOCl$_2$. Treatment of the crude acid chloride with the lithiated amide gave the required carboranyl oxazolidinone in greater than 70% isolated yield overall from allyl carborane.

Direct azidation of the potassium enolate (KHMDS) with triisopropyl benzenesulfonyl azide was unsuccessful due to instability of the carborane cage to disilazide bases. A two step azide introduction *via* diastereoselective bromination followed by azide displacement was then attempted. Optimal conditions for this reaction series were found to consist of enolization by treatment with $TiCl_4$ and diisopropyl ethylamine followed by bromination with NBS. This reaction is highly stereoselective (98:2) as measured by HPLC, ^{13}C and ^1H nmr and the chemical yield (~80%) can probably be increased. Low temperature azide displacement with tetramethylguanidinium azide with inversion at the nascent α-center was completely stereoselective (>99:1 by HPLC) and the desired azide could be isolated in 65% overall yield from the carboranyl oxazolidone.

A continuing problem in characterizing new compounds is unequivocal assignment of ^{13}C nmr peaks. Despite the wide "window" of ^{13}C chemical shifts, two closely spaced peaks of interest are often observed in complex molecules. If the two peaks are fairly close in frequency, correlating assignments to peaks in a precursor molecule's nmr spectrum may not be valid. Two dimensional nmr experiments can often provide the required information but may require long run times. Adaptation of a one dimensional nmr experiment has proven to be quite useful in providing definitive assignments for several of the intermediates in our reaction sequence, and may be more generally useful for chemists working with carboranes. All of the intermediates in the above reaction sequence carry a monosubstituted carborane cage, with one "methine" CH group in the cage, and many of the intermediates also carry one or more functionalized methine carbons whose ^{13}C shifts are very close to the carboranyl CH. Both types of "methines" tend to be found at 60 ± 10 ppm, and absolute assignment of these signals in a simple one dimensional ^{13}C nmr spectrum is impossible. It can be performed, however, using the Attached Proton Test (APT) experiment which is easy, quick and unambiguous. The APT experiment is a multi-phase methodology utilizing one delay that is the inverse of the carbon-proton coupling constant.[12] Since J_{CH} of a carborane CH (~100 H_z) is significantly different from that of most hydrocarbon CH's (~140-160 H_z), discrimination between the two is possible. We have found that when a pulse delay of 7.5 μsec is used, the carborane methine peak intensity is suppressed relative to all other signals, whereas a 10 μsec delay produces a spectrum wherein the carborane methine intensity is enhanced relative to the rest of the signals.

To illustrate this phenomenon, consider the azido imide intermediate in the above pathway. This compound has three methine CH's: the carborane CH, the azide-bearing α center, and the oxazolidinone ring carbon carrying the benzyl group. The ^{13}C spectrum run under normal conditions gives three peaks of nearly equal intensity at 55.05, 59.53 and 59.79 ppm. With a 10 μsec delay, the peak at 59.79 ppm is strongly enhanced whereas the other two are suppressed. With a 7.5 μsec delay, the 59.79 peak is suppressed slightly below the baseline while the other two peaks are actually inverted. The peak at 59.79 ppm is thus assigned to the carborane CH. By comparison of these spectra with the bromoimide and imide precursors, the peak at 55.05 can be assigned to the benzyl-substituted ring carbon and the remaining methine to the α-center methine. Since these APT experiments use a 90°/180° pulse sequence, the FID accumulates much more rapidly than a standard 45° pulse ^{13}C spectrum. Thus, they require less time than a 2-D experiment but yield the same information.

At the azido imide stage, it ws decided to forego the usual methods of oxazolidone template removal in favor of milder conditions due to the instability of the carborane cage to alkaline peroxide. Due to the instability of the carborane cage to alkaline peroxide conditions, it was decided to forego the usual methods of oxazolidone template removal in favor of milder conditions. Transesterification with titanium benzyloxide gave the benzyl azido ester in 90% yield and hydrogenolysis (H_2/Pd or PPh_3) produced the benzyl ester of Car in >90% yield. This transesterification procedure affords any desired ester simply by adjusting the titanium alkoxide. For the first time in our laboratory, it is now possible to produce any ester of either R or S carboranylalanine in $\geq$ 95% enantiomeric excess and in excellent overall chemical yield. If one assumes a 50% yield of allyl carborane from $B_{10}H_{14}$, the overall chemical yield of this multi-step synthesis is approximately 20%. This method is also readily applicable to homologation and to production of either enantiomer of the amino acid ester. Efforts are currently underway in our laboratory to introduce Car into small peptides, as was done earlier by Schwyzer *et. al.*, [13] and it is anticipated that peptides containing (D) or (L)-Car will have both unusual plasma stability and potential selectivity for certain receptor-enriched tumors.

Acknowledgement

The authors gratefully acknowledge the Department of Energy for its support of this work through grant DE-FGO3ER60873-90.

References

1.   G.L. Locher, <u>Am. J. Roentgenol. Radium Ther.</u> 1936, <u>36</u>, 1.
2.   H. Hatanaka, <u>Boron Neutron Capture Therapy for Tumors</u>, Nishimura, 1986.
3.   H.R. Snyder, A.J. Reedy, W. Lennarz, <u>J. Am. Chem. Soc.</u> 1958, <u>80</u>, 835.
4.   Y. Mishima et. al., <u>Lancet</u> 1989, 389.
5.   J.A. Coderre, J.D. Glass, R.G. Fairchild, U. Ray, S.C. Cohen, and I. Fand, <u>Cancer Res.</u> 1987, <u>47</u>, 6377.
6.   J.L. Fauchere, O. Leukhart, A. Eberle and R. Schwyzer, <u>Helv. Chim. Acta</u> 1979, <u>62</u>, 1385.
7.   A. Andersson et. al. in: <u>Progress in Neutron Capture Therapy for Cancer</u>, pp 41-52, B.J. Allen, D.E. Moore, B.V. Harrington (eds.) Plenum Press, New York, 1992.
8.   M.J. O'Donnell and T.M. Eckrich, <u>Tet. Lett.</u> 1978, 4625.
9.   D.A. Evans, F. Urpi, T.C. Somers, J.S. Clark and M.T. Bildeau, <u>J. Am. Chem. Soc.</u> 1990, <u>112</u>, 8215.
10.   G.W. Kabalka, T.M. Sharp and N.M. Goudgaon, <u>J. Am. Chem. Soc.</u> 1989, <u>111</u>, 5930.
11.   H.C. Brown, and R.L. Sharp, <u>J. Am. Chem. Soc.</u> 1966, <u>88</u>, 5851.
12.   S.L. Pratt and J.N. Shoolery, <u>J. Magn. Reson.</u> 1989, <u>46</u>, 535.
13.   R. Schwyzer, K.Q. Do, A.N. Eberle and J.L. Fauchere, <u>Helv. Chim. Acta</u> 1981, <u>64</u>, 2078.

Preparation and Purification of $B_{12}H_{11}{}^{2-}$-S-containing Compounds for Boron Neutron Capture Therapy

D. Gabel, S. Harfst, D. Moller, H. Ketz, T. Peymann, and J. Rösler

DEPARTMENT OF CHEMISTRY, UNIVERSITY OF BREMEN, PO BOX 330 440, D-28334 BREMEN, GERMANY

1 INTRODUCTION

Boron neutron capture therapy (BNCT) requires access to tumor-seeking molecules, which are hydrolytically stable and compatible with physiological systems. In general, these compounds must be water-soluble in order to be transported to the tumor, or they must be solubilized by, e.g., complexation with dextrins.

Recently, a number of compounds with the 1,2-dicarbadodecaborane moiety have been prepared[1-4]. In general, these compounds have been water-insoluble, but could be rendered soluble by degradation to the corresponding nido-carborate.

It is for this reason that we have investigated the use of $B_{12}H_{11}{}^{2-}$-SH as moiety in chemistry directed toward BNCT[5]. Due to its ionic nature, it must be expected that most, if not all, of the compounds containing $B_{12}H_{11}{}^{2-}$-SH will be water-soluble. The boron-attached sulfur of the compound undergoes alkylation and acylation. These reactions can be used for the preparation of suitable boron-containing compounds.

2 ALKYLATION REACTIONS

Alkylation of $B_{12}H_{11}{}^{2-}$-SH usually proceeds to the bis-alkylated sulfonium salt. Only in rare cases can the reaction be stopped at the mono-alkylated product. Usually, the sulfur appears to be rendered more nucleophilic by a carbon-containing substituent and will react with alkyl halides preferentially even in the presence of unreacted $B_{12}H_{11}{}^{2-}$-SH. Thus, with 6-iodo-2,3,4-tri-O-acetyl-methylglucoside (Scheme 1), an S,S-bis-alkylated sugar **1** is obtained. The reaction, carried out in refluxing acetonitrile, did not appear to yield any significant amounts of monosubstituted

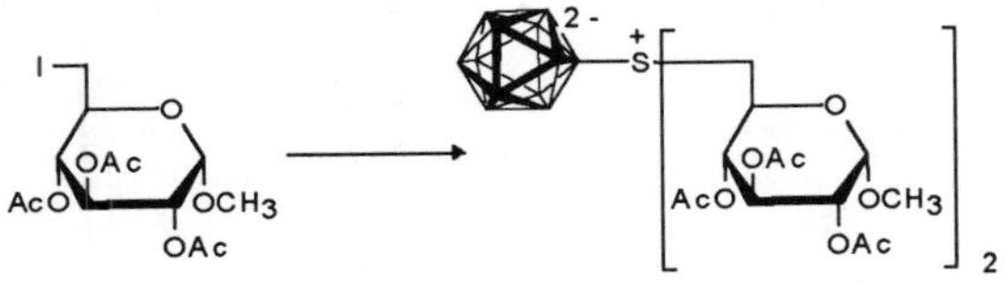

Scheme 1. Synthesis of **1**, the S,S-bis-(tri-O-acetylmethyl-glucoside) of $B_{12}H_{11}{}^{2-}$-SH.

product, as judged by HPLC. For other bromo- and iodoalkyl derivatives, the reaction often takes place at room temperature and over a short time period. The use of the thiolate anion of $B_{12}H_{11}{}^{2-}SH$ (which can be prepared by adding one mole of NaOH and recovering the material by drying) does not prevent the formation of bis-alkylated products.

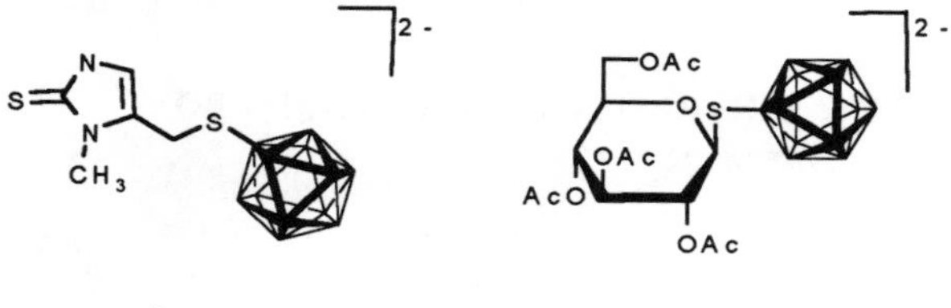

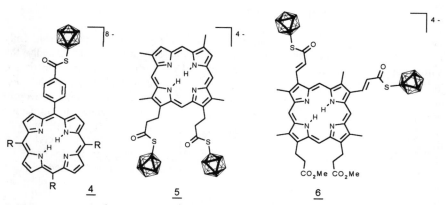

In the case of the 5-bromomethyl derivative of methimazole, only the mono-alkylated product **2** is obtained. This might be due to either steric hindrance or electronic effects of the first substituent. Also in the case of 1-

Scheme 2. Mono-substituted $B_{12}H_{11}{}^{2-}SH$ derivatives **2** and **3**, prepared through alkylation of $B_{12}H_{11}{}^{2-}SH$.

bromo-2,3,4,6-tetra-O-acetyl glucose, a mono-substituted thioglucoside (**3**) is obtained when equimolar amounts of sugar and thiol are reacted.

With the cyanoethyl group, a convenient alkali-labile protection group is available for the general preparation of thioethers and mixed sulfonium salts. The cyanoethyl thioether can be prepared from the bis-cyanoethyl sulfonium salt through treatment with tetramethyl ammonium hydroxide in acetone. Re-alkylation can be performed, and the remaining cyanoethyl group can be removed in an analogous matter. With this reaction sequence, mixed sulfonium salts could be obtained, which were not accessible through other routes.

The sulfur of $B_{12}H_{11}{}^{2-}SH$ can be acylated readily with acyl chlorides to yield surprisingly stable thioesters. This reaction has been utilized to prepare a number of porphyrin derivatives (shown in Scheme 3) where the boron cage is attached to the porphyrin ring system through thioester bonds.

Scheme 3. $B_{12}H_{11}{}^{2-}SH$-containing porphyrins **4**, **5**, and **6**, prepared through reaction of $B_{12}H_{11}{}^{2-}SH$ with the corresponding acyl chlorides.

As expected, the thioester formation with the aromatic chlorocarbonyl group of the meso-substituted porphyrin **4** requires longer reaction times than the aliphatic derivatives **5** and **6**. Compound **6** exhibits an extraordinarily strong ring current, resulting in abnormal shifts for the nitrogen protons to -7.5 ppm.

4 REACTIONS WITH METAL-ORGANIC DERIVATIVES

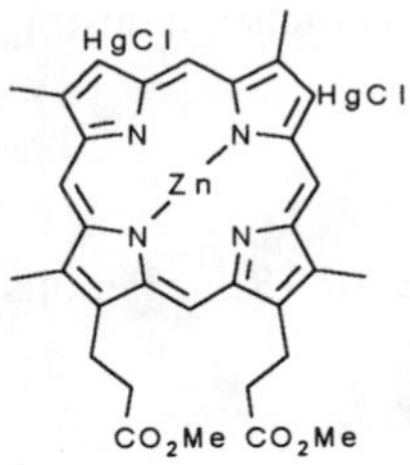

7

Scheme 4. Porphyrin 7

The reaction of the mercury-substituted porphyrin **7** with olefins in the presence of LiPdCl$_3$ proceeds to vinyl-substituted derivatives. This has been utilized before to synthesize a boron-substituted porphyrin[3]. In the presence of B$_{12}$H$_{12}^{2-}$-containing compounds, however, side reactions occur which prevent the formation of any vinyl-substituted derivative. The presence of the borate cage appears to catalyze the decomposition of the intermediate palladium complex, as the de-mercurated product is obtained after the reaction. In ^{11}B-NMR of the reaction mixtures containing B$_{12}$H$_{12}^{2-}$, indications for a substitution reaction on the boron are observed.

5 PURIFICATION AND ANALYSIS

Due to their ionic nature, the compounds obtained are not readily purified to homogeneity. It is problematic to remove the excess of B$_{12}$H$_{11}^{2-}$-SH from the reaction mixture.

It was found that reverse-phase chromatography on RP18 material could be applied successfully. As solvent for preparative chromatography, methanol-water mixtures containing triethylammonium formiate as lyophilizeable buffer was found to be successful. In the case of tetramethylammonium salts of the compounds, low solubility could occasionally present a problem. By exchanging the ammonium ion to a sodium ion, better solubility in the chromatography solvent was achieved.

For analytical purposes, reverse-phase chromatography on Lichrospher RP18 with methanol-water mixtures and 10 mM tetrabutylammonium hydrogen sulfate, adjusted to pH=6.5 with NaOH, proved to be suitable in resolving most of the educts and products. For the porphyrins synthesized, a linear gradient with MeOH/H$_2$O 64:36, 13.5 mM triethylammonium formiate (solvent A) and MeOH (solvent B) between 0 and 100% B , spanning a time period of 25 minutes, was utilized, allowing the separation and elution of widely different porphyrins. Porphyrin **6** (sodium salt) can be separated on silica gel in CH$_3$CN/H$_2$O 97:3 despite its high polarity.

6 CONCLUSIONS

The preparation of B$_{12}$H$_{11}$S^{2-}-containing molecules is possible. These molecules are water-soluble as sodium salts, and hydrolytically stable. Despite the high polarity of the compounds, they can be purified to homogeneity by chromatography. The reactivity of the compounds and their spectroscopic properties are in some cases surprising. Tumor-seeking compounds containing the $B_{12}H_{11}S^{2-}$-moiety could, because of their stability and water-solubility, be good candidates for boron neutron capture therapy.

Acknowledgments

This work has been supported by the Mildred Scheel Stiftung Deutsche Krebshilfe, the Deutsche Forschungsgemeinschaft, and the Fonds der Chemischen Industrie.

References

1. Tjarks, W.; Anisuzzaman, A.K.M.; Liu, L.; Soloway, A.H.; Barth, R.F.; Perkins, D.J.; Adams, D.M. J. Med. Chem. 1992; 35: 1628-1633.

2. Ketz H. ; Tjarks, W.; Gabel D., Tetrahedron Lett, 1990, 31, 4003-4006.

3. Miura, M.; Micca, P.L.; Heinrichs, J.C.; Gabel, D.; Fairchild, R.G.; Slatkin, D.N. Biochem. Pharmacol. 1992; 43: 467-476.

4. Kahl, S.B.; Koo, M.-S. J. Chem. Soc. Chem Commun; 1990: 1769-1771.

5. Gabel, D.; Moller, D.; Harfst, S.; Rösler, J.; Ketz, H. Inorg. Chem. 1993, 32, 2276-2278.

Boron Delivery by Liposomes for BNCT: Development of Lipoidal Boron Compounds

K. Shelly, D. A. Feakes, and M. F. Hawthorne

DEPARTMENT OF CHEMISTRY AND BIOCHEMISTRY, UNIVERSITY OF CALIFORNIA LOS ANGELES, LOS ANGELES, CA 90024, USA

1   INTRODUCTION

The successful boron neutron capture therapy (BNCT) of cancer requires the concentration of significant quantities of the boron-10 isotope within tumor cells. The delivery of water-soluble borane anions, encapsulated within liposomes, has shown great promise in this endeavor.[1] In murine biodistribution experiments, such liposomes have demonstrated the ability to deliver therapeutic amounts of boron to tumors (>30 μg B/g tissue).[2] These tumor levels were attained with relatively small injected doses of boron (<15 mg B/kg body weight) and tumor:blood ratios greater than five. The liposomal delivery of water-soluble encapsulated boron compounds, however, is ultimately limited by factors such as the osmolarity of the borane solution and the small size of tumor-selective liposomes.

Another potential method to accomplish boron delivery with liposomes is to embed lipid-soluble boron species within the phospholipid bilayer membrane of the liposome (Figure 1). Because the mass of the liposome bilayer is much greater than that of the encapsulated species, a significant boron dose may be attained with only a slight alteration of the lipid membrane. For example, the largest quantities of boron which have been encapsulated in liposomes represent only about 5% of the lipid mass. A small perturbation of the lipid bilayer composition could result in significant boron incorporation without adversely affecting the stability or tumor selectivity of the resulting liposome. The combination of hydrophilic and lipophilic boron compounds within the same liposomal formulation could greatly increase the total boron dose provided by liposome administration.

2   SYNTHESIS OF LIPOPHILIC BORON COMPOUNDS

For the production of a useful boron-containing liposome bilayer, the embedded boron compound must be designed to stably pack into the lipid membrane without significant alteration of the membrane properties. This may be accomplished through the synthesis of structural mimics of the phospholipid employed. The basic structural features required are a polar hydrophilic head group to orient the boron species in the membrane and at least one long saturated alkyl chain that can be easily solvated by the hydrophobic interior of the bilayer. The head group that has been employed thus far is based upon the well known polar ion,[3] [$nido$-7,8-$C_2B_9H_{12}$]$^-$. The precursor of this anion ($ortho$-carborane, $closo$-1,2-$C_2B_{10}H_{12}$) presents a variety of routes for the attachment of organic structures. Furthermore, the ion will impart a negative charge to the liposome, and negatively charged liposomes have been found in some cases to enhance tumor selectivity.

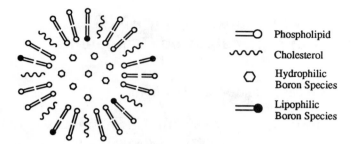

Figure 1. Diagram of the cross-section of a boron-containing liposome.

The synthesis of the first lipophilic boron species examined, [*nido*-7-[CH$_3$(CH$_2$)$_{15}$]-7,8-C$_2$B$_9$H$_{11}$]$^-$ (**2**), is shown in Scheme 1. The formation of hexadecyl-*o*-carborane (**1**) was accomplished by the reaction of 1-octadecyne and decaborane by standard methods.[3] Although the yield of this reaction was somewhat low (30%), improved synthesis methods are under investigation, and the reaction provided a direct route to the immediate precursor of the desired product. Normal base degradation of hexadecyl-*o*-carborane by published methods[3] produced (**2**) in nearly quantitative yield. The transformation of (**2**) to a metallacarborane (**3**) was accomplished with the usual methods.[4]

More effective embedment within the lipid bilayer might be provided by a carborane anion with two hydrophobic alkyl groups. This would enhance the lipophilicity of the boron species, increase the embedment efficiency, and reduce the leakage of the boron compound from the membrane. The synthesis of one such compound, [*nido*-5,6-[CH$_3$(CH$_2$)$_{17}$]$_2$-7,8-C$_2$B$_9$H$_{10}$]$^-$ (**6**), is shown in Scheme 2. Adapting the alkylation procedure of Li et al.,[5] *o*-carborane was iodinated in high yield to produce *closo*-9,12-I$_2$-1,2-C$_2$B$_{10}$H$_{10}$ (**4**). The *bis*-iodocarborane was alkylated with a large excess of the Grignard reagent, CH$_3$(CH$_2$)$_{17}$MgBr, in refluxing THF in the presence of a palladium catalyst. The dialkylated carborane (**5**) was produced in 30% yield. The usual base degradation of this species produced the desired lipophilic boron anion (**6**) in nearly quantitative yield.

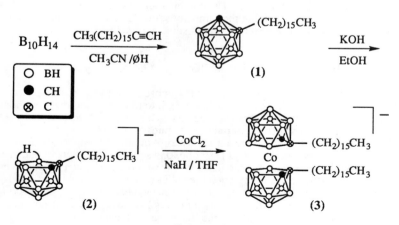

Scheme 1

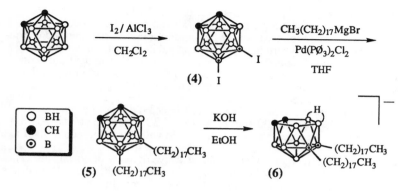

Scheme 2

3 LIPOSOME FORMATION AND MURINE BIODISTRIBUTION

The procedure employed for the preparation of liposomes and the general methods for the biodistribution experiments have been described previously.[1] For the lipophilic boron compounds, the lipid mixture was prepared with the incorporation of 2-4% boron by weight. Liposomes with encapsulated buffer were prepared in the normal fashion, and the embedment efficiency and tissue boron concentrations were determined by inductively coupled plasma-atomic emission spectroscopy.

Liposomes were made from a lipid mixture containing 2.5% boron derived from (2) and encapsulating a normal isotonic saline buffer. The embedment efficiency was 71% and the resulting liposome suspension had a boron concentration of 495 ppm. The biodistribution of these liposomes in mice bearing EMT6 mammary adenocarcinoma tumors is shown in Figure 2A. The mean diameter of the liposomes was 42 nm and the injected dose was 99 µg boron (5.5 mg B/kg body weight). Tumor-boron concentrations were only in the 15-20 ppm range, but these values are recognized as being quite good considering the very small injected dose. Importantly, the tumor values persisted over time, maintaining 16 ppm at 30 hours with a tumor:blood ratio of 5. Also noteworthy is the fact that the carborane-doped liposomes gave only a small uptake of boron in the liver and spleen, the tissues that normally retain boron when provided by many borane anions encapsulated within normal liposomes.

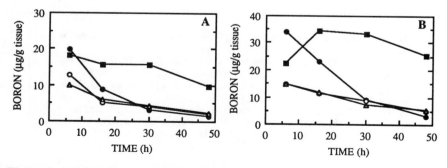

Figure 2. Murine tissue boron concentrations from delivery of lipophilic boron compound (2) by liposomes. Legend: ●=blood; ■=tumor; O=liver; Δ=spleen. (A), liposomes containing (2) and isotonic saline buffer; (B), liposomes containing (2) and hypertonic saline buffer.

In another experiment, liposomes were prepared which contained 2.5% boron from (2) and encapsulating a hypertonic saline buffer (2.2% NaCl). This buffer approximates the osmolarity of the borane salt solutions previously encapsulated in liposomes.[1] The hypertonic solution resulted in liposomes with a much larger diameter, 99 nm. The embedment efficiency was 80%, and the boron concentration of the liposomal suspension was 570 ppm. The biodistribution of these liposomes is shown in Figure 2B for an injected dose of 114 μg boron (6 mg B/kg body weight). The hypertonic liposomes demonstrated greater tumor retention than their isotonic counterparts. Despite the very low injected dose of boron, tumor boron concentrations exceeded 30 ppm at the 16 and 30 hour time points. At 48 hours, the tumor boron concentration is still 25 ppm with a tumor:blood ratio of 8.

The liposome structure is vital for the effective delivery of boron with (2). When injected at a dose of 88 μg boron (5 mg B/kg body weight), free (2) produces tumor boron concentrations of only 5 ppm at 6 hours, decreasing to 2 ppm at 48 hours. Liposomes containing 4% boron from (6) have been produced, but biodistribution results are not yet available.

4 CONCLUSION

Liposomes are an effective boron delivery agent for BNCT. Lipophilic carborane anions have been synthesized which are efficiently incorporated within the liposome bilayer. The resultant membrane retains its ability to stably encapsulate hyperosmotic solutions. Liposomes embedded with the lipophilic carborane anions have exhibited a high degree of boron delivery to tumors *in vivo* at very low injected doses. Studies in progress have indicated that such liposomes, when also containing encapsulated water-soluble borane ions, are capable of the delivery of large amounts of boron to tumors.

ACKNOWLEDGMENTS

The authors wish to thank Teresa A. Krisch of Vestar, Inc. for the murine biodistribution studies and William F. Bauer of the Idaho National Engineering Laboratory for the tissue boron analyses. This research was conducted as part of the BNCT program of the Idaho National Engineering Laboratory and was performed under the auspices of the U.S. Department of Energy (DOE), Office of Energy Research, under DOE Field Office, Idaho, Contract DE-AC07-76ID1570.

REFERENCES

1.   Shelly, K.; Feakes, D.A.; Hawthorne, M.F.; Schmidt, P.G.; Krisch, T.A.; Bauer, W.F. *Proc. Natl. Acad. Sci. USA* **1992**, *89*, 9039.
2.   Feakes, D.A.; Shelly, K.; Hawthorne, M.F., *Proceedings of the Fifth International Symposium for Neutron Capture Therapy for Cancer*, Plenum Publishing, New York..
3.   Hawthorne, M.F., Young, D.C.; Garrett, P.M.; Owen, D.A.; Schwerin, S.G.; Tebbe, F.N.; Wegner, P.A. *J. Am. Chem. Soc.* **1968**, *90*, 862.
4.   Hawthorne, M.F., Young, D.C.; Andrews, T.D.; Howe, D.V.; Pilling, R.L.; Pitts, A.D.; Reintjes, M.; Warren, Jr., L.F.; Wegner, P.A. *J. Am. Chem. Soc.* **1968**, *90*, 879.
5.   Li, J.; Logan, C.F.; Jones, M., Jr. *Inorg. Chem.* **1991**, *30*, 4866.

Synthesis of Monocarbon Carbaboranes with Potential for Application in ^{10}B Neutron Capture Therapy

J. H. Morris, S. Majeed, G. S. Peters, M. D. Spicer, C. Walker, and F. Mair*

DEPARTMENT OF PURE AND APPLIED CHEMISTRY, UNIVERSITY OF STRATHCLYDE, GLASGOW G1 1XL, UK

1 INTRODUCTION

The development of compounds for ^{10}B neutron capture therapy has previously concentrated on dihydroxyboryl groups bonded to a tumour-seeking species, or derivatives of the icosahedral $[closo\text{-}B_{12}H_{12}]^{2-}$ or $closo\text{-}1,2\text{-}C_2B_{10}H_{12}$ with functional substituents.[1] We have sought to extend the range of boron clusters to derivatives of $[closo\text{-}1\text{-}CB_{11}H_{12}]^-$ and its neutral (zwitterionic) analogue $1\text{-}Me_3N\text{-}closo\text{-}1\text{-}CB_{11}H_{11}$, in which either hydrophilic or lipophilic character could be conferred on otherwise closely related species. The availability of substituted derivatives of these monocarbon carbaboranes previously has been limited by the methods available to introduce the substituents.[2,3] In this work we have achieved the controlled substitution of the cluster at boron using low-temperature functionalised boron insertion reactions on *nido*-monocarbon carbaborane precursors,[4-6] and have also examined methods of introducing substituents to the *nido*-precursors prior to the functionalised insertion.[2,7]

2 RESULTS AND DISCUSSION

The *nido*-carbaboranes $[7\text{-}CB_{10}H_{13}]^-$ and $7\text{-}Me_3N\text{-}7\text{-}CB_{10}H_{12}$ were doubly deprotonated (to remove both bridge hydrogens) by either BuLi or NaHBEt$_3$. There was no evidence of the mono-deprotonated intermediates in the ^{11}B n.m.r. spectra of reactions with the reagents in 1:1 molar ratio.

$$7\text{-}L\text{-}7\text{-}CB_{10}H_{12} + 2BuLi = Li_2[7\text{-}L\text{-}7\text{-}CB_{10}H_{10}] + 2BuH$$
$$L = H^- \text{ or } Me_3N$$

$$7\text{-}Me_3N\text{-}7\text{-}CB_{10}H_{12} + 2NaHBEt_3 = Na_2[7\text{-}Me_3N\text{-}7\text{-}CB_{10}H_{10}] + 2H_2 + 2BEt_3$$

Substitution of $7\text{-}Me_3N\text{-}7\text{-}CB_{10}H_{12}$ occurred with $2,4\text{-}(NO_2)_2C_6H_3SCl$ under Friedel-Crafts conditions.[7]

$$7\text{-}Me_3N\text{-}7\text{-}CB_{10}H_{12} + 2,4\text{-}(NO_2)_2C_6H_3SCl \xrightarrow{\text{AlCl}_3}$$
$$4(6)\text{-}\{2,4\text{-}(NO_2)_2C_6H_3S\}\text{-}7\text{-}Me_3N\text{-}7\text{-}CB_{10}H_{11}, \text{ (1).}$$

The anions $[7\text{-}CB_{10}H_{11}]^{2-}$ and $[7\text{-}Me_3N\text{-}7\text{-}CB_{10}H_{10}]^{2-}$ underwent functionalised boron insertion with compounds "RBX$_2$" to yield *closo*-icosahedral products with a functional substituent in position B(2).

*Present Address: Department of Chemistry, University of Cambridge, Cambridge CB2 1EW

In a number of these reactions the anticipated products were obtained:

$[7\text{-L-}7\text{-CB}_{10}H_{10}]^{2-}$ + RBX_2 → $1\text{-L-}2\text{-R-}1\text{-CB}_{11}H_{10}$ + $2X^-$
L = H⁻; R = Ph (2), 4-MeC₆H₄ (3), F (4)
L = Me₃N; R = Ph (5), F (6).

When RBX_2 was BCl_3 or BBr_3 in THF, solvent cleavage occurred and the products isolated were those incorporating the cleaved solvent molecule as the substituent:

$[7\text{-L-}7\text{-CB}_{10}H_{10}]^{2-}$ + BX_3 + THF → $1\text{-L-}2\text{-X(CH}_2)_4\text{O-}1\text{-CB}_{11}H_{10}$
L = H⁻; X = Cl (7)
L = Me₃N; X = Cl (8), Br (9).

When RBX_2 was Me_2NBCl_2 or $^iPr_2NBCl_2$, more complex reactions occurred.[8] In the reactions of $[7\text{-CB}_{10}H_{11}]^{3-}$, the the products isolated respectively were $2\text{-Me}_3\text{N-}1\text{-CB}_{11}H_{11}$, (10), after methylation during work-up, and $2\text{-}^iPr_2\text{NH-}1\text{-CB}_{11}H_{11}$, (11), from adventitious protonation during work-up.

In the reactions of $[7\text{-Me}_3\text{N-}7\text{-CB}_{10}H_{12}]^{2-}$ with Me_2NBCl_2 or $^iPr_2NBCl_2$, remarkable rearrangements occurred which resulted in the isolation of $2\text{-CH}_2\text{Cl-}1\text{-Me}_2\text{NH-}1\text{-CB}_{11}H_{10}$, (12), and $2\text{-}^iPr_2\text{N(H)CH}_2\text{-}1\text{-Me}_2\text{N-}1\text{-CB}_{11}H_{10}$, (13). In each of these, a methyl group was removed from the trimethylamine ligand, a CH_2 group inserted into the B-N bond of the insertion reagent, and the nitrogen was quaternised or underwent nucleophilic substitution by chlorine to give the observed products.

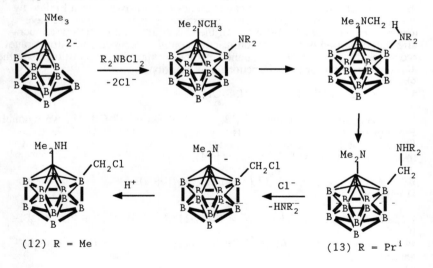

(12) R = Me   (13) R = Pr^i

Further nucleophilic substitution of the terminal bromine in the substituent of (9) was achieved with KSCN to yield 2-(NCS)(CH$_2$)$_4$O-1-Me$_3$N-1-CB$_{11}$H$_{10}$, (14). This reaction demonstrated the potential for further substitution on the substituent side-chain with tumour-seeking species.

The structures of the compounds were established on the basis of their analysis, spectroscopic properties (^{1}H, ^{11}B, and ^{11}B-^{11}B COSY n.m.r.) and X-ray crystallography on selected compounds, some of which are reported elsewhere.[5,6,8]

Studies on the biodistribution of (1) in C57BL6 mice bearing the implanted B16 melanoma established a significant tumour to blood ratio,[7] and therefore further work in this area is now required to synthesise the analogous *closo*-compound from this *nido*-intermediate, and to examine its biodistribution.

3 EXPERIMENTAL

Syntheses

The compounds were synthesised by methods exemplified for compounds (12), (13), (9) and (14).

Preparation of 2-CH$_2$Cl-1-Me$_2$NH-1-CB$_{11}$H$_{10}$, (12).- A soution of 7-Me$_3$N-7-CB$_{10}$H$_{12}$ (0.205 g., 1.07 mmol) in dry THF (10 cm^3) was deprotonated with 1.6M Bu^{n}Li (3 cm^3) to precipitate Li$_2$[7-Me$_3$N-7-CB$_{10}$H$_{10}$]. Freshly distilled Me$_2$NBCl$_2$ (2 cm^3, 15mmol) was added under an inert atmosphere (N$_2$) and the solution stirred for 30 min. to redissolve the precipitate. Volatiles were removed *in vacuo*, the residue extracted with CH$_2$Cl$_2$, and purified by elution with CH$_2$ on a silica gel column. The product, 2-CH$_2$Cl-1-Me$_2$NH-1-CB$_{11}$H$_{10}$, (12), (0.065g, 25.6%) melted at 200°C. Found: C,22.8; H,8.5; N,5.7%. C$_4$H$_{19}$B$_{11}$ClN requires: C,20.4; H,8.1; N,6.0%. The n.m.r spectra comprised: ^{11}B δ(ppm), multiplicity, J$_{B-H}$ (Hz),(rel. intensity),.: -6.78, (s), 1B; -7.30,(d, 143), 1B; -12.36, (d, 147), 2B; -14.63, (d, 109), 4B; -15.38, (d, 138), 3B. ^{1}H δ(ppm), multiplicity, rel. intensity, assignment: 3.35, (s), 2H, CH$_2$Cl; 2.97, (s), 6H, (CH$_3$)$_2$N.

Preparation of 2-^{i}Pr$_2$N(H)CH$_2$-1-Me$_2$N-1-CB$_{11}$H$_{10}$, (13).- A solution of 7-Me$_3$N-7-CB$_{10}$H$_{12}$ (0.23 g., 1.22 mmol) in THF (10 cm^3) was deprotonated with Bu^{n}Li as before, and treated with Pr$^i_2$NBCl$_2$ (2 cm^3, 11.0 mmol) to produce 2-^{i}Pr$_2$N(H)CH$_2$-1-Me$_2$N-1-CB$_{11}$H$_{10}$, (13) (0.07 g, 19.3%) which melted at 191°C. Found: C,41.1; H,10.8; N,9.7%. C$_{10}$H$_{33}$B$_{11}$N$_2$ requires: C,40.0; H,11.1; N,9.3%. The n.m.r. spectra comprised: ^{11}B δ(ppm) (multiplicity, rel. intensity, J$_{B-H}$(Hz)): -7.91 (s, 1B); -9.88 (d, 1B, 139); -15.06 (broad, 9B). ^{1}H: 6.02 (t broad, 1H (NH removed on D$_2$O shake)); 3.92 (septet of doublets, 2H, (CHMe$_2$, collapses to septet on D$_2$O shake)); 2.38 (s, 6H, ((CH$_3$)$_2$N); 1.324 (two doublets, 12H ,((CH$_3$)$_2$CH).

Preparation of 1-Me$_3$N-2-Br(CH$_2$)$_4$O-1-CB$_{11}$H$_{10}$, (9).- A solution of 7-Me$_3$N-7-CB$_{10}$H$_{12}$ (0.2 g.   ) in dry THF was treated under N$_2$ with Bu^{n}Li (ca. 3 cm^3) and Me$_2$SBCl$_3$ (1.2g.   ) added under N$_2$. Volatiles were removed *in vacuo*, methanol (5 cm^3) and a few drops of conc. hydrochloric acid were addedand the solution refluxed for 6 hrs. I was made basic (pH 12) with aqueous NaOH and methanol added dropwise until all solid dissolved. After cooling to 0°C, Me$_2$SO$_4$ (4 cm^3) was added dropwise, and the mixture stirred for 3 hrs; solvent was removed *in vacuo* and the residue extracted into CH$_2$Cl$_2$. After chromatography on silica, the product 1-Me$_3$N-2-Br(CH$_2$)$_4$-1-CB$_{11}$H$_{10}$, (9), was obtained (0.127g., 40%). The n.m.r. spectra comprised: ^{11}B δ(ppm) (multiplicity, rel. intensity); 2.47 (s, 1B); -9.34 (d, 2B); -15.07 (d, 4B), -17.25 (d, 2B); -19.25 (d, 2B); -20.70 (d, 1B). ^{1}H; 3.47 (t, 2H (CH$_2$O)); 3.21 (t, 2H, (CH$_2$Br)); 2.985 (s, 9H, ((CH$_3$)$_3$N)); 1.635 (complex quintet, 2H, (CH$_2$CH$_2$O)); 1.42 (complex quintet, 2H, (CH$_2$CH$_2$Br)).

Preparation of 2-(NCS)(CH$_2$)$_2$O-1-Me$_3$N-1-CB$_{11}$H$_{10}$. (14).- Saturated solutions of (9) (0.127g., 0.39mmol) and KSCN 0.0407g., 0.39 mmol) in acetonitrile were refluxed for 6 hrs. The solution was decanted from the precipitate and evaporated to dryness to yield the product. The n.m.r spectra comprised: ^{11}B δ(ppm) (multiplicity, rel. intensity); 3.18 (s, 1B); -8.09 (d, 1B); -13.96 (d, 4B); -16.95 (d, 4B); -19.38 (d, 1B). ^{1}H; 3.844 (t, 2H); 3.361 (s, 9H); 3.011 (t, 2H); 1.946 (complex quintet, 2H); 1.790 (complex quintet, 2H).

REFERENCES

1.   J. H. Morris, Chem. Br., 1991, 331.
2.   S. A. Khan, J. H. Morris and S. Siddiqui, J. Chem. Soc., Dalton Trans., 1990, 2053.
3.   S. A. Khan, J. H. Morris, M. Harman and M. B. Hursthouse, J. Chem. Soc., Dalton Trans., 1992, 119.
4.   J. H. Morris, S. A. Khan, F. S. Mair and G. S. Peters, Proceedings 4th Int. Symp. on NCT for Cancer, B. J. Allen, D. E. Moore and B. V. Harrington, Eds., Plenum Press, 1992, 289.
5.   F. S. Mair, J. H. Morris, D. F. Gaines and D. Powell, J. Chem. Soc., Dalton Trans., 1993, 135.
6.   F. S. Mair, A. Martin, J. H. Morris, G. S. Peters, F. E. Rayment and M. D. Spicer, Proceedings 5th Int. Symp. on Neutron Capture Therapy, A. H. Soloway, R. F. Barth and D. E. Carpenter, Eds., Plenum Press, 1993, in press.
7.   J. H. Morris, G. S. Peters, E. Koldaeva and R. Spryshkova, manuscript in preparation.
8.   F. S. Mair, A. Martin, J. H. Morris, G. S. Peters and M. D. Spicer, J. Chem. Soc., Chem. Commun., in press.

Carboranyl Amino Acids and Amines with Potential Use in BNCT

S. Sjöberg[1], J. Carlsson[2], P. Lindström[1], and J. Malmquist[1]

[1]DEPARTMENT OF CHEMISTRY, UPPSALA UNIVERSITY, BOX 531,
S-751 21 UPPSALA, SWEDEN
[2]DEPARTMENT OF RADIATION SCIENCES, UPPSALA UNIVERSITY,
BOX 531, S-751 21 UPPSALA, SWEDEN

1. INTRODUCTION

There is much current interest in the synthesis of boron compounds which could be delivered selectively to tumour cells either *per se* or by use of targeting strategies.

Several malignancies such as gliomas, squamous carcinomas and breast cancers have an increased number of epidermal growth factor, EGF, receptors per cell. It has been shown in our laboratories that toxic agents bound to EGF via a dextran spacer can be delivered to cultured glioma cells . We are presently investigating the possibility of loading the target cells with ^{10}B-enriched compounds for BNCT using conjugates with EGF-dextran. For this purpose we have developed methods for the synthesis of amino functionalized carboranes such as amines and carboranyl amino acids with the general formula (1) and (2) respectively.

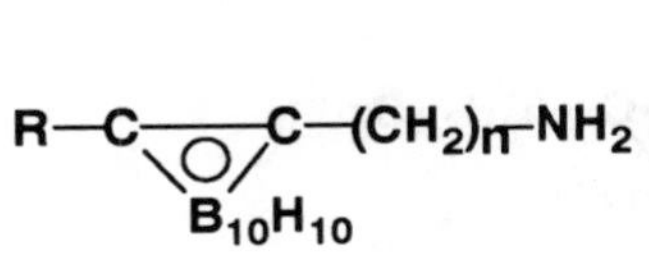

(1) 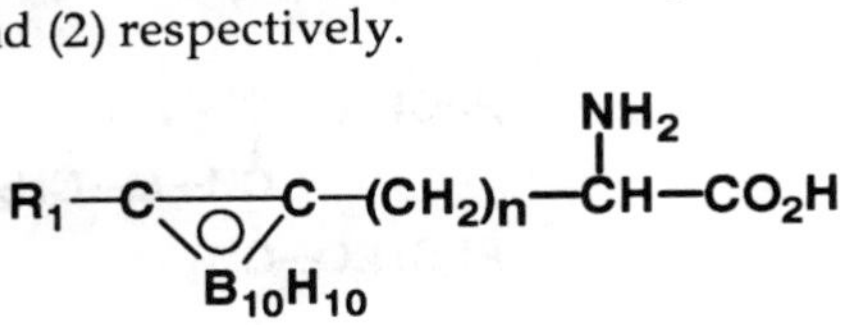   (2)

2. AMINES

Deprotection of *N*-alkyl-substituted phthalimides according to the classical method developed by Garbriel is a well-established method for the synthesis of primary amines. The basic conditions commonly used for deprotection are not suitable for compounds containing the 1,2-dicarba-*closo*-decaboranyl(12) group as the cage will degrade[1].

Several alternative Gabriel reagents allowing milder deprotection conditions are now available and have recently been reviewed[2]. We have used di-*tert*-butyl iminodicarboxylate (HNBoc$_2$) and dibenzyl iminodicarboxylate (HNZ$_2$) for synthesis of carboranyl amines (1). Both iminodicarboxylates can easily be alkylated using the mild phase transfer alkylation conditions described by Brändström[3]. The alkylated iminodicarboxylates are deprotected using hydrogen chloride in diethyl ether and hydrogenolysis for the alkylated HNBoc$_2$ and HNZ$_2$ derivatives respectively.

The synthesis of the hydrogen chloride of 1-(3-aminopropyl)-o-carborane (1a) is shown in Scheme 1 as an example using HNBoc$_2$.[4] Compound (1a) has recently been prepared by an alternative route.[5]

$$ \text{HC}\underset{B_{10}H_{10}}{\overset{}{\diagdown}}\text{C}-(CH_2)_3-Br \quad \xrightarrow{\text{i-ii}} \quad \text{HC}\underset{B_{10}H_{10}}{\overset{}{\diagdown}}\text{C}-(CH_2)_3-NH_3Cl \quad \text{(1a·HCl)} $$

<u>Scheme 1</u> *Reagents*: i, HNBoc$_2$, 1 eq n-Bu$_4$NHSO$_4$, 2 eq. NaOH, H$_2$O, CH$_2$Cl$_2$, reflux; ii HCl(g), Et$_2$O.

In the synthesis of the hydrophilic diol amine (1b), starting with the benzylated diol (3)[6], the Gabriel reagent, HNZ$_2$, is used (Scheme 2).

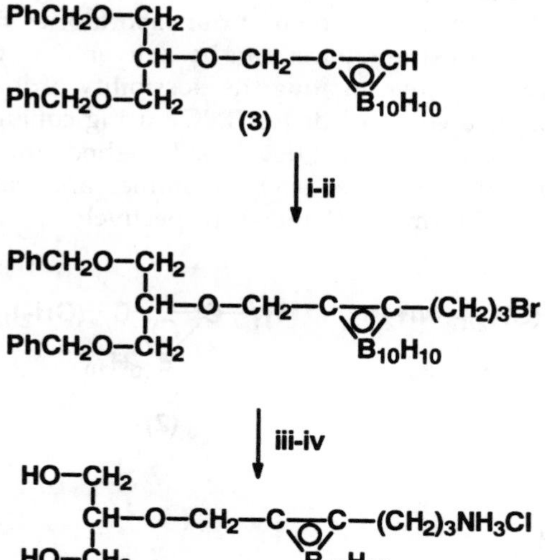

$$
\begin{array}{l}
\text{PhCH}_2\text{O}-\text{CH}_2 \\
\qquad\quad \text{CH}-\text{O}-\text{CH}_2-\text{C}\underset{B_{10}H_{10}}{\diagdown}\text{CH} \\
\text{PhCH}_2\text{O}-\text{CH}_2 \quad \mathbf{(3)}
\end{array}
$$

$$\Big\downarrow \text{i-ii}$$

$$
\begin{array}{l}
\text{PhCH}_2\text{O}-\text{CH}_2 \\
\qquad\quad \text{CH}-\text{O}-\text{CH}_2-\text{C}\underset{B_{10}H_{10}}{\diagdown}\text{C}-(CH_2)_3Br \\
\text{PhCH}_2\text{O}-\text{CH}_2
\end{array}
$$

$$\Big\downarrow \text{iii-iv}$$

$$
\begin{array}{l}
\text{HO}-\text{CH}_2 \\
\qquad\quad \text{CH}-\text{O}-\text{CH}_2-\text{C}\underset{B_{10}H_{10}}{\diagdown}\text{C}-(CH_2)_3NH_3Cl \\
\text{HO}-\text{CH}_2 \\
\mathbf{(1b·HCl)}
\end{array}
$$

<u>Scheme 2</u> *Reagents*: i, BuLi, trimethylene oxide, THF; ii, CBr$_4$, Ph$_4$P, CH$_2$Cl$_2$, 0°C; iii, HNZ$_2$, 1 eq n-Bu$_4$NHSO$_4$, 2 eq. NaOH, H$_2$O, CH$_2$Cl$_2$, reflux; iv, H$_2$, Pd(C), EtOH, HCl(aq).

3. CARBORANYL AMINO ACIDS

Our work on the asymmetric synthesis of carboranyl amino acids have recently been summarised.[7] The amino acids are obtained either via alkylation of the chiral glycine enolate equivalents (4)[8] and (5)[9] or via electrophilic amination of a bornanesultame derivative of a carboranylalkanoic acid according to the procedure by Oppolzer[10].

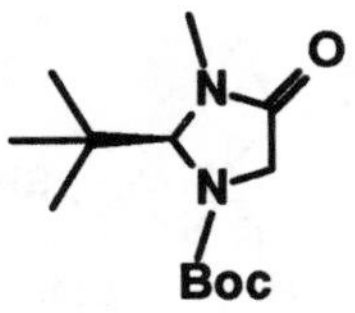

(4) 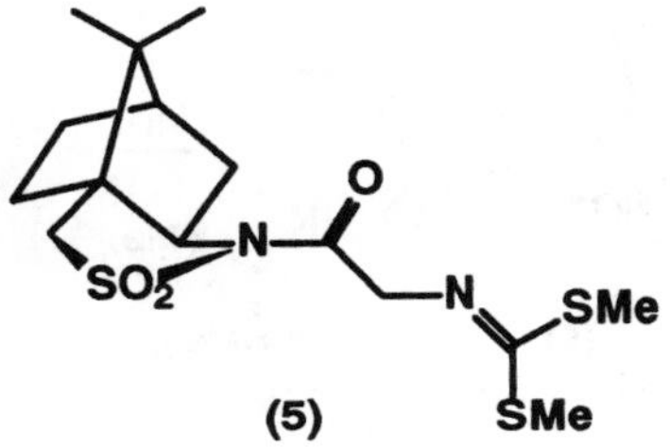   (5)

The alkylation approach is illustrated by three different syntheses of 5-(1,2-dicarba-*closo*-dodecaborane(12)-1-yl)-2-aminopentanoic acid (2a) as shown in Scheme 3 and Scheme 4. All three methods give amino acids with E.P. > 98%.

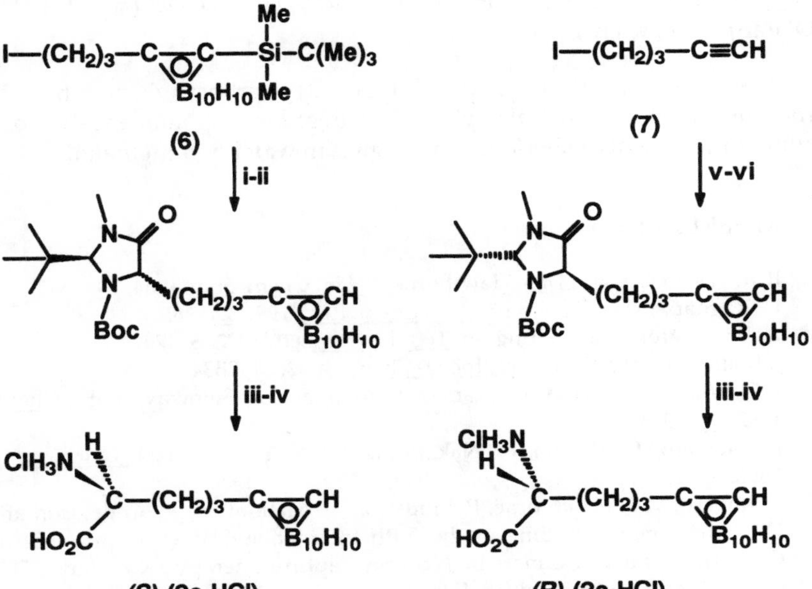

(S)-(2a·HCl)   *(R)*-(2a·HCl)

<u>Scheme 3</u> *Reagents*: i, lithium salt of *(S)*-(4), THF; ii, n-Bu$_4$ F, THF; iii, CF$_3$ CO$_2$ H, CH$_2$ Cl$_2$, r.t.; iv, 3M HCl(aq), toluene, 100°C; v, lithium salt of *(R)*-(4), THF, vi, acetonitrile complex of decaborane, benzene, reflux.

In Scheme 3 *(S)*-2a·HCl is obtained via alkylation of imidazolidinone (4) with the protected iodoalkylcarborane (6) and subsequent deprotection and hydrolysis of the imidazolidinone. *(R)*-2a·HCl is obtained via alkylation of the enantiomer of (4) with 5-iodopentyne (7). In contrast to the imidazolidinone (4) the sultame (5) can be alkylated with iodopropyl carborane (8) without protection of the other cage carbon atom.

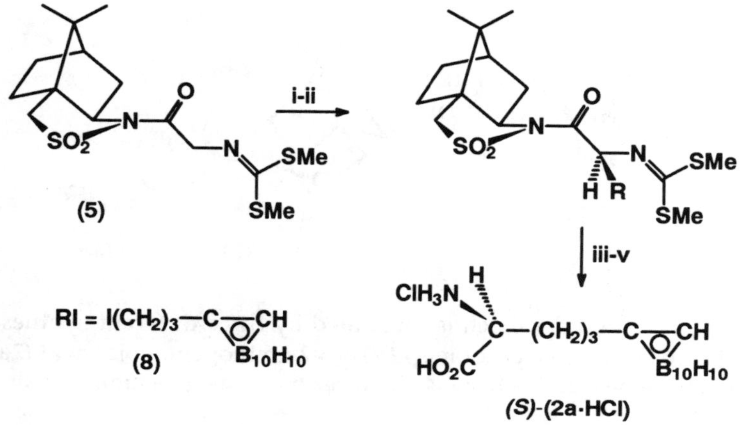

Scheme 4 *Reagents*: i, BuLi, THF; ii, RI, HMPA; iii 1M HCl(aq), THF, iv; LiOH(aq), THF; v, HCL(aq)

The simplest amino acid of type (2), carboranyl alanine, the carboranyl analogue of phenylalanine, degrades spontaneously to a mixture of the diastereomeric *nido*-analogues in water and methanol.[7]

REFERENCES

1.   R.A. Weisboeck and M.F. Hawthorne, *J. Am. Chem. Soc.*, 1964, 29, 1642.
2.   U. Ragnarsson and L. Grehn, *Acc. Chem. Res.*, 1991, 24, 285
3.   A. Brändström and U. Jungren, *Tetrahedron Lett*, 1972, 6, 473.
4.   J. Malmquist and S. Sjöberg, *Inorg. Chem.*, 1992, 31, 2534.
5.   J. G. Wilson, A.K.M. Anisuzzaman, F. Alam and A. Soloway, *Inorg. Chem.*, 1992, 31, 1955.
6.   H. Nemoto,J.G. Wilson, H. Nakamura and Y. Yamamoto, *J. Org. Chem.*, 1992, 57, 435.
7.   S. Sjöberg, M.F. Hawthorne, P. Lindström, J. Malmquist, A. Andersson and O. Pettersson, Proceeding of the Fifth International Meeting on Neutron Capture Therapy of Cancer in 'Neutron Capture Therapy', R.F. Barth, D.E. Carpenter and A.H. Soloway, Eds., Plenum, New York, 1993, in Press.
8.   R. Fitzi and D. Seebach, *Tetrahedron* 1988, 44, 5277.
9.   W. Oppolzer, R. Moretti and S. Thomi, *Tetrahedron Lett.* 1989, 30, 6009.
10.   W. Oppolzer, O. Tamura and J. Deerberg, *Tetrahedron Lett*, 1990, 31, 991.

Carborane-containing Amino Acids as Potential Boron Delivery Agents for Neutron Capture Therapy

Iwona M. Wyzlic and Albert H. Soloway

COLLEGE OF PHARMACY, THE OHIO STATE UNIVERSITY, 500 WEST 12TH AVENUE, COLUMBUS, OH 43210, USA

1 INTRODUCTION

Since BNCT is to be used in the treatment of primary and metastatic brain tumors, it is essential that the boron compounds be capable of crossing the BBB prior to their incorporation into tumor cells. It has been shown that derivatives of L-phenylalanine are transported across the BBB by neutral amino acid transport system.[1] Also recently, a method has been developed to deliver low molecular weight peptides into the brain.[2] These observations suggest that boronated analogues of phenylalanine itself and their peptides modified by replacing aromatic amino acids with highly lipophilic, carborane-containing amino acids might be used to reach and become incorporated into brain cells. We have developed a general, convenient way to synthesize unnatural amino acids containing o-carboranes.[3]

2 SYNTHESIS OF CARBORANE-CONTAINING AMINO ACIDS

O-(o-Carboran-1-ylmethyl)-tyrosine (CBT), p-(o-carboran-1-yl)-phenylalanine (CBPA) and o-carboranyl-alanine (CBA)[4-6] have been synthesized (Figure 1).

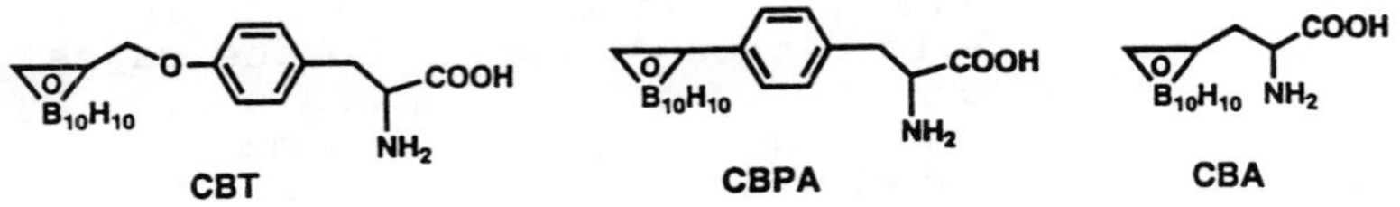

Figure 1 Carborane-containing amino acids

The synthesis of these amino acids involves a phase transfer alkylation of commercially available N-(diphenylmethylene)aminoacetonitrile (1) with appropriate carborane-containing bromide or propargyl bromide (Scheme 1). In the latter case, it is followed

by boronation with a decaborane-acetonitrile complex.
The carborane-containing alkylation products are
hydrolyzed with 6N HCl (Schiff's base) and 70% H_2SO_4
(cyano function). This procedure yields a racemic
mixture of the carborane-containing amino acid with a
good total yield (70-78%) and these are now being
separated into their enantiomers for the synthesis of
peptides. The biological evaluation of these amino acids
is in progress.[7]

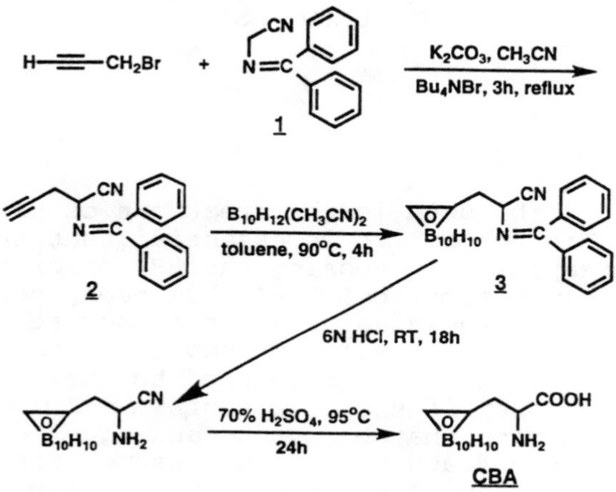

Scheme 1   Synthesis of *o*-carboranylalanine (CBA)

3 SYNTHESIS OF A NEW CARBORANE-CONTAINING KETONE: DI(*o*-CARBORAN-1-YLMETHYL)KETONE

A phase transfer alkylation of N-(diphenylmethylene)aminoacetonitrile **1** with propargyl bromide can
provide either the mono- (**2**, Scheme 1) or dialkylated
product (**4**, Scheme 2). As described above, the
boronation and hydrolysis of **2** produced CBA. The same
procedure applied to compound **4** did not yield **5**. The
boronation of **4** presumbly failed because of steric
hindrance during the formation of two carborane cages.
An alternate approach to compound **5** was to
introduce each carborane cage stepwise (Scheme 2).
Compound **3** was alkylated with propargyl bromide to
obtain product **6**. However, boronation of **6** gave two
products: the expected compound **7** and a second product,
compound **8** arising from dehydrocyanation. This reaction
had been monitored by TLC and showed slow transformation
of **7** into **8** during the boronation. Prolonged reaction
times gave **8** as the predominant product, suggesting that
it was formed by the thermal elimination of hydrogen
cyanide[8,9] from **7**.

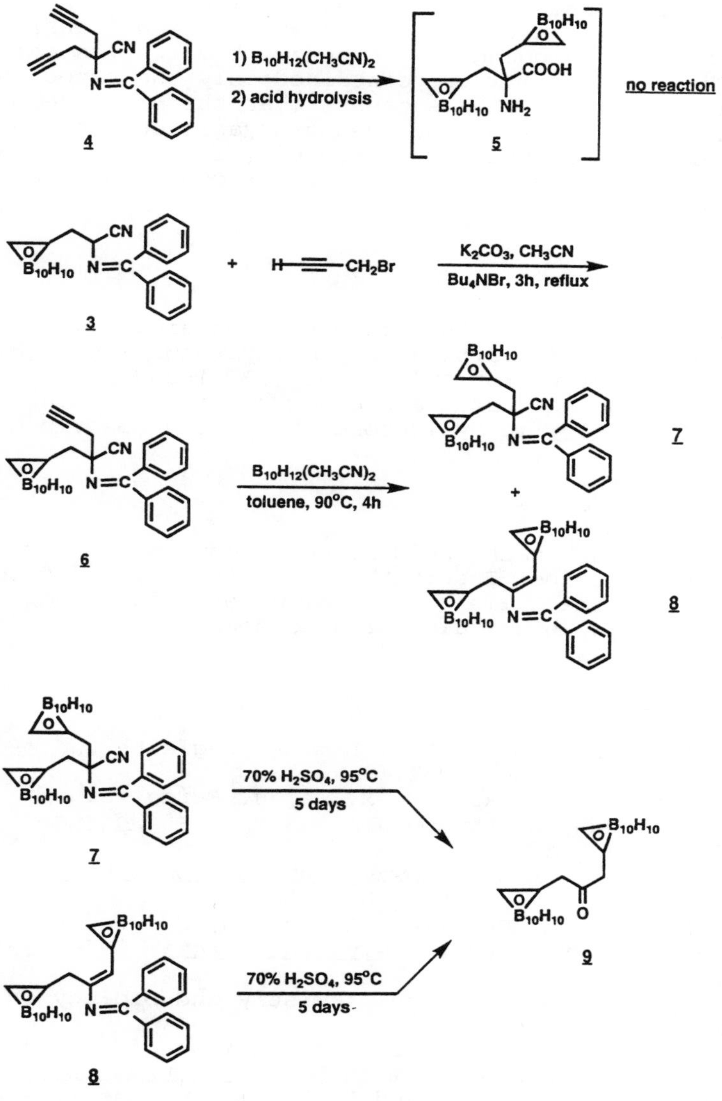

Scheme 2   Synthesis of di(o-carboran-1-ylmethyl)ketone

A new carborane-containing ketone: di(o-carboran-1-ylmethyl)ketone (9) has been synthesized by acid hydrolysis of either 7 or 8. In order to produce 9 (Scheme 2), hydrolytic conditions of 70% H_2SO_4 at 95°C were required. Dehydrocyanation as well as hydrolysis was not observed when the reaction was carried out with 6N HCl at room temperature indicating that hydrogen cyanide elimination was thermally generated.

To determine if there is any influence of the carborane cages on this elimination, a phase transfer alkylation of compound 3 with benzyl bromide was carried out. The reaction mixture contained only the expected alkylated product. This suggests that the presence of two carborane cages promote dehydrocyanation with the transformation of the more hindered 7 into the spatially preferred 8. The crystal structure of compound 8 is being now examined.

4 SUMMARY

The synthesis of the carborane-containig amino acids such as O-(o-carboran-1-ylmethyl)-tyrosine (CBT), p-(o-carboran-1-yl)-phenylalanine (CBPA) and o-carboranylalanine (CBA), and a new boronated ketone, di(o-carboran-1-ylmethyl)ketone, have been described.

ACKNOWLEDGEMENTS

We wish to thank the U.S. Department of Energy and the National Cancer Institute for their support of our research and the Callery Chemical Company for kindly providing us with certain boron compounds.

REFERENCES

1. G.J. Goldenberg, H.-Y.P. Lam, A. Begleiter, *J. Biol. Chem.*, 1979, 254, 1057.
2. N. Bodor, L. Prokai, W.-M. Wu, H. Farag, S. Jonalagadda, M. Kawamura, J. Simpkins, *Science*, 1992, 257, 1698.
3. I.M. Wyzlic, A.H. Soloway, *Tetrahedron Letters*, 1992, 7489.
4. O. Leukart, M. Caviezel, A. Eberle, E. Escher, A. Tunkyi, R. Schweyzer, *Helv. Chim. Acta*, 1976, 59, 2184.
5. S. Sjoberg, Workshop on Chemistry and Biology on BNCT, 1991, Uppsala.
6. P. Radel and S. Kahl, In: proceedings of 5th International Symposium on BNCT, "Neutron Capture Therapy", R.F. Barth and A.H. Soloway (eds.), Plenum Press, New York, London, 1993 (in press).
7. I.M. Wyzlic, A.H. Soloway, R.F. Barth, J. Rotaru, In: proceedings of 5th International Symposium on BNCT, "Neutron Capture Therapy", R.F. Barth and A.H. Soloway (eds.), Plenum Press, New York, London, 1993 (in press).
8. H. Ahlbrecht, W. Raab, C. Vonderheid, *Synthesis*, 1979, 127.
9. S.F. Dyke, E.P. Tiley, A.W.C. White. D.P. Gale, *Tetrahedron*, 1975, 31, 1219.

Synthesis of 5-Carboranyluracil and its N^1-Glycosylation with Natural and Modified Carbohydrates

Yahya El-Kattan[1,2], Naganna M. Goudgaon[1,3], Geraldine Fulcrand[1,3], Dennis C. Liotta[2], and Raymond F. Schinazi[1,3,*]

[1]DEPARTMENT OF PEDIATRICS AND [2]CHEMISTRY, EMORY UNIVERSITY, ATLANTA, GEORGIA, 30322, USA
[3]VETERANS AFFAIRS MEDICAL CENTER, DECATUR, GEORGIA 30033, USA

Introduction

Recently, much attention has been paid to boron neutron capture therapy (BNCT) as a viable binary modality for the treatment of certain malignancies [1, 2]. This combined modality is based on the interaction of boron 10 and a thermal neutron, each relatively innocuous, producing intense ionizing radiation that is confined to single or adjacent cells.

$$^{10}B + {}^1n \ (0.025 \ eV) \longrightarrow {}^7Li + {}^4He + 2.8 \ MeV$$

Previously, Schinazi et al. [3] and Yamamoto et al. [4] reported the synthesis of ^{10}B-nucleosides containing a single boron atom. With the development of practical methods for production of improved thermal neutrons [5], the synthesis of new ^{10}B carriers has become a high priority. The o-carborane moiety, one of the most stable boron clusters, provides an enhancement over the boronic acid function by a factor of 10. Recently, hydroxyalkylated carboranes, glycosyl carboranes, carboranylporphyrins [6], carboranylpeptides [7] and carboranyl nucleosides derivatives have been prepared for their potential use in BNCT [2]. The aim of our work was to synthesize additional boron containing nucleosides that are hydrolytically stable and are substrates for nucleoside kinases found in tumor cells. The targeted molecules were chosen since they may possess the following properties: (1) lipophilicity so they can permeate the blood brain barrier since our primary target is gliomas; (2) boron containing nucleosides will be transported more easily into the nuclei of tumors cells; this requires that the materials be substrates for both cellular kinases and, after triphosphate formation, DNA polymerases; and (3) these compounds will be resistant to catabolism.

Results and Discussion

We synthesized a series of 5-carboranyl substituted nucleosides including compounds modified in the sugar moiety (3'-thia and 3'-oxo)[8] starting from the corresponding 5-iodo nucleosides. One of the compounds developed, CDU, was shown by us to be phosphorylated intracellularly in human lymphocytes thus preventing its rapid egress from cells [9]. This is the first time that a 5-carboranyl nucleoside has been conclusively shown to be a substrate for cellular kinases. The interesting biological results of the carborane containing nucleosides and the need of these compounds for preclinical studies in multigram quantity prompted us to develop alternative chemical syntheses for CDU and related compounds. Our approach was to synthesize the 5-carboranylpyrimidines and then to use novel N-glycosylation technology developed in our laboratories with a variety of natural as well as modified sugars.

We first synthesized the 5-ethynyluracil starting from the 5-iodouracil in 2 steps. We then coupled the protected pyrimidines (acetyl, benzoyl, benzyl, trimethysilyl) with the decaborane using a Lewis acid (propionitrile). Unfortunately, these approaches were ineffective. The successful approach involved conversion of 5-iodouracil to 2,4-

dichloro-5-iodopyrimidine, which on treatment with methanolic NaOMe yielded 2,4-dimethoxy-5-iodopyrimidine (Scheme 1). The latter was coupled with trimethylsilyl acetylene in presence of $(Ph_3P)_2PdCl_2/CuI$, Et_3N, followed by deprotection of the acetylenic group with n-Bu$_4$NF to give 2,4-dimethoxy-5-ethynylpyrimidine (86% yield). Addition of decaborane (as the bispropionitrile adduct) to the alkyne, followed by demethylation using iodotrimethylsilane, gave the desired analytically pure 5-carboranyluracil as a white crystalline solid in an overall 30% yield from 5-iodouracil, m.p. > 280°C (dec); ^1H NMR (DMSO-d_6) δ 11.28 (br s, 2H, NH, D_2O exchangeable), 7.64 (s, 1H, 6-H), 5.90 (s, 1H, carborane proton), 1.2-3.0 (br, 10H, carborane protons). The availability of 5-o-carboranyluracil provided a versatile intermediate for the synthesis of a number of nucleoside analogs.

Scheme 1

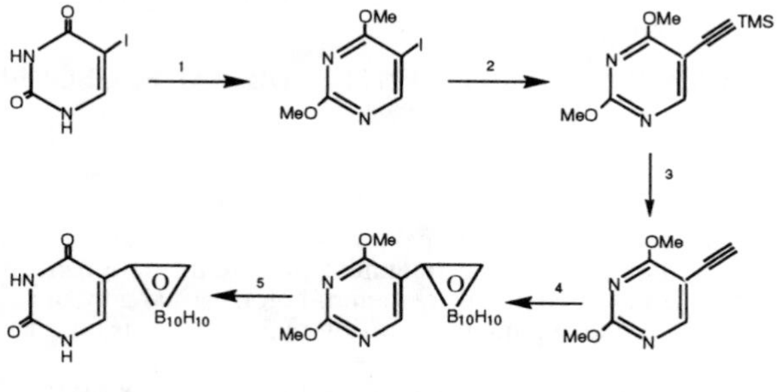

1. POCl$_3$, NaOMe/MeOH; 2. HC≡C-TMS, (Ph$_3$P)$_2$PdCl$_2$/CuI, Et$_3$N; 3. n-Bu$_4$NF/THF; 4. B$_{10}$H$_{12}$(CH$_3$CN)$_2$, toluene, heat; 5. TMSI/CH$_2$Cl$_2$.

The next step consisted of coupling the silylated 5-carboranyluracil with 1,2,3,5-tetra-O-benzoyl-ribofuranose using tin(IV) chloride as a Lewis acid (Scheme 2). The mechanism of the reaction is thought to involve the formation of an oxonium ion at the α face of the sugar moiety [10] which allows the attack of the silylated base on the β face to give the protected nucleoside in excellent yield (95%). After deprotection with sodium methoxide, 1-(β-D-ribofuranosyl)-5-carboranyluracil was obtained.

Scheme 2

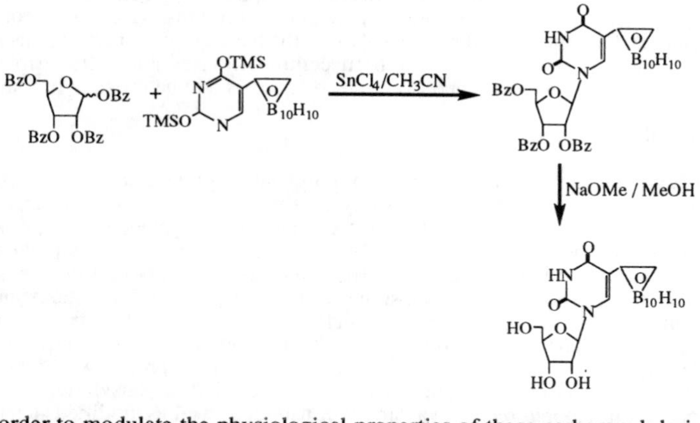

In order to modulate the physiological properties of these carboranyl derivatives, we synthesized a 2',3'-dideoxy-2',3'-unsaturated derivative since such analogs like 2',3'-didehydro-3'-deoxythymidine, D4T [11], have shown selective antiviral activity. For this we applied a strategy developed in our laboratories to control the stereoselectivity of the *N*-glycosylation reaction [12]. The 5-(S)-acetoxy-2-(S)-(*t*-butyldiphenylsilyloxy-methyl)-4-(R)-(2,4,6-triisopropyl)phenylthiotetrahydrofuran was prepared starting from the D-glutamic acid in 4 steps. Coupling of the furan with the silylated base produced the desired protected compound with a good selectivity in a 76% yield (a 20:1 β:α ratio was obtained as determined by ^{1}H NMR) (Scheme 3). Oxidative elimination of the sulfoxide group should lead to the desired compound.

Scheme 3

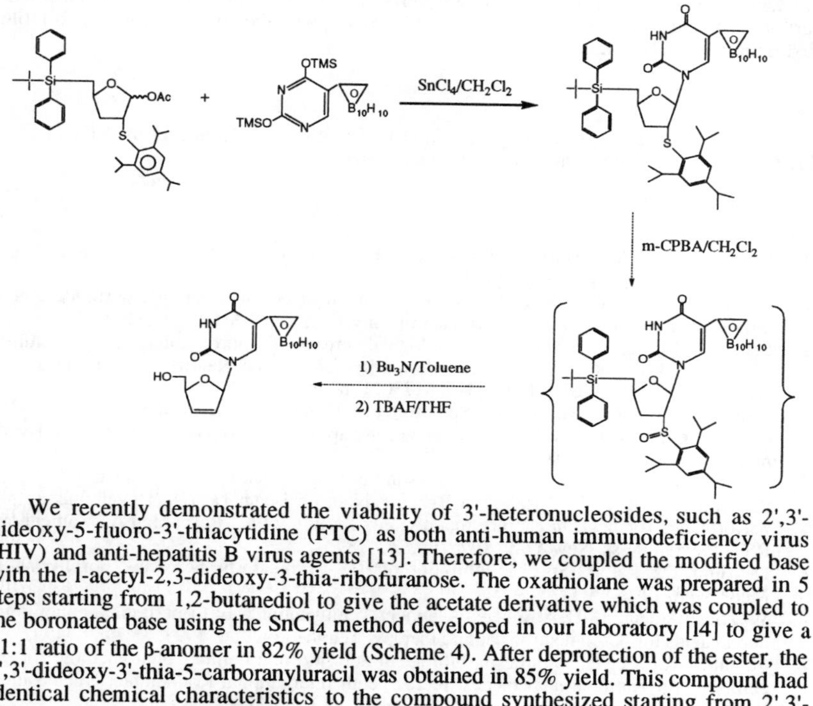

We recently demonstrated the viability of 3'-heteronucleosides, such as 2',3'-dideoxy-5-fluoro-3'-thiacytidine (FTC) as both anti-human immunodeficiency virus (HIV) and anti-hepatitis B virus agents [13]. Therefore, we coupled the modified base with the 1-acetyl-2,3-dideoxy-3-thia-ribofuranose. The oxathiolane was prepared in 5 steps starting from 1,2-butanediol to give the acetate derivative which was coupled to the boronated base using the SnCl$_4$ method developed in our laboratory [14] to give a 31:1 ratio of the β-anomer in 82% yield (Scheme 4). After deprotection of the ester, the 2',3'-dideoxy-3'-thia-5-carboranyluracil was obtained in 85% yield. This compound had identical chemical characteristics to the compound synthesized starting from 2',3'-dideoxy-5-iodo-3'-thiauridine [8,9].

Scheme 4

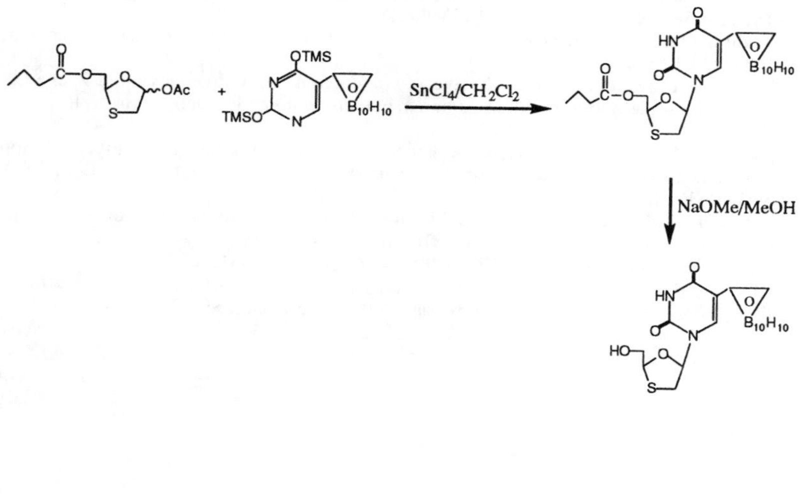

Conclusion

We demonstrated the versatility of this new method to synthesize new nucleosides for BNCT from 5-carboranyluracil. Different conditions for the coupling reaction have been applied to demonstrate that this new 5-o-carboranyl pyrimidine methodology constitutes a facile and efficient way to synthesize different boronated nucleosides analogs bearing modifications in the sugar moiety. We plan to study the potential antiviral and anticancer activity of these compounds and to determine their uptake and egress in clinically relevant cells. The hope is that some of these novel carboranylnucleosides may offer pharmacological advantages over currently available first generation BNCT agents.

Acknowledgments

This work was supported by the National Institutes for Health grant NCI-1RO1-CA-53892 and the Department of Veterans Affairs.

Bibliography

1. Barth, R.F. and A.H. Soloway, Boron neutron capture therapy for cancer - Realities and prospects. Cancer, 1992, 70:2995-3008.
2. Goudgaon, N.M. and R.F. Schinazi, Development of boron containing pyrimidine and nucleoside analogues for neutron capture therapy. Cur. Top. Med. Chem., 1993 (In press).
3. Schinazi, R.F., et al., Synthesis and biological activity of a boron-containing pyrimidine nucleoside. Proc. of The First International Symposium on Neutron Capture Therapy. October 12-13, 1983. BNL Report No. 51730, 1984, pp. 260-265.
4. Yamamoto, Y., T. Seko, and H. Nemoto, New method for the synthesis of boron -10 containing nucleoside derivatives for neutron-capture therapy *via* palladium-catalyzed reaction. J Org Chem, 1989, 54:4734-4736
5. Fairchild, R.G., et al., Installation and testing of a optimized epithermal neutron beam at the Brookhaven Medical Research Reactor. Neutron Beam Design, Development and performance for neutron capture therapy, ed. Harling, O.K., Bernard, J.A., Zamenhof, R.G., Plenum Press, New York, 1990, pp. 185-189.
6. DeCamp, D.L., et al., Specific inhibition of HIV-1 protease by boronated porphyrins. J Med Chem, 1992, 35:3426-3428.
7. Wyzlic, I.M. and A.H. Soloway, A general, convenient way to carborane-containing amino acids for boron neutron capture therapy. Tetrahedron Lett, 1993, 33:7489-7490.
8. Schinazi, R.F., et al., Synthesis, biological activity, and cellular pharmacology of 5-carboranyl pyrimidine nucleosides. p. 28, Tenth International Roundtable: Nucleosides and Nucleotides, September 16-20, 1992, Park City, Utah, 1992.
9. Schinazi, R.F., et al., Synthesis of 5-carboranyl-pyrimidine nucleosides: Evidence for intracellular phosphorylation. p. 11, Fifth International Symposium on Neutron Capture Therapy, September 13-17, 1992, Columbus, Ohio, 1992.
10. Vorbruggen, H. and G. Hofle, On the mechanism of nucleoside synthesis. Chem Ber, 1981, 114:1256-1268.
11. Lin, T.-S, et al., Potent and selective *in vitro* activity of 3'-deoxythymidin-2'-ene (3'-deoxy-2',3'-didehydrothymidine) against human immunodeficiency virus. Biochem Pharmacol, 1987, 36, 2713-2718.
12. Liotta, D.C. and L.J. Wilson, A general method for controlling glycosylation stereochemistry in the synthesis of 2'-deoxyribose nucleosides. Tetrahedron Lett, 1990, 31:1815-1818.
13. Schinazi, R. F., et al., Selective inhibition of human immunodeficiency viruses by racemates and enantiomers of *cis*-5-fluoro-1-[2-(hydroxymethyl)-1,3-oxathiolan-5-yl]cytosine. Antimicrob Agents Chemother. 1992, 36:2423-2431.
14. Choi, W.-B., et al., *In situ* complexation directs the stereochemistry of *N*-glycosylation in the synthesis of oxathiolanyl and dioxolanyl nucleoside analogues. J Am Chem Soc, 1991, 113:9377-9379.

Boronic Acids as Enzyme Models: Catalysis of Oxime Formation

Ling-Hao Niu, Bernarda Frias, Subodh Sharma, and Manfred Philipp

DEPARTMENT OF CHEMISTRY, LEHMAN COLLEGE AND GRADUATE CENTER, CITY UNIVERSITY OF NEW YORK, BRONX, NY 10468, USA

1 INTRODUCTION

Catalysis by the boron family of acids has been an interesting subject in enzyme model studies[2-13]. Unlike model systems based on polymers, these acids are very simple and small molecules that, in some cases, exhibit reversible enzyme-like binding to their substrates prior to the catalytic steps.

Boric acid has been found to be a good catalyst in several investigations. Peer and coworkers observed that boric acid catalyzes the formation of o-(hydroxymethyl)-phenol from the reaction between formaldehyde and phenol[2].

Capon and Ghosh reported the greatly enhanced hydrolysis rate of phenyl salicylate in borate buffers[6]. Okuyama et al. demonstrated that borate accelerates the hydrolysis of S-butyl 2-hydroxy-2-phenylthioacetates by a factor of about 80 at pH 9.0[7]. The catalytic effect of boric acid has also been shown on the formation and hydrolysis of hydroxyl group containing imines[8-11].

Boronic acids possess much higher binding affinity to various alcohols than boric acid[14]. This property makes boronic acids more attractive as potential catalysts.

In experiments done by Letsinger et al., It was observed that 8-quinolineboronic acid effectively accelerates the hydrolysis of chloroethanol to ethyleneglycol[3]. Letsinger and Macheon also used boronoarylbenzimidazole to facilitate the formation of ethers from chloroethanol[5].

Rao and Philipp have recently studied the boronic acid catalyzed hydrolysis of salicylaldehyde imines[12]. These reactions display Michaelis–Menten kinetics. Substrate binding is controlled by the boronic acid ionization. Binding constants are improved by electron-withdrawing substituents on the benzeneboronic acids, whereas the catalytic constants have no pH or electronic dependencies. The mechanism of this catalysis is likely to be an intramolecular transfer of a boron–coordinated hydroxide ion within a borate–substrate complex.

The patent literature has described the boric acid catalyzed hydration of mandelonitrile to form mandelamide[15]. Rao and Philipp investigated the mechanism of this hydration by using substituted benzeneboronic acids[13]. Hydration rates increase continuously with increasing concentration of

benzeneboronic acid. Benzonitrile is not affected by the
presence of boronic acids. This suggests that the hydration of
mandelonitrile depends on the complexation of boronic acids to
its hydroxyl group.

2 RESULTS AND DISCUSSION

Boronic Acid Catalysts

In the presence of 3,5-bis(trifluoromethyl)benzeneboronic
acid, the most effective boronic acid used in this study,
salicylaldehyde oxime formation reaction rates are linearly
dependent on the concentration of boronic acid, hydroxylamine
and salicylaldehyde. This indicates a third-order kinetic
mechanism for the boronic acid catalyzed salicylaldoxime
formation reaction.

The pH profile of this third-order rate constant is bell
shaped with pK_1 of 6.1 and pK_2 of 6.7. The optimum catalytic
activity was observed at pH 6.4 with the limiting third-order
rate constant of 1.15×10^4 M^{-2} sec^{-1}.

In order to determine the origin of the pKs observed in
this pH dependence, other benzeneboronic acids which have
different ionization pKs were selected for the pH study.
Salicylaldoxime formation catalyzed by these other boronic acids
uniformly displayed similar pH dependencies with pK_1 around 6
and pK_2 around 7. The results show that unionized boronic acids
are catalytically active. The pK_1 observed in these pH profiles
reflects the ionization of hydroxylamine. The pK_2 values will be
discussed elsewhere.

The pH profile of 3,5-bis(trifluoromethyl)benzeneboronic
acid profiles was repeated in D_2O buffers. This is to determine
if proton transfer is involved in the rate-determining step of
the reaction. Two-fold lower reaction rate constants were
observed in deuterium oxide. Proton transfer is likely to be
involved in the rate determining step of catalyzed
salicylaldoxime formation.

In a Hammett plot that uses the pH-independent limiting
rate constants, all the boronic acids fall on a straight line
with a slope of +1.5. This shows that benzeneboronic acids with
electron-withdrawing substituents are better catalysts.

Diphenylborinic Acid

When diphenylborinic acid was used as a catalyst, the
reaction rate displayed hyperbolic dependence on both
hydroxylamine and the borinic acid concentrations. These
hyperbolic curves are similar to the one observed in the case of
an enzyme-catalyzed reaction. At low reactant concentration
level (linear region of the hyperbolic curves) this reaction
also follows a third-order kinetic mechanism. K_m and k_{cat} were
calculated by using Eadie-Hofstee plots.

The pH dependence of diphenylborinic acid catalyzed
salicylaldoxime formation is quite different from those observed
in the cases of boronic acid catalysis. This pH profile shows

double-proton ionization with an average pK of 7.5. The limiting third-order rate constant is 2.5×10^4 M^{-2} sec^{-1}. Since the pK of diphenylborinic acid is 6.2[16], the pH dependence suggests that diphenylborinic acid is catalytically active in both ionized and un-ionized forms. The catalytic mechanism involved in this case is clearly different from those used by boric and boronic acids.

When the pH profile was repeated in D_2O. the solvent deuterium isotope effect (1.9) again indicates rate limiting proton transfer.

Boric Acid

A bell-shaped pH dependence was also observed for boric acid catalyzed salicylaldoxime formation. Boric acid catalysis was optimal at pH 7.35 with the limiting third-order rate constant of 76 M^{-2} sec^{-1}. The pK_1 of 6.1 is close to the pK of hydroxylamine. The pK_2 of 8.6 perhaps is due to the ionization of boric acid[17]. The pH profile indicates that neutral boric acid prefers the ionized phenol group of salicylaldehyde for binding in its catalysis. The solvent deuterium isotope effect (1.5) gives a result that indicates, but does not prove the presence of proton transfer in a slow step of the reaction.

pH Dependence of Complexation between Diphenylborinic Acid and Salicylaldoxime

The association of diphenylborinic acid to salicylaldoxime causes the salicylaldoxime band at 304 nm to diminish with simultaneous appearance of a new band around 340 nm. Since diphenylborinic acid has no absorption band in this region, the spectral changes result from the complexation between the borinic acid and salicylaldoxime.

Spectrophotometric titration of the salicylaldoxime-diphenylborinic acid complex shows two pKs. The pK_1 of 5.9 presumably controls the protonation of the phenol oxygen in the complex while the pK_2 of 9.7 perhaps reflects the ionization of the complex's imine group. The complex structure presented here is consistent with that reported in the literature[18].

A hyperbolic saturation curve was observed when the values of net absorbance increase at 340 nm were plotted as a function of diphenylborinic acid concentrations. This relationship is linearized in Eadie-Hofstee plot. From the slopes of this type of plot, the dissociation constants (K_{diss}) of diphenylborinic acid to salicylaldoxime were determined at various pH.

The pH profile of K_{diss} shows best binding at pH 7.3 with the limiting K_{diss} value of 50 M. In the acidic region the limiting K_{diss} is 0.30 mM. pK_1 of 6.2 reflects the ionization of diphenylborinic acid[16]. pK_2 of 8.4 is close to the pK value for the ionization of the phenol group of salicylaldoxime. pK_3 of 5.4 perhaps originates from the ionization of the oxime-diphenylborinic acid complex. The pH dependence suggests that both the neutral and ionized borinic acid complexes with salicylaldoxime. It is clear that the ionized borinic acid displays higher binding affinity.

pH-Dependent Complexation between Salicylaldoxime and Boric and Boronic Acids

Boric and boronic acids also complex with salicylaldoxime resulting in a new absorption band around 320 nm. The complexation between benzeneboronic acid and salicylaldoxime is controlled by two pKs. While pK_1 of 9.1 reflects the ionization of benzeneboronic acid[19], the pK_2 of 8.0 perhaps relates to the ionization of phenolic OH group of the oxime.

Acknowledgements: We thank the PSC-CUNY Research Foundation (RF662164) and the NIH (MARC 5T34GM08182-07 & MBRS 2S06GM08225-07.) The American Chemical Society's Project SEED provided support for S. Sharma.

REFERENCES

1. Data used here can be found in L.-H. Niu, PhD Thesis, City University of New York, 1993.
2. H.G. Peer, Recl. Trav. Chim. Pays-Bas, 1960, 79, 825.
3. R.L. Letsinger, S. Dandegaonker, W.J. Vullo, J.D. Morrison, J. Am. Chem. Soc., 1963, 85, 2223.
4. R.L. Letsinger and J.D. Morrison, J. Am. Chem. Soc., 1963, 85, 2227.
5. R.L. Letsinger and D.B. MacLean, J. Am. Chem. Soc., 1963, 85, 2230.
6. B. Capon and B.C. Ghosh, J. Chem. Soc. B , 1966, 472.
7. T. Okuyama, H. Nagamatsu, T. Fueno, J. Org. Chem., 1981, 46, 1336.
8. H. Matsuda, H. Nagamatsu, T. Okuyama, T. Fueno, Bull. Chem. Soc. Jpn., 1984, 57, 500.
9. J. Hoffman and V. Sterba, Collect. Czech. Chem. Commun., 1972, 37, 2043.
10. H. Nagamatsu, T. Okuyama, T. Fueno, Bull. Chem. Soc. Jpn., 1984, 57, 2502.
11. H. Nagamatsu, T. Okuyama, T. Fueno, Bull. Chem. Soc. Jpn., 1984, 57, 2508.
12. G. Rao, and M. Philipp, J. Org. Chem., 1991, 56, 1505.
13. G.Rao, and M. Philipp, The Bioorganic Chemistry of Enzymatic Catalysis, D'Souza, V. & Feder, J., eds.; CRC Press, pp 129-142, 1992.
14. R. Pizer & L. Babcock, Inorg. Chem. 1977, 16, 1677.
15. M. Wechsberg, and R. Schönbeck, Austrian Patent 358,552, (1978); Chem. Abstr., 1981 94, 120904k.
16. G.N. Chremos, and H.K. Zimmermann, Chim. Chronika, 1963, 28, 103.
17. L. Babcock & R. Pizer, Inorg. Chem., 1980, 19, 56.
18. W. Kliegel, and D. Nanninga, D. Monatshefte für Chemie, 1983, 114(4), 465.
19. J. Juillard, and N. Geugue, C. R. Acad. Paris C, 1967, 264, 259.

The Synthesis of Amine–Borane Adducts of Cyclohexyl-amine, Toluidine, and 3'-Aminodideoxynucleosides and Evaluation of Their Pharmacological Activity

B. S. Burnham[1], S. Y. Chen[1], K. G. Rajendran[1], A. Sood[2,3],
B. F. Spielvogel[2], B. R. Shaw[3], and I. H. Hall[1]

[1]DIVISION OF MEDICINAL CHEMISTRY AND NATURAL PRODUCTS,
SCHOOL OF PHARMACY, THE UNIVERSITY OF NORTH CAROLINA AT
CHAPEL HILL, CB 7360, CHAPEL HILL NC 27599, USA
[2]BORON BIOLOGICALS INCORPORATED, 533 PYLON DRIVE, RALEIGH,
NC 27606, USA
[3]GROSS CHEMISTRY LABORATORY, DUKE UNIVERSITY, DURHAM, NC
27712, USA

A number of the amine-borane adducts of aliphatic,
heterocyclic and aromatic amines, which include analogues
of α-amino acids, peptides, and nucleic acids, have been
previously synthesized. Several amine-boranes of diverse
structures possess antineoplastic[1-5], hypolipidemic[6-8] and
anti-inflammatory[9-10] activities. Active antineoplastic
agents include cyano-, carboxy-, carbomethoxy-, or N-
ethylcarbamoylborane adducts of aliphatic amines[1-3] (e.g.,
trimethylamine-carboxyborane), heterocyclic amines[4] (e.g.,
piperidine-carboxyborane and 4-phenyl-piperidine-carboxy-
borane) and 2'-deoxynucleosides[5] (e.g., 2'-deoxycytidine-
[3]N-cyanoborane and 3',5'-O-bis(triisopropylsilyl)-2'-
deoxyguanosine-[7]N-cyanoborane). Many of these amine-
boranes were also active as hypolipidemic agents, such
as trimethylamine-carbomethoxy-borane[6], 1-methyl-imida-
zole-N-ethylcarbamoylborane[7], N-methylmorpholine-
carboxyborane[7], and 2'-deoxyguanosine-[7]N-cyanoborane[8].
Some of the compounds were active as both antineoplastic
and hypolipidemic agents, for example, trimethylamine-
carboxyborane[1,9] and 3',5'-O-bis(triisopropylsilyl)-2'-
deoxyguanosine-[7]N-cyanoborane[10].

In order to expand the structure-activity relationships
of the aliphatic and heterocyclic amine-boranes, a series
of cyano- and carboxyborane adducts of cylcohexylamine
and toluidine were synthesized. The toluidine-cyano-
boranes can be conveniently prepared by refluxing the
toluidine-hydrochloride (prepared by acidifying a
methanolic solution of the amine with aqueous hydrochloric acid) with sodium cyanoborohydride under inert
atmosphere (See figure 1).

The cyanoborane adducts of cyclohexylamines were synthesized by amine exchange, where excess cyclohexylamine was
refluxed with p-toluidine-cyanoborane, as the source of
cyanoborane, to yield the cyclohexylamine-cyanoborane and
p-toluidine (See figure 2).

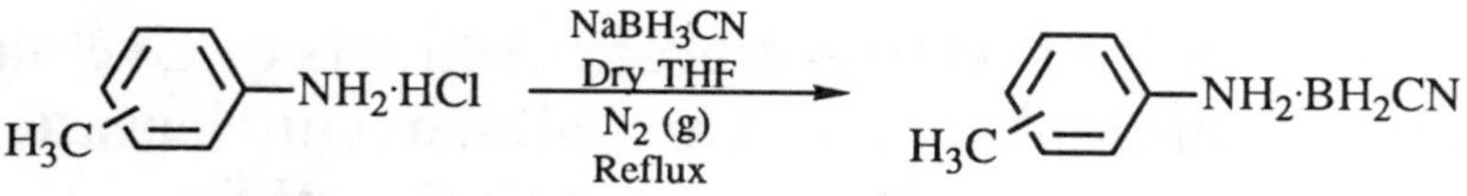

Figure 1. Synthesis of Toluidine Cyanoboranes.

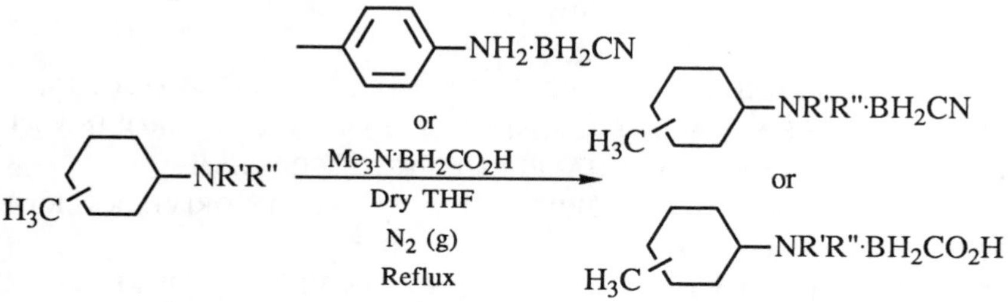

Figure 2. Synthesis of Cyclohexylamine Cyano- or Carboxyboranes.

The cyclohexylamine-carboxyboranes were also prepared by
amine exchange using excess cyclohexylamine and trimethylamine-carboxyborane, as the source of carboxyborane, to
yield the cyclohexylamine-carboxyborane and trimethylamine   (See figure 2).

Since 2'-deoxynucleoside-N-cyanoboranes demonstrated
potent antineoplastic[5] and hypolipidemic[8] activity, and
since the 3'-nitrogen containing nucleosides, 3'amino-
dideoxythymidine and 3'-azidodideoxythymidine have anti-
HIV[11,12] and antineoplastic[13] activity, a series of 3'-amine-
boronated 2',3'-dideoxynucleosides were prepared and
tested for pharmacological activity.   The synthesis of
the 3'-aminodideoxynucleosides proceeds by first
preparing the 5'-protected 2'-deoxynucleoside using
trityl chloride in pyridine.   The next series of steps
includes epimerization and mesylation of the 3'hydroxyl
group, which is then displaced by azide that is subsequently reduced.   This results in the 3'-amino derivative
with the original *erythro* stereochemistry[11,13].   The 5'-
protected nucleoside is mesylated in the 3'-position with
mesyl chloride in pyridine.   The 3'-mesylate (II) is displaced by hydroxide to yield the 5'-protected *threo*
nucleoside (III).   The 3'-epi-hydroxyl is again mesylated, then displaced by azide using lithium or sodium
azide in DMF at 75°C.   The 5'-trityl group is removed by
refluxing in 80% acetic acid.   The azide group is then
reduced by hydrogen gas using 10% palladium on carbon as
the catalyst to yield the 3'-amino-2',3' dideoxynucleo-
side (V).   The 3'-aminecyanoborane adduct (VI) is
prepared by Lewis base exchange using excess
triphenylphosphine-cyanoborane, as the source of cyano-

borane, with the 3'-aminodideoxynucleoside in dry DMF under inert atmosphere at 65°C. (See figure 3).

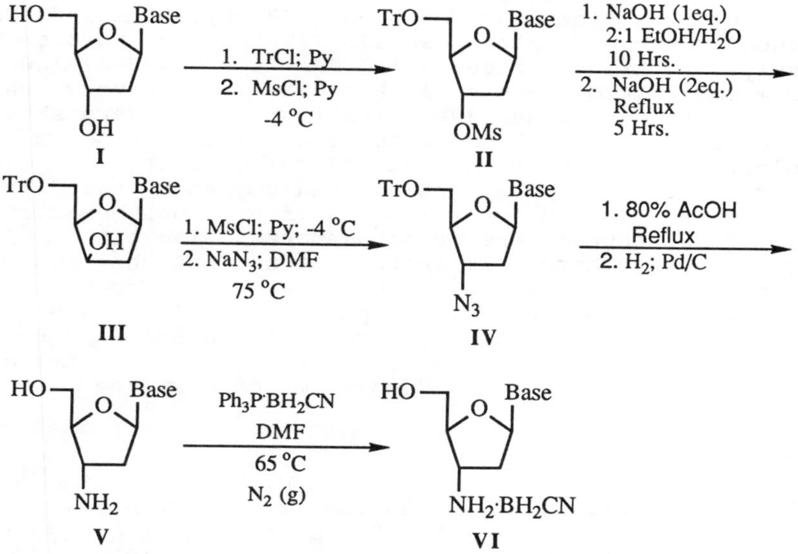

Figure 3. General synthetic route for the preparation of the 3'-amine-cyanoborane-2',3'-dideoxynucleosides. Base = thymidine or uridine.

The two compounds which demonstrated the most potent antineoplastic activity were: N-methylcyclohexylamine-cyanoborane (BSB-12) and 3'-aminecyanoborane-2',3'-dideoxythymidine (B-1). The compounds were active *in vitro* with ED_{50} values less than 4 μg/mL in the suspended cell cultures: murine L_{1210} lymphoid leukemia, human $Tmolt_3$ acute lymphoblastic T-cell leukemia, and HeLa-S^3 suspended cervical carcinoma. The 3'-aminecyanoborane (B-1) was also active in the solid human tumor cell cultures: KB epidermoid nasopharynx and HeLa solid cervical carcinoma. BSB-12 and B-1 were also shown to inhibit the growth (in CF_1 mice at a dose of 8 mg/kg/day, IP) of Ehrlich ascites carcinoma cells *in vivo* 86% and 99%, respectively. *In vitro* cytotoxic mode of action studies in L_{1210} cultured cells showed both compounds inhibited cell growth via inhibition of DNA and RNA syntheses. DNA synthesis was inhibited by reducing the activities of the following enzymes: DNA polymerase-α, IMP dehydrogenase, and dihydrofolate reductase. RNA synthesis was inhibited by the inhibition of mRNA, rRNA, and tRNA polymerase activities.

Some of the amine-boranes were also effective *in vivo* as hypolipidemic agents lowering serum cholesterol and tri-

glyceride levels in CF_1 mice at a dose of 8 mg/kg/day, IP.
The four most potent agents were: 3'-aminecyanoborano-
dideoxythymidine (B-1), 3-methylcyclohexylamine-carboxy-
borane (BSB-7), N-ethylcyclohexylamine-cyanoborane (BSB-
11), and p-toluidine-cyanoborane (BSB-22).   B-1 and BSB-7
reduced serum cholesterol levels 35% and 41%, respectively, and triglycerides 46%.   BSB-11 and BSB-22 lowered
serum cholesterol levels 24% and 22%, respectively, while
reducing triglycerides 39%.   The four agents were shown
to decrease serum LDL cholesterol and increase serum HDL
cholesterol levels in rats at an oral dose of 8 mg/kg/day
after 14 days.   Mode of action studies demonstrated that
all four agents had the following effects on the activities of rat hepatic enzyme homogenates: inhibition of
acetyl CoA synthetase inhibition of neutral cholesterol
esterase and an increase in cholesterol-7α-hydroxylase,
inhibition of phosphatidylate phosphohydrolase, and
inhibition of lipoprotein lipase.   B-1, BSB-7, and BSB-11
also inhibited ATP dependent citrate lyase activity; and
BSB-11 and BSB-22 inhibited acyl CoA carboxylase
activity.

REFERENCES
1.   I.H. Hall, B.F. Spielvogel, K.W. Morse, "Amine-
     Borane Derivatives as Antineoplastic Agents", in
     M.F. Gielen, Ed., Metal-Based Antitumour Drugs,
     Freund Publishing House, Ltd,. London, 1992, 55.
2.   I.H. Hall, C.O. Starnes, B.F. Spielvogel, P. Wisian-
     Neilson, J. Pharm. Sci. 1979, 68, 685.
3.   B.F. Spielvogel, A. Sood, I.H. Hall, R.G. Fairchild,
     P.L. Micca, Strahlenther. Onkol. 1989, 165, 123.
4.   C.K. Sood, A. Sood, B.F. Spielvogel, J.A. Yousef, B.
     Burnham, I.H. Hall, J. Pharm. Sci. 1991, 80, 1133.
5.   A. Sood, B.F. Spielvogel, B.R. Shaw, L.D. Carlton,
     B.S. Burnham, E.S. Hall, I.H. Hall, Anticancer Res.
     1992, 12, 335.
6.   I.H. Hall, B.F. Spielvogel, A. Sood, F.U. Ahmed, S.
     Jafri, J. Pharm. Sci. 1987, 76, 359.
7.   I.H. Hall, A. Sood, B.F. Spielvogel, Biomed.
     Pharmacother. 1991, 45, 333.
8.   I.H. Hall, B.S. Burnham, K.G. Rajendran, S.Y. Chen,
     A. Sood, B.F. Spielvogel, B.R. Shaw, Biomed.
     Pharmacother. 1993, 47, 79-87.
9.   I.H. Hall, C.O. Starnes, A.T. McPhail, P. Wisian-
     Neilson, M.K. Das, F. Harchelroad, Jr., B.F.
     Spielvogel, J. Pharm. Sci. 1980, 69, 1025.
10. K.G. Rajendran, B.S. Burnham, S. Chen, A. Sood, B.
     Spielvogel, B. Shaw, I.H. Hall, J. Pharm. Sci.
     1993 (Submitted).
11. E. De Clerq, J. Med. Chem. 1986, 29, 1561.
12. H. Mitsuya, S. Broder, Nature, 1987, 325, 773.
13. T.S. Lin, W.H. Prusoff, J. Med. Chem. 1978, 21, 109.

Synthesis of Boron Analogues of Biomolecules as Potential New Pharmaceuticals

B. F. Spielvogel[1], A. Sood[1], B. Ramsay Shaw[2], and I. H. Hall[3]

[1]BORON BIOLOGICALS, INC., PO BOX 33489, RALEIGH, NC 27636, USA
[2]P.M. GROSS CHEMICAL LABORATORY, DUKE UNIVERSITY, DURHAM, NC 27708, USA
[3]SCHOOL OF PHARMACY, UNIVERSITY OF NORTH CAROLINA, CHAPEL HILL, NC 27599, USA

1. INTRODUCTION

Two strategies are rapidly emerging in the 90's for the rational design and discovery of new drugs. One strategy involves structures based on small molecule design, while another involves the synthesis of modified oligonucleotides for inhibition of gene expression.

The utilization of small synthetic compounds is becoming increasingly attractive for a number of reasons. During the 1970's and 80's, using the technique of recombinant DNA, therapeutically useful protein drugs were evolved. However, protein-based drugs are still difficult to produce and deliver and also contain unwanted side effects. Thus, one approach gaining popularity is the synthesis of small organic molecules designed to interact with specific enzymes and cellular receptors. Such small molecules are easier to produce and may have fewer side effects.

The other emerging strategy of using modified oligonucleotides to inhibit gene expression is attractive because of the theoretical capability of high affinity and specificity for disease causing nucleic acid targets.[1]

Although both of the above approaches make use chiefly of synthetic organic chemistry, we believe that strategic replacement of carbon atoms by boron in biomolecules offers an entirely new and added dimension to these emerging technologies. Such replacement, which alters the chemical and physical properties of biomolecules, maintains a close structural similarity to the biomolecules and should still permit access and affinity to receptor and enzymatic sites.

2.   SYNTHESIS AND PHYSICAL PROPERTIES OF BORON ANALOGUES

We have reported the synthesis of a number of classes of boronated biomolecules, such as amino acids[2], peptides[3], cholines[4], phosphonates[5], etc. These compounds possess an interesting array of physical and pharmacological properties.

An example of the profound effect of replacement of C by B is on the pKa of the free amino acids. For example, the boron analogue of glycine (ammonia-carboxyborane (H_3NBH_2COOH)) has a carboxyl group pKa of 8.33 compared to 2.4 for glycine.[6] Similarly, the pKa for the ammonium nitrogen deprotonation in $H_3NBH_2CO_2H$ is >11, while for glycine it is 9.7. Thus, while the compounds are similar in size and geometry, they have very different electronic and hydrogen bonding properties.

Boron analogues of biomolecules are generally stable in water and have amphiphilic character, i.e., they are soluble in water as well as most organic solvents and lipids. This amphiphilic character may be useful for the transport of these compounds across cell membranes. Studies with $Me_3NBH_2C(O)NHCH(Ph)C(O)OMe$, a dipeptide of *boro*betaine and phenylalanine, indicate that boronated dipeptides can cross cell membranes without hydrolysis of the amide bond as is found with the majority of non-boronated peptides.[7] Such stability may be useful in overcoming problems associated with peptide or protein drug delivery.

Another class of boronated compounds that have attracted our interest recently is in the area of nucleic acids. Examples of derivatives that have been prepared include oligonucleotides in which a BH_3 group has replaced an oxygen in the phosphodiester backbone.[8]

$$_{5'}O\left(Nu - O - \overset{\overset{O}{\|}}{\underset{\underset{BH_3^-}{|}}{P}} - O\right)_X Nu\text{-}O_{3'}$$

Another series of derivatives have been prepared in which a nitrogen of the purine or pyrimidine base in the nucleoside has been boronated (with BH_2CN).[9]

The above species have shown remarkable hydrolytic stability. The "boronated" phosphodiester backbone[8] is quite resistant to nuclease activity which indicates utility in "antisense" agents. The nucleoside-cyanoboranes also are of sufficient stability[9] to enable pharmacological studies to be carried out.

Recently, two triphosphates (1)(2) have been prepared which are substrates for DNA polymerase showing that boron can be enzymatically incorporated into DNA.[10]

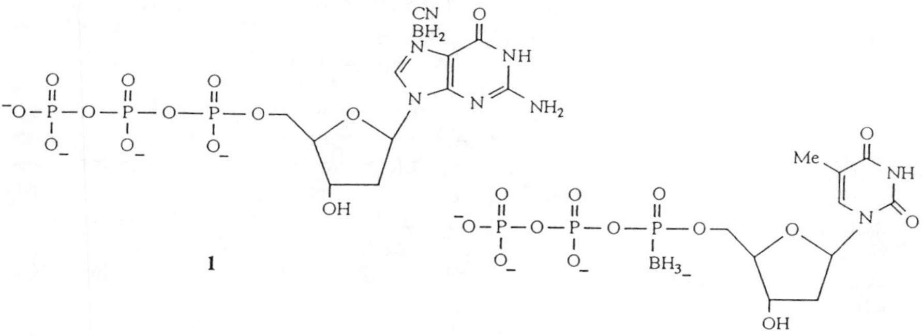

1

2

3. PHARMACOLOGICAL PROPERTIES OF BORON ANALOGUES OF BIOMOLECULES

Antineoplastic Activity

Boron analogues of biomolecules[11], especially amino[12] acid and nucleoside analogues[13], have potent antineoplastic activity. Strong activity has been demonstrated against the growth of Ehrlich ascites carcinoma, Lewis lung carcinoma and B_{16} melanoma in vivo. These compounds are also active in several human cell culture assays. The antitumor activity may be partly attributed to the inhibition of DNA synthesis due to the inhibition of several important enzymes involved in the purine and pyrimidine de novo synthesis. For example, many of these agents inhibit PRPP amidotransferase, IMP dehydrogenase, OMP decarboxylase and dihydrofolate reductase activities.

Hypolipidemic Activity

Analogues of amino acids[14] and related amine-carboxyboranes[15] cause significant reduction of total serum cholesterol and triglyceride levels in rodents. Rat tissue lipids, specifically cholesterol levels, are reduced in the liver and aorta wall, while these levels increased in the feces and bile. The reduction in lipid levels probably is due to the inhibition of key enzyme activities in the de novo synthesis of cholesterol, i.e., HMG CoA reductase, acetyl CoA synthetase, of fatty acids, i.e., citrate lyase, and of triglycerides, i.e., sn-glycerol-3-phosphate acyl transferase and phosphatidylate phosphohydrolase. Another significant effect of boron analogues is on the serum lipoprotein lipid content. These agents effectively reduce cholesterol content in the chylomicron, VLDL, and LDL fractions while raising it in the HDL fraction. Such type of cholesterol modulation has been shown to protect humans against myocardial infarctions. The magnitude of cholesterol increase in the HDL fraction is far superior than clinically-used agents. For example, in rats, borobetaine methyl ester reduces total serum cholesterol by 37% and, at the same time, increases cholesterol content of the HDL fraction to 195% of control. Some heterocyclic amine-carboxyboranes have been shown to increase HDL fraction of cholesterol by 200-300% while decreasing the total cholesterol by 30-40%.

Anti-inflammatory Activity

Another area where these compounds have demonstrated good pharmacological activity is as anti-inflammatory agents. These agents were particularly useful in the inhibition of induced edema, reduction of local pain associated with inflammation, and inhibition of centrally induced pain.[16] Selected agents were also effective in chronic induced arthritis in rats at 2.5 mg/kg/day and active against pleurisy in rats. These derivatives inhibit the activities of lysosomal enzymes from a number of tissues, e.g. PMNs, hepatocytes, leukocytes, with IC_{50} values in the range of 10^{-6} M.

Anti-osteoporosis Activity

Nielson and coworkers[17] at the USDA have studied the effects of dietary boron (as borate) on mineral and hormonal metabolism in postmenopausal women and showed that boron reduced excretion of calcium and magnesium but elevated the serum concentration of 17 β-estradiol and testosterone. Further work[18] has shown that boron supplementation mimics the effects of estrogen therapy and may be affecting calcium and bone metabolism. This work suggests that boron may play a role in the prevention of osteoporosis.

Boron analogues of biomolecules[19,20] also block calcium resorption from the bone. Release of calcium, phosphorus and hydroxyproline levels via urine in mice is decreased by treatment at 8 mg/kg/day for 21 days. During this time, the blood levels remain high. In an *in vitro* assay at 10^{-4} - 10^{-8} M conc. of selected agents, the calcium exchange with the medium from 4-day old pup calvaria bone was decreased. Also, in UMR-106 cultured osteosarcoma cells, calcium resorption was blocked by the agents. These compounds are more active than calcitonin and the bisphosphonate standards in blocking calcium resorption. Concurrent incorporation of calcium into the cell and xyproline into collagen is increased in rat UMR-106 cells, IC-21 macrophages, and Be Sal human osteoporosis cells in the presence of the agents. In addition, an increase in labeled collagen incorporation into cellular collagen is also observed in these cells as well as in the pup calvaria cultures. The exchange of proline to the medium over the next 48 hr is reduced significantly in the presence of drugs. In a lactating rat model in which rats were dosed orally for 17 days at 8 mg/kg/day, these agents increased bone volume, bone weight, bone ash weight and density, while elevating bone and serum calcium levels.

Acute Toxicity Studies of Boron Analogues

The boron analogues of many of the biomolecules which we have prepared, such as amino acids, peptides, nucleosides, etc., are considered to be of low toxicity. The LD_{50} values in mice range from 500 mg/kg to 1800 mg/kg I.P. Mouse organ weights, hematological parameters, clinical chemistry values and major tissue morphology are generally unchanged.[21]

Biodistribution and metabolism studies are currently under investigation with select boron analogues. Since one ultimate metabolite or decomposition product of the boron biomolecules may be boric acid or borate, the conclusions reached at the first International Symposium on the Health Effect of Boron and Its Compounds[22] are of particular relevance for the potential use of boron compounds as pharmaceuticals. A low order of toxicity for boric acid and borates in humans seems consistent with the results of many studies. Also, no evidence for accumulation of boron in humans exists.

References

1. Crooke, S.T. and Lebleu, B., "Antisense Research and Applications," CRC, Boca Raton, FL, 1993.
2. Spielvogel, B.F., Boron Chemistry-4, 1UPAC, R.W. Parry, G. Kodama, eds., Peragamon Press, NY 1980, 199..
3. Sood, A., Sood, C.K., Spielvogel, B.F., and Hall, I.H., Eur. J. Med. Chem. 1990, 25, 301.
4. Spielvogel, B.F., Ahmed, F., and McPhail, A.T., Inorg. Chem. 1986, 25, 4395.
5. Sood, A., Sood, C.K., Hall, I., and Spielvogel, B.F., Tetrahedron 1991, 47, 6915.

6. Scheller, K.H., Martin, R.B., Spielvogel, B.F. and McPhail, A.T., Inorg. Chim. Acta 1982, 57, 227.
7. Elkins, A.L., Cho, M., Shrewsbury, R.B., Sood, A., Spielvogel, B.F., Hall, I.H. and Miller III, M.C., Abstracts, p. 103, Eighth Intl. Meeting on Boron Chemistry, Knoxville, TN July 11-15, 1993.
8. Sood, A., Shaw, B.R. and Spielvogel, B.F., J. Amer. Chem. Soc. 1990, 112, 9000.
9. Sood, A., Spielvogel, B.F., and Shaw, B.R., J. Amer. Chem. Soc. 1989, 111, 9234.
10. Spielvogel, B.F., Sood, A., Powell, W., Tomasz, J., Porter, K. and , Shaw, B.R., Proceedings of the 5th Intl. Symposium on Neutron Capture Therapy, Plenum Pub. Corp. New York, NY in press, 1993.
11. Hall, I.H., Spielvogel, B.F. and Morse, K.W., Metal Based Antitumor Drugs, M.F. Gielen, ed., Freund Publishing House, 1992, 55.
12. Hall, I.H., Sood, A., Anti-Cancer Drugs 1990, 1, 133.
13. Sood, A., Spielvogel, B.F., Shaw, B.R., Carlton, L.D., Burnham, B.S., Hall, E.S. and Hall, I.H., Anticancer Research 1992, 12, 335.
14. Sood, A., Sood, C.K., Spielvogel, B.F., Hall, I.H., Wong, O.T., Mittakanti, M. and Morse, K., Archiv. der Pharmazie 1991, 324, 423.
15. Hall, I.H., Sood, A. and Spielvogel, B.F., Biomed & Pharmacother 1991, 45, 333.. et. al
16. Hall, I.H. et al. Acta Pharm. 1990, 2, 387.
17. Nielsen, F.H., Hunt, C.D., Mullen, L.M. and Hunt, J.R., FASEB J. 1987, 1, 394-397.
18. Nielsen, F.H. et al. International Symposium on the Health Effects of Boron and Its Compounds 1992, Session 4, 1.
19. Hall, I.H., Chen, S.Y., Rajendran, K.G., Sood, A., Spielvogel, B.F. and Shih, J., Envir. Health Persp. 1993, in press.
20. Rajendran, K.G., Sood, A., Spielvogel, B.F. and Hall, I.H., Bone Mineral Metab., Submitted.
21. Hall, I.H., Reynolds, D.J., Chang, J., Spielvogel, B.F., and Griffing, T.S., Archiv. der Pharmazie 1991, 324, 573.
22. Proceedings, Intl. Symp. on the Health Effects of Boron and Its Compounds, Envir. Health Persp. in press.

Radiopharmaceutical Synthesis via Boronated Polymers

G. W. Kabalka, J. F. Green, M. M. Goodman, J. T. Maddox, and S. J. Lambert

DEPARTMENT OF CHEMISTRY, THE UNIVERSITY OF TENNESSEE, KNOXVILLE, TENNESSEE 37996, USA

1. INTRODUCTION

A significant portion of our research is focused on the development of new techniques for rapidly preparing medical imaging agents containing short-lived, positron-emitting isotopes such as carbon-11, nitrogen-13, and oxygen-15.[1] These nuclides, along with fluorine-18, have important applications in a tomographic imaging technique known as positron emission tomography (PET). There are nearly 100 institutions throughout the world that carry out PET research. The University of Tennessee is generally recognized as the first institution to offer PET on a clinical basis. The value of PET rests on its ability to evaluate biological processes *in vivo*. The difficulties encountered in the synthesis of PET pharmaceuticals are a consequence of the very short half-lives of many of the medically important nuclides. As an example, carbon-11 and oxygen-15 have half-lives of 20 minutes and two minutes, respectively. Clearly, there is little time to synthesize complex reagents from the radiolabeled precursors (normally oxides) generated by the medical cyclotrons and accelerators.

When we initiated our research program in PET radiopharmaceuticals, we focused on developing new reactions which incorporated the desired isotope at, or near, the end of the synthetic sequence. In this manner, a significant portion of the life time of the element could be relegated to medical and imaging protocols. [As a general rule, the radiopharmaceutical is effective for approximately five half-lives, inclusive of synthesis, administration, and imaging.] Our studies have generally utilized organoboranes as reactive precursors to the pharmaceuticals.[2] Interestingly, prior to our studies, little had been known about the use of organoboranes for isotope incorporation; only the isotopes of hydrogen had been incorporated into organic molecules via organoborane

reactions. It occurred to us that many other reactions
were feasible based on the utility of organoboranes in
organic syntheses. As an example, it had been
established that a number of functional groups could
replace boron in a regiospecific and stereospecific
fashion:

$$
\begin{array}{ccc}
 & R-I & \\
 & \uparrow & \\
R-OH \longleftarrow & R_3B & \longrightarrow \quad R-\overset{\displaystyle O}{\underset{\displaystyle H}{\big\langle}} \\
 & \downarrow & \\
 & R-NH_2 &
\end{array}
$$

2.   RESULTS AND DISCUSSION

We initiated our studies by investigating the
rates of the various organoborane reactions in order to
evaluate their potential use in syntheses involving
short-lived isotopes. Interestingly, few investigators
had focused on this aspect of organoborane chemistry.
Fortunately, our studies revealed that many of the
reactions were applicable, with certain modifications,
to PET syntheses. Many of the modifications involved
converting relatively unreactive anionic reagents into
more electropositive species. As an example, it was
well known that halogens react with organoboranes to
provide the corresponding halogenated reagents;
however, radiohalogens are generally obtained in their
halide states. We initiated a study which focused on
the rapid, *in situ* conversion of halides to more
reactive electropositive species in the presence of
organoboranes.[2] This study led to the development of
a wide variety of radiohalogenations that could be
utilized in medicine.

During the past decade, we developed a number of
new synthetic sequences which allowed us to prepare PET
pharmaceuticals from medically useful isotopes such as
carbon-11, nitrogen-13, oxygen-15, and a variety of
radiohalogens.[2] These developments, and those of other
boron chemists, have contributed to the syntheses of a
variety of important pharmaceuticals for use in PET
imaging . These include oxygen-15 labeled butanol
which is currently being used to quantitate blood flow
in humans and 17-α-iodovinylestradiol which is now
being evaluated as tumor specific imaging agents for
breast cancer patients.[3,4] [These agents are
illustrated on the next page.]

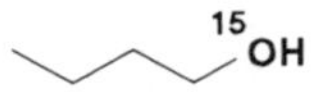

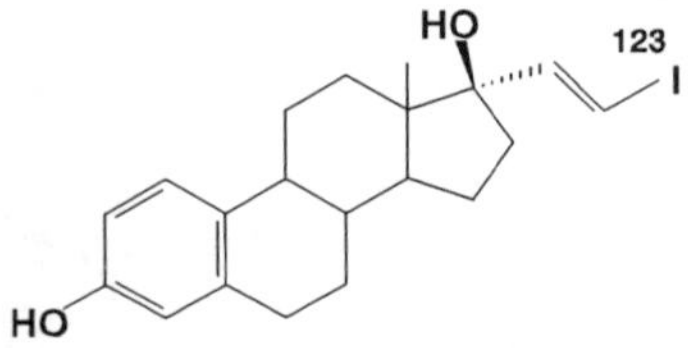

Throughout the course of our studies it has become apparent that the use of simple trialkylboranes in these reactions might not be ideal. As an example, the formation of nitrogen-13 labeled putrescine from the corresponding trialkylborane results in the formation of the dialkylborinic acid byproduct. Through the use of "disposable" alkyl groups such as methyl and phenyl, we were able to maximize the utility of the desired functionally substituted alkyl group.[5]

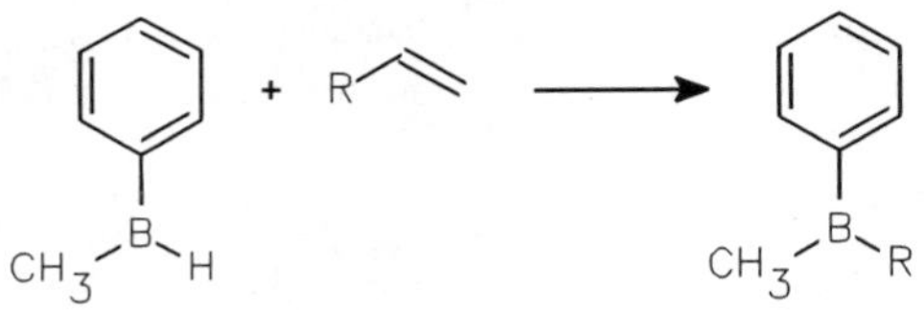

Unfortunately, this modification does little to simplify the isolation and purification of the desired pharmaceuticals. As a consequence, we have begun to investigate the use of boronated surfaces as potential radiopharmaceutical precursors. The concept is quite straightforward; the boronated surfaces are contained in a column fitted with appropriate valving and receptacles. The reactive isotopes are injected onto the column, followed by other reactive intermediates (generally, oxidizers); the byproducts and the desired product are isolated by sequentially washing the column with sterile saline solutions.

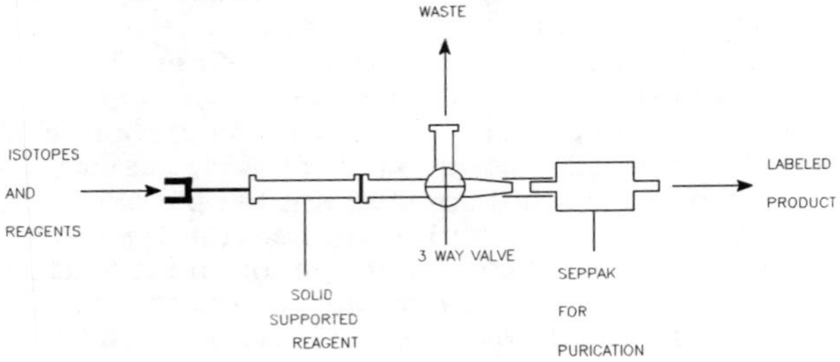

We are currently studying two different boronated surfaces: boronated polymers and boronated aluminas. The boronated polymers have been used to prepare oxygen-15 labeled butanol via the preparation of a boronated polystyrene agent.[6]

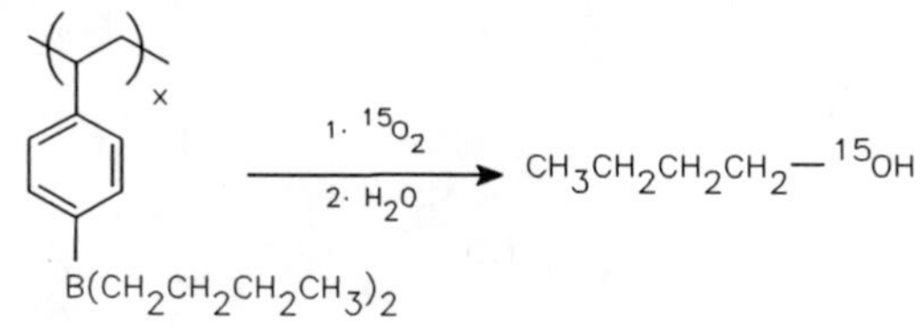

The yields of oxygen-15 labeled butanol using the polymeric reagents are competitive with standard tributyl-borane reactions but the product isolation is vastly simplified. More recently, we have modified the procedure to utilize the methyl group as a sacrificial group and have applied the methodology to the synthesis of a number of clinically important nitrogen-13 labeled amines including putrescine, dopamine, and γ-aminobutyric acid.[7]

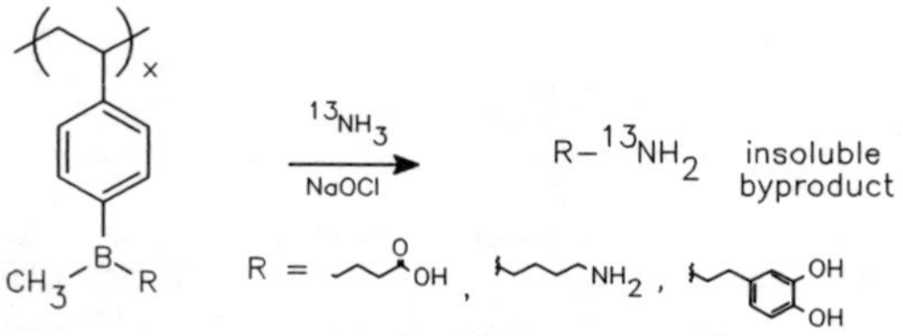

Alumina can also be utilized to catalyze organic reactions. This is an area which we have been investigating for some time.[8] The fact that both Lewis Acid and Lewis Base sites exist simultaneously and in close proximity on the alumina surface often provides new types of chemistry. As an example, a surface which is modified by substituting deuterated water for normal water will substitute deuterons for hydrogens on an enolizable site in an organic molecule as it passes over the surface.[9] Coating alumina with reactive substrates, such as boron tribromide, often leads to entirely new chemistry.[10] In an earlier study, we noted that organoboranes react with sodium iodide on a TLC plate in the absence of added oxidant;[11] presumably, sufficient oxygen is bound to the surface to oxidize the iodide to a more electrophilic iodine reagent. We recently reported that boronic acids bound to alumina surfaces react differently than their solution counterparts.[12] In essence, the E/Z ratio can be varied by modifying the quantity of water hydrated to the surface.

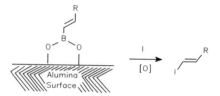

This study has been expanded to include bromination reactions. In these studies the ratio of E/Z vinyl bromide varies dramatically from the in solution counterparts. We are continuing the application of surface bound boron reagents to radiopharmaceutical syntheses. It is clear from our current investigations that these reactions have important implications in radiopharmaceutical syntheses as well as more traditional organic syntheses.

ACKNOWLEDGEMENT

We wish to thank the Department of Energy for support of this research.

REFERENCES

1. E. Buonocore, K. Hubner, J. Collmann and G. W. Kabalka, 'Clinical Positron Emission Tomography', Mosby Yearbook, Hanover, MD, 1991

2. G. W. Kabalka, R. S. Varma, <u>Tetrahedron</u>, 1989, <u>45</u>, 1989.

3. G. W. Kabalka, R. M. Lambrecht, M. Sajjad, J. S. Fowler, S. S. Kunda, G. W. McCollum and R. Macgregor, <u>Int. J. Appl. Rad. Isot.</u>, 1985, <u>35</u> 853.

4. G. W. Kabalka, <u>Acc. Chem. Res.</u>, 1984, <u>17</u>, 215.

5. Z. Wang, C. Narayana, P. P. Wadgaonkar and G. W. Kabalka, <u>J. Organomet. Chem.</u>, 1992, <u>440</u>, 243.

6. G. W. Kabalka, J. F. Green and G. McCollum, <u>J. Labelled Compd. Radiopharm.</u>, 1989, <u>24</u>, 76.

7. G. W. Kabalka, J. F. Green, Z. Wang and M. M. Goodman, <u>J. Labelled Compd. Radiopharm.</u>, 1989, <u>24</u>, 90.

8. G. W. Kabalka, R. M. Pagni, S. Bains, G. Hondrogiannis, M. Plesco, R. Kurt, D. Cox and J. Green, <u>Tetrahedron Asymmetry</u>, 1992, <u>2</u>, 1283.

9. G. W. Kabalka, R. M. Pagni, P. Bridwell, E. Walsh and H. M. Hassaneen, <u>J. Org. Chem.</u>, 1981, <u>46</u>, 1513.

10. S. Bains, J. Green, L. C. Tan, R. N. Pagni and G. W. Kabalka, <u>Tetrahedron Lett.</u>, 1992, <u>33</u>, 7475.

11. T. E. Boothe, R. D. Finn, M. M. Vora, A. Emran, P. Kothari and G. W. Kabalka, <u>J. Labeled Compd.</u>, 1985, <u>22</u>, 1109.

12. W. R. Sponholtz, J. F. Green, L. C. Lee, R. M. Pagni and G. W. Kabalka, <u>J. Org. Chem.</u>, 1991, <u>56</u>, 5700.

CARBORANE CHEMISTRY

Host–Guest Chemistry of Anion Complexation by Macrocyclic Multidentate Lewis Acids

M. Frederick Hawthorne

DEPARTMENT OF CHEMISTRY AND BIOCHEMISTRY, UNIVERSITY OF CALIFORNIA AT LOS ANGELES, LOS ANGELES, CA 90024, USA

1. INTRODUCTION

In contrast to the extraordinary achievements of cation complexation in host-guest chemistry,[1] only very recently has anion complexation by compounds containing electron-deficient atoms such as boron,[2] mercury,[3,4] tin[5] and silicon[6] received attention, even though anion-inclusion complexes were reported as early as 1968.[7] It was first reported that halide ions can be encapsulated in "in-in" isomers of bicyclic diazaalkane hosts via electrostatic and hydrogen bond interactions. Since then, several protonated and charged spherical tricyclic cryptands and tricyclic tertiary ammonium cages have been synthesized[7] as well as diprotonated planar sapphyrin.[8] These hosts bind halide ions through electrostatic and hydrogen bond interactions. However, recently reported Lewis acid hosts which employed tin, mercury, silicon and boron atoms as binding sites interact with anion guests through coordinate bond formations which offers new prospects for selective molecular recognition, anion transport and the catalytic activation of electron-rich organic and inorganic substrates. The synthesis and application of Lewis acid hosts in anion recognition has initiated studies of coordinate bonding in host-guest chemistry.

Linear mercuric compounds, with two empty mercury p-orbitals, exhibit Lewis acidity towards a variety of bases and nucleophiles. Organomercury(II) compounds form anionic complexes with halide or pseudohalide ions in solution.[9] However, no anionic complexes were isolated in the solid state because of their low stability. It has long been recognized that multiple complementary interactions between host and guest species is of vital importance in determining the stability of host-guest complexes (e.g. the chelate effect). It is thus not unusual that host-guest systems containing at least two binding centers have been designed and synthesized. In addition to multiple-binding, preorganization of the host plays another

key role in determining the stability of host-guest
complexes and the macrocyclic and cryptate effects have been
invoked to explain the high stability of macrocyclic and
cryptate host-guest complexes. It is obvious that the
design and synthesis of preorganized multidentate Lewis
acid hosts is the key to the discovery of efficient hosts
which can recognize and transport anionic species.
Presented here is a description of our recent work on the
design, synthesis and structural characterization of a new
class of carborane-supported macrocyclic multidentate Lewis
acid hosts such as (1) and (2).[10]

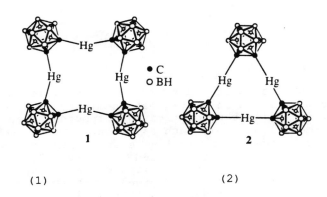

(1)   (2)

2. DESIGN AND SYNTHESIS OF LEWIS ACID HOSTS

To design and synthesize macrocyclic Lewis acid
hosts with Hg(II) as binding sites, we employed 1,2-
carborane as the supporting structure for the host for
two reasons. First, the $1,2-C_2B_{10}H_{12}$ cage is electron-
withdrawing in character, when examined from a C-vertex,
which would greatly enhance the acidity of attached
Hg(II) centers. Secondly, the icosahedral carborane 1,2-
$C_2B_{10}H_{12}$ contains CH vertices which are easily lithiated
by n-BuLi to give $1,2-Li_2-1,2-C_2B_{10}H_{10}$[11] which facilitate
synthesis. The reaction of mercury(II) halides with
$1,2-Li_2-1,2-C_2B_{10}H_{10}$ results in the formation of the host-
guest species, $(1)\cdot XLi$ (X = Cl⁻, Br⁻) or $(1)\cdot I_nLi_n$ (n = 1,
2) (13a,b). When $Hg(OAc)_2$ is employed in the reaction,
trimeric (2) is formed (13d), as shown in Scheme 1.
Macrocyclic multidentate Lewis host (1),
[12]mercuracarborand-4, is a charge-reversed analogue of
[12]crown-4 with a preorganized structure. Similarly,
(2), [9]-mercuracarborand-3, is the electrophilic
analogue of [9]-crown-3. The direct synthesis of halide
ion compplexes of [12]mercuracarborand-4 is significant
because it efficiently produces a macrocyclic
multidentate Lewis acidic host, (1), in a single step.
It is believed that halide ions function as templates
which direct the assembly of the host macrocycle, since
cyclic trimer (2) is formed in the absence of halide ion

(Scheme 1). Therefore, the formation of cyclic trimer (2) would appear to be both kinetically and thermodynamically more favored than the tetramer (1) in the absence of templates. It is also observed that the reaction of $1,2-Li_2-1,2-C_2B_{10}H_{10}$ with mercuric halides gives higher yields of cyclic tetramer than the corresponding reaction with HgOAc, which gives cyclic trimer. The presence of halide ions in the cyclic tetramer product further strengthens the claim that halide ions exercise a template effect in the formation of mercuracarborands. The template effect of metal ions in promoting certain cyclization reactions has been previously recognized.[12]

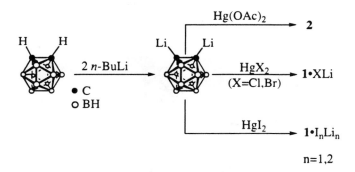

Scheme 1

X-ray diffraction studies of (1)·ClLi and (1)·I_2(AsPh$_4$)$_2$ revealed that halide ions are tetra-coordinated by four mercury atoms as shown in Figure 1 and 2, respectively. The Cl$^-$ is essentially inside the cavity of the cyclic host and the Hg-Cl distance of 2.944 (2) Å is significantly shorter than the van der Waals distance. Iodide ion is too large to fit into the cavity of (1). Two iodide ions are found both above and below the host plane with Hg-I distances ranging from 3.277 to 3.774 Å. The location of Cl$^-$ in (1) maximizes the cooperative interaction of four empty mercury p-orbitals with two filled p-orbitals of the chloride ion. The chloride ion therefore binds to four mercury atoms *via* two 3c-2e bonds, so does the iodide ion in (1)·I_2^{2-}. The chloride ion complex is the only reported structure which contains square planar chlorine.

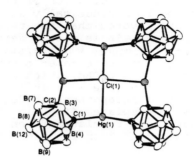

Figure 1 ORTEP presentation of (1)·Cl⁻.

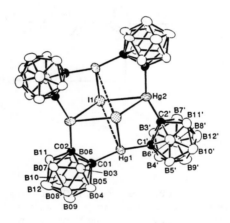

Figure 2 ORTEP presentation of (1)·I₂²⁻.

The trimeric cycle, (2), also binds halide ions and
organic solvents such as telrahydrofuran (THF) and
acetonitrile. The chloride ion complex is obtained by
the addition of LiCl in H₂O to a diethyl ether solution of
(2). Lithium chloride is transfered from the aqueous
phase to organic phase upon formation of (2)·ClLi. The
species (2)·Cl⁻ could be detected by negative ion FAB mass
spectroscopy although suitable crystals have not yet been
obtained for x-ray diffraction studies. Because of the
linearity of mercury coordination in (2), it is possible
for (2) to coordinate more than two molecules of
acetonitrile as seen in the structures of (2)·(CH₃CN)₅ and
(2)·(CH₃CN)₃ shown in Figure 3.

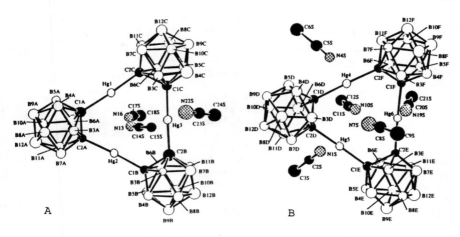

Figure 3 The structures of (2)·(CH₃CN)₃ (A) and (2)·(CH₃CN)₅ (B).

At this point in our studies, only halide ion complexes of (1) had been isolated and the question remained as to the stability of (1) in the absence of halide ion guests. It was found that (1)·I₂Li₂ reacts with AgOAc to produce (1) and insoluble AgI . Species (1) crystallized from THF as (1)·(THF)₄.₂H₂O and its structure is shown in Figure 4. The buckled structure of (1) facilitates the coordination of THF to the mercury centers and the primary mercury coordination in (1) is less distorted from linearity.

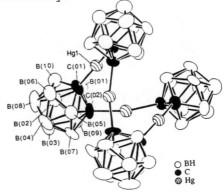

Figure 4 The molecular structure of (1)·(THF)₄ with THF molecules removed for clarity.

Host (1) is soluble in most electron-donor solvents and complexes with these solvent molecules. These solvent interactions mitigate the potential usefulness of (1) in the homogeneous catalysis of organic reactions. In order to eliminate this competitive binding of solvent molecules, we have synthesized the octaethyl deriverative of [12]mercuracarborand-4 (3) as shown in Scheme 2. The

$9,12\text{-}Et_2\text{-}1,2\text{-}C_2B_{10}H_{10}$ precursor can be easily lithiated and the resulting dilithio reagent reacts with HgI_2 to form $(3)\cdot I_2Li_2$ and $(3)\cdot ILi$ in a 2:1 ratio. We have also prepared the 3-phenyl substituted carborane and the iodide ion complexes of the corresponding tetrameric cycles, $(4)\cdot ILi$[10f] (Scheme 3). The formation of tetrameric cycles appears to be universal with derivatized carboranes. All of these derivatized host-guest complexes have higher solubilities in organic solvents than those of (1). The host (3), obtained from the reaction of AgOAc with $(3)\cdot I_2Li_2$, is an air-stable white solid soluble in benzene, toluene, chloroform and methylene chloride. When (3) was crystallized from $CHCl_3$, solvent molecules were not coordinated to (3), as revealed by X-ray diffraction analysis. Like $(1)\cdot(THF)_4\cdot 2H_2O$, (3) has a buckled structure as shown in Figure 5. In principle, four structural isomers of the tetraphenyl species (4) could be formed with phenyl substitutents arrayed about the periphery of the host cavity. The isomer of $(4)\cdot ILi$ shown in Figure 6 is the only product isolated from the reaction of $1,2\text{-}Li_2\text{-}3\text{-}Ph\text{-}1,2\text{-}C_2B_{10}H_9$ and HgI_2, as observed by HPLC. This synthesis is significant since it demonstrates the facility with which structural modifications of the cavity of conformationally immobile mercuracarborand species can be made and portends the creation of chiral complexes and hosts. Preliminary work indicates that chloride ion complexes of three isomeric mercuracarborand complexes of (4) are formed, when $HgCl_2$ is employed.

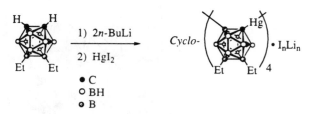

Scheme 2

Scheme 3

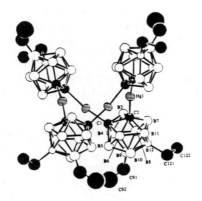

Figure 5 The molecular structure of (3).

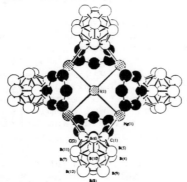

Figure 6 The molecular structure of (4)·I⁻.

3. Mercury-199 NMR STUDIES

Mercury-199 NMR spectroscopy has been employed in studies of biological systems and the interactions of organomercurials with a variety of bases and ligands.[12,13] The ^{199}Hg nucleus has a spin quantum number of I=1/2 and a moderately large natural abundance (16.9%). The extreme sensitivity of ^{199}Hg chemical shifts to its immediate enviroment make it very useful in the study of the systems we have described here, especially so since mercuracarborands and their anion complexes have very similar ^{13}C and ^{11}B NMR spectra.

The ^{199}Hg resonances of mercuracarborands and their complexes range from -1300 ppm for (1) in THF to -710 ppm for (3)·I₂Li₂ in acetone. The resonances of (1), (2) and (3) are among display the highest field chemical shifts reported for diorganomercurials at -1300 in THF, -1221 in CDCl₃ and -1364 ppm in acetone-d_6, respectively. Complex (1)·ClLi has a resonance at -1077 ppm, (1)·BrLi -1010 ppm, (1)·ILi -810 ppm and (1)·I₂Li₂ -714 ppm in acetone solution, respectively. Similarly, (3)·ILi and (3)·I₂Li₂

have chemical shifts at -774 and -674 ppm in acetone
respectively. The downfield shifts upon complexation of
halide ions to hosts (1) are 130 ppm, 200 ppm and 390 ppm
for Cl^-, Br^- and I^- ions, respectively. Similar downfield
^{199}Hg chemical shifts were also observed in $RHgX$ and HgX_2
systems, when $RHgX_2^-$ and HgX_3^- are formed.[9] It has been
known that organomercurials exhibit large solvent and
concentration dependence in their ^{199}Hg NMR spectra.
However, anion complexes of (1) and (3) have ^{199}Hg
chemical shifts which are essentially independent of
concentration and solvents at room temperature. We
believe that the solid state structures of $(1) \cdot X_nLi_n$ (X =
Cl, Br, n=1; X=I, n=1,2) and $(3) \cdot I_n{}^{n-}$ (n = 1,2) are
largely intact in solution with a planar host and its Hg
atoms bending towards the center of the tetramer cycle.
Since the carborane cages and halide ions virtually
surround the mercury binding sites and it is therefore
unlikely that solvent molecules could coordinate to Hg
atoms of the tetramer cycle.

Previous studies of halide ion exchange and
complexation of mercuric complexes have shown that fast
exchange was always observed when compared to the NMR
time scale. The observed peak is then the weighted
average of the chemical shifts of several species present
in the solution. However, the mixtures of $(1) \cdot ClLi$ and
$(1) \cdot BrLi$ in acetone exhibit two resonances which
correspond to the two individual complexes. Apparently,
the exchange between the chloride and bromide ions
complexed to (1) is too slow to be detected by ^{199}Hg NMR.
Similarly, a mixture of $(1) \cdot ILi$ and $(1) \cdot I_2Li_2$ displays two
individual resonances. The fact that the characteristic
^{199}Hg chemical shift of each of these complexes in
solution is independent of solvent and concentration
allows us to study halide ion exchange and association
with (1).

The conversion of (1) to $(1) \cdot I_2{}^{2-}$ was studied as
suggested above. The incremental addition of nBu_4NI to an
acetone-d_6 solution of (1) results in the formation of
$(1) \cdot I^-$ and then $(1) \cdot I_2{}^{2-}$ after more then one equivalent of
I^- was added. When two equivalents of I^- was added, only
one resonance, corresponding to $(1) \cdot I_2{}^{2-}$, remains. When
more than two equivalents of n-tetrabutyl ammonium iodide
were added to the acetone-d_6 solution of (1), no further
change in the spectrum was observed, indicating that (1),
in agreement with the solid state structure of $(1) \cdot I_2{}^{2-}$,
is not capable of hosting more than two iodide ions.
This fact suggests that the solid state structure of
$(1) \cdot I_2{}^{2-}$ is largely maintained in solution. It is also
suggests that $(1) \cdot I_2{}^{2-}$ is sufficiently stable that its
exchange with bulk halide ions does not occur on the NMR
time scale.

4 SUPRAMOLECULAR CHEMISTRY

Supramolecular chemistry, an extension of host-guest chemistry, has attracted increasing attention because of its relationship to molecular recognition, reactivity and catalysis, transport processes, molecular assemblies and the design of devices based upon supramolecular phenomenona. Recent work in this area has demonstrated that the successful construction of supramolecular systems is often based on the design and creation of microenvironments with specific features which include the steric and electronic complementarity between host and guest, and the preorganization of those features responsible for associative interactions.

Closo-$B_{10}H_{10}^{2-}$ (5) has long been known to be nucleophilic and to form covalent bonds with Lewis acids such as Ag^+, Cu^+ and Hg^{2+}.[16] Similarly, we have studied the intermolecular interaction of *closo*-$B_{10}H_{10}^{2-}$ with [12]mercuracarborand-4 (1) and its octaethyl derivative (3) in organic solvents (acetone, CH_2Cl_2 and CH_3CN) using ^{11}B and ^{199}Hg NMR spectroscopy. The ^{11}B NMR spectrum of (5) in CH_2Cl_2 exhibited two doublets at 0.83 and -27.3 ppm with a 1:4 ratio which corresponds to axial and equatorial boron atoms, respectively. Upon the addition of one equivalent of (1) or (3) in CH_3CN/CH_2Cl_2 (1:1), the doublet at -27.3 ppm splits into a broad singlet at -23.3 ppm and a sharp doublet of equal intensity at -26.9 ppm, which can be explained by the formation of adducts of (5) with (1) or (3) in which four equatorial BH groups of (5) bind to four mercury atoms in (1) or (3) by forming B-H-Hg three center-two election bonds.

The structure of supramolecular ion (3)·(5)$_2$$^{4-}$ was deduced from an X-ray diffraction analysis (Figure 7). The supramolecular ion consists of a planar macrocyclic tetradentate Lewis acid host and two bicapped square antiprismatic borane cages. The bonding between (5) and (3) is a four-point-bonding interaction between the host and the guest. The borane dianion binds to (3) through four B-H--Hg intercations. The eight lone-pairs of the B-H σ bonds are donated to the eight empty p-orbitals of the four Hg centers. Apparently, these interactions are weak. However, this supramolecule is stable in acetonitrile solution, as revealed by ^{11}B NMR and ^{199}Hg NMR. The stability of the supramolecular ion reported here therefore must arise from the cooperative and complementary interactions of the host and guest, noting that planar (1) and (3) both have C_{4h} symmetry and (5) has D_{4d} symmetry.

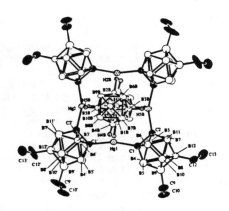

Fig. 7   The molecular structure of $(3) \cdot (5)_2{}^{4-}$.

ACKNOWLEDGEMENT   This work was supported by National
Science Foundation under the grant CHE-91-11437.

REFERENCES

1.   (a) Cram, D. J. *Science* 1983, 219 , 1177; (b)
     Vögtle, F.; Weber, E. 'Host-Guest Complex Chemistry/
     Macrocycles', Springer-Verlag: Berlin, 1985.
2.   (a) Katz, H. E. *Organometallics* 1987, 6, 1134; (b)
     Katz, H. E. *J. Am. Chem. Soc.* 1986, 108, 7640.
3.   (a) Wuest, J. D.; Zacharie, B. *Organometallics* 1985,
     4, 410; (b) Beauchamp, A. L.; Olivier, M. J.; Wuest,
     J. D.; Zacharie, B. J. Am. Chem. Soc. 1986, 108, 73.
4.   (a) Shur, V. B.; Tikhonova, I. A.; Yanovsky, A. I.;
     Struchkov, Y. T.; Petrovskii, P. V.; Panov.; S. Yu.;
     Furin, G. G.; Vol'pin, M. E. *J. Organomet. Chem.*,
     1991, 418, C29. (b) Shur, V. B.; Tikhonova, I. A.;
     Dolgushin, F. M.; Yanovsky, A. I.; Struchkov, Y. T.;
     Volkonsky, A. Yu.; Solodova, E. V.; Panov.; S.
     Yu.; Petrovskii, P. V.; Vol'pin, M. E. *J. Organomet.
     Chem.*, 1993, 443, C19.
5.   (a) Newcomb, M.; Horner, J. H.; Blanda, M. T.;
     Squattrito, P. J. *J. Am. Chem. Soc.* 1989, 111, 6294.
     (b) Jurkschat, K.; Kuivila, H. G.; Liu, S.; Zubieta,
     J. *Organometallic* 1989, 8, 2755;   (c) Jurkschat, K.;
     Rühlemann, A.; Tzschach, A. *J. Organomet. Chem.*, 1990,
     381, C53.
6.   (a) Jung, M. E.; Xia, H. *Tetrahedron Lett.*, 1988, 29,
     297; (b) Tamao, K.; Hayashi, T.; Ito, Y.; Shiro, M. *J.
     Am. Chem. Soc.*, 1990, 112, 2422.
7.   (a) Park, C. H.; Simmons, H. E. *J. Am. Chem. Soc.*,
     1968, 90, 2431. (b) Dietrich, B.; Lehn, J.-M.;
     Guilhem, J.; Pascard, C. *Tetrahedron Lett.*, 1989,
     4125. (f) Schmidtchen, F. P.; Gleich, A.; Schummer, A.
     Pure & Appl. Chem., 1989, 61, 1535.

8. Sessler, J. L.; Cyr, M.; Furuta, H.; Kral, V.; Mody, T.; Morishima, T.; Shionoya, M.; Weghorn, S. *Pure & Appl. Chem.*, 1993, <u>65</u>, 393 and references therein.

9. Goggin, P. L.; Goodfellow, R. J.; Hurst, N. W. *J. Chem. Soc. Dalton* 1978, 561.

10. (a) Yang, X.; Knobler, C. B.; Hawthorne, M. F. *Angew. Chem. Int. Ed. Engl.*, 1991, 30, 1507. (b) Yang, X; Knobler, C. B.; Hawthorne, M. F. *J. Am. Chem. Soc.*, 1992, <u>114</u>, 380. (c) Yang, X.; Johnson, S. E.; Kahn, S.; Hawthorne, M. F. *Angew. Chem. Int. Ed. Engl.*, 1992, <u>31</u>, 893. (d) Yang, X.; Zhang, Z.; Knobler, C. B.; Hawthorne, M. F. *J. Am. Chem. Soc.* 1993, <u>115</u>, 193. (e) Yang, X.; Knobler, C, B.; Hawthorne, M. F. *J. Am. Chem. Soc.*, 1993, <u>115</u>, 4904. (f) Zheng, Z.; Yang, X.; Knobler, C, B.; Hawthorne, M. F. *J. Am. Chem. Soc.*, 1993, <u>115</u>, 5320.

11. Grimes, R. N. 'Carboranes', Academic Press: New York; 1970; p.66.

12. Healy, M. D. S.; Rest, A. J. *Adv. Inorg. Chem. Radiochem.*, 1978, <u>21</u>, 1.

13. Sudmerier, J. L.; Perkins, T. G. *J. Am. Chem. Soc.*, 1977, <u>99</u>, 7732.

14. Sen, M. A.; Wilson, N. K.; Ellis, P. D.; Odom, J. D. *J. Magn. Resonance*, 1975, <u>19</u>, 323.

15. (a) Godfrey, P. D.; Heffernan, M. L.; Kerf, D. F. *Aust. J. Chem.*, 1964, <u>17</u>, 701. (b) Bach, R. D.; Vardhan, H. B. *J. Org. Chem.*, 1986, <u>51</u>, 1609.

16. Middaugh, R. L. 'Boron Hydride Chemistry' Muetterties, E. L., Ed.; Academic Press: New York, 1975; p 273.

Synthesis and Structural Characterizations of New Polyhedral Di-, Tri- and Tetra-carbon Carboranes

Kai Su, Patrick J. Carroll, and Larry G. Sneddon*

DEPARTMENT OF CHEMISTRY, UNIVERSITY OF PENNSYLVANIA, PHILADELPHIA, PA 19104, USA

We have previously shown that the carbon in a nitrile group is susceptible to nucleophilic attack by polyhedral borane anions and these reactions can result in either CN or monocarbon cage-insertion products in high yields.[1] These results suggested that borane or carborane anions might be sufficiently nucleophilic to attack positive sites in other polarized triple bonds, such as those found in alkynes containing strongly electron withdrawing substituents, to form either mono- or di-carbon insertion products. We now describe the use of this new synthetic strategy to produce important new classes of di-, tri- and tetra-carbon carbaboranes that have been inaccessible using conventional synthetic procedures.

Monocarbon Insertions: New Tricarbaboranes.[2]

The $arachno$-6,8-$C_2B_7H_{12}^-$ anion reacts with cyanoacetylene, methyl propiolate or 3-butyn-2-one in THF solution to yield, following protonation, a series of new tricarbaboranes of general formula $arachno$-6-(RCH_2)-5,6,7-$C_3B_7H_{12}$ in yields ranging from 51 to 82%.

$$arachno\text{-}C_2B_7H_{12}^- + HC{\equiv}CR \xrightarrow[\text{(2) H}^+]{\text{(1) RT}} arachno\text{-}6\text{-}(RCH_2)\text{-}5,6,7\text{-}C_3B_7H_{12} \ (\mathbf{1})$$

R = CN (**1**), C(O)OMe (**2**), or C(O)Me (**3**)

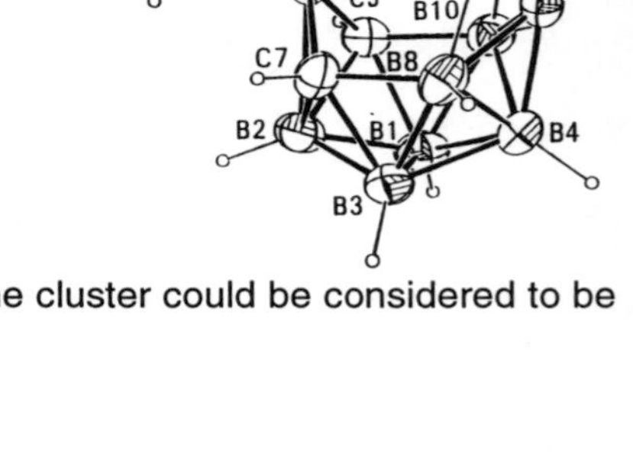

Single crystal X-ray studies have shown that these compounds, consistent with their 26 skeletal-electron counts, adopt 10-vertex arachno cage geometries based on an icosahedron missing two vertices, with the carbon atoms occupying adjacent positions on the open 6-membered face. Examination of the interatomic cage distances suggests a reduced bonding interaction between C6 and B2 and largely localized two-center single bonds between the C6 carbon and C5 and C7. Thus, the cluster could be considered to be composed of both "classical" electron-precise and "nonclassical" electron-deficient components. In the limit where C6 and B2 are nonbonding, then, instead of being considered part of the cluster framework, C6 might be viewed as a carbon-carbon bridging exopolyhedral substituent on the starting carborane framework, i.e. *arachno*-$\mu_{6,8}$-RCH-6,8-$C_2B_7H_{11}$.

1 was found to undergo cage degradation with aqueous ammonium hydroxide to give the new hypho tricarbaborane, *hypho*-1-(NCCH$_2$)-1,2,5-$C_3B_6H_{12}^-$ (**4**).

$$\textit{arachno}\text{-6-(NCCH}_2\text{)-5,6,7-C}_3\text{B}_7\text{H}_{12} \xrightarrow{\text{NH}_4\text{OH}}$$

$$\textit{hypho}\text{-1-(NCCH}_2\text{)-1,2,5-C}_3\text{B}_6\text{H}_{12}^- \quad \textbf{(2)}$$

A single crystal X-ray study has confirmed the structure shown in the ORTEP drawing. If the anion is considered a tricarbaborane cluster, then the compound would contain 26 electrons and fall into the 9-vertex 2n+8 hypho electronic class (n = no. of cage atoms) and should adopt a structure based on an icosahedron missing three vertices. Such a cage geometry has been observed for the isoelectronic

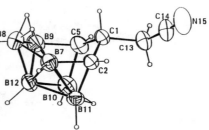

analogues, *hypho*-1-(CH$_2$)-2,5-$S_2B_6H_8$[3] and *hypho*-1,2,5-(η^6-C_6Me_6)RuClS$_2B_6H_9$.[4] Alternatively, the (NCCH$_2$)CH- group may be viewed as an exo-polyhedral substitutent on a C_2B_6-cluster, i.e. *hypho*-$\mu_{2,5}$-(NCCH$_2$)CH-2,5-$C_2B_6H_{11}^-$. The (NCCH$_2$)CH- fragment could then be considered to bond to the two cage-carbons by two conventional 2-center, 2-electron bonds. In this view, **4** would be considered a carbon-bridged derivative of the previously known[5] dicarbaborane anion *hypho*-7,8-$C_2B_6H_{13}^-$, in which two endo-hydrogens are replaced by the carbon-bridging (NCCH$_2$)CH- group.

Dicarbon Insertion Reactions: Syntheses of Di- and Tetra-Carbon Carboranes.

Dicarbon insertions were observed in the reactions of *nido*-5,6-$C_2B_8H_{11}^-$ with methyl propiolate, and $B_{10}H_{13}^-$ with cyanoacetylene which resulted in the production of the isoelectronic *arachno*-tetracarbaborane, *arachno*-8-(CH$_3$OC(O))-7,8,9,10-$C_4B_8H_{13}$ (**5**) and *arachno*-dicarbaborane, *arachno*-8-(NC)-7,8-$C_2B_{10}H_{14}^-$ (**6**), clusters, respectively.

$$\textit{nido}\text{-C}_2\text{B}_8\text{H}_{11}^- + \text{HC}\equiv\text{C(O)OMe} \xrightarrow[\text{(2) H}^+]{\text{(1) RT}}$$

$$\textit{arachno}\text{-8-(CH}_3\text{OC(O))-7,8,9,10-C}_4\text{B}_8\text{H}_{13} \quad \textbf{(3)}$$

$$\text{B}_{10}\text{H}_{13}^- + \text{HC}\equiv\text{CCN} \longrightarrow \textit{arachno}\text{-8-(NC)-7,8-C}_2\text{B}_{10}\text{H}_{14}^- \quad \textbf{(4)}$$

Consistent with their arachno 30 skeletal-electron counts, both compounds adopt similar structures based on a bicapped hexagonal antiprism missing two vertices.

In both compounds two-carbon insertion of the acetylenic carbons into the cage framework has occurred to produce the tetra- and di-

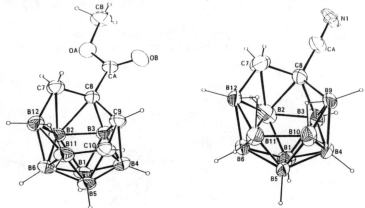

carbaboranes, respectively. The observed cage structures are consistent with that found for the only previously structurally characterized example of a 12-vertex arachno-cluster, $[\sigma-(\eta^5-C_5H_5)Co(\eta^4-C_5H_4)^+]-(CH_3)_4-arachno-C_4B_8H_8^-$.[6]

The results presented above have clearly demonstrated that terminal acetylenes containing strongly electron withdrawing groups are activated toward reaction with various polyhedral borane anions and that such reactions can ultimately lead to monocarbon or dicarbon insertions. Although the mechanisms of the insertion reactions have not yet been studied in detail, the results of previous studies[7] of the reactions of polarized acetylenes with organic nucleophiles suggest several key steps in the reaction sequences that are important in both the monocarbon and dicarbon insertion reactions. The triple bond in cyanoacetylene, methyl propiolate or 3-butyn-2-one, is strongly polarized by the cyano or carbonyl groups with the terminal carbon having a partial positive charge. As a result, these acetylenes undergo attack at the terminal carbons by organic nucleophiles, such as alcohols or mercaptans.[7a] Subsequent rearrangements in these organic systems can then lead to the incorporation of either one or two carbons into a ring system.

As shown in **Scheme 1**, a reaction sequence based on those previously suggested[7a] for the formation of cyclic organics may be proposed to account for the carbon-insertion reactions. The initial step in both the monocarbon and dicarbon insertion reactions would again involve the nucleophilic attack of the anions at the positive terminal-carbon of the alkynes to generate anionic sigma-vinyl intermediates with the negative charge localized on the internal vinyl carbon. Once formed, this intermediate can then react, depending upon the nature of the polyhedral anions, in two different ways ultimately leading to either monocarbon or dicarbon insertions. Thus, as shown by path **A**, intramolecular transfer of one acidic proton to the vinylic carbon anion, followed by a cyclization reaction could occur to incorporate the terminal carbon into the cluster. In this case, the internal-carbon of the olefin would be converted to an exopolyhedral methylene. Alternatively, as shown in path **B**, the formation of a new bond between the internal carbon and an electrophilic site in the cluster could occur resulting in cyclization at the internal-carbon and dicarbon insertion. In this case, the terminal-carbon would be incorporated as a cage(bridging)-CH_2 fragment.

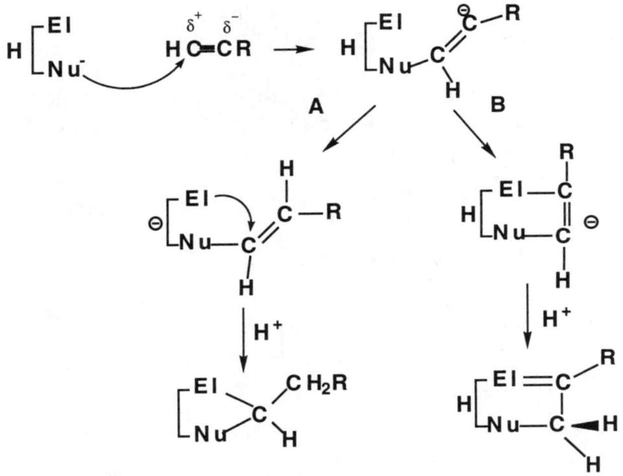

Scheme 1

Monocarbon insertions were observed in the reactions of *arachno*-6,8-$C_2B_7H_{12}^-$ with cyanoacetylene, methyl propiolate or 3-butyn-2-one. A reaction sequence consistent with that given in path **A** is illustrated below:

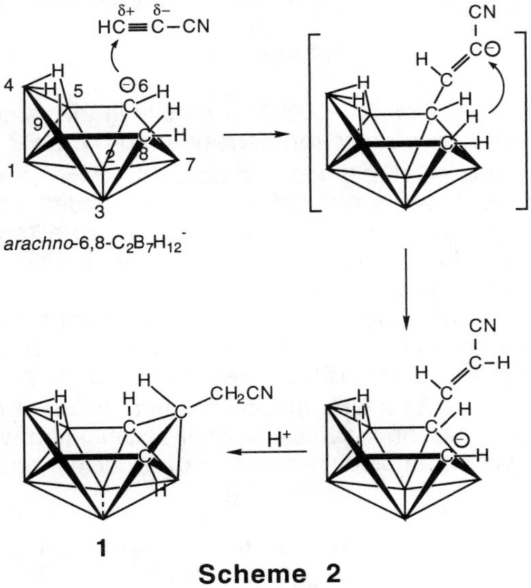

arachno-6,8-$C_2B_7H_{12}^-$

Scheme 2

Previous studies[8] have shown that in the *arachno*-6,8-$C_2B_7H_{12}^-$ anion, the negative charge is localized on a cage carbon. Nucleophilic attack of the *arachno*-6,8-$C_2B_7H_{12}^-$ anion at the cyanoacetylene terminal carbon, accompanied by hydrogenation of the acetylenic bond, should then produce a carbon-substituted sigma-vinyl anionic product. Such an anion was, in fact, isolated as the sole product of the reaction before the acidification step and the spectral data are consistent with the structure proposed in the figure. Subsequent acidification of this intermediate then results in monocarbon insertion to produce the tricarbaborane **1** containing the exopolyhedral $NCCH_2$-substituent at C6.

A reaction sequence leading to dicarbon insertion which is based on that of path **B** is illustrated in **Scheme 3** for the reaction of 5,6-$C_2B_8H_{11}^-$ with methyl propiolate.

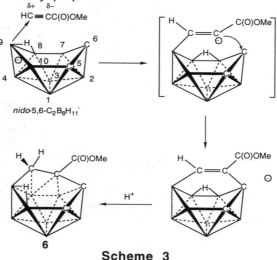

Scheme 3

Deprotonation of *nido*-5,6-$C_2B_8H_{12}$ results in the removal of one bridge-proton and enhanced electron density at both the B8-B9 and B9-B10 edges. Thus, the initial reaction of *nido*-5,6-$C_2B_8H_{11}^-$ with methyl propiolate should result in the production of a B9-bonded sigma-vinyl anionic intermediate which could then rearrange to an alkenyl-bridged compound. Based on the ^{11}B NMR spectrum of the reaction mixture, such a compound appeared to form, but it was too unstable to be isolated and characterized. In accord with path **B**, acidification of the reaction mixture then resulted in the formation of the dicarbon insertion product, *arachno*-8-($CH_3OC(O)$)-7,8,9,10-$C_4B_8H_{13}$ (**6**), in which the terminal carbon is incorporated in the cage as a CH_2 group. In the reaction of *nido*-$B_{10}H_{13}^-$ with cyanoacetylene, carbon insertion without acidification was observed, presumably because intramolecular transfer of an acidic bridging-hydrogen occurs.

Whether mono- or dicarbon insertion is observed in these reactions (i.e. path **A** or **B**) appears to depend largely on the structure and the reactivity of the polyhedral anion. The fact that the products derived from the reactions of *arachno*-6,8-$C_2B_7H_{12}^-$ are tricarbaboranes may arise as a result of both the acidity of the *endo*-C8H proton and the proximity of the C8 carbon to the sigma-vinyl group, thereby making addition to the olefin possible. On the other hand, in 5,6-$C_2B_8H_{11}^-$ and $B_{10}H_{13}^-$, an additional electrophilic site (i.e. C6, B9) exists on the opposite side of these molecules, which facilitates the two-carbon insertions.

In conclusion, the results described in this paper have demonstrated an important new synthetic pathway by which mono- or dicarbon insertions into polyhedral boron clusters may be attained. Furthermore, the use of this approach has been shown to allow the syntheses of a number of new types of carborane clusters, most notably the *arachno*-tricarbaboranes and the *arachno*-12-vertex, di- and tetracarboranes, not accessible by conventional carborane-forming reactions. This synthetic method may now result in an even wider array of new carboranes as it is applied to reactions with other polyhedral borane anions. We are presently exploring these possibilities along with the chemistry of the many unique carborane clusters that have already resulted.

Acknowledgements. We thank the National Science Foundation for the support of this research.

References.

1   S. O. Kang, G. T. Furst and L. G. Sneddon, Inorg. Chem. 1989, 28, 2339-2347.
2   K. Su, B. A. Barnum, P. J. Carroll and L. G. Sneddon, J. Am. Chem. Soc. 1992, 114, 2730-2731.
3   (a) S. O. Kang and L. G. Sneddon, J. Am. Chem. Soc. 1989, 111, 3281-3289. (b) S. O. Kang and L. G. Sneddon, "Electron Deficient Boron and Carbon Clusters"; G. A. Olah, K. Wade and R. E. Williams, eds., Wiley, New York, 1991, pp 195-213.
4   K. Mazighi, P. J. Carroll and L. G. Sneddon, Inorg. Chem. 1992, 31, 3197-3204.
5   T. Jelinek, J. Plesek, S. Heřmánek and B. Stíbr, Main Group Met. Chem. 1987, 10, 387-388.
6   R. N. Grimes, J. R. Pipal and E. Sinn, J. Am. Chem. Soc. 1979, 101, 4172-4180.
7   (a) F. Bohlmann and E. Bresinsky, Chem. Ber. 1964, 97, 2109-2117. (b) J. I. Dickstein and S. I. Miller, "The Chemistry of the Carbon-Carbon Triple Bond" S. Patai, ed., Wiley, New York, 1978, Chapter 19, pp 813-955.
8   (a) F. N. Tebbe, P. M. Garrett and M. F. Hawthorne, J. Am. Chem. Soc. 1966, 88, 607-608. (b) F. N. Tebbe, P. M. Garrett and M. F. Hawthorne, J. Am. Chem. Soc. 1968, 90, 869-879.

Carboranes Derived from *nido*-4,5-C$_2$B$_6$H$_{10}$: Syntheses of *closo*-2,3-C$_2$B$_5$H$_7$ and *arachno*-5,6-C$_2$B$_7$H$_{12}^-$

Joseph W. Bausch, Patrick J. Carroll, and Larry G. Sneddon*

DEPARTMENT OF CHEMISTRY, UNIVERSITY OF PENNSYLVANIA,
PHILADELPHIA, PA 19104, USA

We recently reported[1] the high yield synthesis of the carborane anion *nido*-4,5-C$_2$B$_6$H$_9^-$ from the reaction of acetonitrile with *arachno*-4,5-C$_2$B$_7$H$_{12}^-$:

$$arachno\text{-}4,5\text{-}C_2B_7H_{12}^- \ + \ CH_3CN \ \xrightarrow{\Delta}$$
$$nido\text{-}4,5\text{-}C_2B_6H_9^- \ + \ 1/3\ Et_3B_3N_3H_3 \quad (1)$$

This anion is an example of an 8-vertex, 20 skeletal-electron, nido cluster system. However, a single-crystal X-ray study of Bu$_4$N$^+$*nido*-4,5-C$_2$B$_6$H$_9^-$ (1) showed an arachno-type structure. This was the first structural confirmation of this geometry for the parent carborane system. Subsequent protonation using HCl yielded the neutral carborane *nido*-4,5-C$_2$B$_6$H$_{10}$ (2) that was structurally characterized using NMR spectroscopy in conjunction with ab initio and IGLO chemical shift calculations (Figure 1). These studies showed an arachno type structure, with the two bridging-hydrogens being fluxional around the four borons of the open face.

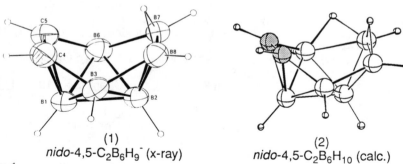

(1)
nido-4,5-C$_2$B$_6$H$_9^-$ (x-ray)

(2)
nido-4,5-C$_2$B$_6$H$_{10}$ (calc.)

Figure 1

We have now found that this *nido*-4,5-C$_2$B$_6$H$_{10}$ carborane and its *nido*-4,5-C$_2$B$_6$H$_9^-$ anion can serve as useful precursors to both smaller and larger cage systems. Based on the proposed geometry (2), it appeared likely *nido*-4,5-C$_2$B$_6$H$_{10}$ would be susceptible to loss of BH$_3$, and this was accomplished by either thermal or chemical degradation. Thermal degradation was achieved by slowly passing the *nido*-4,5-$C_2B_6H_{10}$ through a hot tube heated at 350 °C under vacuum:

$$\textit{nido-}4,5\text{-}C_2B_6H_{10} \xrightarrow[\text{flow system}]{350\ ^\circ C} \textit{closo-}2,3\text{-}C_2B_5H_7 + 1/2\ B_2H_6 \quad (2)$$

The resulting product *closo*-2,3-$C_2B_5H_7$ (3) was isolated in a -78 °C trap in ~60% yield. Smaller amounts of other closo-carboranes (eg., 2,4-$C_2B_5H_7$ and 1,7-$C_2B_6H_8$) were also produced. The carborane was characterized by ^{11}B, ^{1}H, and ^{13}C NMR spectroscopy in conjunction with ab initio and IGLO chemical shift calculations (Figure 2). The calculated pentagonal bipyramid structure (3) is consistent with its 16 skeletal-electron count. This is the first isolation and characterization of this parent closo-carborane.

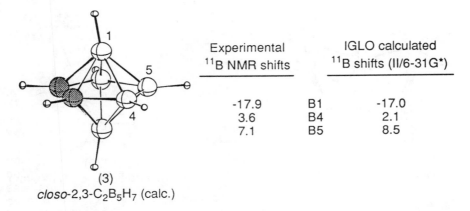

	Experimental ^{11}B NMR shifts		IGLO calculated [11]B shifts (II/6-31G*)
	-17.9	B1	-17.0
	3.6	B4	2.1
	7.1	B5	8.5

(3)
closo-2,3-$C_2B_5H_7$ (calc.)

Figure 2

Chemical degradation of *nido*-4,5-$C_2B_6H_{10}$ was carried out by reaction with an alkyne:

nido-4,5-$C_2B_6H_{10}$ + alkyne ⟶
(2)

(3)

(4)
proposed structure

closo-2,3-$C_2B_5H_7$ + R$_3$B ⟵ Δ
(3)

The *nido*-4,5-$C_2B_6H_{10}$ undergoes hydroboration to form a stable exopolyhedral (2-alkenyl)-B-substituted $C_2B_6H_{10}$ (4). The ^{11}B NMR data combined by the IGLO calculated chemical shifts indicate a static structure, in contrast to the fluxional *nido*-4,5-$C_2B_6H_{10}$ (2). Upon heating, (4) degrades to give *closo*-2,3-$C_2B_5H_7$ in ~35% yield. An equivalent amount of 1-(2-alkenyl)-*closo*-2,3-$C_2B_5H_6$ is also formed.

The *nido*-4,5-C$_2$B$_6$H$_9^-$ anion, like the neutral *nido*-4,5-C$_2$B$_6$H$_{10}$, is also prone to loss of BH$_3$, as exemplified by its reactions with transition metals. For example, the reaction of *nido*-4,5-C$_2$B$_6$H$_9^-$ with CpCo(CO)I$_2$ or (C$_6$Me$_6$)$_2$RuCl$_2$ gave (C$_6$Me$_6$)RuC$_2$B$_5$H$_7$ and the previously reported CpCoC$_2$B$_5$H$_7$.[2]

The addition of BH$_3$·THF to a THF solution of K$^+$*nido*-4,5-C$_2$B$_6$H$_9^-$ gave the new carborane anion K$^+$*arachno*-5,6-C$_2$B$_7$H$_{12}^-$ in quantitative yield (Figure 3). The structure has been characterized by ^{11}B, ^{1}H, and ^{13}C NMR spectroscopy (along with an ab initio/IGLO theoretical study) as well as a single crystal X-ray diffraction study. This anion (5) is the third isomer of this carborane system. The two previously known *arachno*-4,5-C$_2$B$_7$H$_{12}^-$ and *arachno*-6,8-C$_2$B$_7$H$_{12}^-$ have structures based upon *i*-B$_9$H$_{15}$, but this new isomer *arachno*-5,6-C$_2$B$_7$H$_{12}^-$ is the first heteroborane analog of *n*-B$_9$H$_{15}$.[3]

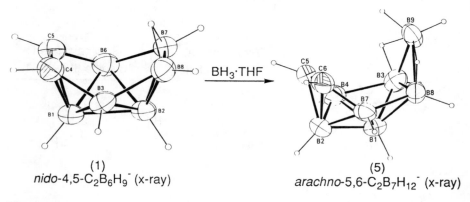

(1)
nido-4,5-C$_2$B$_6$H$_9^-$ (x-ray)

(5)
arachno-5,6-C$_2$B$_7$H$_{12}^-$ (x-ray)

Figure 3

The reactivity of *nido*-4,5-C$_2$B$_6$H$_{10}$ with hydride reagents has also been explored. Sodium or potassium hydride reacts with *nido*-4,5-C$_2$B$_6$H$_{10}$ to give *nido*-4,5-C$_2$B$_6$H$_9^-$. However, when "super hydride" is employed (LiBEt$_3$H), the main product is the previously reported[4] *arachno*-C$_2$B$_6$H$_{11}^-$ (6) arising from hydride addition to the cage.

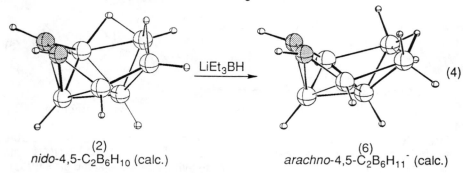

(2)
nido-4,5-C$_2$B$_6$H$_{10}$ (calc.)

(6)
arachno-4,5-C$_2$B$_6$H$_{11}^-$ (calc.)

The chemistry of *closo*-2,3-C$_2$B$_5$H$_7$ is also being explored (Scheme 1). For example, the reaction of *closo*-2,3-C$_2$B$_5$H$_7$ with LiEt$_3$BH gave *nido*-

$3,4\text{-}C_2B_5H_8^-$ (7) as evidenced by the ^{11}B, ^{1}H, and ^{13}C NMR data in conjunction with ab initio and IGLO chemical shift calculations. This nido electron-count carborane is isoelectronic and isostructural with the previously structurally characterized $nido\text{-}3,4\text{-}Et_2C_2B_5H_5^-$ and $exo\text{-}6\text{-}(Me_3P^+\text{-}CH_2)\text{-}3,4\text{-}Et_2\text{-}nido\text{-}3,4\text{-}C_2B_5H_5^-$.[5,6] The carborane $nido\text{-}2,3\text{-}C_2B_4H_8$ (8) can be synthesized from the reaction of $closo\text{-}2,3\text{-}C_2B_5H_7$ with TMEDA followed by water.[7] The TMEDA cage-opens the $closo\text{-}2,3\text{-}C_2B_5H_7$ to give a nido 7-vertex structure containing a bound TMEDA ligand. Presumably, the water serves as a nucleophile to remove the boron with the attached TMEDA ligand and to protonate the resulting C_2B_4 framework to give $nido\text{-}2,3\text{-}C_2B_4H_8$ (8).

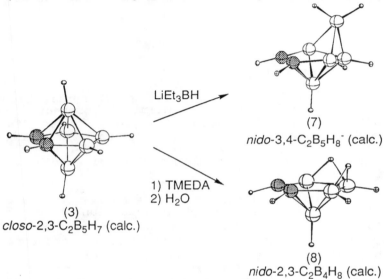

LiEt$_3$BH

(7)
$nido\text{-}3,4\text{-}C_2B_5H_8^-$ (calc.)

1) TMEDA
2) H$_2$O

(3)
$closo\text{-}2,3\text{-}C_2B_5H_7$ (calc.)

(8)
$nido\text{-}2,3\text{-}C_2B_4H_8$ (calc.)

Scheme 1

We thank the National Science Foundation for support of this research.

REFERENCES

1. S.O. Kang, J.W. Bausch, P.J. Carroll, and L.G. Sneddon, J. Am. Chem. Soc., 1992, 114, 6248.
2. G.J. Zimmerman, L.W. Hall, and L.G. Sneddon, Inorg. Chem., 1980, 19, 3642.
3. R.W. Dickerson, P.J. Wheatley, P.A. Howell, and W.N. Lipscomb, J. Chem. Phys., 1957, 27, 200.
4. T. Jelínek, B. Stíbr, S.Hermánek, and J. Plesek, J. Chem. Soc., Chem. Comm., 1989, 804.
5. J.S. Beck, W. Quintana, and L.G. Sneddon, Organometallics, 1988, 7, 1015.
6. K. Su, P.J. Fazen, P.J. Carroll, and L.G. Sneddon, Organometallics, 1992, 11, 2715.
7. R.N. Grimes, Pure Appl. Chem., 1991, 63, 369.

O-Carboranes from Ynol Ethers

Christophe Morin* and Mintradeo Ramburrun

LABORATOIRE D'ÉTUDES DYNAMIQUES ET STRUCTURALES DE LA
SÉLECTIVITÉ, DÉPARTEMENT CHIMIE RECHERCHES, BÂTIMENT 52,
UNIVERSITÉ DE GRENOBLE, BP 53X, 38041 GRENOBLE, FRANCE

1 INTRODUCTION

There has been interest in biological applications of
boron clusters [1] since the clinical boroneutrotherapy
results obtained by Hatanaka's group [2] with disodium
undecahydromercaptododecaborate. Dicarba-*closo*-dodecarbo-
ranes, in particular ortho-carboranes [3], have been found to
be appropriate boron-rich moieties for incorporation into
various organic molecules.

A number of C-linked carborane derivatives have been
prepared, essentially through the addition of decaborane
to substituted acetylenes [4-6]. Although B-hydroxy [7,8] and C-
hydroxy [9-12] carboranes have been obtained, apparently no
carboranyl ether has been prepared efficiently [13], which
could reflect a lack of adequate preparative methods. Since
a number of C-carboranyl deivatives of carbohydrates have
already been described [14-18], the preparation of a
carboranyl ether in this field seemed appropriate to
demonstrate the feasibility of the approach.

2 RESULTS AND DISCUSSION

The addition of decaborane to an ynol ether was
considered a potentially appealing approach to such
carboranyl ethers (scheme 1) .

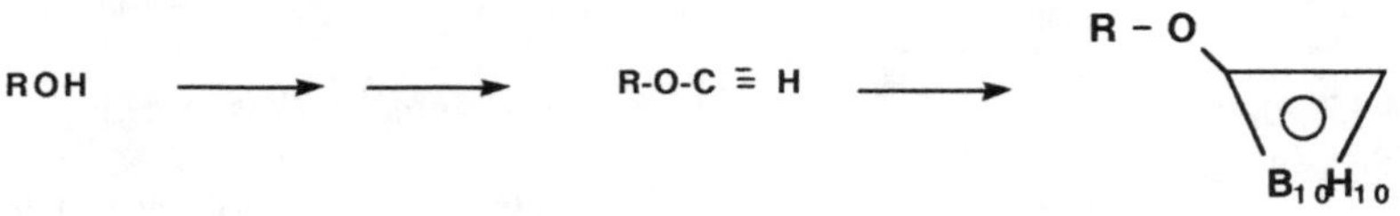

ROH $\longrightarrow$ $\longrightarrow$ R-O-C $\equiv$ H $\longrightarrow$ R-O / O / $B_{10}H_{10}$

Scheme 1 : the ynol ether approach to carboranyl ethers

Among the available syntheses of ynol ethers, the procedure of Greene et al.[19] seemed particularly attractive since it allows a one-pot conversion from the alcohol. Thus, diacetone-D-glucose (1), was converted to (2) as follows :

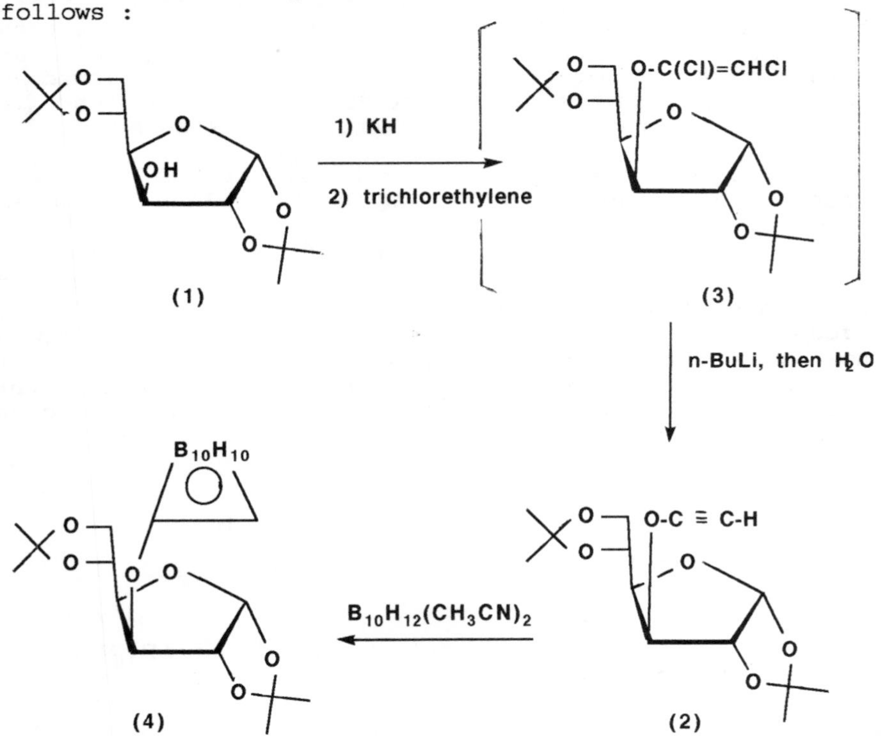

(1)　　　　　　　　　　　　　　　　　(3)

n-BuLi, then H_2O

(4)　　　　　　　　　　　　　　　　　(2)

After reaction of the preformed alkoxide (KH / THF - rt, 1 h) with trichlorethylene at -78 °C (then rt over 2 h) to produce intermediate (3) which needed not be isolated, base-induced elimination of chlorine (n-BuLi at -78 °C then rt over 2 h) smoothly afforded (2) (v acetylenic = 2140 cm^{-1}) after chromatography on buffered silica gel.

The addition of the preformed acetonitrile complex [20] of decaborane [21] under standard conditions then gave the desired carboranyl ether. The structure of (4) is supported, in particular, by the resonance of the carborane carbon directly bound to the oxygen which is considerably downfield (δ = 115.6 ppm) in comparison with C-substituted carboranes [22]. Furthermore in its ^{1}H-decoupled ^{11}B nmr spectrum (4) displays resonances at -4.0, -10.6 (shoulder), -11.8, -13.6 (shoulder) and -14.5 ppm; in its HRMS, the expected boron cluster is evidenced.

The yield of this reaction was moderate (25-30 %), reproducibly reached after 12 to 14 h reflux time. In

that (4)   appeared to be stable, and as every hydroxyl
function borne in (3) is protected (carboranes are known
not to be obtainable in presence of free hydroxyl groups),
the   strong polarisation of the ynol ether moiety itself
could be having a deleterious effect in the carborane
formation. Deprotection of both acetals of (4) was then
achieved under standard acidic conditions to yield 3-0-
ortho-carboranyl-D-glucose. This molecule can be viewed as
the simplest possible carboranyl derivative of glucose, in
that the boron cage is linked <u>directly</u> to the sugar.

In order to assay the flexibility of the synthetic
scheme, the same   process was next applied to other
carbohydrate derivatives.

Commercially available (5) and (6), available in
three steps from methyl α-D-glucopyranoside [23] (and in
which benzyl ethers are protecting the other hydroxyl
groups), were   selected since the reaction sequence would
now involve primary hydroxyl groups. The transformation of
(5) and (6) to carborane ethers (7) and (8) *via* the
corresponding ynol ethers (which were isolated) could be
similarly accomplished.

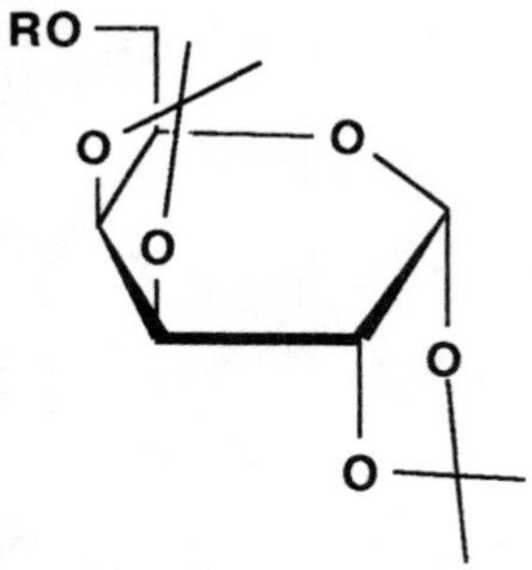

(5) : R = H

(7) : R =

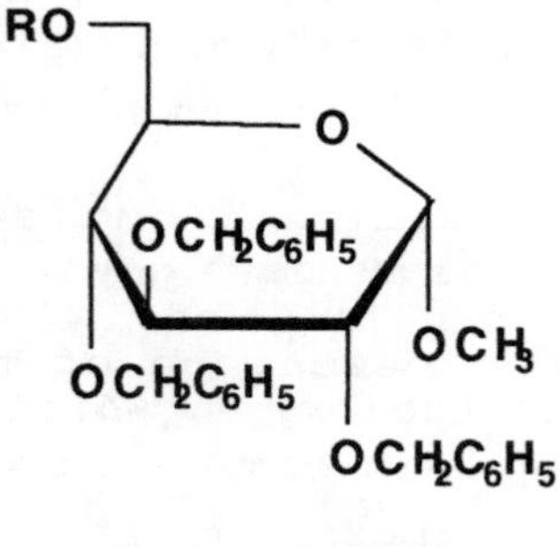

$B_{10}H_{10}$

(6) : R = H

(8) : R =

$B_{10}H_{10}$

It is therefore anticipated that the above process
will allow access to various ether derivatives of
carboranes, thus expanding the interest of the organic and
bioorganic chemistry of this class of boron clusters.

<u>Acknowledgements</u> : Dr. J. Ulrich, Institut de Biologie Structurale, Grenoble is gratefully thanked for HRMS data.

<u>References and note</u>.

1. M.F.Hawthorne, <u>Pure Appl. Chem.</u>, 1991, <u>63</u>, 327.
2. H.Hatanaka, W.H.Sweet, K.Sano and F.Ellis, <u>Pure Appl. Chem</u>, 1991, <u>63</u>, 373.
3. V.I.Bregadze, <u>Chem. Rev.</u>, 1992, <u>92</u>, 209.
4. T.L.Heying, J.W.Ager Jr., S.L.Clark, D.J.Mangold, H.L.Goldstein, M.Hillman, R.J.Polak, J.W.Szymanski, <u>Inorg. Chem</u>., 1963, <u>2</u>, 1089.
5. M.M.Fein, J.Bobinski, N.Mayes, N.Schwartz, M.S.Cohen, <u>Inorg. Chem</u>., 1963, <u>2</u>, 1111.
6. Z.I.Zakharkin, V.I.Stanko, V.A.Brattsev, Y.A. Chapovskii, O.Y.Okhlobystin, <u>Izv.Akad. Nauk SSSR Ser</u>. <u>Khim.</u> (Engl.Transl.) 1963, 2074.
7. V.I.Stanko, V.A.Brattsev, N.N.Ovsyannikov,T.P. Klimova, <u>Zh. Obsch. Khim.</u> (Engl.transl) 1974,<u>44</u>,2441.
8. V.A.Brattsev, S.P. Knyazev, G.N.Danilova, T.N Vostrikova, V.I.Stanko, <u>Zh. Obsch.Khim</u>.(Engl. Transl.) 1976, <u>46</u>, 2752.
9. L.Zakharkin, <u>Zh. Obsch. Khim</u>.(Engl.transl), 1969, <u>39</u>, 1856.
10. L.I.Zakharkin,G.G.Zhigareva, <u>Zh.Obsch.Khim</u> (Engl. transl),1970, <u>40</u>, 2318.
11. L.I.Zakharkin,G.G.Zhigareva, <u>Izv. Akad. Nauk SSSR,</u> (Engl.transl), 1970, 2153.
12. V.I.Stanko, Y.V..Gol'tyarin <u>Zh. Obsch. Khim</u>.,(Engl. transl) 1971, <u>41</u>, 2053.
13. V.Gregor, S.Hermanek, J.Plesek,<u>Coll. Czech Chem.</u> <u>Commun</u>., 1968, <u>33</u>, 980.
14. J.L.Maurer, A.J.Serino, M.F.Hawthorne, <u>Organometallics</u> 1988, <u>7</u>, 2519.
15. J.L.Maurer, F.Berchier, A.J.Serino, C.B.Knobler, M.F.Hawthorne, <u>J.Org. Chem</u>., 1990, <u>55</u>, 838.
16. A.K.Anisuzzaman, F.Alam, A.H.Soloway, <u>Polyhedron</u>, 1990, <u>9</u>, 891.
17. W.Tjarks, A.K.M.Anisuzzaman, L.Liu, A.H.Soloway, R.F.Barth, D. J.Perkins, D.M.Adams, <u>J. Med. Chem.</u>, 1992, <u>35</u>, 1628.
18. W.Tjarks, A.K.Anisuzzaman, A.H.Soloway, <u>Nucleosides</u> <u>Nucleotides</u>, 1992, <u>11</u>, 1765.
19. A. Moyano, F. Charbonnier, A.E. Greene, <u>J. Org.</u> <u>Chem.</u>, 1987, <u>52</u>, 2919.
20. R.Schaeffer, <u>J. Amer. Chem. Soc.</u>, 1957, <u>79</u>, 1006.
21. This compound is highly toxic, flammable and therefore should be handled with appropriate care.
22. O.A.Subbotin, T.V.Klimova, V.I.Stanko, Y.A.Ustynyuk, <u>Zh. Obsch. Khim.</u> (Engl.transl.) 1979,<u>49</u>,363.
23. F.Ramirez, S.B.Mandal, J.F.Marecek,<u>J. Org.Chem</u>.,1983, <u>48</u>, 2008.

Carborane Derivatives and Polymer Synthesis

W. Clegg[1], H. M. Colquhoun[2], R. Coult[3], M. A. Fox[3], W. R. Gill[3], P. L. Herbertson[3], J. A. H. MacBride[3], and K. Wade[3]

[1]CHEMISTRY DEPARTMENT, NEWCASTLE UNIVERSITY, NEWCASTLE NE1 7RU, UK
[2]NORTH WEST WATER PLC., THE HEATH, RUNCORN WA7 4QE, UK
[3]CHEMISTRY DEPARTMENT, DURHAM UNIVERSITY, DURHAM DH1 3LE UK

INTRODUCTION

The icosahedral carboranes 1,2-, 1,7- and 1,12-$C_2B_{10}H_{12}$ are chemical building blocks of high boron content and remarkable thermal and chemical stability, resistant to attack by acids and oxidising agents and generally inactive towards biological systems. Despite their cost, such properties make them suitable for various specialised applications. These include the incorporation of large concentrations of boron atoms in tumour-seeking drugs for boron neutron capture therapy (BNCT),[1] and the synthesis of polymers for high temperature[2] or neutron shielding[3] purposes or for firing to form ceramics related to boron carbide.[4] Their exceptional hydrophobic character, and the unusual solubility characteristics of their ionic derivatives,[5] together with their ability to form <u>nido</u> anionic species with remarkable ligand properties by treatment with nucleophiles, has led to their use in metal complexing agents for solvent extraction, particularly of fissionable metals where neutron capture is desirable,[6] in radiochemical drugs[7] and in new catalysts.[8] The non-linear optical (NLO) properties of selected derivatives are also attracting attention.[9]

Our own recent interest in such systems has focussed on the synthesis of new types of thermally stable or electronically interesting polymers incorporating icosahedral carboranes, and of ceramic materials related to boron carbide accessible through precursor polymers.[2,10] Particular attention has been paid to polymers of the poly-ether-ketone (PEK) or poly-ether-ether-ketone (PEEK) type in which <u>ortho</u>, <u>meta</u> or <u>para</u> carborane units and phenylene rings are linked through such functional groups. We are also exploring related oxygen-free systems, including macrocycles, in which carborane icosahedra are linked through benzene or other aromatic rings, and are investigating the capacity of carborane icosahedra to interact electronically with sources of π-bonding electrons attached to their carbon atoms. Our findings are summarized below.

PEEK- AND PEK-TYPE CARBORANE POLYMERS

Studies on model systems ROR' or RCOR' containing carboranyl groups R attached directly to ether or ketone units showed them to be susceptible to hydrolytic cleavage of their C-O or C-CO bonds.[2] We have therefore, in our polymer work, used phenylene spacers to separate the carborane icosahedra from the ether or ketone units. Illustrative routes to representative polymers are shown in Figure 1. For ortho-carborane systems $1,2-C_2R_2B_{10}H_{10}$, it was possible to attach suitable substituents to an alkyne before forming the carborane polyhedron.[10] For meta- and para-systems $1,7-$ and $1,12-C_2R_2B_{10}H_{10}$, it proved to be more convenient to attach suitable substituents R to the icosahedron by exploiting the reactivity of copper derivatives $Cu_2C_2B_{10}H_{10}$ towards aryl iodides in the presence of pyridine.[11] Polymers were made by acylation reactions between appropriate carboxylic acids and phenoxyphenyl derivatives in triflic acid as both solvent and catalyst, from which the soluble product polymers were recovered by pouring into water.

Figure 1   Routes to PEEK- and PEK-type carborane polymers

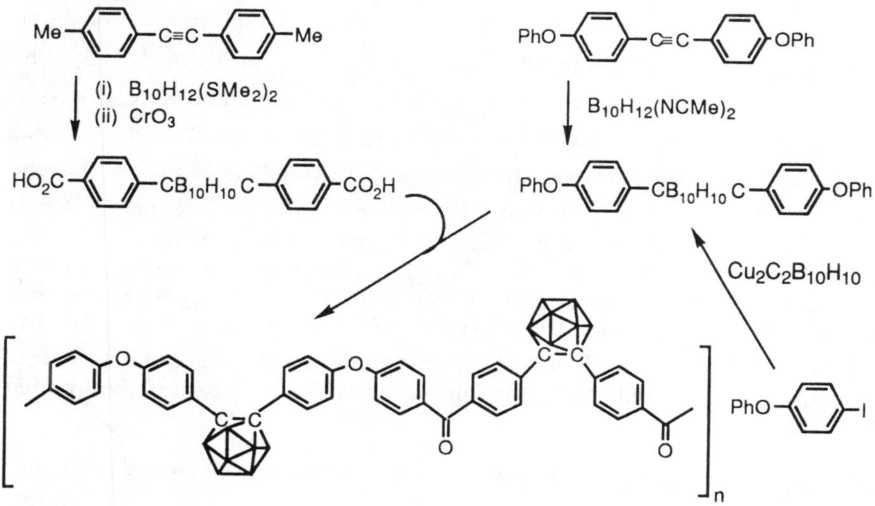

Products could be cast as films from CH_2Cl_2 or $CHCl_3$ solution, and were generally amorphous, with Tg in the region 180-220 °C. Thermogravimetric analysis showed that evolution of volatiles began at about 350-400 °C without melting, the total weight loss even at 1000 °C being only some 10-15%, suggesting that inert atmosphere pyrolysis might lead to a ceramic residue of boron carbide dispersed in a graphite matrix.

EXO PI-BONDING IN DERIVATIVES OF ORTHO-CARBORANE

Following our discovery[12] that deprotonation of the C-hydroxy ortho-carborane $1\text{-Ph-}2\text{-OH-}1,2\text{-}C_2B_{10}H_{10}$ gave an anion RO^- in which the negative charge was delocalised into the carborane cage, to which the oxygen atom was attached by a <u>double</u> bond (length 1.25Å), we have prepared and characterised several other systems that offered scope for related <u>exo</u> π-bonding (Figure 2).

Figure 2   Ortho carborane systems showing exo π-bonding

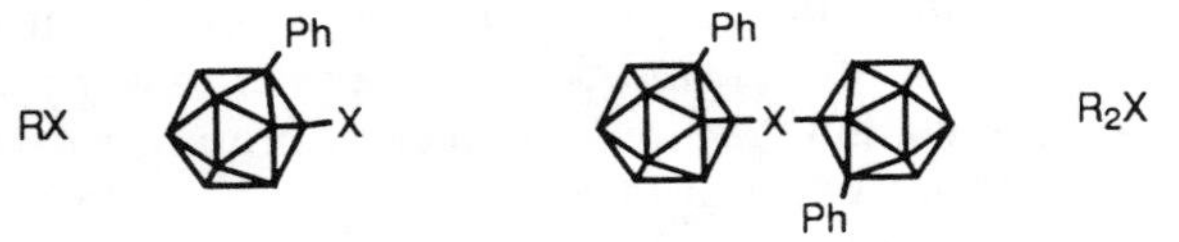

Compound	RO^-	RS^-	R_2S	R_2NH	R_2N^-
d(C---C)/Å	2.00	1.86	1.76	1.80	1.99
d(C⁝X)/Å	1.25	1.73	1.79	1.40	1.35

They include the sulphur analogue of the original anion,[13] RS^-, the bis(carboranyl)sulphide R_2S and amine R_2NH, and the anion R_2N^- (R=1-Ph-$1,2\text{-}C_2B_{10}H_{10}$).[14] All have relatively short <u>exo</u> C---X bonds and long C1---C2 bonds, and the CXC' planes in the systems R_2X (X=S, NH, N⁻) show that the <u>exo</u> π-bonding uses the p AO on C2 that lies in the C1-C2-X plane. Apart from revealing a hitherto unprecedented capacity of carbon atoms to form a multiple bond to one atom whilst bonding simultaneously to four or five other atoms (when the skeletal C---C distance increases to <u>ca</u> 2.0Å, this link hardly counts as a bond) these systems effectively show that carborane polyhedra attached directly to sources of π-bonding electrons can in principle participate in extended conjugative interactions that may lead to exploitable bulk properties.

MACROCYCLES INCORPORATING CARBORANE ICOSAHEDRA

During an exploration of the feasibility of preparing polymers in which carborane polyhedra were linked through arylene rings, we have also developed synthetic routes to macrocyclic systems in which <u>meta</u>-carborane units were linked by 1,3-disubstituted benzene rings or 2,6-disubstituted pyridine rings (Figure 3).[15] Such systems promise to have a rich chemistry themselves.

Figure 3   Carboranes linked through benzene rings

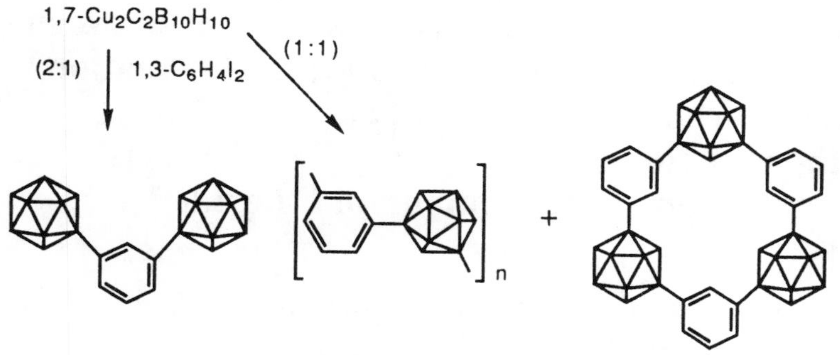

$1,7\text{-}Cu_2C_2B_{10}H_{10}$

(2:1) | $1,3\text{-}C_6H_4I_2$ (1:1)

Acknowledgement We thank SERC, ICI plc and North West Water plc for financial support.

REFERENCES

1.   R.F. Barth, A.H. Soloway and R.G. Fairchild, <u>Cancer Res.</u>, 1990, <u>50</u>, 1061; M.F. Hawthorne, <u>Pure Appl. Chem.</u>, 1991, <u>63</u>, 327; K. Shelly, D.A. Feakes, M.F. Hawthorne, P.G. Schmidt, T.A. Krisch and W.F. Bauer, <u>Proc. Natl. Acad. Sci. USA</u>, 1992, <u>89</u>, 9039.
2.   D.A. Brown, H.M. Colquhoun, J.A. Daniels, J.A.H. MacBride, I.R. Stephenson and K. Wade, <u>J. Mater. Chem.</u>, 1992, <u>2</u>, 793 and references cited therein.
3.   J. Plesek, <u>Chem. Rev.</u>, 1992, <u>92</u>, 269.
4.   L.G. Sneddon, M.G. Mirabelli, A.T. Lynch, P.J. Fazen, K. Su and J.S. Beck, <u>Pure Appl. Chem.</u>, 1991, <u>63</u>, 407.
5.   V. Skarda, J. Rais and M. Kyrs, <u>J. Inorg. Nucl. Chem.</u>, 1979, <u>41</u>, 1443.
6.   E. Makrlik and P. Vanura, <u>Talanta</u>, 1985, <u>32</u>, 423; M. Kyrs, J. Plesek, J. Rais and E. Makrlik, <u>Czech Patent</u> 211, 942, 1982; <u>Chem. Abstr.</u>, 1985, <u>98</u>, 115595p.
7.   R.J. Paxton, B.G. Beatty, M.F. Hawthorne, A. Varadarajan, L.E. Williams, F.L. Curtis, C.B. Knobler. J.D. Beatty and J.E. Shively, <u>Proc. Natl. Acad. Sci. USA</u>, 1991, <u>88</u>, 3387.
8.   M.F. Hawthorne in J.F. Liebman, A. Greenberg and R.E. Williams (Eds.) "Advances in Boron and the Boranes," VCH Publishers, New York, 1988, Chapter 10, p. 225.
9.   D.M. Murphy, D.M.P. Mingos, J.L. Haggitt, H.R. Powell, S.A. Westcott, T.B. Marder, N.J. Taylor and D.R. Kanio, <u>J. Mater. Chem.</u>, 1993, <u>3</u>, 139.
10. H.M. Colquhoun, J.A. Daniels, I.R. Stephenson and K. Wade, <u>Polym. Commun.</u>, 1991, <u>32</u>, 272.
11. R. Coult. M.A. Fox. W.R. Gill, P.L. Herbertson, J.A.H. MacBride and K. Wade, <u>J. Organomet. Chem.</u>, 1993 in press.
12. D.A. Brown, W. Clegg, H.M. Colquhoun, J.A. Daniels, I.R. Stephenson and K. Wade, <u>Chem. Commun.</u>, 1987, 889.
13. R. Coult. M.A. Fox. W.R. Gill, K. Wade and W. Clegg, <u>Polyhedron</u>, 1992, 2717.
14. W. Clegg, R. Coult, M.A. Fox, W.R. Gill, J.A.H. MacBride and K. Wade, unpublished work.
15. W. Clegg, W.R. Gill, J.A.H. MacBride and K. Wade, <u>Angew. Chem.</u>, 1993 in press.

Carborane Product Analyses, by Means of IGLO/NMR and GIAO/NMR, of Reactions between *closo*-Carboranes and Fluoride Ion; the Formation of a New Carborane Ion, [*arachno*-2,6-C₂B₆H₁₁]⁻, from the Reaction of *closo*-1,6-C₂B₇H₉ with THF Solutions of Tetrabutylammonium Fluoride Ion

Thomas Onak*, Dan Tran, James Tseng, Martin Diaz, Sergio Herrera, and Joachin Arias

DEPARTMENT OF CHEMISTRY, CALIFORNIA STATE UNIVERSITY, LOS ANGELES, LOS ANGELES, CA 90032, USA

1   INTRODUCTION

Previous studies in our laboratories[1] demonstrated that fluoride ion can displace higher halogens in appropriately substituted closo-2,4-C₂B₅H₇ derivatives. More recently we have discovered that it is not necessary to have a halogen attached to the cage in order to encourage reactions of carboranes with fluoride ion. This report discusses the results of such reactions with a number of parent closo-carboranes. Some of the products of these reactions do not lend themselves to easy isolation. So it was important to find a reliable method of establishing their structures in the reaction solutions.

The novel *ab-initio*/IGLO/NMR method has been applied with some considerable success in the prediction of ¹¹B chemical shifts from MO optimized structures of known compounds, and for the utilization of calculated shifts for unique structure assignments.[2] The number of successful 'experimental vs calculational' correlations that have now been made strongly suggests that structural assignments based on the *ab-initio*/IGLO/NMR method is quickly approaching a confidence level that rivals modern-day X-ray diffraction determinations of molecular structures. If the IGLO calculated values from a suggested structure (optimized by appropriate *ab-initio* MO methods) match the experimental data reasonably well, it gives considerable credence to such a suggested structure. We recently applied this method to the conjectured structures for the product(s) from a number of reactions, including those from carborane/fluoride reactions, and these are the subject of the present study. Our calculational studies also employ the GIAO[3] method for corroborative structure/NMR correlations.

2 RESULTS AND DISCUSSION

Fluoride ion, in aprotic solvents, such as acetonitrile or tetrahydrofuran, has been found to be effective in the (partial) cage opening of *closo*-carboranes such as $1,6$-$C_2B_4H_6$, $2,4$-$C_2B_5H_7$, $1,7$-$C_2B_6H_8$, $1,6$-$C_2B_7H_9$, $1,10$-$C_2B_8H_{10}$, $1,2$-$C_2B_{10}H_{12}$, and $1,7$-$C_2B_{10}H_{12}$. In some cases new carborane ions, and in at least one instance a fluoro carborane derivative, have been obtained. The reactions usually proceed at a moderate rate at ambient temperatures, and nearly quantitative conversions to carborane anions are observed.

In the case of the (near) octahedral *closo*-$1,6$-$C_2B_4H_6$ the product is the $[5$-F-*nido*-$2,4$-$C_2B_4H_6]^-$ ion, the pentagonal pyramidal framework of which contains a fluorine atom terminally attached to one of the two equivalent basal borons of the parent $[nido$-$2,4$-$C_2B_4H_7]^-$ ion. This structural assignment is in agreement with ^{11}B, ^{1}H and ^{19}F NMR results.[4] Additionally, the IGLO ^{11}B NMR shifts, as calculated by Bausch[5] on an *ab-initio* optimized structure of this anion,[4] agree very well with the experimental chemical shifts.

In each of the three carboranes $2,4$-$C_2B_5H_7$, $1,2$-$C_2B_{10}H_{12}$, and $1,7$-$C_2B_{10}H_{12}$,[4] a *nido*-carborane is produced that is expected from a removal of single skeletal boron. Specifically, the action of fluoride ion on the pentagonal bipyramidal *closo*-$2,4$-$C_2B_5H_7$ produces the $[nido$-$2,4$-$C_2B_4H_7]^-$ ion, a product expected from the removal of a single high-coordination boron. This reaction is analogous to the removal of a single vertex of *closo*-$2,4$-$C_2B_5H_7$ by amides[6] and the earlier reported removal of a single boron vertex from *closo*-$1,2$- and $1,7$-$C_2B_{10}H_{12}$ using hydroxide ion, or alkoxide ion, in alcohol.[7] The reaction of fluoride ion with both the $1,2$- and $1,7$- isomers of the (near) icosahedral *closo*-$C_2B_{10}H_{12}$ mimics the action of the $[OH]^-$/ROH reagent combination in that the products are the same, the $7,8$- and $7,9$- isomers, respectively, of the $[nido$-$C_2B_9H_{12}]^-$ ion. It is to be noted that the conversion of *closo*-$1,7$-$C_2B_{10}H_{12}$ to $[nido$-$7,9$-$C_2B_9H_{12}]^-$ appears to be much more facile[4] when carried out with tetrabutylammonium fluoride in THF than with the $[OH]^-$/ROH reagent combination.

Reaction of tetrabutylammonium fluoride in tetrahydrofuran with *closo*-$1,6$-$C_2B_7H_9$ results in the removal of a single boron and yields the $[arachno$-$2,6$-$C_2B_6H_{11}]^-$ ion, Figure 1. The structure for this ion is derived from 1D ^{11}B NMR, 2D ^{11}B-^{11}B NMR spectra, and *ab-initio*/IGLO/NMR and *ab-initio*/GIAO/NMR results, as well as an intercorrelation

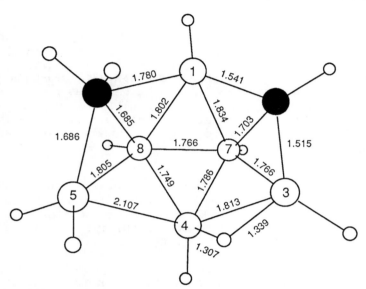

<u>**Figure 1**</u> The 6-31G* geometry-optimized (vibrationally stable) structure for the [*arachno*-2,6-C$_2$B$_6$H$_{11}$]$^-$ ion; bond distances are given in Å.

of all of these results for self-consistency. The ^{11}B NMR data denotes a structure with C$_1$ symmetry; i.e., all six ^{11}B resonances, in an area ratio of 1:1:1:1:1:1, are unique. Five of the six borons are each attached to a single terminal hydrogen and the remaining boron is attached to two terminal hydrogens. The proton NMR indicates the presence of one bridging hydrogen at δ^{11}B = -3.24 ppm. The skeletal structure proposed for the new ion is related to the known C$_2$B$_6$H$_{10}$,[8] but unlike the latter molecule the carbons of the [*arachno*-2,6-C$_2$B$_6$H$_{11}$]$^-$ ion are in *low* coordination (i.e. 3k)[9] non-adjacent 2,6- positions, a single bridging hydrogen positioned between the borons 3 and 4, a BH$_2$ group located at position 5, and a CH$_2$ group located at position 6.

Geometry optimization of the [*arachno*-2,6-C$_2$B$_6$H$_{11}$]$^-$ ion at the 6-31G* level of MO theory gives a structure, Figure 1, which is vibrationally stable. The four shortest of the nine B-B bond distances in the 6-31G* optimized structure are the B4-B8, B7-B8, B4-B7 and B3-B7 bonds. The first three of these give the strongest cross peaks in the ^{11}B-^{11}B 2D NMR and the last gives a medium strength cross peak, Figure 2. Additional ^{11}B-^{11}B 2D NMR cross peaks are observed for B1-B8, B5-B8 and B1-B7. The only ^{11}B-^{11}B cross peaks that are not observed between adjacent borons are for the two bonds B3-B4 and B4-B5; and these two B-B bonds are calculated to be the longest in the molecule-ion.

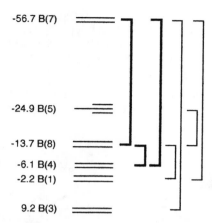

Figure 2. 2D ^{11}B-^{11}B NMR interactions (strong lines represent stronger interactions) for the [*arachno*-2,6-$C_2B_6H_{11}$]$^-$ ion.

Both IGLO calculations at the DZ level, and GIAO calculations at the 6-31G level, performed on the 6-31G* optimized [*arachno*-2,6-$C_2B_6H_{11}$]$^-$ ion predict ^{11}B and ^{13}C NMR chemical shifts that are exceptionally close to those obtained experimentally. The IGLO and GIAO results are plotted against the ^{11}B experimental data in Figure 3, and it is visually obvious that excellent linear correlations are obtained.

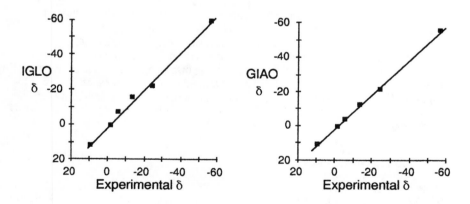

Figure 3 Graphical correlations between IGLO (dz//6-31G*) vs experimental, and GIAO (6-31G//6-31G*) vs experimental, ^{11}B chemical shifts for the [*arachno*-2,6-$C_2B_6H_{11}$]$^-$ ion. Best fit linear relationships: δ(IGLO) = 1.056*δ(exp) + 1.42 (r^2 = 0.993), standard deviation 1.9 ppm; δ(GIAO) = 1.008*δ(exp) + 2.56 (r^2 = 0.999), standard deviation 0.9 ppm.

The reaction of *closo*-1,10-$C_2B_8H_{10}$ with fluoride ion produces the [*arachno*-2,6-$C_2B_6H_{11}$]⁻ in smaller yield than obtained from the *closo*-1,6-$C_2B_7H_6$; in addition, some other cage products are also obtained that are presently under study.

Fluoride ion reacts with *closo*-1,7-$C_2B_6H_8$ to give as the major product a [$C_2B_5H_{10}$]⁻ ion, Figure 4,

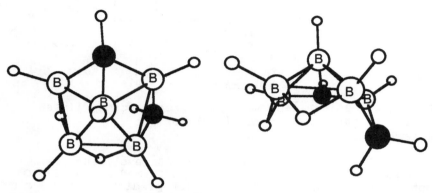

Figure 4 Two perspectives of the MP2/6-31G* geometry-optimized structure for the [*arachno*-$C_2B_5H_{10}$]⁻ ion.

on the basis of GIAO (6-31G//MP2/6-31G*) ¹¹B NMR analyses. This product bears little resemblance to the types of product(s) found from the reactions of *closo*-1,7-$C_2B_6H_8$ with other Lewis bases such as hydride and trimethylamine.[10] The framework of the [*arachno*-$C_2B_5H_{10}$]⁻ ion can be considered to be that of a pentagonal B_5C pyramid with a CH_2 group bridging two adjacent borons in the pentagonal base. It is related to the geometry of the [*arachno*-2,6-$C_2B_6H_{11}$]⁻ ion by the simple replacement of the BH_2 group in the latter, Figure 1, with a bridging hydrogen.

ACKNOWLEDGMENT

The authors wish to thank the NSF, CHE-8922339 and CHE-9222375, for partial support of this project. M.D, S.H. and J.A. thank the MBRS-NIH program for partial support. We also thank California State University, Sacramento, CA, for access to the Multiflow Trace (NSF Grant CHE-8822716) minisupercomputer facilities. We also wish to thank M. Schindler for permission to use the IGLO program designed by W. Kutzelnigg and M. Schindler, and to thank P. Pulay for the use of the GIAO program.

REFERENCES

1. B. Ng, T. Onak and K. Fuller, <u>Inorg. Chem.</u>, 1985, <u>24</u>, 4371; B. Ng and T. Onak, <u>J. Fluorine Chem.</u>, 1985, <u>27</u>, 119.

2. See M. Bühl, P. v.-R. Schleyer, <u>J. Am. Chem. Soc.</u>, 1992, <u>114</u>, 477, and references therein; W. Kutzelnigg, U. Fleischer, and M. Schindler, in 'NMR, Principles and Progress', Vol. 23, Springer Verlag: Berlin, 1990, pgs 165-262; J.W. Bausch, G.K.S. Prakash, M. Bühl, P. v.-R. Schleyer and R.E. Williams, <u>Inorg. Chem.</u>, 1992, <u>31</u>, 3060; S.O. Kang, J.W. Bausch, P.J. Carroll and L.G. Sneddon, <u>J. Am. Chem. Soc.</u>, 1992, <u>114</u>, 6248; J.W. Bausch, G.K.S. Prakash, R.E. Williams, <u>Inorg. Chem.</u>, 1992, <u>31</u>, 3763. A.M. Mebel, O.P. Charkin, M. Bühl and P. v. R. Schleyer, <u>Inorg. Chem.</u>, 1993, <u>32</u>, 463; A.M. Mebel, O.P. Charkin, and P. v. R. Schleyer, <u>Inorg. Chem.</u>, 1993, <u>32</u>, 469; M.L. McKee, M. Buhl and P.v. R. Schleyer, <u>Inorg. Chem.</u>, 1993, <u>32</u>, 1712.

3. K. Wolinski, J.F. Hinton, P. Pulay, <u>J. Am. Chem. Soc.</u>, 1990, <u>112</u>, 8251; R. Ditchfield, <u>Mol. Phys.</u>, 1974, <u>27</u>, 789.

4. H. Tomita, H. Luu and T. Onak, <u>Inorg. Chem.</u>, 1991, <u>30</u>, 812.

5. J. Bausch, footnote 14 in reference 4.

6. Z.J. Abdou, F. Gomez, G. Abdou and T. Onak, <u>Inorg. Chem.</u>, 1988, <u>27</u>, 3679; T. Onak, J. Tseng, M. Diaz, D. Tran, J. Arias, S. Herrera and D. Brown, <u>Inorg. Chem.</u>, 1993, <u>32</u>, 487.

7. See M.F. Hawthorne, D.C. Young, P.M. Garrett, D.A. Owen, S.G. Schwerin, F.N. Tebbe and P.A. Wegner, <u>J. Amer. Chem. Soc.</u>, 1968, <u>90</u>, 862, and references therein.

8. A.J. Gotcher, J.F. Ditter and R.E. Williams, <u>J. Amer. Chem. Soc.</u>, 1973, <u>95</u>, 7514.

9. R.E. Williams, in 'Electron Deficient Boron and Carbon Clusters'; Olah, G.A., Wade, K., Williams, R.E., Eds.: Wiley: New York, 1991, pgs 11-93; R.E. Williams, <u>Chem. Rev.</u> 1992, <u>92</u>, 177-207, and references therein.

10. T. Onak, J. Tseng, D. Tran, S. Herrera, B. Chan, J. Arias and M. Diaz, <u>Inorg. Chem.</u>, 1992, <u>31</u>, 3910.

Boron-Carboranyl Derivatives of Phosphorus

V. I. Bregadze, V. Ts. Kampel, L. V. Ermanson, V. A. Antonovich, and N. N. Godovikov

A. N. NESMEYANOV INSTITUTE OF ORGANO-ELEMENT COMPOUNDS, RUSSIAN ACADEMY OF SCIENCES, VAVILOV STR. 28, B-334, MOSCOW 117813, RUSSIA

1 INTRODUCTION

Since 1976 a new field of carborane chemistry, the chemistry of boron-metallated carboranes, has been developed[1-4]. In these compounds the usually labile boron-nontransition metal σ-bond is stabilized due to B atom's incorporation into carborane polyhedron. A great variety of B-carboranyl derivatives of nontransition metals has been prepared by direct mercuration of carboranes[1] followed by the substitution of mercury atom with nontransition metals under an action of metal halides as electrophiles[4]:

$$C_2H_2B_{10}H_{10} \quad + \quad (CF_3CO_2)_2Hg \quad \longrightarrow \quad CF_3CO_2HgB_{10}H_9C_2H_2$$
$$\text{(1)} \qquad\qquad\qquad \text{(2)} \qquad\qquad\qquad\qquad \text{(3)}$$

$$(C_2H_2B_{10}H_9)_2Hg \quad + \quad EX_n \quad \longrightarrow \quad C_2H_2B_{10}H_9HgEX_{n-1}$$
$$\text{(4)} \qquad\qquad\qquad \text{(5)} \qquad\qquad\qquad \text{(6)}$$

$E = Ga, In, Tl, Sn, As, Sb$
$n = 3, 4$

However, B-carboranyl derivatives of phosphorus could not be prepared by this method. The attempts to obtain them failed using all other traditional methods of synthesis of organophosphorus compounds, like the use of B-halogencarboranes in the Arbusov type reaction, the phosphorylation of carboranes with PCl_3 in the presence of $AlCl_3$, the oxidative chlorophosphorylation of carboranes, and the interaction of o-carboran-3-yldiazonium with PCl_3. The only known carborane derivative with B-P bond has been prepared by a photochemical reaction of bis(m-carboran-9-yl)bromonium tetrafluoroborate with triphenylphosphine[5]:

$$(9-m-C_2H_2B_{10}H_9)_2Br^+BF_4^- \quad + \quad PPh_3 \xrightarrow[\text{56}^{\circ}\text{C, 20 h}]{\text{UV, acetone}}$$

$$\text{(7)} \qquad\qquad\qquad \text{(8)}$$

$$\longrightarrow \quad C_2H_2B_{10}H_9PPh_3^+BF_4^-$$

$$\text{(9)}$$

A purpose of this work was the development of method of synthesis of B-carboranyl derivatives of phosphorus, a determination of their structure and reactivity, an estimation of the electronic influence of B-carboranyl group on phosphorus atom, and a comparison of the properties of B- and C-carboranyl derivatives of phosphorus.

2 CARBORANES WITH B-P BOND

We developed a method of synthesis of B-carboranyl derivatives of phosphorus based on photochemical decomposition of B-carboranyl derivatives of mercury[6] in the presence of trimethylphosphite[7] or phosphorus trichloride[8].

Photolysis of bis(m-carboran-9-yl)mercury and bis(p-carboran-2-yl)mercury in trimethylphosphite leads to the formation of dimethyl esters of (m-carboran-9-yl) and (p-carboran-2-yl)phosphonic acids[7]:

$$(m(p)-C_2H_2B_{10}H_9)_2Hg \quad + \quad (CH_3O)_3P \xrightarrow{\text{UV}}$$

$$\text{(10) (11)} \qquad\qquad \text{(12)}$$

$$\longrightarrow \quad (m(p)-C_2H_2B_{10}H_9P(O)(OCH_3)_2 \quad + \quad C_2H_2B_{10}H_{10} \quad + \quad Hg$$

$$\text{(13) (14)} \qquad\qquad\qquad \text{(15)} \qquad\qquad \text{(16)}$$

The course of the reaction was shown to depend on the nature of the carboranyl group. The yield of the m-carborane derivative is 5% and of the p-isomer is 60% while in the case of bis(o-carboran-9-yl)mercury the phosphonate does not form. Since steric factors in o-, m- and p-carboranyl radicals are similar, the different yields were ascribed to the inductive effects of B-carboranyl groups. This apparently affects the formation of the intermediate phosphoranyl radical, $[C_2H_2B_{10}H_9P(OCH_3)_3]\cdot$. The formation of such type radicals by the addition of the boron-centered carboranyl radical to phosphite under its reaction with bis(carboranyl)mercury was proved by the ESR method[9]. The stability of the phosphoranyl radical $RP(OCH_3)_3\cdot$ is known to depend primarily on the inductive effect of the group R : the more the electronegativity of R, the

more stable is the phosphoranyl radical[10]. It is known
that o- and m-carboran-9-yl groups are strong donors
and the corresponding phosphoranyl radicals are
apparently unstable. The p-carboran-2-yl radical has
negligible donor properties, resulting in a maximum
yield of dimethyl(p-carboran-2-yl)phosphonate.

To prepare o-carboranyl derivatives of phosphorus
and to increase a yield of m-carboranyl derivatives we
decided to use PCl_3 instead of $(CH_3O)_3P$ expecting the
formation of the more stable intermediate phosphoranyl
radical, $[C_2H_2B_{10}H_9PCl_3]\cdot$. Indeed, photolysis of
bis(m(o)-carboran-9-yl)mercury in phosphorus trichloride results in the formation of the corresponding
(m(o)-carboran-9-yl)dichlorophosphines[8]:

$$(m(o)-C_2H_2B_{10}H_9)_2Hg \;+\; PCl_3 \xrightarrow{UV}$$

    (10) (17)   (18)

$$\longrightarrow \quad m(o)-C_2H_2B_{10}H_9PCl_2 \;+\; m(o)-C_2H_2B_{10}H_9HgCl$$

        (19) (20)

Oxidation of (19) and (20) with sulphuryl chloride
in CCl_4 gives (B-carboranyl)phosphonic acid dichlori-
des[8]. Action of methanol on m-carboran-9-ylphosphonic
acid dichloride (24) results in the formation of
dimethyl ester of m-carboran-9-ylphosphonic acid (13)
described previously[7].

$$(19) \; (20) \;+\; SO_2Cl_2 \xrightarrow{CCl_4} m(o)-C_2H_2B_{10}H_9P(O)Cl_2$$

               (23)   (24) (25)

$$(24) \;+\; CH_3OH \xrightarrow{Et_3N} (13)$$

               (26)

The presence of the B-P bond in B-carboranyl derivatives of phosphorus was confirmed by ^{11}B and ^{31}P NMR
(Table 1) spectral data and X-ray diffraction.

Table 1   Parameters of ^{31}P NMR spectra of B-P
            compounds

Compound	δ ^{31}P ppm	J([31]P-[11]B) Hz	J([31]P-[10]B) Hz
(19)	177.1	30	–
(20)	176.3	30	–
(24)	56.5	272	91
(25)	51.3	280	91
(13)	37.4	305	101

     Dimethyl (p-carboran-2-yl)phosphonate is the first icosahedral carborane derivative with the B-P σ-bond to be studied structurally[7]. The phosphorus atom is in a tetrahedral configuration with the P-B bond length of 1.903(5)Å. In the crystal there is a weak intermolecular hydrogen bond C-H...O=P between one of the carborane nucleus C-H bonds and a phosphoranyl group. The possibility of the existence of such a hydrogen bond has also been suggested on the basis of the IR spectra of B-carboranylphosphonates. The comparison of the spectra of the solid samples with the ones in CCl_4 solution shows the high frequence shift of ν_{CH} ($\Delta\nu_{CH}$=20-75 cm^{-1} and ν_{PO} ($\Delta\nu_{PO}$= 20-35 cm-1) bands[7].

     σ^{ϕ}-Constants of the B-carboranyl groups were estimated on the basis of $\Delta\nu_{OH}$ dependence on $\Sigma\sigma^{\phi}$ of H-complexes of p-fluorophenols with phosphoryl compounds[11]. σ^{ϕ}-Constants of p-carboran-2-yl group and m-carboran-9-yl group were found to be -1.68 and -1.81, respectively[7]. It is of interest that, among all known substituents on the phosphorus atom, the m-carboran-9-yl group is the strongest electron donor with respect to this atom while o-carboran-1-yl group is one of the strongest acceptors (σ^{ϕ}=2.0)[12].

3 CARBORANES WITH B-Hg-P MOIETY

Compounds of the type $(RO)_2P(O)HgX$ with P-Hg bond are known only with strong electron acceptor groups at mercury atom, X=Hal,CN,OCOMe,Ph and $P(O)(OR)_2$. We intended to synthesize compounds of such type containing electron donor B(9)-carboranyl group at mercury atom – namely, the compounds with B-Hg-P moiety. Interaction of sodium diethylphosphite with (m-carboran-9-yl)mercury chloride in THF at room temperature results in the formation of the first compounds of the new type with B-Hg-P moiety[13]:

$$m\text{-}C_2H_2B_{10}H_9HgCl \quad + \quad (EtO)_2PONa \quad \xrightarrow{\text{THF}}$$

     (21)   (27)

$$\xrightarrow{} m\text{-}C_2H_2B_{10}H_9HgPO(OEt)_2$$

                       (28)

     Compound (28) was shown to disproportionate in benzene or THF solutions at room temperature in darkness to give the corresponding symmetrical compounds[13]:

$$2 \ m-C_2H_2B_{10}H_9HgPO(OEt)_2 \longrightarrow$$

$$(28)$$

$$\longrightarrow \ (m-C_2H_2B_{10}H_9)Hg \ + \ [(EtO)_2PO]_2Hg$$

$$(10) \hspace{3cm} (29)$$

This equilibrium was studied by the ^{31}P NMR. The highest content of the symmetrical compound (29) was 8.3% (after two weeks). Mixing of equimolecular amounts of the symmetrical compounds (10) and (29) in benzene, the minimal percentage (8.8%) of unsymmetrical compound (28) was reached in two days. Interaction of bis(carboranyl)mercury compounds with $[(MeO)_2PO]_2Hg$ also lead to the same type of equilibrium with the preferable formation (84-97%) of unsymmetrical compounds[13] :

$$R_2Hg \ + \ [(R'O)_2PO]_2Hg \longrightarrow RHgPO(OR')_2$$

$$R \ = \ m-C_2H_2B_{10}H_9; \ R' \ = \ Me \ (30)$$
$$R \ = \ o-C_2H_2B_{10}H_9; \ R' \ = \ Et \ (31), \ Me \ (32)$$

The presence of the B-Hg-P moiety in compounds (28), (30)-(32) was confirmed by ^{31}P and ^{11}B NMR spectroscopy data (Tables 2,3) as well as X-ray diffraction. It is known that $(RO)_2PO$ fragment can be linked with metal atom via either oxygen or phosphorus atoms, to give a structure of phosphite[14] or phosphonate[15], respectively. In ^{31}P NMR spectrum of the compound (28) in benzene an intensive quadruple signal at 161.9 ppm was observed due to $^{31}P-^{11}B$ coupling. Satellite quadruplet caused by spin-spin interaction with the ^{199}Hg nucleus was also detected. Thus the considerable down field shift of ^{31}P and the decrease of $J(^{31}P-^{199}Hg)$ in comparison with the symmetrical compound (29) were observed.

Table 2   Parameters of ^{31}P NMR spectra of compounds
           (28-32)

Compound	Solvent	^{31}P ppm	$J(^{31}P-^{199}Hg)$ Hz	$J(^{31}P-^{11}B)$ Hz
(28)	THF	160.6	3350	210
(28)	C_6H_6	161.9	3240	202
(29)	C_6H_6	108.4	7374	–
(30)	THF	163.14	3240	190
(31)	THF	163.59	2700	220
(32)	THF	163.95	2670	200

Table 3 Parameters of ^{11}B NMR spectrum of
m-$C_2H_2B_{10}H_9HgP(O)(OEt)_2$ in C_6H_6

^{11}B (ppm)	J (Hz)	Integral	
8.2	195 (B-P)	1	B(9)
	953 (B-Hg)		
-5.3	167 (B-H)	2	B(5,12)
-8.9	165 (B-H)	1	B(10)
-12.0	160 (B-H)	4	B(4,6,8,11)
-16.0	185 (B-H)	2	B(2,3)

These data are in agreement with the observation that the increase of electron acceptor properties of X in $(RO)_2P(O)HgX$ results in the considerable up-field shift of ^{31}P and the increase of $J(^{31}P-^{199}Hg)$[15]. On the other hand when Hg and P atoms not linked together in these compounds the chemical shift practically does not depend on the nature of X[16]. Thus, ^{31}P NMR data of the compounds (28),(30)-(32) confirm the existence of B-Hg-P fragment, and the chemical shift of ^{31}P signal to the weak field and the decrease of $J(^{31}P-^{199}Hg)$ are caused by the electron donor properties of B(9)-carboranyl group. An additional confirmation of the existence of B-Hg-P fragment can be found in the ^{11}B NMR spectral data (Table 3)

IR-spectra of o(m)-$C_2H_2B_{10}H_9HgP(O)(OR)_2$ contain all bands typical for the carboranyl group and dialkyl-posphonate fragment. However, $\nu(P=O)$ band in the spectra of m-$C_2H_2B_{10}H_9HgP(O)(OMe)_2$ is 40 cm^{-1} lower than in the spectra of m-$C_2H_2B_{10}H_9P(O)(OMe)_2$. This decrease is probably due to the P=O...Hg coordination.

In fact, an X-ray study[17] of the molecule of (m-carboran-9-yl)(diisoprpopoxyphosphoryl)mercury, (i-PrO)$_2$P(O)HgB$_{10}$H$_9$C$_2$H$_2$ (33), shows that molecules in the crystal are united into the centrosymmetrical dimer associates due to the coordination bond Hg...O which equals 2.629(7)Å. This distance is essentially shorter than the sum of van der Waals radii of mercury (1.5Å) and oxygen (1.4Å) atoms. Molecule (33) is the first compound with σ-bond between a mercury atom and a boron atom of the carborane cage to be studied structurally. The Hg-B bond length is 2.13(1)Å, and it is close to the value 2.16Å found for the similar B-Hg bond in the molecule of bis(p-carboran-2-yl)mercury studied by diffraction in the gas phase[18]. The Hg-B bond length in (33) is close to the sum of covalent radii of mercury (1.30Å) and boron (0.81Å) atoms. The Hg-P bond length in (33) is 2.503(4)Å.

The photolysis of (m-carboran-9-yl)(diethoxyphos-phoryl)mercury in acetone or benzene solution leads to the formation of bis(m-carboran-9-yl)mercury, metallic mercury and a complex mixture of products from reaction of the phosphoryl radical[19]:

$$m-C_2H_2B_{10}H_9HgPO(OEt)_2 \xrightarrow{\text{UV}}$$

(28)

$$\longrightarrow \quad m-C_2H_2B_{10}H_9Hg\cdot \quad + \quad \cdot PO(OEt)_2$$

$$\downarrow$$

$$(m-C_2H_2B_{10}H_9)_2Hg \quad + \quad Hg$$

(10)   (16)

The formation of these radicals was proved by the ESR method. Depending on the type of trap used, spectra of adducts derived from either m-carboranylmercury radical or phosphonyl radical were recorded. The first one gave adduct with o-quinone, the latter gave adduct with 4-methyl-2,4,6-tri-tert-butylcyclohexa-2,5-dien-1-one.

Electrochemical properties of (B-carboranyl) (dialkoxyphosphoryl) mercury compounds were studied on a platinum electrode and a redox electrochemical mechanism was proposed.[20]

REFERENCES

1.   V.I. Bregadze, V.Ts. Kampel and N.N. Godovikov, J.Organometal.Chem., 1976, 112, 249.
2.   V.I. Bregadze, V.Ts. Kampel, A.Ya. Usiatinsky and N.N. Godovikov, J.Organometal.Chem., 1978, 154, C1.
3.   V.I. Bregadze, V.Ts. Kampel and N.N. Godovikov, J.Organometal.Chem., 1978, 157, C1.
4.   V.I. Bregadze, V.Ts. Kampel, A.Ya. Usiatinsky and N.N. Godovikov, Pure & Appl.Chem., 1991, 63, 835.
5.   V.V. Grushin, T.P. Tolstaya, I.N. Lisichkina, Yu.K. Grishin, T.M. Shcherbina, V.Ts. Kampel, V.I. Bregadze and N.N. Godovikiov, Izv. Akad. Nauk SSSR, Ser. Khim., 1983, 472
6.   B.L. Tumansky, V.Ts. Kampel, S.P. Solodovnikov, V.I. Bregadze and N.N. Godovikov, Izv. Akad. Nauk SSSR. Ser. Khim., 1985, 2644
7.   V.I.Bregadze, V.Ts. Kampel, E.I.Matrosov, V.A. Antonovich, A.I. Yanovsky, Yu.T. Struchkov, N.N. Godovikov and M.I. Kabachnik, Dokl. Akad. Nauk SSSR, 1985, 285, 1127
8.   V.Ts. Kampel, V.I. Bregadze, L.V. Ermanson, V.A. Antonovich, E.I. Matrosov, N.N. Godovikov and M.I Kabachnik, Metalloorg. Khim., 1992, 5, 1024

9.   B.L. Tumansky, V.Ts. Kampel, V.I. Bregadze, N.N.
     Bubnov, S.P. Solodovnikov, A.I. Prokof'ev, E.S.
     Kozlov, N.N. Godovikov and M.I. Kabachnik, <u>Izv.
     Akad. Nauk SSSR. Ser. Khim.</u>, 1986, 458

10. Ya.A. Levin and E.I. Vorkunova, "Homolytical
     chemistry of phosphorus", Nauka, Moscow, 1978

11. E.I. Matrosov, E.E. Nifant'ev, A.A. Krjuchkov and
     M.I.Kabachnik, <u>Izv. Akad. Nauk SSSR. Ser. Khim.</u>,
     1976, 530

12. A.N. Degtyarev, N.N. Godovikov, V.I. Bregadze,
     E.I. Matrosov, T.M. Shcherbina and M.I. Kabachnik,
     <u>Izv. Akad. Nauk SSSR. Ser. Khim.</u>, 1978, 2099

13. V.Ts. Kampel, L.V. Ermanson, V.I. Bregadze, V.A.
     Antonovich, E.I. Matrosov and N.N. Godovikov,
     <u>Metalloorg. Khim.</u>, 1993, <u>6</u>, 82

14. G.R.Van der Berg, D.H. Platenbury, H.P. Benschop,
     <u>Chem. Commun.</u>, 1971, 606

15. J. Eichbichler and P. Peringer, <u>Inorg. chim. acta</u>,
     1980, <u>43</u>, 121

16. C. Glidewel, <u>Inorg. chim. acta</u>, 1978, <u>27</u>, 129

17. A.I. Yanovsky, Yu.T. Struchkov, V.Ts. Kampel, L.V.
     Ermanson, N.N. Godovikov and V.I. Bregadze,
     <u>Metalloorg. Khim.</u>, 1993, <u>6</u>, 88

18. V.S. Mastryukov, A.A. Remorova, A.V. Golubinsky,
     M.V. Popik, L.V. Vilkov, V.Ts. Kampel and V.I.
     Bregadze, <u>Metalloorg. Khim.</u>, 1991, <u>4</u>, 132 [English
     translation in <u>Organometal. Chem. USSR</u>, 1991, <u>4</u>,
     69]

19. B.L. Tumansky, V.Ts. Kampel, L.V. Ermanson, N.N.
     Bubnov, S.P. Solodovnikov, A.I. Prokof'ev and N.N.
     Godovikov, <u>Metalloorg. Khim.</u>, 1991, <u>4</u>, 941
     [English translation in <u>Organometal. Chem. USSR</u>,
     1991, <u>4</u>, 462]

20. R.D. Rakhimov, K.P. Butin, L.V. Ermanson, V.Ts.
     Kampel, N.N. Godovikov and V.I. Bregadze,
     <u>Metalloorg. Khim.</u>, 1991, <u>4</u>, 823 [English
     translation in <u>Organometal. Chem. USSR</u>, 1991, <u>4</u>,
     400]

METALLABORANE CHEMISTRY

Studies of the Triosmium Carbonyl Borylidyne Cluster $(\mu\text{-H})_2Os_3(CO)_9(\mu_3\text{-BCO})$

S. G. Shore

DEPARTMENT OF CHEMISTRY, THE OHIO STATE UNIVERSITY,
COLUMBUS, OH 43210, USA

1 INTRODUCTION

Metal ketenylidene clusters possess a rich and diverse chemistry. In this laboratory we are developing the chemistry of the carbonyl borylidyne cluster $(\mu\text{-H})_3Os_3(CO)_9$-$(\mu_3\text{-BCO})$, **I**, an analogue of the ketenylidene cluster $(\mu\text{-H})_2Os_3(CO)_9(\mu_3\text{-CCO})$. Results of these studies are provided in this report.

<u>Hydroboration Reactions</u>
From the hydroboration of $(\mu\text{-H})_2Os_3(CO)_{10}$, the tetrahedral cluster **I** is produced (Reaction (1) (Figure 1a).[1,2] In

$$(\mu\text{-H})_2Os_3(CO)_{10} + 1/2B_2H_6 \xrightarrow{\quad CH_2Cl_2/Me_2O \quad} (\mu\text{-H})_3Os_3(CO)_9(\mu_3\text{-BCO}) + H_2 \quad (1)$$

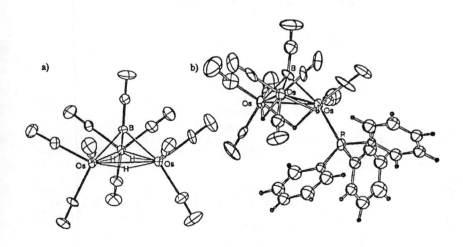

<u>Figure 1</u> a) Structure of $(\mu\text{-H})_3Os_3(CO)_9(\mu_3\text{-BCO}$, **I**. b) Structure of $(\mu\text{-H})_3Os_3(CO)_8(PPh_3)(\mu_3\text{-BCO})$, **II**.

the formation of **I**, a BH_3 unit adds to the Os_3 triangle, transfers hydrogen to it and effectively inserts into the OsCO bond in the process of capping the Os_3 unit. The B-C distance is 1.469 (15) Å. A related carbonyl borylidyne $(\mu-H)_3Os_3(CO)_8(PPh_3)(\mu_3-BCO)$, **II**, is obtained[2] from the hydroboration of $(\mu-H)_2Os_3(CO)_9(PPh_3)$; it is structurally similar to **I** (Figure 1b). Experimental evidence and details in support of the proposed reaction pathway for the formation of **I** are described elsewhere.[1]

Displacement of Carbon Monoxide from Carbonyl Borylidynes

Complexes **I** and **II** are remarkable molecules. Unlike many transition metal carbonyl clusters, they are not fluxional on the NMR time-scale up to their decomposition temperatures, ca. 90 °C.[1,2] Furthermore, no detectable exchange of carbon monoxide gas with carbon monoxide in ^{13}C enriched **I** and **II** occurs up to 1,000 psi and 50 °C for three days. On the other hand the carbonyl on the boron site is readily displaced by PMe_3 at room temperature. In the case of **I** the carbon monoxide at the apical site is exclusively replaced by PMe_3 to form $(\mu-H)_3Os_3(CO)_9(\mu_3-BPMe_3)$, **III**, at room temperature within 1 day when the molar ratio PMe_3:**I** is $\leq 1:1$ (Reaction (2)).

$$(\mu-H)_3Os_3(CO)_9(\mu_3-BCO) + PMe_3 \xrightarrow{CH_2Cl_2}$$
$$(\mu-H)_3Os_3(CO)_9(\mu_3-BPMe_3) + CO \quad (2)$$

The reaction of **II** with PMe_3, however, gives a mixture of **III** and $(\mu-H)_3Os_3(CO)_8(PPh_3)(\mu_3-BPMe_3)$, **IV**, in a ratio of 1.5:1 - 2.0:1 over the temperature range 15 - 40 °C. These products are formed in parallel, concurrent reactions and the products are produced in a constant ratio as the reaction progresses (Reactions (3a) and (3b)).[2]

$$(\mu-H)_3Os_3(CO)_8(PPh_3)(\mu_3-BCO) + PMe_3 \xrightarrow{CH_2Cl_2}$$
$$(\mu-H)_3Os_3(CO)_8(PPh_3)(\mu_3-BPMe_3) + CO \quad (3a)$$

$$(\mu-H)_3Os_3(CO)_8(PPh_3)(\mu_3-BCO) + PMe_3 \xrightarrow{CH_2Cl_2}$$
$$(\mu-H)_3Os_3(CO)_9(\mu_3-BPMe_3) + PPh_3 \quad (3b)$$

Studies[2] of kinetics of reactions of PMe_3 with **I** and **II** indicate that they are associative in nature, being first order in cluster and first order in PMe_3. Rate constants and activation parameters are summarized in Table I. Several pathways have been considered as possible routes for substitution Reactions (2), (3a), and (3b). Since no experimental method appears to be available for preferentially labeling complexes **I** and **II** with ^{13}CO, additional experimental information that might assist in choosing a mechanistic pathway is not available.

One pathway that has been considered and rejected involves addition of PMe_3 to a basal osmium atom followed

Table 1 Rate Constants and Activation Parameters for Reactions (2), (3a), and (3b) at 293 °K

Reaction	$k, M^{-1}s^{-1} \times 10^3$	$\Delta H^{\ddagger}$, kcal/mol	$\Delta S^{\ddagger}$, cal/mol-deg
(2)	2.56 ± 0.17	18.0 ± 0.8	-8.8 ± 2.6
(3a)	2.27 ± 0.05	19.1 ± 0.8	-5.3 ± 2.6
(3b)	1.42 ± 0.03	17.8 ± 0.6	-11.0 ± 2.0

by migration of the PMe_3 to the boron atom, displacing the apical CO. This pathway is considered to be unlikely in the present case, since the enthalpies of activation for Reactions (2), (3a), and (3b) are not significantly different. The position of PPh_3 at an axial position on an osmium atom (Figure 1b) is expected to hinder, significantly, the approach of PMe_3 with consequent increase in the enthalpies of activation for Reactions (3a) and (3b) compared to that for Reaction (2).

A second pathway that has been considered and rejected involves initial attack of the PMe_3 at the carbon atom of the carbonyl attached to boron to form an adduct followed by a concerted exchange between CO and PMe_3. One difficulty with this proposal resides in the substantial steric hindrance caused by the arrangement of six of the nine carbonyls that are disposed upward in the direction of the apical boron (Figures 1a and 1b). Furthermore, the low entropy of activation observed compared to that observed for mononuclear substitution reactions with a transition state of higher coordination number than that of the ground state (ca. -25 cal/mol-deg) suggests that there is appreciable rearrangement in the structure of the activated complex in the current system.[3]

A reaction pathway that we favor involves a cluster opening step by adding PMe_3 to **I** or **II** to form an intermediate with a "butterfly" structure followed by a subsequent cluster reclosing step to eject CO with the formation of **III** or **IV**. If PPh_3 is eliminated in the cluster reclosing step, **III** is formed. Scheme 1 represents these pathways for the formation of **III** and **IV** in the reaction of **II** with PMe_3 (Reactions (3a) and (3b)). In this scheme PMe_3 adds to the cluster to form one of two possible isomeric "butterfly" intermediates. In the reformation of the Os_3B tetrahedral core either CO or PPh_3 is eliminated to yield respectively either **III** or **IV**. Statistically it is twice as favorable to eliminate CO than PPh_3 thereby accounting for a ratio of **III:IV** that approaches 2:1 with increasing temperature. Route (b) shown in this scheme is also applicable to the formation **III** in the reaction of **I** with PMe_3.

A reaction pathway with an intermediate of "butterfly geometry" has also been proposed in the CO substitution reaction of $Ir_4(CO)_{12}$ by trialkyl phosphines.[4] This reaction pathway appears to be operative in several systems in which intermediates with open structures have been isolated or detected.[5]

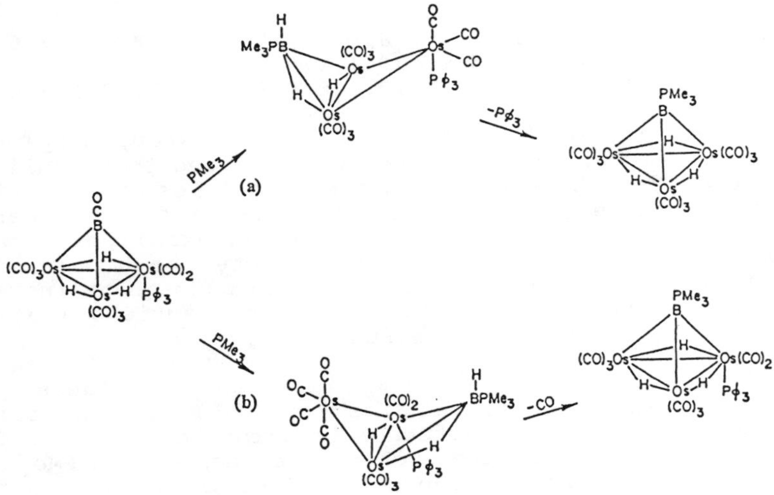

Scheme 1

Reactions of I with Lewis Acids

Vinylidene cluster analogues. The unique carbonyl in I is reduced to a methylene group by $THFBH_3$ to form $(\mu-H)_3Os_3(CO)_9(\mu_3-\eta^2-BCH_2)$, **V**, (Reaction (4)).[6a,b] Its structure (Figure 2a) resembles that of the vinylidene cluster

$$(\mu-H)_3Os_3(CO)_9(\mu_3-BCO) + THFBH_3 \longrightarrow$$
$$(\mu-H)_3Os_3(CO)_9(\mu_3-\eta^2-BCH_2) + B_3H_3O_3 \quad (4)$$

$(\mu-H)_2Os_3(CO)_9(\mu_3-\eta^2-CCH_2)$.[7] However the B-C bond is canted 60° from the perpendicular with respect to the Os_3 plane. This is significantly larger than observed for the analogous C-C bond in structurally characterized vinylidene clusters (40-50°).[8] The two B-H-Os bridges in the structure probably force the BCH_2 unit to an extreme tilt angle compared to the vinylidene complexes. The extreme tilt angle implies that the compound could also be described as a methylene-bridged complex. However, the "short" B-C distance, 1.498 (15) Å, compared to observed B-C single bond distances that are ca. 0.1 Å longer[9a,b] suggest partial double bond character and the relatively long Os-C distance, 2.325 (17) Å, favors the vinylidene analogy.

Formation of **V** is believed to occur through initial coordination of BH_3 to the oxygen atom of the carbonyl to give $(\mu-H)_3Os_3(CO)_9(\mu_3-BCOBH_3)$ followed by transfer of two BH hydrogens to the carbon atom. Elimination of H-B-O as the boroxine trimer, $B_3H_3O_3$, would then result in the formation of **V**. Deuterium labeling experiments indicate that reduction of the CO occurs with no apparent scrambling of B-H, C-H, and Os-H-Os hydrogen atoms.

Boron trihalides react with **I** to form vinylidene cluster analogues $(\mu\text{-H})_3Os_3(CO)_9(\mu_3\text{-CBX}_2)$ (X = Cl, B), **VI**, in which the boron and carbon have exchanged positions (Reaction (5)).[6b,c] The reaction of $^{10}BCl_3$ with **I** does not

$$(\mu\text{-H})_3Os_3(CO)_9(\mu_3\text{-BCO}) + BX_3 \longrightarrow$$
$$(\mu\text{-H})_3Os_3(CO)_9(\mu_3\text{-CBX}_2) + 1/3\ B_3X_3O_3 \quad (5)$$
$$X = Cl,\ Br$$

involve interchange of ^{10}B with the ^{11}B in the cluster. Therefore, the formation of **VI** appears to involve intramolecular exchange of the boron and carbon atoms of **I**. The structure of **VI** (Figure 3b) differs from that of **V** not only in that the carbon positions are reversed, but

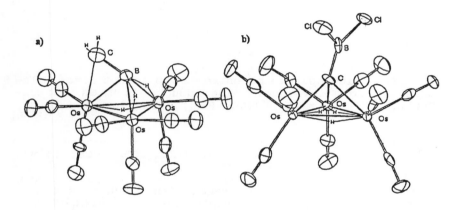

<u>Figure 2</u> a) Structure of $(\mu\text{-H})_3Os_3(CO)_9(\mu_3\text{-}\eta^2\text{-BCH}_2)$, **V**. b) Structure of $(\mu\text{-H})_3Os_3(CO)_9(\mu_3\text{-CBX}_2)$ (X = Cl, Br), **VI**.

also the C-B bond in **VI** is canted only 15° from perpendicularity to the Os_3 plane. As in the case of **V**, the B-C distance 1.47 (2) Å is "short" compared to observed[9a,b] single bond B-C distances.

Scheme 2 presents a proposed pathway by which **VI** is formed. Upon interacting with the unique carbonyl oxygen, the boron trihalide is a sufficiently strong electron withdrawing agent to reduce the bond order of the carbon oxygen bond causing it to move to a bridging site. Movement of the carbonyl ligand into the μ_3 site exposes the boron and results in successive halogen atom transfer from the reagent boron to the cluster boron. Compound **VI** is produced when X-B-O is eliminated as $B_3X_3O_3$. That reaction of **I** with BH_3 differs from its reactions with boron trihalides is attributed[6b] to the relatively stronger Lewis acidities of the trihalides toward oxygen donors than BH_3.[10]

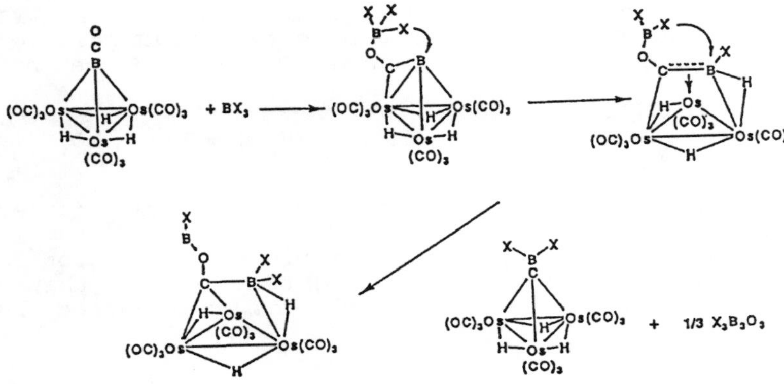

Scheme 2

The tricoordinate boron in **VI** can accept donor molecules to form Lewis acid-base adducts $(\mu-H)_3Os_3(CO)_9(\mu_3-CBCl_2L)$, **VII**, (L = NMe$_3$, PMe$_3$, PPh$_3$) (Reaction (6)).[6b]

$$(\mu-H)_3Os_3(CO)_9(\mu_3-CBCl_2) \; + \; L \; \xrightarrow{\;-40\ °C\;} \; (\mu-H)_3Os_3(CO)_9(\mu_3-CBCl_2L) \quad (6)$$

$$L = NMe_3, \; PMe_3, \; PPh_3$$

However, above −10 °C the trimethylene adduct is transformed to the salt $[NMe_3H][(\mu-H)_2Os_3(CO)_9(\mu_3-CBCl_2)]$ through deprotonation of the cluster by the amine.

Alkyne Cluster Analogues

Reaction of **I** with B-Cl-9BBN and BPhCl$_2$. An alkyne cluster analogue $(\mu-H)_3Os_3(CO)_9[\mu_3-\eta^2-C(OBC_8H_{14})B(Cl)]$, **VIII**, is formed from the reaction **I** with B-Cl-9BBN ($C_8H_{14}BCl$) (Reaction (7)).[6b,d] The structure of **VII** (Figure 3a) reveals that in the formation of this compound the

$$(\mu-H)_3Os_3(CO)_9(\mu_3-BCO) \; + \; C_8H_{14}BCl \; \xrightarrow{\hspace{2cm}}$$
$$(\mu-H)_3Os_3(CO)_9[\mu_3-\eta^2-C(OBC_8H_{14})B(Cl)] \quad (7)$$

unique carbonyl of **I** moves to a μ_3-site that caps two osmium atoms and the boron atom whereas the chlorine atom of the B-Cl-9BBN moves to the boron of the cluster. This compound is an alkyne cluster analogue in which a BH group takes the place of a carbon atom. The B-C bond is oriented nearly parallel (within 10°) to an Os-Os bond. It adopts the $\mu^3-\eta^2$ bonding mode that occurs for the C-C bond in alkyne cluster analogues.[11] The bond distance is 1.46 (2) Å, comparable to that in **I**, **V**, and **VI** and between the vaules for a B-C single bond, ca. 1.6 Å[9a,b] and a B-C double bond, 1.361 (5) Å[9c].

Another alkyne cluster analogue $(\mu-H)_3Os_3(CO)_9[\mu_3-\eta^2-C\{OB(Ph)Cl\}B(Cl)]$, **IX**, is formed in the reaction of **I** with $BPhCl_2$ (Reaction (8)).[6b] The proposed structure of **IX**

$$(\mu-H)_3Os_3(CO)_9(\mu_3-BCO) + BPhCl_2 \longrightarrow$$
$$(\mu-H)_3Os_3(CO)_9[\mu_3-\eta^2-C\{OB(Ph)Cl\}B(Cl)] \quad (8)$$

(Figure 3b) is related to that of **VIII** and is based upon spectroscopic data (IR, NMR).

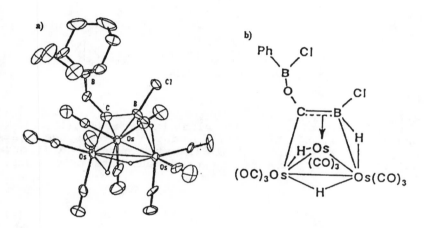

Figure 3 a) Structure of $(\mu-H)_3Os_3(CO)_9[\mu_3-\eta^2C(OBC_8H_{14})-B(Cl)]$, **VIII**. b) Proposed structure of $(\mu-H)_3Os_3(CO)_9[\mu_3-\eta^2-C\{OB(Ph)Cl\}B(Cl)]$, **IX**.

Complexes **VIII** and **IX** are related to the proposed intermediate in Scheme 2 in the reaction of **I** with BX_3 (X = Cl, Br). They react with boron trihalides to form compounds of type **VI**, vinylidene cluster analogues (Reactions (9) and (10)).[6b]

$$(\mu-H)_3Os_3(CO)_9[\mu_3-\eta^2-C(OBC_8H_{14})B(Cl)] + BX_3 \longrightarrow$$
$$(\mu-H)_3Os_3(CO)_9(\mu_3-CBClX) + 1/3B_3X_3O_3 + \underline{B}-X-9BBN \quad (9)$$
$$X = Cl, Br$$

$$(\mu-H)_3Os_3(CO)_9[\mu_3-\eta^2-C\{OB(Ph)Cl\}B(Cl)] + BCl_3 \longrightarrow$$
$$(\mu-H)_3Os_3(CO)_9(\mu_3-CBCl_2) + 1/3B_3Cl_3O_3 + BPhCl_2 \quad (10)$$

In scheme 2 the initial intermediate is an alkyne cluster analogue like **VIII** and **IX**. In subsequent steps halogen is transferred to the boron atom of the cluster. In Reactions (9)and (10) the boron halide provides the additional halogen. A proposed sequence is given in Scheme 3.

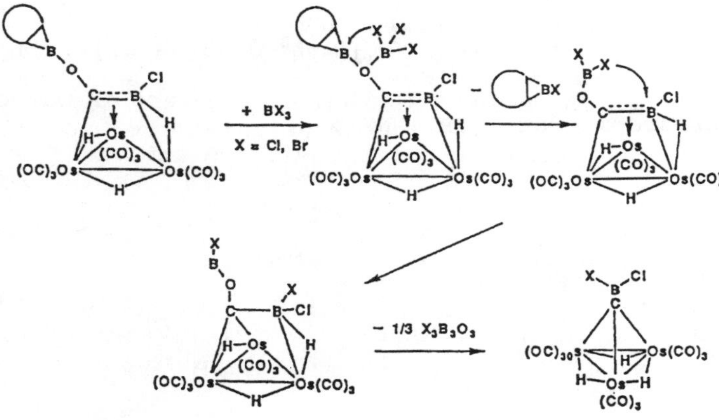

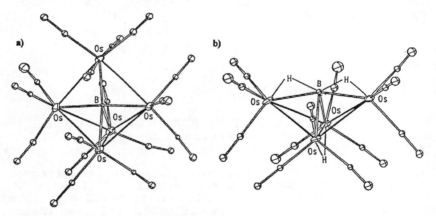

Formation of Osmaboride Clusters from I

The first examples of osmaboride clusters, $HOs_5(CO)_{16}B$, **X**, and $HOs_4(CO)_{12}BH_2$, **XI**, were produced through the thermolysis of **I** at 110 °.[12a] In the structure of **X** (Figure 4a) the five Os atoms define a bridged "butterfly" metal

a)

b)

Figure 4 a) Structure of $HOs_5(CO)_{16}B$, **X**. b) Structure of $HOs_4(CO)_{12}BH_2$, **XI**.

framework. The boron atom is encapsulated in this framework, bonded to all five osmium atoms. The hydrogen atom was not located, but it is believed to be on the cluster surface, possibly bridging the two osmium atoms that form the hinge of the "butterfly". The overall molecular geometry of this cluster resembles the pentaosmium carbonyl carbide cluster $Os_5(CO)_{16}C$[12b] containing a carbon atom encapsulated in the Os_5 core. Complex **XI** like $Os_5(CO)C$ is a 76 valence electron system. From the Wade, Williams, and Rudolph electron counting rules[13a-c], this compound can be considered to be an <u>arachno</u> cluster that is derived from a pentagonal bipyramid from which non-adjacent equatorial vertices are removed.

The molecular structure of **XI** (Figure 4b) contains four Os atoms forming a "butterfly" cluster core with the boron atom residing midway between the osmium atoms that define the wing tips. This molecule is isostructural with ruthenium[14a] and iron[14b] analogues and is considered to be an <u>arachno</u>, four atom cluster with an interstitial boron or alternatively a 62 valence electron complex in which the BH_2 ligand contributes five electrons on the basis of the skeletal electron-pair theory.[13a] Although the bridging hydrogens were not located from the X-ray data, ^{11}B and ^{1}H NMR spectra indicate that they are located at the positions shown in Figure 4b. Complex **XI** is deprotonated by KH in ether solvents (Reactions (11) and (12)).[12a] In these reactions, deprotonation occurs at the

$$HOs_4(CO)_{12}BH_2 + KH \longrightarrow K[HOs_4(CO)_{12}BH] + H_2 \quad (11)$$

$$K[HOs_4(CO)_{12}BH] + KH \longrightarrow K_2[HOs_4(CO)_{12}B] + H_2 \quad (12)$$

the Os-H-B bridges, which is consistent with results from deprotonation studies of $HRu_4(CO)_{12}BH_2$[14a] and confirms, further, the predictions of Fehlner[15] concerning the deprotonation of these clusters.

Acknowledgement. This work was supported by the National Science Foundation.

REFERENCES

1. a) S. G. Shore, D.-Y. Jan, L.-Y. Hsu, and W.-L. Hsu, *J. Am. Chem. Soc.*, 1983 <u>105</u>, 5923. b) D.-Y. Jan, D. P. Workman, L.-Y. Hsu, J. A. Krause, and S. G. Shore, *Inorg. Chem.*, 1992, <u>31</u>, 5123.
2. D.-Y. Jan, <u>PhD Thesis</u>, The Ohio State University, 1985.
3. a) C. Y. Chang, C. E. Johnson, T. G. Richmond, Y. T. Chen, W. C. Trogler, and F. Basolo, *Inorg. Chem.*, 1981, <u>20</u>, 3137. b) Q.-Z. Shi, T. G. Richmond, W. C. Trogler, and F. Basolo, *J. Am. Chem. Soc.*, 1984, <u>106</u>, 71.
4. D. C. Sonneberger, J. D. Atwood, *Inorg. Chem.*, 1981, <u>20</u>, 3243.
5. a) G. Hutter, J. Schneider, H.-D. Muller, G. Mohr, J. von Seyerl, and L. Wohlfahrt, *Angew. Chem. Int. Ed. Engl.*, 1979, <u>18</u>, 76. b) L. J. Farrugia, J. A. K. Hward, P. Mitroprachachon, J. L. Spencer, F. G. A. Stone, and P. Woodward, *J. Chem. Soc., Chem. Commun.*, 1978, 260. c) L. J. Farrugia, M. Green, D. R. Hankey, A. G. Orpen, and F. G. A. Stone, *J. Chem. Soc., Chem. Commun.*, 1983, 310.
6. a) D.-Y. Jan and S. G. Shore, <u>Organometallics</u>, 1987, <u>6</u>, 428. b) D. P. Workman, D.-Y. Jan, and S. G. Shore,

Inorg. Chem., 1990, 29, 3518. c) D.-Y. Jan, L.-Y. Hsu, D. P. Workman, and S. G. Shore, Inorg. Chem. , 1987, 6, 1984. d) Workman, D. P, H.-B. Deng, and S. G. Shore, Angew. Chemie, Int. Ed. Engl. , 1990, 29, 3518.

7. a) A. J. Deeming and M. Underhill, J. Chem. Soc., Chem. Commun., 1973, 277. b) A. J. Deeming and M. Underhill, J. Chem. Soc., Dalton Trans., 1974, 1415.

8. a) R. Dodsworth, T. Dutton, B. F. G. Johnson, J. Lewis, and P. R. Raithby, Acta Crystallogr., 1989, C45, 707. b) A. A. Aradi, F. W. Grevels, C. Krueger, E. Raabe, Organometallics, 1988, 7, 812. c) D. Seyferth, J. B. Hoke, M. Cowie, and A. D. Hunter, J.Organomet. Chem., 1988, 346, 91. d) T. Albietz, W. Bernhardt, C. von Schnering, E. Roland, H. Bantel, and H. Vahrenkamp, Chem. Ber., 1987, 120,141. e) Roland, E., B. Wolfgang, and H. Vahrenkamp, ibid., 1985, 110 , 2858 (1985).

9. a) D. J. Saturnino, M. Yamauchi, W. R. Clayton, W. R. Nelson, and S. G. Shore, J. Am. Chem. Soc. , 1975, 97, 6063. b) L.-Y. Hsu, J. F. Marategui, K. Niedenzu, and S. G. Shore, Inorg. Chem., 1987, 26, 143. c) R. Boese, P. Paetzold, A. Tapper, and R. Ziembinski, Chem. Ber., 1989, 122, 1057.

10. A. Fratiello, T. P. Onak, and R. E. Schuster, J. Am. Chem. Soc. , 1968, 90, 1194.

11. a) E. Sappa, A. Tripicchio, and P. Braunstein, Chem. Rev. 1983, 83, 203 and references therein. b) P. R. Raithby, and M. J. Rosales, Adv. Inorg. Chem. Radiochem. , 1985, 29, 169.

12. a) J. H. Chung, D. Knoeppel, D. McCarthy, A. Columbie, and S. G. Shore, Inorg. Chem., 1993, 32, 3391. b) B. F. G. Johnson, J. Lewis, W. J. H. Nelson, J. N. Nicholls, J. Puga, P. R. Raithby, M. J. Rosales, M. Schroder, and M. D. Vargas, J. Chem. Soc. Dalton Trans., 1983, 2447.

13. a) K. Wade, Adv. Inorg. Chem. Radiochem. , 1976, 18, 1,. b) R. E. Williams, Inorg Chem. , 1971, 10, 210, (1971). c) R. W. Rudolph, and W. R. Pretzer, Inorg. Chem. , 1972, 11, 1974.

14. a) F.-E. Hong, D. A. McCarthy, J. P. White, III, C. E. Cottrell, and S. G. Shore, Inorg. Chem. , 1990, 28, 3284. b) K. S. Wong, W. R. Scheidt, T. P. Fehlner, J. Am. Chem. Soc., 1982, 104, 1111. c) T. P. Fehlner, C.E. Housecroft, W. R. Scheidt, K. S. Wong, Organometallics., 1983, 7, 2302.

15. T. P. Fehlner, Polyhedron , 1990, 9, 1955.

Routes to Transition Metal–Boron Bonds. Some Principles and Practices

Thomas P. Fehlner*, A. K. Bandyopadhyay, Chang-Soo Jun, Yasushi Nishihara, and Kathryn J. Deck

DEPARTMENT OF CHEMISTRY AND BIOCHEMISTRY, UNIVERSITY OF NOTRE DAME, NOTRE DAME, IN 46556, USA

1 INTRODUCTION

Interest in the chemistry associated with compounds containing transition metal-carbon bonds arises in part from the ability to perturb the chemistry associated with the carbon fragment in useful ways. In principle, a transition metal, when bonded to a fragment containing boron, can also be used to modify the chemistry of the boron fragment. This idea extends to main group fragments in general.[1] Although there is a substantial body of synthetic organometallic chemistry, albeit in continuous development, the situation with boron is not as well developed either in terms of structural types or chemistry. In the following some of our approaches to the formation of compounds containing the metal-boron bond and some of the ways in which the transition metal center perturbs the structure and reactivity of the boron fragments are summarized. Although space does not allow a detailed comparison, the three systems discussed possess interesting similarities and differences to the analogous organometallic compounds.

Discrete Boride Clusters

We have demonstrated in previous work that the reaction of $BH_3 \cdot L$ with a low temperature source of $Fe(CO)_x$ fragments in the appropriate stoichiometric ratio leads to the cluster $HFe_3(CO)_9(H_3BH)$ in good yield.[2] The product can be cleanly deprotonated and reaction of $[Fe_3(CO)_9(H_3BH)]^-$ with $Fe_2(CO)_9$ leads quantitatively to $[HFe_4(CO)_{12}(HB)]^-$.[3] This tetrairon cluster in turn reacts quantitatively with $[Rh(CO)_2Cl]_2$ to yield $1,2\text{-}[Rh_2Fe_4(CO)_{16}B]^-$ which rearranges cleanly into $1,6\text{-}[Rh_2Fe_4(CO)_{16}B]^-$ (Figure 1a).[4] The latter exhibits an octahedral metal framework with a centered, six coordinate boron atom and is structurally similar to other main group atom centered octahedral metal clusters. The IR absorptions associated with the interstitial boron atom have been identified and compared with those for an analogous carbide cluster. The electronic structure of the discrete borides, e.g., $[Rh_2Fe_4(CO)_{16}B]^-$, particularly with regard to their unusual ^{11}B NMR spectroscopic behavior, has been described and compared with similar properties of a carbide cluster.[5] In an investigation of the mechanism of the boride cluster isomerization process, evidence has been presented for a Lewis base promoted process in which rapid, reversible base coordination precedes isomerization as well as ligand substitution. The fact that the isomerization process is more rapid than ligand substitution for PMe₂Ph as the entering ligand demonstrates a special role for the promoter base (Figure 2).[6] Protonation of 1,6-[Rh₂Fe₄(CO)₁₆B]⁻ results in metal cluster disproportionation and the formation of H₂Rh₃Fe₃(CO)₁₅B (Figure 1b).

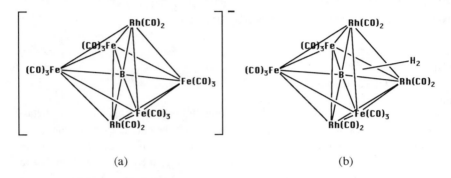

(a)   (b)

<u>Figure 1</u>   Cluster core structures of 1,6-[Rh₂Fe₄(CO)₁₆B]⁻ and H₂Rh₃Fe₃(CO)₁₅B

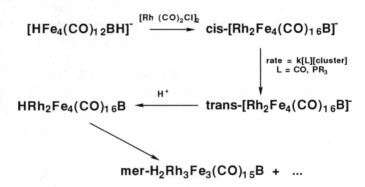

<u>Figure 2</u>   Schematic reaction pathway for the formation and transformation of discrete borides

Carbonyl Cobaltaboranes

The relationship of the geometric and electronic structures of cluster compounds is expressed in its most simple form in the cluster electron counting rules.[7,8]   An emphasis on stable geometries for a given electron count inevitably hides the metal and ligand identities of a given fragment. Also this focus causes relatively little attention to be paid to the differences in structure or chemistry due to element variation at constant cluster electron count. Although external ligands are known to have an effect on cluster structure[9,10], the effects of changing metals while keeping the same type of external ligands is not as well explored. The following shows that metallaboranes formed from borane and cobalt carbonyl fragments can be synthesized. Although there are the expected similarities, the differences between the borane-cobalt carbonyl and borane-iron carbonyl compounds are striking.

There are only a few examples of metallaboranes or carboranes in which a $Co(CO)_x$ rather than $Fe(CO)_3$ or CpCo fragments occupies one or more cluster vertices.[11-14] This is striking considering the important role of, e.g., $Co_2(CO)_8$, in organometallic chemistry. Some earlier work on the reaction of $Co_2(CO)_8$ with $BH_3 \cdot THF$ demonstrated reaction (1), however, the cobaltaborane was not stable with respect to THF cleavage and consequent degradation above 0°C.[15] Although this metal substituted borane was not crystallographically characterized, the structure of a phosphine derivative, $(CO)_2(\eta^1\text{-dppm})Co(\mu\text{-dppm})BH_2$, which has since been published[16], provides geometric parameters for the structure postulated in our work. Unfortunately, $(CO)_4CoBH_2 \cdot THF$ was of no use in building larger clusters. We

$$Co_2(CO)_8 + 2BH_3 \cdot THF \rightarrow 2(CO)_4CoBH_2 \cdot THF + H_2 \qquad (1)$$

have now found that the reaction of $Co_2(CO)_8$ with $BH_3 \cdot SMe_2$ at elevated temperatures produces $B_2H_4Co_2(CO)_6$ as well as minor amounts of other cobaltaboranes including $(CO)_4CoBH_2 \cdot SMe_2$. $B_2H_4Co_2(CO)_6$ is a volatile, very air sensitive, orange compound, which is liquid at room temperature. Up until now it has been isolated only in low yield but has been unambiguously characterized spectroscopically. The low yield is due in part to large losses in purification due to unoptimized procedures for handling this difficult compound and in part to the formation of large amounts of $Co_4(CO)_{12}$. The spectroscopic data are consistent with the structure shown in Figure 3a. The location of the CoHB hydrogen shown is based on variable temperature ^1H NMR and spin saturation transfer experiments. $B_2H_4Co_2(CO)_6$ is strictly isoelectronic with $B_2H_6Fe_2(CO)_6$ and has an analogous structure.

One of the larger clusters formed has also been isolated from the reaction mixture. The compound, $Co_5(CO)_{14}BBH$, has been characterized spectroscopically and its ^{11}B NMR spectrum exhibits resonances consistent with a BH vertex and a boride-like boron atom. The polyhedral electron count of 80 is consistent with a trigonal prismatic cluster with 1 boron and 5 cobalt vertices and an interstitial boron atom, an example of which has been reported recently.[17] However, the crystallographic result (Figure 3b) shows a structure that ostensibly bears little resemblance to this prediction. It contains a B_2H fragment ($d_{BB} = 1.85\text{Å}$) surrounded by a square array of 4 cobalt atoms yielding a distorted octahedron with one of the BCo_2 faces thereby generated capped with a $Co(CO)_3$ fragment. The observed structure can be generated formally from a B atom centered, BCo_5 trigonal

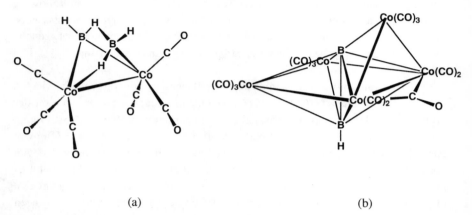

(a)   (b)

<u>Figure 3</u> Structures of $B_2H_4Co_2(CO)_6$ and $Co_5(CO)_{14}BBH$

prismatic cluster by taking the BH vertex and bending it around to cap the rectangular Co_4 face. This model is more consistent with the structural metrics and electron counting rules. The origin of this curious structure is thought to lie in the geometric constraints imposed by the differing Co and B radii in a mixed metal-boron cluster.[18] Although presently obtained in very low yield we suspect that $Co_5(CO)_{14}BBH$ results from the reaction of $Co_2(CO)_8$ with $B_2H_4Co_2(CO)_6$ but this point remains to be established. Thus, the use of cobalt carbonyl fragments allows access to new and unusual metallaborane structures not easily predicted by the existing electron counting rules.

<u>Pentamethylcyclopentadienyl Cobaltaboranes</u>

Due largely to the work of Grimes et al, an extensive series of cobaltaboranes containing CpCo fragments have been generated utilizing pentaborane as the borane source.[19] Curiously, only a few have ferraborane analogs. Thus, we have sought such analogs via the reaction of monoborane adducts with cyclopentadienyl cobalt fragments. We began some time ago with the reaction of $CpCo(PPh_3)_2$ and $BH_3 \cdot THF$. This resulted in the characterization of several cobaltaboranes of unusual composition and structure.[20] The synthetic problem associated with this system is two-fold. First, the presence of copious quantities of the byproduct $R_3P \cdot BH_3$ in the product mixture created difficulties in separation. Second, many of the products contained a PPh_x fragment which, although interesting, suggested our objective could only be attained in the absence of phosphines.

Hence, we sought a related approach in which the metallaboranes and byproducts can be separated on the basis of solubility. The work of Messerle et al has shown that halo-compounds react with borohydrides to yield metallaboranes which are analogs of $B_2H_6Fe_2(CO)_6$, e.g., $B_2H_6Cp*_2Ta_2Br_2$.[21] This has led us to investigate the reaction of the paramagnetic dimer $[Cp*CoCl]_2$, $Cp* = C_5Me_5$,[22,23] with BH_4^- as a potential route to metallaboranes. The target molecules were $B_2H_6Co_2Cp*_2$ and higher nuclearity cluster analogs of our ferraboranes.

The reaction of $[Cp*CoCl]_2$ with $LiBH_4$ results in a single final product, formed in good yield, after about 20 hours at room temperature. It can be conveniently isolated from the salt byproduct by filtration and crystallized. However, paramagnetic impurities are only removed by low temperature chromatography. The product is a diamagnetic molecule that has been fully characterized spectroscopically as well as by a single crystal X-ray diffraction study as the metallaborane, $Cp*_3Co_3(\mu_3-HBH)_2$ (Figure 4a). This cluster is analogous to $Cp_3Co_3(\mu_3-PPh)(\mu_3-BPh)$.[24] (Figure 4b) and constitutes the first example of a metallaborane analog of a biscarbyne complex $Cp_3Co_3(\mu_3-CH)_2$ (Figure 4c). It is an important addition to our "stable" of metallaboranes which are isoelectronic with organometallic clusters but, as yet, has no ferraborane analog.

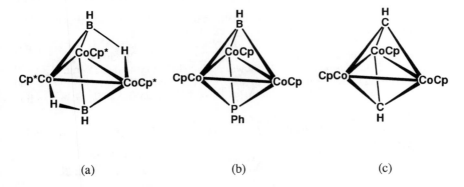

(a)   (b)   (c)

Figure 4 Structures of three analogous M_3E_2 closo clusters.

It is pertinent to note that Theopold et al demonstrated that the major products of the reaction of $[Cp*CoCl]_2$ with AlH_4^- are the paramagnetic hydrides $[Cp*_3Co_3(\mu-H)_3(\mu_3-H)]$ and $[Cp*_2Co_2(\mu-H)_3]$.[25] Now it is well known that borohydrides can also be used to make hydrides via elimination of BH_3. Thus, we find that some of the other products in the reaction of $[Cp*CoCl]_2$ with borohydride can be formed in high yield by the reaction of the dimer directly with $BH_3\cdot THF$. In this manner nido-$1-[\eta^5-Cp*Co]-2-[\eta^4-Me_5C_5HCo](\mu-H)B_3H_7$, which eliminates H_2 to form nido-$2,4-[\eta^5-Cp*Co]_2B_3H_7$, and arachno-$1-[\eta^5-Cp*Co]B_4H_{10}$, which eliminates H_2 to form nido-$1-[\eta^5-Cp*Co]B_4H_8$, have been prepared and characterized.

ACKNOWLEDGMENT
The support of the National Science Foundation is gratefully acknowledged.

REFERENCES

1.   T. P. Fehlner, 'Inorganometallic Chemistry', Plenum, New York, 1992.
2.   X. Meng, A. K. Bandyopadhyay, T. P. Fehlner and F.-W. Grevels, J. Organomet. Chem., 1990, 394, 15.
3.   C. E. Housecroft and T. P. Fehlner, Organometallics, 1986, 5, 379.
4.   R. Khattar, J. Puga, T. P. Fehlner and A. L.Rheingold, J. Am. Chem. Soc., 1989, 111, 1877.
5.   R. Khattar, T. P. Fehlner and P. T. Czech, New J. Chem. , 1991, 15, 705.
6.   A. K. Bandyopadhyay, R. Khattar, J. Puga, T. P. Fehlner and A. L. Rheingold, Inorg. Chem. , 1992, 31, 465.
7.   K. Wade, Adv. Inorg. Chem. & Radiochem. , 1976, 18, 1.
8.   D. M. P. Mingos and R. L.Johnston, Structure and Bonding, 1987, 68, 29.
9.   D. M. P. Mingos, In 'Inorganometallic Chemistry', T. P. Fehlner, Ed.; Plenum: New York, 1992, p 179.
10.   D. M. P. Mingos, Inorg. Chem., 1982, 21, 464.
11.   S. G. Shore, J. Ragaini, T. Schmitkons, L. Barton, G. Medford and J. Plotkin, In 'Abstr. IMEBORON-4', 1979; p 36.
12.   R. L. Sneath, J. L. Little, A. R. Burke and L. J. Todd, Chem. Commun., 1970, 693.
13.   G. Schmid, V. Bätzel, G. Etzrodt and R. Pfeil, J. Organomet. Chem., 1975, 86, 257.
14.   D. M. Schubert, C. B. Knobler, P. A. Wegner and M. F. Hawthorne, J. Am. Chem. Soc., 1988, 110, 5219.
15.   J. D. Basil, A. A. Aradi, N. K. Bhattacharyya, N. P. Rath, C. Eigenbrot, T. P. Fehlner, Inorg. Chem. 1990, 29, 1260.
16.   D. J. Elliot, C. J. Levy, R. J. Puddephatt, D. G. Holah, A. N. Hughes, V. R. Magnuson and I. M. Moser, Inorg. Chem., 1990, 29, 5014.
17.   C. E. Housecroft, D. M. Matthews, A. L. Rheingold and X. Song, Chem. Commun., 1992, 842.
18.   C. S. Jun, T. P. Fehlner and A. L. Rheingold, J. Am. Chem. Soc., 1993, 115, 4393.
19.   R. N. Grimes, Pure & Appl. Chem., 1982, 54, 43.
20.   J. Feilong, T. P. Fehlner and A. L. Rheingold, J. Organomet. Chem., 1988, 348, C22.
21.   C. Ting and L. Messerle, J. Am. Chem. Soc., 1989, 111, 3449.
22.   U. Kölle, F. Khouzami and B.Fuss, Angew. Chem. Int. Ed. Engl., 1982, 21, 131.
23.   U. Kölle, F. Khouzami and B. Fuss, Angew. Chem. Suppl., 1982, 230.
24.   J. Feilong, T. P. Fehlner and A. L. Rheingold, Angew. Chem. Int. Ed. Engl., 1988, 27, 424.
25.   J. L. Kersten, A. L.Rheingold, K. H.Theopold, C. P.Casey, R. A. Widenhoefer and C. E. C. A. Hop, Angew. Chem. Int. Ed. Engl., 1992, 31, 1341.

Organometallacarborane Clusters in Synthesis: Recent Advances and New Directions

Russell N. Grimes

DEPARTMENT OF CHEMISTRY, UNIVERSITY OF VIRGINIA,
CHARLOTTESVILLE, VA 22901, USA

1. INTRODUCTION

The small nido-carboranes $2,3\text{-}RR'C_2B_4H_6$, which have a pyramidal cage structure with a C_2B_3 open face, are the starting point for an area of synthetic organometallic chemistry whose scope seems almost without limit, and whose potential power in "designer chemistry" is becoming increasingly apparent.[1] Until recently, the development of this chemistry was somewhat constrained by safety-imposed limits on the synthesis of $RR'C_2B_4H_6$ from B_5H_9 and alkynes. The highly exothermic nature of this reaction, which was usually conducted without solvent in the presence of a Lewis base, limited it to a scale of 1-2 g of isolated carborane product.[2] However, recent findings have radically altered this picture. Following the discovery by Spencer and Cendrowski-Guillame that the reaction can be run efficiently in dilute THF solution at room temperature,[3] we found that the process can be scaled up and employed with concentrated (2 M) solutions of B_5H_9 in cold diethyl ether, affording 50 g or more of the C,C'-diethylcarborane in a simple and relatively safe procedure (Scheme I).[4]

Scheme I

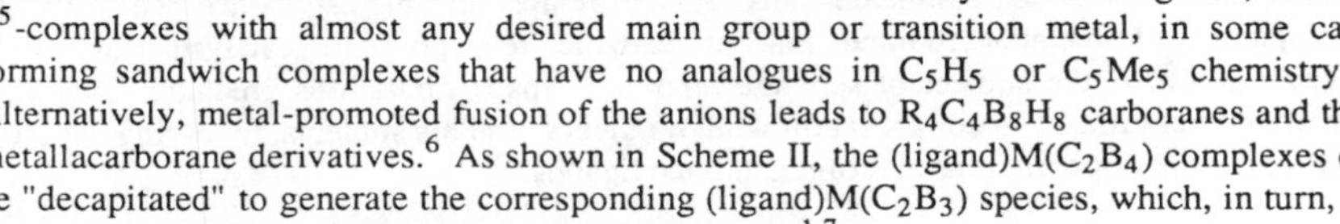

$$Et-C\equiv C-Et \; + \; B_5H_9 \; + \; Et_3N \; \xrightarrow[0°]{Et_2O}$$

0.8 mole   0.8 mole   0.8 mole

(50 g)

$Et_2C_2B_4H_6$
Isolated yield
~ 50 g (50%)

Other alkynes can be used in place of 2-butyne. Bridge-deprotonation of the carborane, via treatment with 1 or 2 equivalents of n-butyllithium in THF, generates the carborane monoanion or dianion. These ions are extraordinarily versatile ligands, forming η^5-complexes with almost any desired main group or transition metal, in some cases forming sandwich complexes that have no analogues in C_5H_5 or C_5Me_5 chemistry.[1,5] Alternatively, metal-promoted fusion of the anions leads to $R_4C_4B_8H_8$ carboranes and their metallacarborane derivatives.[6] As shown in Scheme II, the (ligand)$M(C_2B_4)$ complexes can be "decapitated" to generate the corresponding (ligand)$M(C_2B_3)$ species, which, in turn, are the springboard for a whole host of sandwich systems.[1,7]

Other aspects of metallacarborane chemistry have been developed in several laboratories and were very recently reviewed by Saxena and Hosmane.[8] Related metal-sandwich chemistry based on organoboron ring systems (e.g., C_3B_2, C_4B, C_4B_2) has been extensively explored in the laboratories of Siebert[9] and Herberich.[10] Our own efforts have centered on the role of small metallacarborane sandwich complexes in organometallic chemistry, as reagents for organic chemistry and as building blocks for assembling large organometallic systems having useful electronic, magnetic, or optical properties. In a series of papers and reviews we have described methods for preparing and alkylating (ligand)$M(C_2B_3)$ synthons, especially $Cp^*Co(Et_2C_2B_3H_5)$ ($Cp^* = \eta^5\text{-}C_5Me_5$), and their use in construction of multidecker sandwich and linked-sandwich systems.[1,5] Here I will

Scheme II

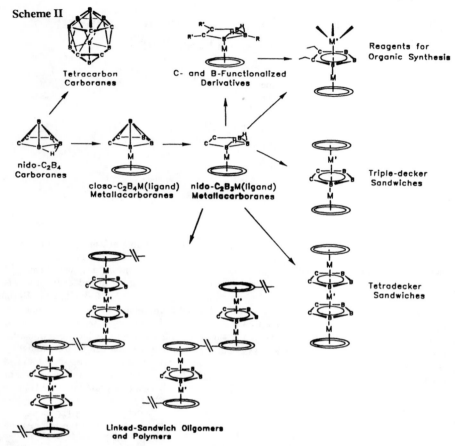

summarize some of our recent work that addresses three aspects of this chemistry: (1) tailoring via introduction of functional groups, (2) studies of the electronic properties of selected systems, and (3) synthesis of metallocene analogues and mixed-sandwich metallocenium metallacarboraneide salts. A fourth area of interest, concerned with the development of metallacarborane Fischer carbene reagents and catalysts for olefin polymerization, is the subject of separate reports.[11] Space limitations preclude discussion here of another project, in collaboration with W. Siebert and his students, involving "hybrid" metal complexes of carborane and organoborane ligands.[12]

2. DERIVATIZATION OF *NIDO*-(ligand)M(C_2B_3) COMPLEXES

Substituent groups on the metallacarborane synthons serve two purposes: they enable the synthesis of linked systems via intermolecular reactions, and they modify the chemistry of the host metallacarborane through electronic or steric effects. In Cp*Co($Et_2C_2B_3H_5$), substitution at B3 (the middle boron) occurs easily by electrophilic attack of alkyl and organomethyl groups on the bridge-deprotonated carborane anion,[13a] but the introduction of electron-withdrawing groups requires other approaches. For example, the B3-acetyl derivative was obtained only via a vinyl ester intermediate (Scheme III).[13b] Mono- and di-B-halo derivatives are readily generated by treatment of the neutral complexes with N-halosuccinimides.[13b]

Scheme III

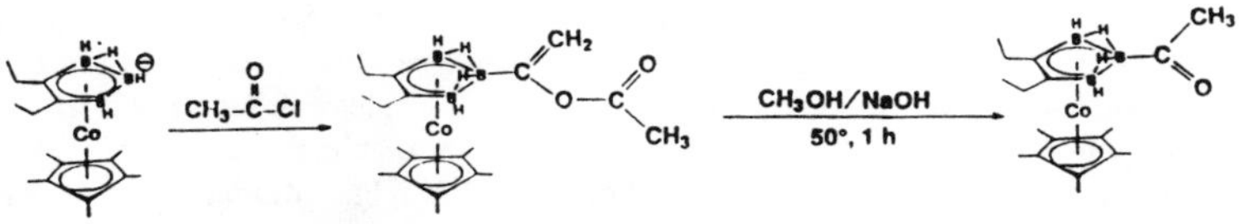

$CH_3-\overset{O}{\underset{}{C}}-Cl$ $\xrightarrow{\begin{array}{c}CH_3OH/NaOH\\50°, 1\ h\end{array}}$

Alternatively, the substituents on the ring *carbons* can be varied by desilylating closo-$Cp*Co(SiMe_3)_2C_2B_4H_4$, lithiating the resulting parent (CH) species, displacing lithium via electrophilic attack, and decapitating to form the nido complex (Scheme IV).[14]

Scheme IV

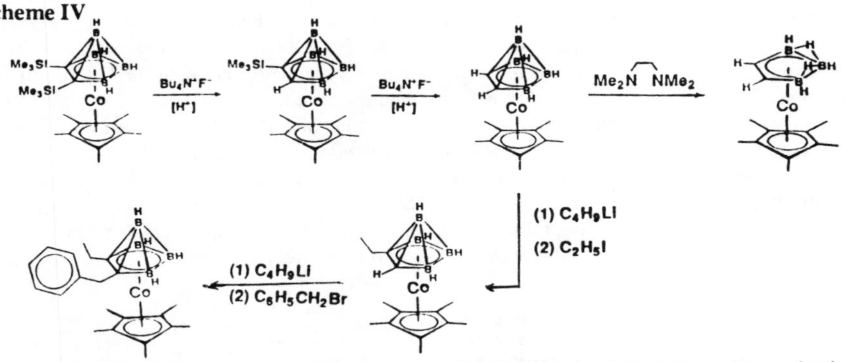

$\xrightarrow[[H^+]]{Bu_4N^+F^-}$ $\xrightarrow[[H^+]]{Bu_4N^+F^-}$ $\xrightarrow{Me_2N\frown NMe_2}$

(1) C_4H_9Li
(2) C_2H_5I

(1) C_4H_9Li
(2) $C_6H_5CH_2Br$

Several examples serve to illustrate the role of derivatized complexes in synthesis. Reactions of monoanion synthons with organic dihalides give bis(metallacarborane) compounds linked at boron or carbon, as desired (Scheme V).[14] By combining the methods for B- and C-alkylation, the possibilities for designed synthesis are considerably extended; a recent example is the preparation of a pentamethylcarboranyl analogue of $Cp*_2Co^+$ ion, described below.

Scheme V

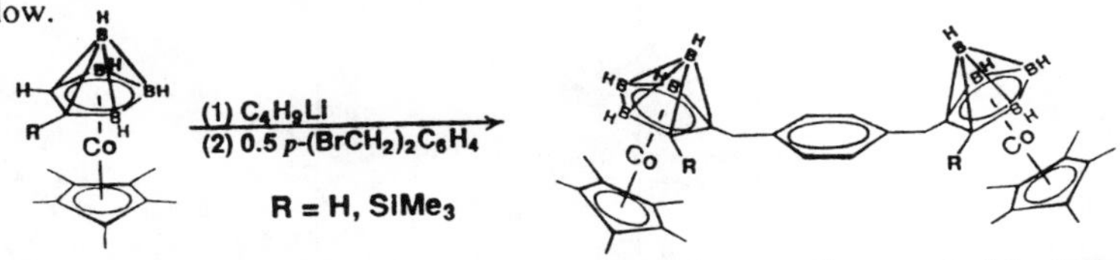

$\xrightarrow{\begin{array}{c}(1)\ C_4H_9Li\\(2)\ 0.5\ p\text{-}(BrCH_2)_2C_6H_4\end{array}}$

$R = H,\ SiMe_3$

Formation of tetradecker sandwich complexes from double-decker $Cp*Co(Et_2C_2B_3H_4\text{-}X)$ synthons[7a] is controlled in part by the nature of the X substituent, the most facile reactions occurring when X is most electron-withdrawing, e.g., Cl, Br, acetyl (Scheme VI).

Scheme VI

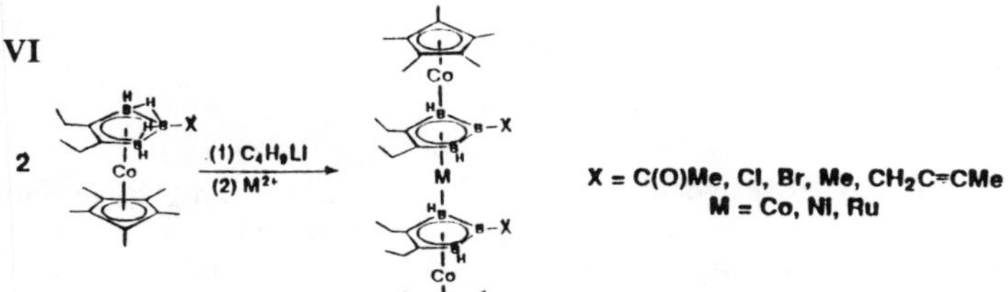

2 $\xrightarrow{\begin{array}{c}(1)\ C_4H_9Li\\(2)\ M^{2+}\end{array}}$

$X = C(O)Me,\ Cl,\ Br,\ Me,\ CH_2C=CMe$
$M = Co,\ Ni,\ Ru$

3. ELECTRONIC STUDIES OF METALLACARBORANE SANDWICHES

Studies of Fe(III)-arene and Ru(III)-arene Complexes. In collaboration with W. E. Geiger and his students, (arene)$M(Et_2C_2B_4H_4)$ mixed-sandwich species were shown to undergo reversible electrochemical oxidation to the M(III) (d^5) cation and (for the Fe system) reduction to the Fe(I) (d^7) monoanion.[15] The cations and anions persist in solution, and the $(arene)Fe(Et_2C_2B_4H_4)^+$ cations, whose ESR spectra indicate similarity to ferrocenium ion, are nearly unprecedented examples of stable Fe(III) arene sandwich compounds. These findings demonstrate the remarkable power of the carborane ligands to stabilize such systems via attenuation of high metal oxidation states in the electron-delocalized cluster framework.

Mononuclear $Cp^*Fe^{III}(Et_2C_2B_4H_4)$ and $Cp^*Fe^{II}H(Et_2C_2B_4H_4)$ ferracarboranes, and the corresponding phenylene-linked dinuclear species, undergo chemical and electrochemical interconversions that have been studied via ESR, magnetic measurements, and multinuclear NMR spectroscopy (Scheme VII).[12] Stepwise reduction of green paramagnetic $Cp^*Fe^{III}(Et_2C_2B_4H_4)$ to its orange diamagnetic anion in THF solution was followed by ^1H and ^{13}C NMR spectroscopy, and the resulting correlation diagrams, plotted as chemical shift vs. mole fraction of reduced species, allowed assignment of all signals in the paramagnetic spectra (extreme left in each diagram below). The slopes of the lines provide measurements of the sensitivity of the electronic environment of each nucleus to the addition of electron density; thus, the cage carbon atoms (line 2 in the ^{13}C diagram) are much more shielded in the anion than in the neutral species. This complex is a rare example of a species whose ESR and NMR spectra are both obtainable.

Scheme VII

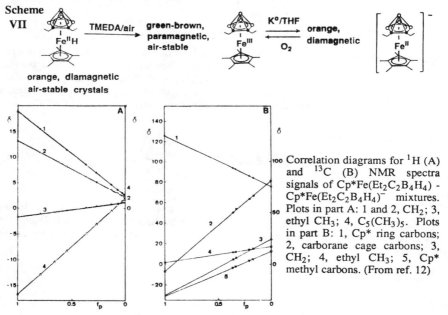

Correlation diagrams for ^1H (A) and ^{13}C (B) NMR spectra signals of $Cp^*Fe(Et_2C_2B_4H_4)$ - $Cp^*Fe(Et_2C_2B_4H_4)^-$ mixtures. Plots in part A: 1 and 2, CH_2; 3, ethyl CH_3; 4, $C_5(CH_3)_5$. Plots in part B: 1, Cp* ring carbons; 2, carborane cage carbons; 3, CH_2; 4, ethyl CH_3; 5, Cp* methyl carbons. (From ref. 12)

The indenylferracarborane system $(\eta^6-C_9H_8)Fe^{III}(Et_2C_2B_4H_4)$ exhibited unusual chemistry.[16] On deprotonation and warming, the iron migrated from the C_6 to the C_5 ring; the resulting anion underwent reversible oxidation and protonation at Fe, as shown in Scheme VIII. The electronic transformations in these species have been explored in detail via cyclic voltammetry, NMR, and ESR spectroscopy.

Addition of nickel occurred not at the indenyl ligand, as intended, but instead on the cage, resulting in cage expansion to form 8-vertex clusters (Scheme VIII).[16] Thus, this complex features a dual personality, exhibiting behavior typical of both polyhedral boranes and organotransition-metal sandwich complexes.

Scheme VIII

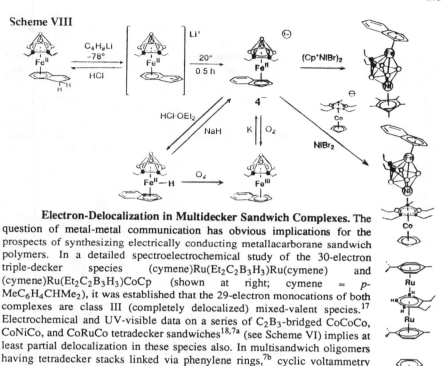

Electron-Delocalization in Multidecker Sandwich Complexes. The question of metal-metal communication has obvious implications for the prospects of synthesizing electrically conducting metallacarborane sandwich polymers. In a detailed spectroelectrochemical study of the 30-electron triple-decker species (cymene)Ru(Et$_2$C$_2$B$_3$H$_3$)Ru(cymene) and (cymene)Ru(Et$_2$C$_2$B$_3$H$_3$)CoCp (shown at right; cymene = *p*-MeC$_6$H$_4$CHMe$_2$), it was established that the 29-electron monocations of both complexes are class III (completely delocalized) mixed-valent species.[17] Electrochemical and UV-visible data on a series of C$_2$B$_3$-bridged CoCoCo, CoNiCo, and CoRuCo tetradecker sandwiches[18,7a] (see Scheme VI) implies at least partial delocalization in these species also. In multisandwich oligomers having tetradecker stacks linked via phenylene rings,[7b] cyclic voltammetry indicates that stack-to-stack communication is blocked, evidently for steric reasons.[18] However, we have prepared multi-tetradecker systems bridged by planar fulvalene groups[7b] in which inter-stack delocalization is anticipated.

4. METALLOCENE ANALOGUES

The planar C$_2$B$_3$H$_5^{4-}$ ligand and its substituted derivatives are isosteric and isoelectronic with η^5-C$_5$H$_5^-$ and η^5-C$_5$Me$_5^-$, so that the carborane ring can function as a surrogate for cyclopentadieneide and thus allow the synthesis of mixed-ligand and mixed-sandwich compounds. These "pseudo-metallocenes" are potential precursors to novel types of stacked-sandwich electroactive or magnetic materials. A recent example from our laboratory is the preparation and structural characterization of the mixed-ligand neutral complex Cp*Co(C$_2$B$_3$Me$_5$H$_2$), depicted in Figure 1.[19] Structural and spectroscopic comparison of this decamethyl species with the isoelectronic decamethylcobaltocenium ion, Cp*$_2$Co$^+$, has allowed direct measurement of the effects of replacing ring carbons with boron atoms.

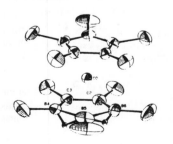
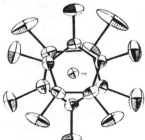

Figure 1. Structure of Cp*Co(C$_2$B$_3$Me$_5$H$_2$), with hydrogens omitted (from ref. 19).

In related work, we have prepared the cobaltocenium metallacarboraneide mixed-sandwich salts Cp_2Co^+ $(Et_2C_2B_4H_4)_2M^-$ ($M = Co^{III}$, Fe^{III}) and Cp_2Co^+ $(Et_2C_2B_3H_4-X)Co^{III}(Et_2C_2B_4H_4)^-$ ($X = H$, Me), the first compounds combining discrete metallocenium cations and metallacarboranide anions (Scheme IX).[20] The crystal packing of the orange diamagnetic cobalt(III) complex revealed pairwise alignment of Cp_2Co^+ and $(Et_2C_2B_4H_4)_2Co^-$ ions, an association that is also seen in the FAB mass spectrum. The dark red paramagnetic Fe(III) salt has a very different crystal packing, currently under study, that features a very long unit cell with three cation-anion pairs in the asymmetric unit.[20] This new class of ionic complexes has a relationship to metallocenium salts of planar organic ligands such as TCNQ (7,7,8,8-tetracyanoquinodimethane) and TCNE (tetracyanoethylene), and stacking to facilitate electron transfer is a distinct possibility that is under investigation.

Scheme IX

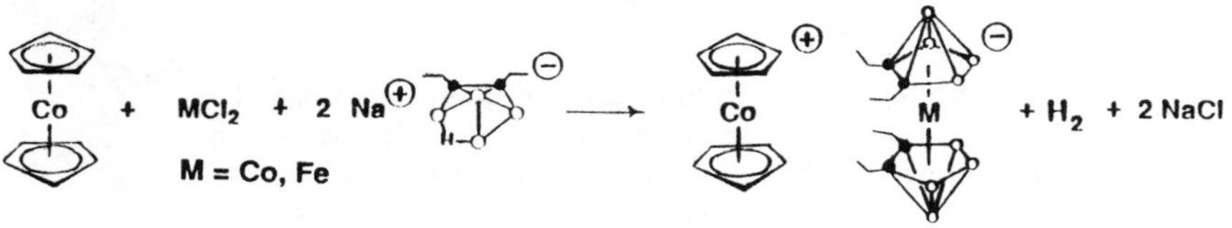

$$M = Co, Fe$$

REFERENCES

1. R. N. Grimes, *Chem. Rev.* **1992**, *92*, 251, and references therein.
2. R. B. Maynard, L. I. Borodinsky, and R. N. Grimes, *Inorg. Synth.* **1983**, *22*, 211.
3. S. M. Cendrowski-Guillame and J. T. Spencer, *Organometallics* **1992**, *11*, 969.
4. K. E. Stockman, X. Meng, B. Gangnus, and R. N. Grimes, unpublished work.
5. R. N. Grimes, in *Electron-Deficient Boron and Carbon Clusters*, G. A. Olah, K. Wade, and R. E. Williams, Eds.; John Wiley and Sons, New York, 1991, Chapter 11, pp. 261-285.
6. (a) R. N. Grimes, *Adv. Inorg. Chem. Radiochem.* **1983**, *26*, 55. (b) R. N. Grimes, in *Advances in Boron and the Boranes [Mol. Struct. Energ. Vol. 5]*, J. F. Liebman, A. Greenberg, R. E. Williams, Eds., VCH Publishers, Inc., New York, 1988, Chapter 11, pp. 235-263, and references therein.
7. (a) K. W. Piepgrass, X. Meng, M. Hölscher, M. Sabat, and R. N. Grimes, *Inorg. Chem.* **1992**, *31*, 5202. (b) X. Meng, M. Sabat, and R. N. Grimes, *J. Am. Chem. Soc.*, in press.
8. A. K. Saxena and N. S. Hosmane, *Chem. Rev.* **1993**, *93*, 1081.
9. W. Siebert, *Angew. Chem. Internat. Edit. Engl.* **1985**, *24*, 943 and *Pure Appl. Chem.* **1987**, *59*, 947.
10. G. E. Herberich, in *Comprehensive Organometallic Chemistry*; G. Wilkinson, F. G. A. Stone, E. Abel, E., Eds., Pergamon Press, Oxford, 1982, Chapter 5.3.
11. (a) K. E. Stockman, M. Sabat, M. G. Finn, and R. N. Grimes, *J. Am. Chem. Soc.*, **1992**, *114*, 8733. (b) K. E. Stockman, M. Sabat, M. G. Finn, and R. N. Grimes, Eighth International Meeting on Boron Chemistry, Knoxville, TN, July 1993.
12. M. Stephan, J. H. Davis, Jr., X. Meng, K. P. Chase, J. Hauss, U. Zenneck, H. Pritzkow, W. Siebert, and R. N. Grimes, *J. Am. Chem. Soc.* **1992**, *114*, 5214, and references therein.
13. (a) J. H. Davis, Jr., M. D. Attwood, and R. N. Grimes, *Organometallics* **1990**, *9*, 1171. (b) K. W. Piepgrass and R. N. Grimes, *Organometallics* **1992**, *11*, 2397. (c) K. W. Piepgrass, K. E. Stockman, M. Sabat, and R. N. Grimes, *Organometallics* **1992**, *11*, 2404.
14. M. A. Benvenuto and R. N. Grimes, *Inorg. Chem.* **1992**, *31*, 3897.
15. J. W. Merkert, W. E. Geiger, M. D. Attwood, R. N. Grimes, *Organometallics* **1991**, *10*, 3545.
16. A. Fessenbecker, M. Stephan, R. N. Grimes, H. Pritzkow, U. Zenneck, and W. Siebert, *J. Am. Chem. Soc.* **1991**, *113*, 3061.
17. J. W. Merkert, J. H. Davis, Jr., W. E. Geiger, and R. N. Grimes, *J. Am. Chem. Soc.* **1992**, *114*, 9846.
18. J. R. Pipal and R. N. Grimes, submitted for publication.
19. M. A. Benvenuto, M. Sabat, and R. N. Grimes, *Inorg. Chem.* **1992**, *31*, 3904.
20. X. Meng, S. Waterworth, M. Sabat, and R. N. Grimes, *Inorg. Chem.*, in press.

Mixed-Ligand Organoborane/Carborane Metal Complexes

W. Siebert

ANORGANISCH-CHEMISCHES INSTITUT, IM NEUENHEIMER FELD 270,
69120 HEIDELBERG, GERMANY

INTRODUCTION

In 1984 we discovered that the bis(η^5-2,3-dihydro-1,3-diborole)nickel complex **1a** undergoes a unique sequence of capping and stacking reactions which led to the first examples of mixed-ligand complexes **3a** and **4a** (n = 1-8) with the new 2,3,5-tricarbahexaborane **5a** and the 2,3-dihydro-1,3-diborole heterocycle **7a**[1]. These findings initiated studies of joining the two separately developed areas of metallacarboranes[2,3] and π-complexes of boron heterocycles[4]. By incorporating carboranyl ligands into bridging and terminal positions of triple- and tetra-decker complexes the HOMO/LUMO of isoelectronic species will be altered, and a stabilization effect of the carboranyl ligands in boron heterocycle π-complexes is expected. In this account results will be presented which have mostly been obtained in cooperation with the groups of R.N. Grimes, L.G. Sneddon and D.M.P. Mingos by combining our 2,3-dihydro-1,3-diborolylmetal π-complex chemistry with that of the carboranyl ligands $C_2B_3H_5$, $C_2B_4H_6^{2-}$, $C_3B_7H_{10}^-$ and $C_2B_9H_{11}^{2-}$.

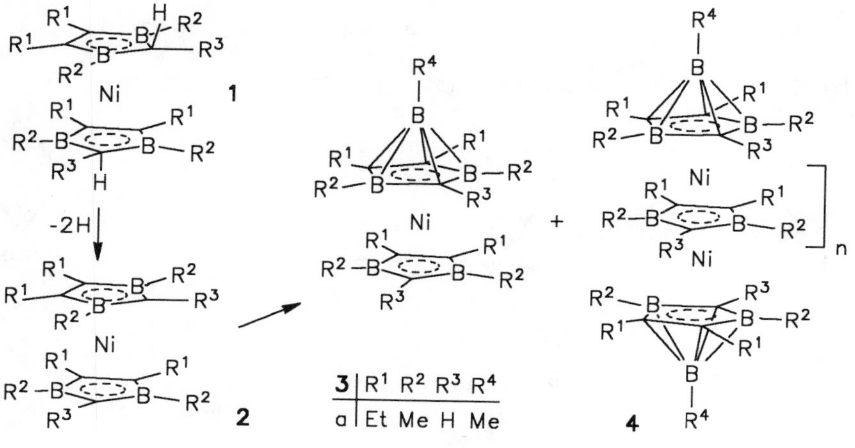

3	R^1	R^2	R^3	R^4
a	Et	Me	H	Me

2,3,5-TRICARBAHEXABORANYLMETAL COMPLEXES

Transformation **1a** into **3a** and **4a** (n = 1-8)

The 2,3-dihydro-1,3-diborole heterocycle 7^5 exhibits remarkable ligand properties toward 14 VE complex fragments such as $(C_5H_5)Co$, $(RC_6H_5)Fe$, $(CO)_3Fe$ and (**7**)Ni. In the complexes the hydrogen at the C2 carbon of **7** forms a C-H-B 3c/2e bond, which allows a 4c/4e interaction C-H-B/C-Co-B. The C-H bond is weak as indicated by a low $^{13}C-^1H$ coupling of J = 80 Hz and facile deprotonation. Presumably in the first step of the 2,3,5-tricarbahexaboranyl formation hydrogen is eliminated from **1a** to yield the 16 VE bis(2,3-dihydro-1,3-diborolyl)nickel intermediate **2a**. Because of its electron-deficiency **2a** interacts with either **1a** or **2a** and by a transfer of a methylborandiyl fragment the $(\eta^5-2,3,5-$tricarbahexaboranyl)(2,3-dihydro-1,3-diborolyl)nickel complex **3a** is formed. In addition to the capping, stacking reactions also occur as indicated by the formation of a mixture of the bis($\eta^5-2,3,5$-tricarbahexaboranyl)(μ, $\eta^5-2,3$-dihydro-1,3-diborolyl)nickel oligodecker complexes **4a** (n = 1-8). The triple-, tetra-, and pentadecker complexes (n = 1-3) could be separated by HPLC, whereas the higher oligodeckers **4a**, n = 4-8, were identified by FAB mass spectra.

Designed Synthesis of 2,3,5-Tricarbahexaboranes **5**

The nido 2,3,5-tricarbahexaborane **5** and its 2,3,4-isomer **6**, first studied by R.N. Grimes et al.[6], are isolobal with cyclopentadiene. In **6** the extra hydrogen bridges the basal boron atoms and (**6**-H) functions as a 5e donor in $[(6-H)Mn(CO)_3]$. The extra hydrogen in **5** is bonded to the basal C5 carbon and forms a weak C-H-B 3c/2e bond with an adjacent boron. An early attempt to synthesize **5** from the 2,3-dihydro-1,3-diborole **7b** and B_2Cl_4 was only partly successful. We could not identify the expected derivative of **5** after cis-addition of B_2Cl_4 to **7b** and reaction of $7b(BCl_2)_2$ with $AlEt_3$ to yield $7b(BEt_2)_2$. It was hoped that $7b(BEt_2)_2$ would lose BEt_3 with formation of **5**. However, a reaction between

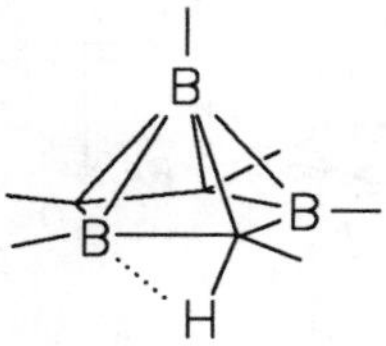

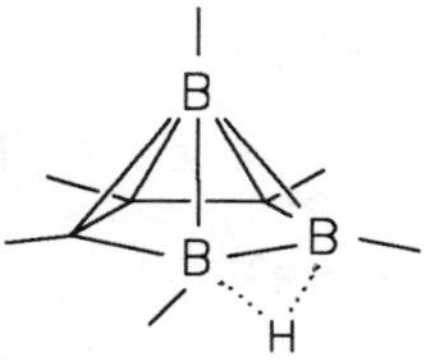

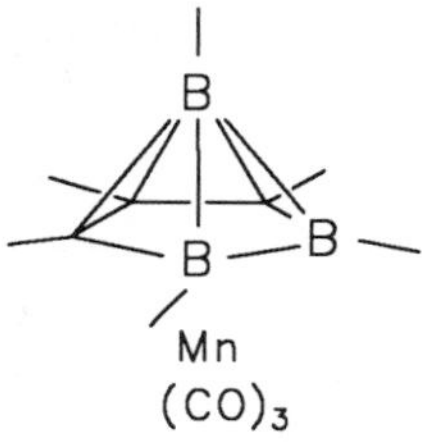

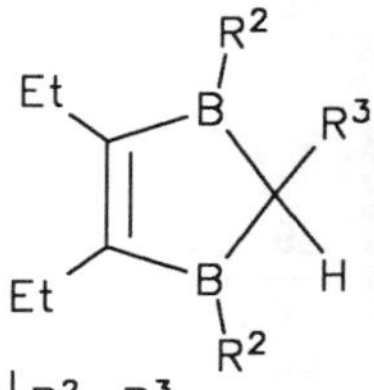

	R^2	R^3
a	Me	H
b	Et	Me

$$\text{8} \xrightarrow[\text{-BEt}_3]{(\text{Et}_2\text{BH})_2} \text{5}$$

8	R^1	R^2
b	Et	Me
c	Et	Et

5	R^1	R^2	R^3	R^4
b	Et	Me	iPr	Et
c	Et	Et	iPr	Et

7b $(\text{BEt}_2)_2$ and $\text{Mn}_2(\text{CO})_{10}$ led to $(5-\text{H})\text{Mn}(\text{CO})_3$ in very low yield, identified by MS. Hydroboration of 1,3-dihydro-1,3-diborafulvenes **8b,c** with diethylborane yields **5b,c** under mild conditions[7]. Long reaction time causes exchange of the boron alkyl groups as demonstrated by GC-MS analysis of the reaction products of **5b** and $\text{Ni}(\eta^3-\text{C}_3\text{H}_5)_2$. Hydroborations of 4,5-bis(isopropyliden)-1,3-diborolanes **9a,b** lead to the diisopropyl derivatives **5c,d** and unexpectedly to the 4,5-diisopropyl-1,3-diborole **7c**.

The constitution of the carboranes **5b-e** follows from spectroscopic data: the proton at the C5 carbon shows in the ^1H-NMR a broad signal at $\delta = -1.34$ (**5b,c**) and -1.37 (**5d,e**). In the ^{11}B-NMR two signals in the ratio 1:2 are observed: $\delta = -36$ and 23 (**5b,c**) and -36 and 22 (**5d,e**) for the apical boron and for the basal boron atoms. The oily carboranes **5b,c** are heat-sensitive; **9b-e** are deprotonated by BuLi, MeLi or NaH to yield light-yellow anions $(5-\text{H})^-$, soluble in hexane. These anions react with $[(\text{C}_3\text{H}_5)\text{NiBr}]_2$ to form the red complexes $(5-\text{H})\text{Ni}(\text{C}_3\text{H}_5)$. Their ^{11}B-NMR spectra are very similar to those of **3a** and **4a**; the basal boron atoms exhibit a shift of $\delta = 10$ and the apical boron of $\delta = -3$. Attempts to replace the allyl group in $(5-\text{H})\text{Ni}(\text{C}_3\text{H}_5)$ by 2,3-dihydro-1,3-diborole **7** did not yield **3**. Independently, **4a** was obtained by oxidative cleavage of the triple-decker **3a** ($n = 1$) in dimethoxyethane. The resulting cation $[(5\text{a}-\text{H})\text{Ni}]^+$ reacts with COD to give $[(5\text{a}-\text{H})\text{Ni}(\text{COD})]^+$. Surprisingly the bis(2,3,5-tricarbahexaboranyl)nickel **10**, a paramagnetic nickelocene analogue, is formed in the reaction of $\text{Ni}(\text{C}_2\text{H}_4)_3$ with **7a**.

$$\text{9} \xrightarrow[\text{-BEt}_3]{(\text{Et}_2\text{BH})_2} \text{5} + \text{7c}$$

9	R^2
a	Me
b	Et

5	R^1	R^2
d	iPr	Me
e	iPr	Et

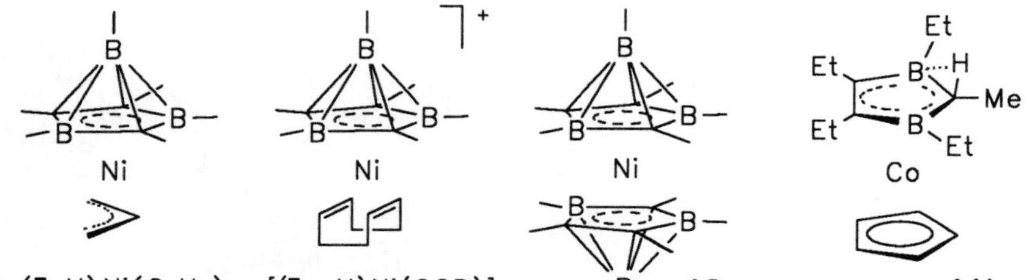

(5-H)Ni(C₃H₅)   [(5a-H)Ni(COD)]   B 10   11b

The compounds **3a**, **4a** (n = 1), [(5a-H)Ni(COD)]$^+$ and **10** have been studied by X-ray structure analyses[1,8,9]. In the carboranylnickel complexes the 2,3,5-tricarbahexaboranyl ligands function as 5e ligands.

2,3-DICARBAHEXABORANYL-1,3-DIBOROLYL-METAL COMPLEXES

The controlled displacement of cyclooctatriene in $(C_8H_{10})Fe[(RC)_2B_4H_4]$ (R = Et, Bz) by the 18 VE sandwich **11b** leads to the triple-decker complexes **12a,b**[10]. The paramagnetic compounds undergo apex BH-removal ("decapitation") on treatment with $(Me_2NCH_2)_2/H_2O$ to yield **13a,b**. A general approach to **15a,b** and **16a** is provided by the reaction between (11b-H)$^-$, $[(RC)_2B_4H_4]^{2-}$ (R = Et, Bz) and Co^{2+}/Rh^{3+}. **15** and **16** are 30 VE triple-deckers, which are characterized by NMR, MS, electrochemistry and an X-ray structure analysis (**16a**)[11]. Apex-removal from **15a,b** leads to **14a,b**. The anion (14a-H)$^-$ reacts with organometal fragments to give the first mixed-ligand (C_3B_2/C_2B_3) tetradecker complexes **17a,b**, **18a,b** and **19a**[12]. The X-ray diffraction study of **18a** confirms the tetradecker structure, in which the rings are planar, but not quite parallel, resulting a slighly "bent" tetradecker. (11b-H)$^-$ and the anion $[(C_5Me_5)Fe(EtC)_2B_4H_4]^-$ react with Ni^{2+} to yield the trinuclear complex **20a**[13].

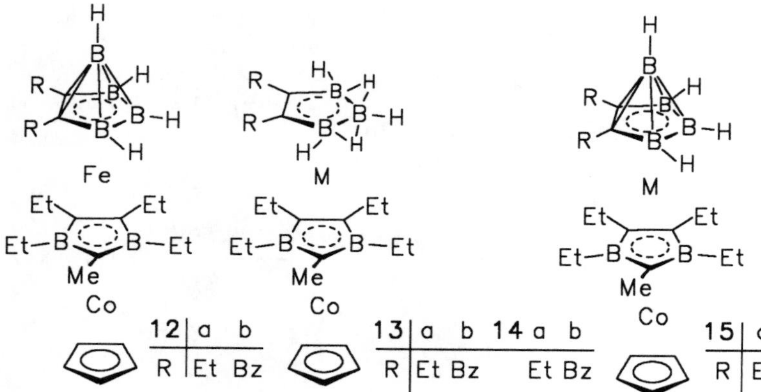

12	a	b
R	Et	Bz

13	a	b
R	Et	Bz
M	Fe	Fe

14	a	b
	Et	Bz
	Co	Co

15	a	b	16a
R	Et	Bz	Et
M	Co	Co	Rh

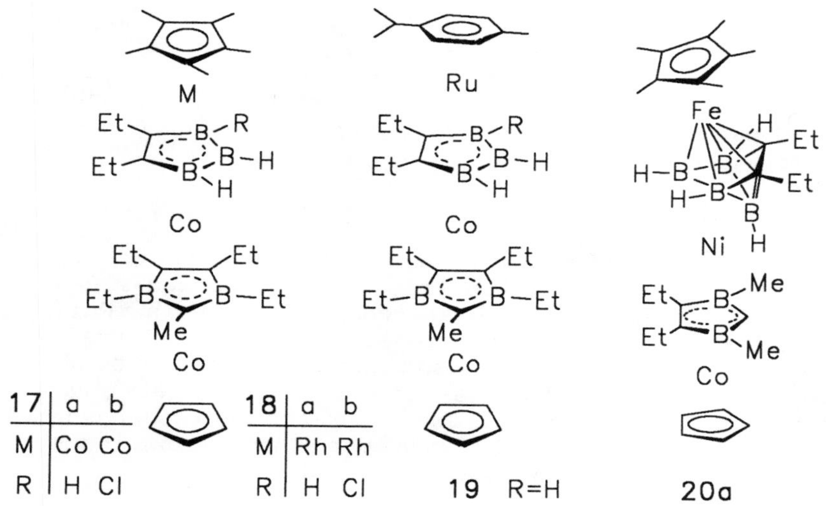

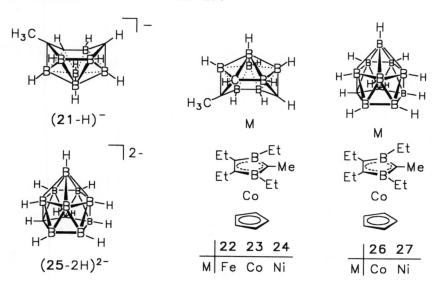

C_3B_7 and C_2B_9 CARBORANYL-METAL-1,3-DIBOROLYL COMPLEXES

Our efforts to synthesize other "hybrid" complexes have led to **22-24** in which the sandwich anion (**11b**-H)⁻ and the tricarbadecaborane anion (**21**-H)⁻ are complexed by Fe^{2+}, Co^{2+}, and Ni^{2+} to give the dinuclear species[14]. The C_3B_7M clusters have 24, 25 and 26 framework electrons, which indicates a nido structure for **24**, a closo arrangement for **22** and an intermediate structure for **23**. The X-ray diffraction analyses confirm nido structures for **24** and for **23**.

(21-H)⁻

M

M

(25-2H)²⁻

	22	23	24
M	Fe	Co	Ni

	26	27
M	Co	Ni

The synthesis of the two novel triple-decker compelxes **26** and **27** is achieved by reacting the anions **(11b**-H)$^-$ and **(25**-2H)$^{2-}$ with Co^{2+} and Ni^{2+}, respectively[15]. X-ray structure analyses reveal that in the CoCo complex **26** the Co to C$_2$B$_3$-ring distance is shorter than that of **27**, the latter having one electron in an antibonding MO.

ACKNOWLEDGEMENTS

I wish to thank my coworkers for their valuable contributions to develop this field of mixed-ligand organoborane/carborane metal complexes. This was made possible through the cooperations with R.N. Grimes, L.G. Sneddon and D.M.P. Mingos and their coworkers, to whom we owe our thanks. The research reported here was generously supported by the Deutsche Forschungsgemeinschaft (SFB 247), the Fonds der Chemischen Industrie, the State of Baden-Württemberg, and the BASF AG.

REFERENCES

1. T.Kuhlmann, H.Pritzkow, U.Zenneck, W.Siebert, *Angew.Chem.Int.Ed.Engl.* 1984, 23, 965.
2. G.B.Dunks, M.F.Hawthorne in E.L.Muetterties, *Boron Hydride Chemistry*, Academic Press, New York 1975, 383.
3. R.N. Grimes, *Coord. Chem.Rev.* 1979, 28, 47; R.N. Grimes, *Chem. Rev.* 1992, 92, 251.
4. W.Siebert, *Adv. Organmet. Chem.* 1980, 18, 301.
5. W.Siebert, *Angew.Chem. Int.Ed.Engl.* 1985, 24, 943.
6. R.N.Grimes, C.L.Bramlett, R.L.Vance, *Inorg.Chem.* 1968, 7, 1066.
7. A.Feßenbecker, A.Hergel, R.Hettrich, V.Schäfer, W.Siebert, *Chem.Ber.* 1993, in press.
8. J.Zwecker, T.Kuhlmann, H.Pritzkow, W.Siebert, U.Zenneck, *Organometallics* 1988, 7, 2316.
9. J.Zwecker, H.Pritzkow, U.Zenneck, W.Siebert, *Angew.Chem.Int.Ed.Engl.* 1986, 25, 1099.
10. M.D.Attwood, K.K.Fonda, R.N.Grimes, G.Brodt, D.Hu, U.Zenneck, W.Siebert, *Organometallics* 1989, 8, 1300.
11. A.Feßenbecker, M.D.Attwood, R.F.Bryan, R.N.Grimes, M.K.Woode, M.Stephan, U.Zenneck, W.Siebert, *Inorg. Chem.* 1990, 29, 5157.
12. A.Feßenbecker, M.D.Attwood, R.N.Grimes, M.Stephan, H.Pritzkow, U.Zenneck, W.Siebert, *Inorg.Chem.* 1990, 29, 5164.
13. A.Feßenbecker, M.Stephan, R.N.Grimes, H.Pritzkow, U.Zenneck, W.Siebert, *J.Am.Chem.Soc.* 1991, 113, 3061.
14. W.Weinmann, H.Pritzkow, W.Siebert, B.Barnum, L.G.Sneddon, in preparation.
15. J.M.Forward, D.M.P.Mingos, W.Siebert, J.Hauß, H.R.Powell, *J.Chem.Soc. Dalton Trans.* 1993, 1783.

Novel Reactivities of *closo* Group 11 Metallacarboranes

Y.-W. Park[1], J. Kim[2], S. Kim[3], and Youngkyu Do[1,*]

[1]DEPARTMENT OF CHEMISTRY AND CENTER FOR MOLECULAR SCIENCE, KOREA ADVANCED INSTITUTE OF SCIENCE AND TECHNOLOGY, TAEJON 305-701, KOREA
[2]DEPARTMENT OF CHEMISTRY, KONGJU NATIONAL UNIVERSITY, KONGJU 314-701, KOREA
[3]LUCKY CENTRAL RESEARCH INSTITUTE, TAEJON 302-343, KOREA

Many research works on metallacarborane chemistry have been directed toward metal–metal bond formation with the aid of dicarbollide anion 7,8-*nido*-$C_2B_9H_{11}^{2-}$ and its homologues.[1] Particularly, the involvement of a heterodinuclear monoanionic metallacarborane with a Tl–group 6 metal bond in the formation of multinuclear metallacarborane systems[2,3] has prompted us to synthesize new heterodinuclear metallacarboranes containing a thallium ion and to utilize them as synthons.[4] Described herein are brief accounts of the use of thallium salts of group 11 metallacarboranes as direct complexation reagents as well as cage transfer reagents.

A set of group 11 carborane complexes $Tl[(PPh_3)MC_2B_9H_{11}]$ (M = Cu, **1a**; Ag, **1b**) was prepared and utilized as synthons in the synthesis of heterobimetallic metallacarboranes as outlined in Scheme 1. The off-white microcrystalline compound **1a** was isolated quantitatively from an equimolar slurry of $Tl_2C_2B_9H_{11}$ and $Cu(PPh_3)_3Cl$ in DMF and characterized by several means. Spectroscopic data suggest that in solution the compound **1a** has the *closo* nature with non-interacting thallium cation whereas it may interact with the carborane cage in solid state. The compound **1b** which is insoluble in most organic solvents was generated analogously *in situ* and used in subsequent reactions.

Direct complexation of $[M'(PPh_3)]^+$ with $[(PPh_3)CuC_2B_9H_{11}]^-$, observed in the reaction systems of **1a**/CuCl/PPh$_3$ in CH$_2$Cl$_2$ and **1a**/AgBr/2PPh$_3$ in CH$_2$Cl$_2$, afforded bimetallic complexes $[4,8\text{-}exo\text{-}\{M'(PPh_3)\}\text{-}4,8\text{-}(\mu\text{-}H)_2\text{-}3\text{-}(PPh_3)\text{-}3,1,2\text{-}closo\text{-}$

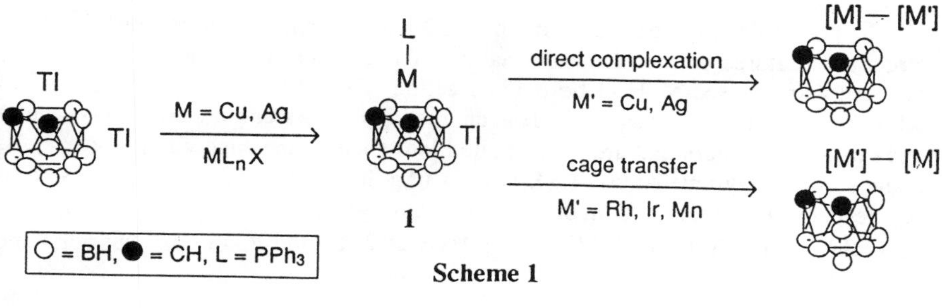

$O = BH, \bullet = CH, L = PPh_3$

Scheme 1

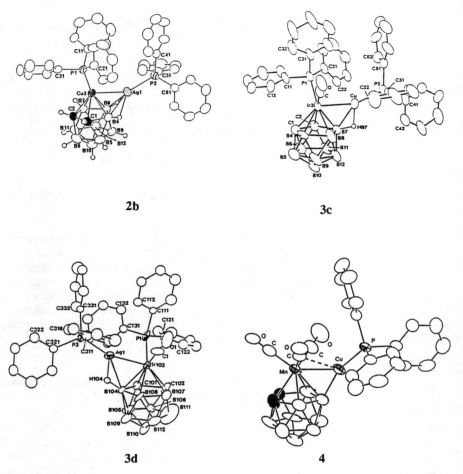

2b   3c

3d   4

CuC$_2$B$_9$H$_9$)] (M' = Cu, 2a; Ag, 2b), respectively. On the other hand, a set of reactions of 1/M'(CO)(PPh$_3$)$_2$Cl (M'=Rh, Ir) gave rise to cage-transferred dinuclear complexes [4-*exo*-{M(PPh$_3$)}-4-(μ-H)-3-(PPh$_3$)-3-(CO)-3,1,2-*closo*-M'C$_2$B$_9$H$_{10}$]   (M'−M = Rh−Cu, 3a; Rh−Ag, 3b; Ir−Cu, 3c; Ir−Ag, 3d) containing bonding interaction between group 9 and 11 metals. The cage transfer was also observed in the reaction of 1a/Mn(CO)$_5$Br, yielding [8-*exo*-{Cu(PPh$_3$)}-8-(μ-H)-3,3,3-(CO)$_3$-3,1,2-*closo*-MnC$_2$B$_9$H$_{10}$] (4).

The molecular structures of 2b, 3c, 3d and 4, shown above, reveal unique structural features of the *closo-exo* nature of two heterometal centers with close proximity and closed M−H−B bridge interaction. The compound 2b is isostructural with the well characterized[5] dinuclear cupracarborane 2a and contains *closo*-Cu(I) and *exo*-Ag(I) centers that are incorporated into a carborane cage via the open C$_2$B$_3$ plane and two asymmetric Ag−H−B bridges (Ag−B: 2.395, 2.509; Ag−H: 2.28, 2.34 Å), respectively, resulting in intermetallic Cu−Ag distance of 2.622(1) Å. The observation of v(B−H) of bridging Ag−H−B at 2316 and 2346 cm^{-1} supports the

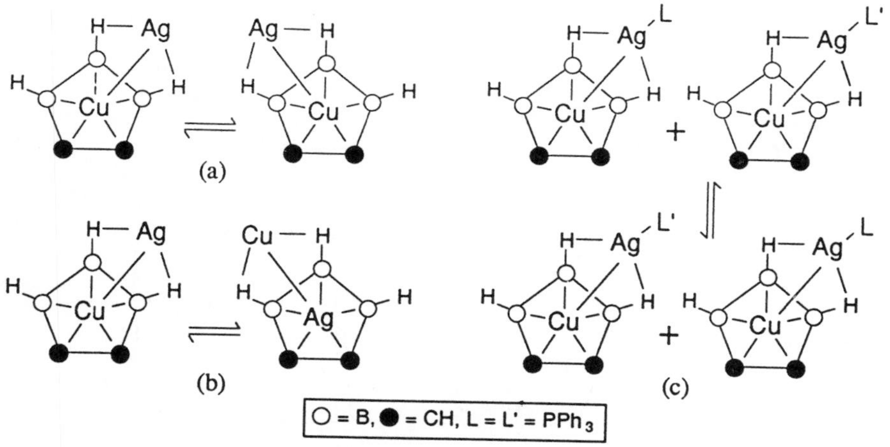

Scheme 2. Three conceivable fluxional motions for the compound **2b**: (a) wigwag motion, (b) *closo/exo* exchange, and (c) intermolecular phosphine exchange.

presence of the bridge interactions.

Among three conceivable fluxional motions for the compound **2b** in solution, outlined in Scheme 2, the ^{1}H, ^{11}B, and ^{31}P NMR spectra support the presence of the wigwag type motion as low as 203 K and of the intermolecular phosphine exchange above 263 K in solution phase of **2b** whereas the homodinuclear complex **2a** is known to undergo both wigwag and *closo/exo* exchange motions.[5] For **2b**, the phosphorus line splitting of Ag–*P*Ph$_3$ due to the coupling with ^{107}Ag and ^{109}Ag becomes coalesced above 263 K, indicating the presence of intermolecular phosphine exchange, a rather common property observed for silver complexes of phosphine.[6] It can be noted that the value of 1.157 which is the ratio of coupling constants $^1J(^{109}$Ag–P$) = 648$ to $^1J(^{107}$Ag–P$) = 587$ is in good agreement with the theoretical prediction of 1.150.

The crystal structure of **3d** contains two enantiomers as independent molecules in the asymmetric unit and shown above is the R form based on the CIP sequence rule.[7] In the compounds **3c** and **3d**, the [M(PPh$_3$)]$^+$ (M = Cu or Ag) unit interacts with *closo*-IrC$_2$B$_9$ cage via a closed M–H–B bridge bond involving asymmetric boron atom in the open C$_2$B$_3$ plane and Ir–M interaction (Ir–Cu, 2.526(1) Å; Ir–Ag, 2.633(1) and 2.649(1) Å), giving rise to C_1 solid symmetry. The ^{1}H and ^{11}B{^{1}H} (5) NMR spectra of the compounds **3** reveal that in solution only the Ir–Ag system maintains structural integrity while the others show pseudo C_s solution symmetry owing to fluxionalities. The ^{31}P NMR spectra (6; 298 K for M–Cu and 203 K for M–Ag systems) of **3** rule out the possibility of *closo/exo* exchange motion.

The X-ray analysis on **4** reveals several unique solid structural features. The *closo* nature of Mn(CO)$_3$ fragment is manifested as zero value of the slipping parameter Δ.[8] The [Cu(PPh$_3$)]$^+$ unit interacts with the *closo*-MnC$_2$B$_9$ cage via an exopolyhedral σ-Cu–B bond at the symmetric boron atom and a Mn–Cu bond

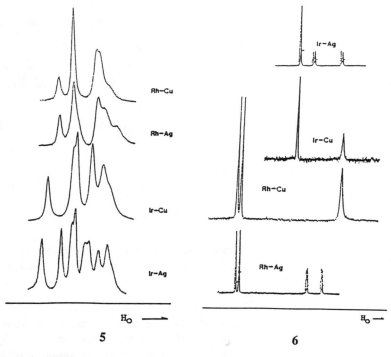

5   6

(2.563(1) Å). The copper atom interacts very weakly with one of carbonyl carbon atom (Cu–C, 2.436(4) Å), resulting in a weak semi-bridging CO with Mn–C–O angle of 172.7(4)°. In solution, the observations of one carboranyl proton signal and five distinct signals with 1:2:2:1:3 ratio suggest the C_s nature of the C_2B_9 cage.

Support of this work by the Korea Science and Engineering Foundation is gratefully acknowledged.

REFERENCES

1. A. K. Saxena and N. S. Hosmane, Chem. Rev., 1993, 93, 1081.

2. Y. Do, C. B. Knobler and M. F. Hawthorne, J. Am. Chem. Soc., 1987, 109, 1853.

3. J. Kim, Y. Do, Y. S. Sohn and M. F. Hawthorne, J. Organomet. Chem., 1991, 418, C1.

4. J. Kim, S. Kim and Y. Do, J. Chem. Soc., Chem. Commun., 1992, 938.

5. (a) Y. Do, H. C. Kang, C. B. Knobler and M. F. Hawthorne, Inorg. Chem., 1987, 26, 2348. (b) H. C. Kang, Y. Do, C. B. Knobler and M. F. Hawthorne, Inorg. Chem., 1987, 27, 1716.

6. S. Attar, N. W. Alcock, G. A. Bowmaker, J. S. Frye, W. H. Bearden and J. H. Nelson, Inorg. Chem., 1991, 30, 4166 and references cited therein.

7. (a) R. S. Cahn, C. Ingold and V. Prelog, Angew. Chem., Int. Ed. Engl., 1966, 5, 385. (b) V. Prelog and G. Helchen, Angew. Chem., Int. Ed. Engl., 1982, 21, 567.

8. (a) D. M. P. Mingos, M. I. Forsyth and A. J. Welch, J. Chem. Soc., Chem. Commun., 1977, 605. (b) D. M. P. Mingos, M. I. Forsyth and A. J. Welch, J. Chem. Soc., Dalton Trans., 1978, 1363.

Polyhedral Metallaborane and Metallaheteroborane Chemistry. Aspects of Cluster Flexibility and Cluster Fluxionality

John D. Kennedy[1] and Bohumil Štíbr[2]

[1]SCHOOL OF CHEMISTRY, THE UNIVERSITY OF LEEDS, LEEDS LS2 9JT, UK
[2]INSTITUTE OF INORGANIC CHEMISTRY, THE ACADEMY OF SCIENCES OF THE CZECH REPUBLIC, 25068 ŘEŽ NEAR PRAGUE, THE CZECH REPUBLIC

1. Introduction

Metallaboranes and metallaheteroboranes exhibit a fascinating variety of equilibria, some quite dynamic. Some of these dynamic processes occur very readily indeed, which implies transition states and intermediates with very similar energies to those of the ground-state conformers. By modification of cluster constituents and cluster substituents, models for these intermediates can be stabilised, either electronically or sterically, and isolated. This can generate interesting chemistry. In particular, new bonding and structural patterns may be revealed. The systematisation of chemical observations into related patterns is one of the important steps by which the science develops. For the further development of polyhedral boron chemistry there is therefore some importance in the recognition, description, and delineation of any new patterns that may emerge from exploratory work. In this paper we develop this context by the examination of some aspects of experimentally observed behaviour, principally in nine-, ten-, and eleven-vertex boron-containing cluster chemistry, although it seems likely that the general perceptions follow through to twelve- and eight-vertex cluster species also.

2. Background

Simple structural and electronic patterns associated with the observed cluster structures of the basic boranes, and of related species such as the dicarbaboranes, were recognised and classified some twenty years ago by Williams and Wade,[1,2] and others also contributed soon afterwards.[3,4] The basic cluster structural patterns are based on the series of most symmetrical closed triangulated polyhedral, known as closed deltahedra, and these are said to have *closo* geometries. Such a delta-hedron that lacks one vertex has *nido* cluster geometry, and conventionally it is the vertex of highest connectivity that is missing. If it has an additional missing vertex, conventionally adjacent to the first, then it has *arachno* cluster geometry. Removal of a third vertex generates *hypho* geometry. At the simplest level, the electronic patterns are based on electron-counting schemes. If electrons that are associated with the cluster bonding are counted up according to a set of criteria known as Wade's Rules,[2] then an n-vertex *closo* cluster notionally has $2n + 2$ cluster bonding electrons, an n-vertex *nido* has $2n + 4$, an n-vertex *arachno* has $2n + 6$, and an *n*-vertex *hypho* has $2n + 8$. The electron counts can be used to predict geometries. These geometric and electron-counting formalisms work surprisingly well, and not only for clusters containing boron. They are now well exemplified in the general literature, and they have played and will continue to play a key role in the development of the extensive and still expanding cluster chemistry of boron.

However, exceptions to these patterns have long been recognised. Well-known examples here include B_8H_{12} which has an *arachno* rather than a *nido* eight-vertex geometry,[5] the *closo*-structured neutral perhaloboranes B_nX_n that have $2n$ rather than $2n + 2$ skeletal electrons,[6] and the two metallaboranes $[(C_5H_5)_4Co_4B_4H_4]$ and $[(C_5H_5)_4Ni_4B_4H_4]$ which also have essentially *closo* cluster structures but which have respectively $2n$ and $2n + 4$ cluster electrons associated with them.[7] Most of these long-recognised examples have the conventional geometries of the Williams-Wade formalism, but are anomalous in terms of their formal cluster electron-counts. By contrast, some more recently reported anomalies include examples which differ geometrically from the classical Williams-Wade patterns, in that they exhibit closed and open deltahedra of regular shape that have different geometries to those associated with conventional *closo, nido, arachno, etc.*[8,9,10,11] This paper addresses some recent experimental work and perceptions that impinge upon this general area.

3. Aspects of fluxionality

Fluxionality pervades polyhedral boron chemistry. It is readily examined by NMR spectroscopy. The first fluxional compound to be recognised was a metalla-borane.[12] Atoms of the same element that exchange position among different chemical environments will give a single NMR resonance line if the rate of exchange is substantially faster than $\pi\Delta v/2^{1/2}$, but will show separate lines if the rate is substantially slower. As the rate of exchange increases from the limiting slow case, the separate lines broaden, coalesce, and then sharpen again to give the limiting fast-exchange single resonance. For two equally populated sites, at coalescence the activation energy $\Delta G^{\ddagger}$ at the experimental temperature T is given by $\Delta G^{\ddagger} = -RT\ln(\pi\Delta vh/2^{1/2}kT)$. Fluxionalities that cannot be stopped on the NMR time-scale down to about -100°C imply $\Delta G^{\ddagger}$ values of < *ca.* 30 kJ mol^{-1}. This puts an upper limit of this magnitude on the differences in energy among the ground state, transition states, and intermediates involved in such a fluxional process. Consequences of this in structural interpretation include the following: *(a)* crystal packing forces can easily approach this sort of value so that a crystallographically determined molecular structure may not represent the most stable conformer; *(b)* calculations that purport to demonstrate relative stabilities in such systems must be conducted at sufficiently high levels that calculated differences in energy of these sorts of magnitude not only emerge, but also are meaningful.

4. Aspects of ten-vertex cluster behaviour

The *closo* metalladicarbaborane $[2-(\eta^5-C_5Me_5)-2,1,6-RhC_2B_7H_9]$ together with its $2-\{Ru(\eta^6-C_6Me_6)\}$ and $2-\{RhH(PPh_3)_2\}$ analogues are fluxional between enant-iomers with activation energies $\Delta G^{\ddagger}$ of *ca.* 58, 45 and 60 kJ mol^{-1} respectively.[13]

The assigned NMR spectra suggest that the fluxionality occurs *via* a linked double diamond-square-diamond (DSD) process involving intermediates with square-faced *isonido* structures. These must be energetically close (< *ca.* 50 kJ mol^{-1}) to the *closo* structures, and thereby amenable to stabilisation by use of appropriate cluster constituent and substituent. Ten-vertex *isonido* is derived geometrically from eleven-vertex *closo* by the removal of a vertex of cluster connectivity four, rather than by the removal of the six-connectivity vertex that generates the conventional ten-vertex *nido* geometry. Actual *isonido* compounds are themselves difficult to come by, but we have isolated and identified some {IrB$_9$} examples, *e.g.* [(PPh$_3$)$_2$IrB$_9$H$_{10}$(PPh$_3$)] and [Cl(PPh$_3$)-μ-(Ph$_2$PC$_6$H$_4$)IrB$_9$H$_5$(OEt)(PPh$_3$)].[8,14]

Completion of a single DSD process from *closo* through intermediate *isonido* generates a closed cluster of formal C_{3v} symmetry. This *isocloso* shape derives geometrically from the complete capping of a nine-vertex *arachno* shape by a six-connectivity vertex. The *isocloso* type is well recognised , with [Fe$_2$C$_2$B$_6$},[15] {RuC$_2$B$_7$},[15] {IrB$_9$},[16] {RuB$_9$},[17] {Os B$_9$},[18] and {WB$_9$}[19] compounds structurally characterised, and with {FeB$_9$},[20] {RhB$_9$},[21] and {ReB$_9$}[18] species also recorded. The *closo - isonido - isocloso* DSD sequence forms the basis of a structural continuum. A plot of the diagonals *versus* the fold angle for the four atoms involved in the DSD process for *closo, isonido,* and *isocloso* {IrB$_9$} cluster compounds illustrates this.[8] Here, interestingly, many ten-vertex *closo* compounds show partial opening to *isonido* in the ground state.[9] Known compounds generate a continuous progression from *closo* to *isonido*, but there is a gap between *isonido* and *isocloso*. It is not certain whether this is a function of the chemistry that happens to have emerged so far, or whether there is a catastrophic structural "flip" as the fold angle passes through zero.

5. Aspects of eleven-vertex cluster behaviour

The conventional eleven-vertex *closo* species [B$_{11}$H$_{11}$]$^{2-}$ and [CB$_{10}$H$_{11}$]$^-$ exhibit a very facile fluxionality, suggesting that open-faced structures are energetically very close to the classical eleven-vertex C_{2v} closed deltahedron.[22] Compounds such as [1-(η^6-C$_6$Me$_6$)-1,2,4-RuC$_2$B$_8$H$_8$Me$_2$}] and [1,1,1-H(PPh$_3$)$_2$-1,2,4-IrC$_2$B$_8$H$_{10}$] of conventional Wadian *closo* eleven-vertex electron count in fact exhibit quadrilateral open faces and open *isonido* geometry.[11,23] This open face is not hydrogen-bridged, but compounds such as [(C$_5$Me$_5$)$_2$RhHIrB$_9$H$_{10}$][24] and [(C$_5$Me$_5$)$_2$Rh$_2$-B$_9$H$_8$Cl(NHEt$_2$)][25] have open-face hydrogen bridges (two and one respectively). Taken successively, these last two species, followed by the eleven-vertex {Rh$_2$B$_9$} subcluster of [(C$_5$Me$_5$)RhB$_9$(SMe$_2$)H$_{10}$RhB$_9$H$_7$(SMe$_2$)],[26] and then by [(PMe$_3$)$_2$H(PPh$_3$)-μ-(Ph$_2$PC$_6$H$_4$)PtIrB$_9$H$_{10}$],[27] exhibit steps along a continuum from *isonido* to pentagonally open-faced classical *nido*. This particular sequence entails a successive increase in the intermetal distance. A compound such as [2-(C$_6$Me$_6$)-"*nido*"-2,8,10-OsC$_2$B$_8$H$_8$-8,10-Me$_2$],[28] still of formal *closo* electron count, thence demonstrates further opening to hexagonally open-faced eleven-vertex "*isoarachno*" geometry. The eleven-vertex *isonido* and *isoarachno* shapes both derive from the twelve-vertex *isocloso* icosahedron (Section 7 below) by the removal of either a four-connected or a six-connected vertex respectively. The *isoarachno* opening is reminiscent of the anomalous opening observed in the ten-vertex formally *nido* $[1-(C_6Me_6)-1-RuB_9H_{13}]$,[29] of which the basic cluster structure derives from the removal of adjacent six- and five-connected vertices from the twelve-vertex *isocloso* shape.

6. Aspects of nine-vertex cluster behaviour

Many nine-vertex classical *closo* compounds also in fact show partial opening towards a quadrilaterally faced *isonido* shape.[9] Isomeric $[(CO)(PPh_3)_2IrCB_7H_8]$ and $[(CO)(PPh_3)HIrCB_7H_7(PPh_3)]$ thence demonstrate completely open *isonido* structures.[10] This *isonido* geometry is generated by the removal of a four-connectivity vertex from the C_{3v} ten-vertex *isocloso* shape. Continuation of the DSD sequence here generates nine-vertex *isocloso* geometry of C_{2v} symmetry. This is derived geometrically by the complete capping of an eight-vertex *arachno* shape by a six-connectivity vertex. Structurally it is exemplified so far only by $[(PMe_3)_2HIrB_8H_7Cl]$,[30] although $[(PPh_3)_3HReB_8H_8]$ has been tentatively identified by NMR spectroscopy.[18] Again a plot of diagonal versus fold angle for the four DSD atoms illustrates the continuum.[10] Again there is a gap between *isonido* and *isocloso*, but again this aspect of chemistry is as yet very underinvestigated.

Another sequence terminating in the nine-vertex *isocloso* structure is that of *arachno* - *nido* - *isocloso* cluster closure formally represented by the structures of $[(CO)(PMe_3)_2HIrB_8H_{11}Cl]$,[31] $[(CO)(PMe_3)_2IrB_8H_{11}]$,[32] and $[(PMe_3)_2H-IrB_8H_7Cl]$.[30] Here the metal centre shows successively a three-, a four-, and a six-connectivity to a basic $\{B_8\}$ *arachno* shape;[3] this possibly reflects a successive two-, three- and four-orbital involvement of the metal centre with the cluster. If this last hypothesis holds, then *arachno* - *nido* - *isocloso* represents an equally fundamental systematic closure sequence to that of the classical *arachno* - *nido* - *closo* progression; in this context it should be noted that the closer relationship between *arachno* and *nido*, and a corresponding much more distant relationship between classical *closo* and these two, has recently been pointed out.[34] From these alternative viewpoints,[33,34] it is interesting that the formal successive addition of electron pairs in the *isocloso* - *nido* - *arachno* cluster-opening sequence actually results in the successive removal of iridium-based atomic orbitals from the cluster bonding scheme, rather than addition of electrons to the cluster bonding itself.

7. Aspects of eight- and twelve-vertex cluster behaviour

Fewer results are available in twelve- and eight-vertex boron chemistry, but corresponding *isocloso* and *isonido* behaviour is already either suggested or apparent from experimental results. In twelve-vertex chemistry $[\{(PEt_3)_2Pt\}-(\mu-CO)_2-WC_2B_9H_8Me_2(CH_2C_6H_4-4-Me)]$[35] exhibits an "*isocloso*" icosahedral structure derived from classical *closo* icosahedral by a diamond-square-diamond process, and $[(C_5Me_5)RhC_2B_9H_9Ph_2]$[36] exhibits a quadrilaterally open-faced "*isonido*" structure intermediate between these two extremes. Here also $[(PMe_2Ph)_2Pt_2B_{10}H_{10}]$ of classical *closo* geometry exists in equilibrium with its (reasonably presumed) *isocloso* isomer,[37] emphasizing the energetic similarities of the two cluster types when the cluster constituents are apposite. The twelve-vertex *isocloso* structure is the parent for *isonido* and *isoarachno* shapes of lower vertex number by vertex-removal as discussed above. Conversely, the twelve-vertex *isonido* derives by vertex-removal from the thirteen-vertex *closo* structure that also has vertices of four- and six-connectivity. In eight-vertex chemistry $[B_8H_8]^{2-}$, $[CB_7H_8]^-$, $[CB_7H_7I]^-$, and $[CB_7H_6I_2]^-$ all have classical dodecahedral *closo* structures in the solid state in salts examined so far,[38] but they are all very fluxional in solution, which must imply open-face isomers of very similar energies. This in turn implies that suitable cluster substituents or constituents will readily stabilise open clusters in this eight-vertex system also. It may be that relatively simple modifications, such as protonation,[39] or crystallisation with a small polarising countercation, can achieve this in the ionic species just mentioned.

8. Conclusions

The experimental results summarised above show that many classically *closo* species in twelve-, eleven-, ten-. nine-, and also probably eight-vertex boron cluster chemistry exhibit partial opening towards square-faced *isonido* structures, and with suitable cluster constituents and substituents a complete opening can be induced to generate stable *isonido* species. These openings may be induced sterically, electronically, or sometimes by crystal-packing forces if the energetic differences are small. Completion of this DSD process gives *isocloso* structures, which again are exemplified experimentally. Geometrical removal of vertices from *isocloso* can generate the *isonido*, and also "*isoarachno*", geometries, and examples of many of these have also been isolatable. There are obvious parallels here to the vertex-removal concepts involved in the classical Williams-Wade[1,2] structural patterns. In accord with these parallels it is possible[10] to start the construction of a structural paradigm for these "*iso*" clusters that seems to be analogous to those drawn up, for example by Rudolph,[3] for the more classical borane cluster shapes. It is emphasised that these new cluster shapes have their own symmetries and (as judged by NMR) their own electronic structures, and are not local distortions of the classical types. It is also emphasised that in general these "*iso*" shapes are not merely artefacts of kinetic control. Among the many alternative structure-energy surfaces in the multidimensional structure-energy continuum they represent alternative energy minima that are stabilised by the incorporation of cluster constituents other than the main-group carbon and boron centres that are generally taken to exemplify borane cluster chemistry. Sometimes these other surfaces are close to, or can overlap with or cross over, the classical Williams-Wade surface, and then alternative minima are available to simpler borane or carbaborane clusters. These conditions then generate, for example, the electron-count anomalies referred to in Section 2 above, and the observed fluxionalities[22,40] in species such as $[B_8H_8]^{2-}$ and $[B_{11}H_{11}]^{2-}$.

There are interesting bonding questions to pose and to answer here.[41,42] and further experimental work is needed to distinguish and test among possible electronic models as a prelude to rigorous theoretical work. We have started to tackle this in the eleven-vertex systems of classical[1,2] *closo* shape, for which we have so far observed[43] no fewer than five distinct and successively more compact cluster bonding modes within the constraints of the formal C_{2v} symmetry: (*a*) bridged *arachno* with a two-orbital metal-to-cluster interaction as in the $[Ph_2PB_{10}H_{12}]^-$ anion,[44] (*b*) bridged *nido*, also with a two-orbital interaction, as in

$[Me_2SnC_2B_8H_{10}]$,[43,45] *(c)* true *closo* with a three-orbital interaction as in $[SnC_2B_8H_{10}]^{46}$ and $[\{P(OMe)_3\}_2PtC_2B_8H_{10}]$,[45] and then *(d)* and *(e)* two more compact types that may involve four-orbital metal-to-cluster interactions, *viz.* as in successively more compact $[(PMe_2Ph)_2PtC_2B_8H_9Ph]^{43,47}$ and $[(C_5Me_5)Ir\text{-}B_{10}H_{10}]$.[48] This progression emphasises the variation in contribution that one particular vertex can make to a cluster bonding scheme, which is sometimes open to some discussion.[41,42,49] In the $\{PtC_2B_8\}$ clusters, for example, a change of metal ligand or cluster substituent can induce a change in metal oxidation state, either electronically,[43] or sterically.[45] The fluxional molecule $[(PMe_2Ph)_2Pt\text{-}C_2B_8H_9Me]^{43}$ shows that in this system the difference in energy between the formal oxidation states of platinum(II) and platinum(IV) is $< ca.$ 60 kJ mol^{-1}. This difference is associated with a concomitant fundamental change in the nature of the metal-to-cluster interaction within the same C_{2v} cluster geometry. Obviously it is important[8,41] to accommodate the possibilities of such variations in metal-vertex behaviour when setting up the bases of any molecular orbital schemes[49] which attempt to account for these new structure types.

9. Acknowledgements

Contribution no. 40 from the Řež-Leeds Anglo-Czech Polyhedral Collaboration (ACPC). We have pleasure in acknowledging the hard work and enthusiasm of all other past and present full, associate, and honorary members of the ACPC, and in thanking the Academy of Sciences of the Czech republic, the UK SERC, Borax Research Ltd., and the Royal Society for support of work in this general area.

10. References

1.   R. E. Williams, *Inorg. Chem.*, 1971, **10**, 210; *Adv. Inorg. Chem. Radiochem.*, 1978, **18**, 64.
2.   K. Wade, *J. Chem. Soc., Chem. Commun.*, 1971, 792; *Adv. Inorg. Chem. Radiochem.*, 1978, **18**, 1.
3.   R. W. Rudolph and W. R. Pretzer, *Inorg. Chem.*, 1972, **11**, 1974; R. W. Rudolph, *Accounts Chem. Res.*, 1976, **9**, 446.
4.   D. M. P. Mingos, *Nature (Phys. Sci.)*, 1972, **99**, 236.
5.   R. E. Enrione, F. P. Boer, and W. N. Lipscomb, *J. Am. Chem. Soc.*, 1964, **86**, 1451; *Inorg. Chem.*, 1964, **3**, 1659.
6.   See, for example, J. A. Morrison, in *Advances in Boron and the Boranes*, Eds. J. L. Liebman, A. Greenberg, and R. E. Williams, VCH, Weinheim and New York, 1988, Ch. 8, pp. 151-189, and references cited therein.
7.   J. R. Bowser, A. Bonny, J. R. Pipal, and R. N. Grimes, *J. Am. Chem. Soc.*, 1979, **101**, 6229; J. R. Pipal and R. N. Grimes, *Inorg. Chem.*, 1979, **18**, 257.
8.   J. Bould, J, D. Kennedy, and M. Thornton-Pett, *J. Chem. Soc., Dalton Trans.*, 1992, 563;
9.   K. Nestor, B. Štíbr, J. D. Kennedy, M. Thornton-Pett, and T. Jelínek, *Collect. Czech. Chem. Commun.*, 1992, **57**, 1262.
10. B. Štíbr, J. D. Kennedy, M. Thornton-Pett, E. Drdáková, T. Jelínek, and J. Plešek, *Collect. Czech. Chem. Commun.*, 1992, **57**, 1439; B. Štíbr, J. D. Kennedy, E. Drdáková, and M. Thornton-Pett, *J. Chem. Soc., Dalton Trans.*,

1993, in press (results on *isonido* {IrCB$_7$} clusters).

11. K. Nestor, X. L. R. Fontaine, N. N. Greenwood, J. D. Kennedy, J. Plešek, B. Štíbr, and M. Thornton-Pett, *Inorg. Chem.*, 1989, **28**, 2219.

12. R. A. Ogg and J. D. Ray, *Disc. Farad. Soc.*, 1955, **19**, 239.

13. M. Bown, T. Jelínek, B, Štíbr, S. Heřmánek, X. L. R. Fontaine, N. N. Greenwood, J. D. Kennedy, and M. Thornton-Pett, *J. Chem. Soc., Chem. Commun.*, 1988, 974; K.Nestor, M. Murphy, B. Štíbr, T. R. Spalding, X. L. R. Fontaine, M. Thornton-Pett, and J. D. Kennedy, *Collect. Czech. Chem. Commun.*, 1993, **58**, 1555.

14. J. E. Crook, N. N. Greenwood, J. D. Kennedy, and W. S. McDonald, *J. Chem. Soc., Chem. Commun.*, 1981, 933; J. Bould, P. Brint, J. D. Kennedy, and M. Thornton-Pett, *Acta Cryst., Sect. C*, 1990, **46**, 1010.

15. K. P. Callahan, W. J. Evans, F. Y. Lo, C. E. Strouse, and M. F. Hawthorne, *J. Am. Chem, Soc.*, 1975, **97**, 296; C. W. Jung, R. T. Baker, C. B. Knobler, and M. F. Hawthorne, *J. Am. Chem. Soc.*, 1980, **102**, 5782.

16. J. Bould, N. N. Greenwood, and J. D. Kennedy, *J. Chem. Soc., Dalton Trans.*, 1990, 1451.

17. J. E. Crook, M. Elrington, N. N. Greenwood, J. D. Kennedy, M. Thornton-Pett, and J. D. Woollins, *J. Chem. Soc., Dalton Trans.*, 1985, 2407; E. J. Ditzel, X. L. R. Fontaine, N. N. Greenwood, J. D. Kennedy, and M. Thornton-Pett, *J. Chem. Soc., Chem. Commun.*, 1989, 1115,

18. R. S. Coldicott, Ph.D. Thesis, University of Leeds, 1993; see also R. S. Coldicott, elsewhere in this book.

19. I. Macpherson, Ph.D. Thesis, University of Leeds, 1988; see also pp. 229-230 in J. D. Kennedy, *Boron Chemistry; Proceedings of the Sixth International Conference on Boron Chemistry (IMEBORON VI)*, Ed. S. Hermánek, World Scientific, Singapore, 1987, 207.

20. R. P. Micciche, J. J. Briguglio, and L. G. Sneddon, *Inorg. Chem.*, 1984, **23**, 3992.

21. H. Fowkes, unpublished results, University of Leeds, 1985.

22. R. E. Wiersema and M. F. Hawthorne, *Inorg. Chem.*, 1973, **12**, 785; E. I. Tolpin and W. N. Lipscomb, *J. Am. Chem. Soc.*, 1973, **95**, 2384.

23. M. Bown, X. L. R. Fontaine, N. N. Greenwood, J. D. Kennedy, and M. Thornton-Pett, *Organometallics*, 1987, **6**, 2254; M. Bown, X. L. R. Fontaine, N. N. Greenwood, J. D. Kennedy, and M. Thornton-Pett, *J. Chem. Soc., Dalton Trans.*, 1990, 3039.

24. K. Nestor, X. L. R. Fontaine, N. N. Greenwood, J. D. Kennedy, and M. Thornton-Pett, *J. Chem. Soc., Chem. Comm.*, 1989, 455.

25. E. J. Ditzel, Zhu Sisan, and M. Thornton-Pett, unpublished results, University of Leeds, 1989.

26. E. J. Ditzel, X. L. R. Fontaine, N. N. Greenwood, J. D. Kennedy, and M. Thornton-Pett, *J. Chem. Soc., Chem. Commun.*, 1989, 1262.

27. J. Bould, J. E. Crook, J. D. Kennedy, and M. Thornton-Pett, *Inorg. Chim. Acta*, 1992, **203**, 193.

28. M. Bown, X. L. R. Fontaine, N. N. Greenwood, J. D. Kennedy, and M. Thornton-Pett, *J. Chem. Soc., Chem. Commun.*, 1987, 1650.

29. M. Bown, X. L. R. Fontaine, N. N. Greenwood, J. D. Kennedy, and P. MacKinnon, *J. Chem. Soc. Chem. Commun.*, 1987, 817.

30. J. Bould, J. E. Crook, N. N. Greenwood, J. D. Kennedy, and W. S. McDonald, *J. Chem. Soc., Chem. Commun.*, 1982, 346.

31. J. Bould, J. E. Crook, N. N. Greenwood, and J. D. Kennedy, *J. Chem. Soc., Dalton Trans.*, 1984, 1903.

32. J. Bould, N. N. Greenwood, and J. D. Kennedy, *J. Chem. Soc., Dalton Trans.*, 1984, 2477.

33. See, for example, pp. 249-253 in J. D. Kennedy, *Prog. Inorg. Chem.*, 1986, **34**, 211, and references cited therein.

34. M. J. Moore and P. Brint, *J. Chem. Soc., Dalton Trans.*, 1993, 427.

35. M. J. Attfield, J. A. K. Howard, A. N. de Jelfs, C. M. Nunn, and F. G. A. Stone, *J. Chem. Soc., Dalton Trans.*, 1987, 2219; see also N. Carr, D. F. Mullica, E. L. Sappenfield, and F. G. A. Stone, *Organometallics*, 1992, **11**, 3697.

36. Z. G. Lewis, and A. J. Welch, *J. Organomet. Chem.*, 1992, **430**, C45.

37. Y. M. McInnes and M. Thornton-Pett, *Abstracts Sixth International Meeting on Boron Chemistry (IMEBORON VI), Bechyně, The Czech Republic, June 1987*, p 55, Abstract no. CA 16; J. D. Kennedy, Y. M. McInnes, and M. Thornton-Pett, 1993 (to be) submitted (results on closed {Pt_2B_{10}} clusters).

38. T. Jelínek, B. Štíbr, J. D. Kennedy, and M. Thornton-Pett, *J. Chem. Soc., Dalton Trans.*, 1993, (to be) submitted (results on closed {CB_7} clusters), and references cited therein.

39. J. W. Bausch, G. K. Surya Prakash, and R. E. Williams, *Inorg. Chem.*, 1992, **31**, 3763; M. Bühl, A. M. Mebel, O. P. Charkin, and P. v. R. Schleyer, *Inorg. Chem.*, 1992, **31**, 3769.

40. E. L. Muetterties, *Tetrahedron*, 1974, **30**, 1595; E. L. Muetterties, E. L. Hoel, C. G. Salentine, and M. F. Hawthorne, *Inorg. Chem.*, 1975, **14**, 950; E. L. Muetterties, R. J. Wiersema, and M. F. Hawthorne, *J. Am. Chem. Soc.*, 1973, **95**, 7520.

41. See, for example, R. T. Baker, *Inorg. Chem.*, 1986, **25**, 109, and J. D. Kennedy, *Inorg. Chem.*, 1986, **25**, 111, together with references cited therein.

42. See, for example, pp. 223-225 in A. A. Aradi and T. P. Fehlner, *Adv. Organomet. Chem.*, 1990, **30**, 189.

43. J. D. Kennedy, B. Štíbr, T. Jelínek, X. L. R. Fontaine, and M. Thornton-Pett, *Collect. Czech. Chem. Commun.*, 1993, in press (results on C_{2v} {MC_2B_8} clusters)

44. M. Thornton-Pett, M. A. Beckett, and J. D. Kennedy, *J. Chem. Soc., Dalton Trans.*, 1986, 303.

45. J. D. Kennedy, K. Nestor, B. Štíbr, M. Thornton-Pett, and G. S. A. Zammit, *J. Organomet. Chem.*, 1992, **437**, C1.

46. K. Nestor, B. Štíbr, T. Jelínek, and J. D. Kennedy, *J. Chem. Soc., Dalton Trans.*, 1993, 1661.

47. J. D. Kennedy, B. Štíbr, M. Thornton-Pett, and T. Jelínek, *Inorg. Chem.*, 1991, **30**, *4481*.

48. K. Nestor, V. Petříček, M. Thornton-Pett, and J. D. Kennedy, to be submitted (results on closed {$(C_5Me_5)IrB_9$} and {$(C_5Me_5)IrB_{10}$} clusters).

49. R. L. Johnston, D. M. P. Mingos, and P. Sherwood, *New J. Chem.*, 1991, **15**, 831, and references cited therein.

Syntheses and Structures of Transition Metal Complexes of Lewis Base Adducts of Borane and Borylene

M. Shimoi[1], K. Katoh[2], M. Uruichi[2], S. Nagai[1], and H. Ogino[2]

[1]THE UNIVERSITY OF TOKYO, COLLEGE OF ARTS AND SCIENCES, KOMABA 3-8-1, MEGURO, TOKYO 153, JAPAN
[2]TOHOKU UNIVERSITY, FACULTY OF SCIENCE, SENDAI 980, JAPAN

1   INTRODUCTION

Neutral Lewis base adducts of monoborane, $BH_3 \cdot L$, have not been known to coordinate to any metal ions, although isoelectronic tetrahydroborate, BH_4^-, is known to be a versatile ligand. On the other hand, $B_2H_4 \cdot 2PMe_3$, an electron precise borane, has been known to act as a uni- or bidentate ligand.[1,2] We reported that photochemical reactions of group 6 metal carbonyls were very useful in the preparation and isolation of complexes containing weakly coordinating $B_2H_4 \cdot 2PMe_3$.[2] These results prompted us to try to photochemical synthesis of $BH_3 \cdot L$ complexes (L = Lewis base). Tertiary phosphine- and amine-boranes afforded $[M(CO)_5(BH_3 \cdot L)]$ complexes (M = Cr, W), in which $BH_3 \cdot L$ is coordinated to the metal through a single M-H-B linkage. Dehydrogenation of secondary and primary amine-boranes took place by the photochemical reaction of $[M(CO)_6]$ to form aminoboranes or borazine derivatives. Reaction of $B_2H_4 \cdot 2PMe_3$ with $[Co_2(CO)_8]$ afforded a dinuclear complex bridged with a phosphine adduct of borylene.

2. SYNTHESES AND STRUCTURES OF $[M(CO)_5(BH_3 \cdot L)]$

Ultraviolet irradiation of $M(CO)_6$ (M = Cr, W) in toluene in the presence of $BH_3 \cdot L$ (L = PMe_3, PPh_3, NMe_3) followed by recrystallization from pentane afforded $[M(CO)_5(BH_3 \cdot L)]$:

$$[M(CO)_6] + BH_3 \cdot L \xrightarrow{h\nu} [M(CO)_5(BH_3 \cdot L)] + CO \qquad (1)$$

The corresponding molybdenum complexes could not be isolated under similar reaction conditions, although their formation was confirmed by NMR spectroscopy. Figures 1 and 2 show X-ray structures of $[Cr(CO)_5(BH_3 \cdot PMe_3)]$ at -60°C and $[W(CO)_5(BH_3 \cdot PMe_3)]$ at -50°C, respectively. Trimethylphosphine-borane adopts staggered conformation with respect to the B-P axis and coordinates to the metal through a single M-H-B linkage: M-H 1.97 (11), 2.21 (13), M···B 2.79 (1), 2.87 (1) Å, and M-H-B 129 (10), 114 (9) °

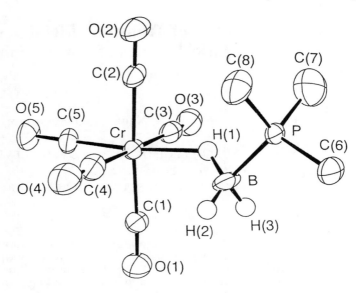

<u>Figure 1</u> X-ray Structures of $[Cr(CO)_5(BH_3 \cdot PMe_3)]$ at $-60\ °C$

for Cr and W complexes, respectively. Although hydrogen
positions were not determined by the X-ray study of
$[Cr(CO)_5(BH_3 \cdot NMe_3)]$ and $[W(CO)_5(BH_3 \cdot PPh_3)]$, the $M \cdots B$
interatomic distances clearly show the unidentate
coordination mode in these complexes. ^1H NMR spectra of
$[M(CO)_5(BH_3 \cdot L)]$ (M = Cr, W) show only one BH proton
signal, 4–6 ppm to higher field than the free $BH_3 \cdot L$
indicating that the $BH_3 \cdot L$ ligand also acts as a unidentate
ligand in solution but there is a fast exchange process
of coordination site between the bridging and terminal BH
protons. The activation energy of this exchange process
should be very small, since in $[W(CO)_5(BH_3 \cdot PMe_3)]$ the
exchange is rapid even at $-100\ °C$. Further photolysis of
$[M(CO)_5(BH_3 \cdot L)]$ did not afford complexes containing
bidentate $BH_3 \cdot L$, although the only known BH_4^- complexes of
group 6 metal carbonyls are $[M(CO)_4(BH_4)]^-$ in which BH_4^-
acts as a bidentate ligand.

      3   PHOTOCHEMICAL DEHYDROGENATION OF AMINE–BORANES
WITH ONE OR TWO N–H GROUPS IN THE PRESENCE OF $[M(CO)_6]$

Photochemical reaction of borane adducts of secondary
amines in the presence of $[M(CO)]_6$ undertook
dehydrogenation of amine–borane to give dimers of
aminoboranes:

$$BH_3 \cdot NHR_2 \xrightarrow{\quad h\nu / [M(CO)_6] \quad} 1/2[BH_2NR_2]_2 + H_2 \atop R = CH_3, C_2H_5 \qquad (2)$$

      The reaction did not take place in the absence of
$[M(CO)_6]$. The reaction proceeded even after the UV

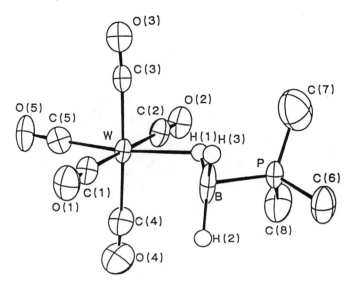

<u>Figure 2</u> X-ray Structures of $[W(CO)_5(BH_3 \cdot PMe_3)]$ at -50 °C

irradiation was shut off. These results indicate that the reaction is catalyzed by photo-induced $[M(CO)_5]$ or $[M(CO)_4]$ species, such as $[M(CO)_5(BH_3 \cdot NHR_2)]$ or $[M(CO)_4(BH_3 \cdot NHR_2)]$.

The similar reaction took place for primary amine-boranes to give borazine derivatives:

$$BH_3 \cdot NHR_2 \xrightarrow{h\nu / [M(CO)_6]} 1/3[BHNR]_3 + 2H_2 \qquad (3)$$
$$R = CH_3, \ C(CH_3)_3$$

These results show that photochemical reactions of amine-boranes may provide versatile processes for the preparation of boron-nitrogen compounds.

4 SYNTHESIS OF A COMPLEX BRIDGED WITH TRIMETHYL-PHOSPHINE ADDUCT OF BORYLENE

The reaction of $B_2H_4 \cdot 2PMe_3$ with $[Co_2(CO)_8]$ at -15 °C afforded $[Co_2(CO)_7BHPMe_3]$, the first borylene bridged dinuclear complex, liberating carbon monoxide and trimethylphosphine-borane:

$$[Co_2(CO)_8] + B_2H_4 \cdot 2PMe_3 \longrightarrow$$
$$[Co_2(CO)_7BHPMe_3] + CO + BH_3 \cdot PMe_3 \qquad (4)$$

Figure 3 shows the X-ray structure of $[Co_2(CO)_7BHPMe_3]$. Two cobalt atoms are bridged by carbonyl and BHPMe$_3$ groups: Co(1)-B 2.112 (9), Co(2)-B 2.108 (11), Co(1)-Co(2) 2.486 (2), B-P 1.921 (10), B-H 1.02 (7) Å. ^{11}B NMR spectrum shows a broad signal at 17.5 ppm. This compound can be regarded as an inorganic counterpart of carbene

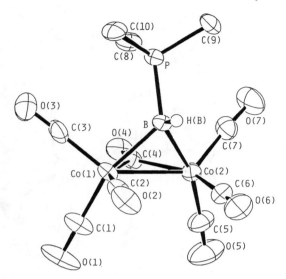

<u>Figure 3</u>   X-ray Structure of $[Co_2(CO)_7BHPMe_3]$

bridged complex, because $BHPMe_3$ ligand is isoelectronic with methylene.

The reaction scheme is closely related to the borane cage expansion reactions with $B_2H_4 \cdot 2PMe_3$ such as:[3,4]

$$B_2H_6 + B_2H_4 \cdot 2PMe_3 \longrightarrow B_3H_7 \cdot PMe_3 + BH_3 \cdot PMe_3 \qquad (5)$$

$$B_4H_8 \cdot PH_3 + B_2H_4 \cdot 2PMe_3 \longrightarrow$$
$$B_5H_9 \cdot PMe_3 + BH_3 \cdot PMe_3 + PH_3 \qquad (6)$$

These results show the correlation between the reactivity of metal clusters and that of borane clusters.

REFERENCES

1.   S.A. Snow, M. Shimoi, C.D. Ostler, B.K. Thompson, G. Kodama, and R.W. Parry, <u>Inorg.Chem.</u>, 1983, <u>35</u>, 511.
2.   K. Katoh, M. Shimoi, and H. Ogino, <u>Inorg.Chem.</u>, 1992, <u>31</u>, 670.
3.   M. Kameda and G. Kodama, <u>J.Am.Chem.Soc.</u>, 1980, <u>102</u>, 3647.
4.   G. Kodama, in 'Advances in Boron and the Boranes' J. F. Liebman, A. Greenberg, and R.E. Williams, Eds; VCH 1986, p 105.

Some Reactions of Unsaturated Molecules with Metallaboranes

Richard S. Coldicott

SCHOOL OF CHEMISTRY, UNIVERSITY OF LEEDS, LEEDS LS2 9JT, UK

1 INTRODUCTION

Polyhedral boron-containing clusters should, in principle, be able to function as two-electron sinks or donors via *arachno* $\rightleftharpoons$ *nido* and *nido* $\rightleftharpoons$ *closo* processes. By combining electronically flexible boron-frame matrices with a transition-element centre, then any flexibility of co-ordination mode and valence state of the metal atom could produce species that undergo interesting and unusual redox chemistry in their reactions, for example with unsaturated molecules. Although the reaction chemistry of non-metallated borane/heteroborane species has been reasonably extensively studied, progress in the equivalent reaction chemistry of the metallaboranes has hitherto been slow and is far less developed.

The reactions of MeNC with the three mutually related metallaboranes [6-(η^5-C_5Me_5)-*nido*-6-RhB$_9$H$_{13}$],[1] [6-(η^5-C_5Me_5)-*nido*-6-IrB$_9$H$_{13}$],[2] and [6-(η^6-C_6Me_6)-*nido*-6-RuB$_9$H$_{13}$][3] have previously been examined and various interesting products ranging from straightforward 6,9-*bis*(ligand) adducts to novel *isocloso*-clustered products have been isolated. Since novel products were obtained from these processes it was envisaged that, by using metal centres that have more labile phosphine ligands, a more extensive redox chemistry might develop.

The *nido*-6-iridadecaborane [6,6,6-H(PPh$_3$)$_2$-*nido*-6-IrB$_9$H$_{13}$] (1) and its *ortho*-cyclophenylated analogue [6,6-H(PPh$_3$)-6,5-μ-(Ph$_2$P-*ortho*-C_6H_4)-*nido*-6-IrB$_9$H$_{12}$] (2) are known readily to lose dihydrogen when heated, to give [1,1-H(PPh$_3$)-1,2-μ-(Ph$_2$P-*ortho*-C_6H_4)-*isocloso*-1-IrB$_9$H$_8$] (3).[4] Since this loss of hydrogen occurs readily it was anticipated that these metallaboranes might undergo interesting redox processes with unsaturated molecules. On this basis they have been used as metallaborane substrates in the reactions described below. In addition, the reactions of the related[5] rhenaborane [6,6,6,6-H(PMe$_2$Ph)$_3$-*nido*-6-ReB$_9$H$_{13}$] (4) and osmaborane [6,6,6-(PMe$_2$Ph)$_3$-*nido*-6-OsB$_9$H$_{13}$] (5) have also been investigated, in order to compare the chemistry across the third row of the transition-metal series.

2 REACTIONS OF METALLABORANES WITH UNSATURATED MOLECULES

CS$_2$ has been used in metallaheteroborane chemistry to yield products in which a reduced {η^2-S_2CH} group can bridge *(a)* two boron atoms, [as in the [3,3'-*commo*-*bis(closo)*-Co(1,2-$C_2B_9H_{11}$)$_2$]$^-$ complexes[6,7] and the equivalent iron complexes[8]], *(b)* adjacent metal and boron atoms,[9] and *(c)* in some cases it is bound solely to the metal atom.[10,11] CS$_2$ was thereby thought to be an interesting reagent for initial examination.

Compound **1** reacts with CS_2 to yield black [2-(PPh$_3$)-2,6;2,9-(μ-S$_2$CH)$_2$-*closo*-2-IrB$_9$H$_6$-10-(PPh$_3$)] (compound **6**; 20%), characterised by NMR spectroscopy and a single-crystal X-ray diffraction analysis (Fig.1). Interesting features include the two {μ-S$_2$CH} metal-boron bridging groups. The similar reaction of CS_2 with compound **2** yields black [2,6;2,9-(μ-S$_2$CH)$_2$-2,1-μ-(Ph$_2$P-*ortho*-C$_6$H$_4$)-*closo*-2-IrB$_9$H$_5$-10-(PPh$_3$)] (compound **7**; 38%), the *ortho*-cyclised analogue of compound **6**. Compounds **6** and **7** are both examples of {*closo*-2-MB$_9$} species and prior to this work few examples of this particular cluster type were known.[12-14] *Isocloso* compound **3** also reacts with CS_2 to yield compound **7** (54% yield), now via a unique *isocloso*-to-*closo* cluster transformation.

Compounds **1**, **2** and **3** react in similar ways with acetic acid. The heating of compound **1** in a refluxing solution of CH_2Cl_2 containing CH_3CO_2H-$(CH_3CO)_2O$ (4:1 v/v) affords the yellow acetate-bridged compound [2,2-(CO)(PPh$_3$)-2,9-(μ-O$_2$CMe)-*closo*-2-IrB$_9$H$_7$-10-(PPh$_3$)] (**8**) in 11% yield. The structure (Fig.1) was established by X-ray diffraction analysis and shows a similar {*closo*-2-IrB$_9$} cage with an acetate group bridging Ir(2) and B(9). The presence of the Ir-bound CO unit implies that a cleavage of acetic acid molecules may be occurring. The equivalent reaction of compound **2** gives yellow [2-(CO)-2,9-(μ-O$_2$CMe)-2,1-μ-(Ph$_2$P-*ortho*-C$_6$H$_4$)-*closo*-2-IrB$_9$H$_6$-10-(PPh$_3$)] (**9**), the *ortho*-cyclised analogue of compound **8**, in 9% yield, as the sole product, but a reflux period of 40 hours is needed. Shorter reaction times (<24 hours) afford three yellow products: compound **9** (9% yield), [2-(CO)-2-(PPh$_3$)-2,1-μ-(Ph$_2$P-*ortho*-C$_6$H$_4$)-*closo*-2-IrB$_9$H$_6$-10-(PPh$_3$)] (compound **10**; 16%), and [2,9-(μ-O$_2$CMe)-2-(PPh$_3$)-2,1-μ-(Ph$_2$P-*ortho*-C$_6$H$_4$)-*closo*-2-IrB$_9$H$_6$-10-(PPh$_3$)] (compound **11**; 11%). *Isocloso* compound **3** reacts similarly to give compounds **10** and **11** (6% yield each) over a reflux period of 4 hours. The apparent similarities between reactions with CS_2 and CH_3CO_2H have previously been noted in the reactions of [3,3`-*commo*-bis(*closo*)-Co(1,2-C$_2$B$_9$H$_{11}$)$_2$]$^-$ complexes,[6-8] where related products with 8,8`-{μ-S$_2$CH} and 8,8`-{μ-O$_2$CMe} bridges have been formed.

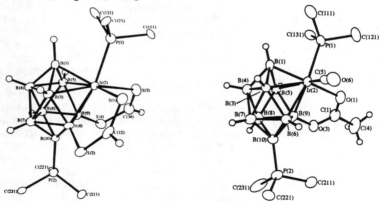

Figure 1. Molecular structures of compound **6**, [2-(PPh$_3$)-2,6;2,9-(μ-S$_2$CH)$_2$-*closo*-2-IrB$_9$H$_6$-10-(PPh$_3$)], and compound **8**, [2,2-(CO)(PPh$_3$)-2,9-(μ-O$_2$CMe)-*closo*-2-IrB$_9$H$_7$-10-(PPh$_3$)].

In contrast, the equivalent reactions of the *nido*-6-rhenadecaborane (**4**) and *nido*-6-osmadecaborane (**5**) with CS_2 and CH_3CO_2H produce different types of products. Compound **4** reacts with CS_2 under reflux conditions, in a sealed tube, to afford green [1-(PMe$_2$Ph)-1,2;1,3-(μ-S$_2$CH)$_2$-*isocloso*-1-ReB$_9$H$_6$-6-(PMe$_2$Ph)] (compound **12**; 11%) and its 5-(PMe$_2$Ph) isomer (compound **13**; 4%). Compounds **12** and **13** are the first characterised {*isocloso*-1-ReB$_9$} compounds and each shows two metal-to-boron {μ-S$_2$CH} bridging groups. The complementary reaction with CH$_3$CO$_2$H gives purple [6,6,6,6-H(PMe$_2$Ph)$_3$-*nido*-6-ReB$_9$H$_{12}$-9-(OCOMe)] (compound **14**; 21%; Fig.2), a simple ligand-substituted derivative of compound **4**. Compound **5** reacts with CS$_2$ in a heated toluene solution, in a sealed tube, to produce orange [5,5-(PMe$_2$Ph)$_2$-5,1-(μ-S$_2$CH)-*nido*-5-OsB$_9$H$_{12}$] (compound **15**; 9%). This product has an unusual *nido*-5-osmadecaborane structure and a {μ-S$_2$CH} bridge from Os(5) to B(1). Its formation presumably involves a "vertex-flip" process associated with the attack of CS$_2$. The reaction of compound **5** with CH$_3$CO$_2$H-(CH$_3$CO)$_2$O in refluxing toluene yields orange [6,6,6-(PMe$_2$Ph)$_3$-*nido*-6-OsB$_9$H$_{12}$-2-(OCOMe)] (compound **16**; 18%). Formation of compounds **14** and **16** with their monodentate acetate groups contrasts notably with the iridadecaborane reactions which yield bridged products.

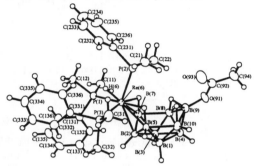

Figure 2. The molecular structure of compound **14**, [6,6,6,6-H(PMe$_2$Ph)$_3$-*nido*-6-ReB$_9$H$_{12}$-9-(OCOMe)].

The reactions with organyl isocyanides have also given some new and interesting products. The iridium compound **1** reacts with MeNC in refluxing CH$_2$Cl$_2$ to yield colourless [6,6,6,6-H(MeNC)$_2$(PPh$_3$)-*arachno*-6-IrB$_9$H$_{11}$-*exo*-9-(PPh$_3$)] (compound **17**; 35%) and [6,6,6,6-H(MeNC)$_2$(PPh$_3$)-*arachno*-6-IrB$_9$H$_{11}$-*exo*-9-(MeNC)] (compound **18**; 42%; Fig.3). The similar reaction of *ortho*-cycloboronated compound **2** afforded yellow [2,2,2-(MeNC)(PPh$_3$)$_2$-*closo*-2-IrB$_9$H$_8$-6-(MeNC)] (compound **19**; 20%), colourless [6,6,6-H(MeNC)(PPh$_3$)-6,5-μ-(Ph$_2$P-*ortho*-C$_6$H$_4$)-*arachno*-6-IrB$_9$H$_{10}$-*endo*-9-(MeNC)] (compound **20**; 13%) and [6,6,6-H(MeNC)(PPh$_3$)-6,5-μ-(Ph$_2$P-*ortho*-C$_6$H$_4$)-*arachno*-6-IrB$_9$H$_{10}$-*exo*-9-(MeNC)] (compound **21**; 30%). Compounds **17**, **18**, **20** and **21** are all formed via formal cluster reduction processes and their structures were assigned by comparison with related species. Compound **19** is of interest, as it is believed to be the first product formed via a novel decycloboronation process which may be favoured in that it may assist cluster oxidation by effective dihydrogen sequestration.

[6,6,6,6-H(PMe$_2$Ph)$_3$-*nido*-6-ReB$_9$H$_{13}$] (**4**) yields orange [5,5,5,5-H$_2$(PMe$_2$Ph)$_2$-*nido*-5-ReB$_9$H$_{12}$-2-(PMe$_2$Ph)] (compound **22**; 12%) and pale orange [1,1,1-(MeNC)(PMe$_2$Ph)$_2$-*isocloso*-1-ReB$_9$H$_8$-5-(PMe$_2$Ph)] (compound **23**; 6%) in a similar reaction with MeNC in refluxing toluene. The "vertex-flip" process which converts compound **4** to compound **22** is a general process,which occurs independently of the presence of MeNC, when the reaction is carried out solely in refluxing toluene. Interestingly the analogous reaction of the osmadecaborane (**5**) with ^{t}BuNC gives novel pale yellow [2,2,2-(PMe$_2$Ph)$_3$-*nido*-2-OsB$_4$H$_8$] (compound **24**; 10%), and the first {*isocloso*-1-OsB$_9$} species, yellow [1,1,1-H(PMe$_2$Ph)$_2$-*isocloso*-1-OsB$_9$H$_8$-5-(PMe$_2$Ph)] (compound **25**; 15%; Fig.3) which was structurally characterised by X-ray diffraction analysis. Compound **24** is obviously formed via a cluster degradation process, but the nature of this is unknown.

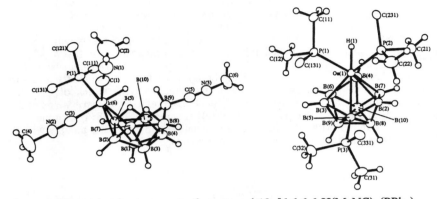

Figure 3. The molecular structures of compound **18**, [6,6,6,6-H(MeNC)$_2$(PPh$_3$)-*arachno*-6-IrB$_9$H$_{11}$-*exo*-9-(MeNC)], and compound **25**, [1,1,1-H(PMe$_2$Ph)$_2$-*isocloso*-1-OsB$_9$H$_8$-5-(PMe$_2$Ph)].

3 ACKNOWLEDGEMENTS

I am indebted to John Kennedy for his enthusiastic supervision of this work and his stream of ideas which prompted my research. I also thank Mark Thornton-Pett for his X-ray crystallographic expertise, Darshan Singh for mass spectrometry, and the University of Leeds for funding and facilities.

4 REFERENCES

1. X.L.R. Fontaine, N.N. Greenwood, J.D. Kennedy, P. MacKinnon and M. Thornton-Pett, J. Chem. Soc., Dalton Trans., 1988, 2809
2. K. Nestor, X.L.R. Fontaine, N.N. Greenwood, J.D. Kennedy, Zhu Sisan and M. Thornton-Pett, J. Chem. Soc., Dalton Trans., 1989, 1465
3. E.J. Ditzel, X.L.R. Fontaine, N.N. Greenwood, J.D. Kennedy, Zhu Sisan and M. Thornton-Pett, J. Chem. Soc., Chem. Commun., 1989, 1762
4. J. Bould, N.N. Greenwood and J.D. Kennedy, J. Chem. Soc., Dalton Trans., 1990, 1451
5. M.A. Beckett, N.N. Greenwood, J.D. Kennedy and M. Thornton-Pett, J. Chem. Soc., Dalton Trans., 1986, 795
6. M.R. Churchill, K. Gold, J.N. Francis and M.F. Hawthorne, J. Am. Chem. Soc., 1969, 91, 1222
7. M.R. Churchill and K. Gold, Inorg. Chem., 1971, 10, 1928
8. J.N. Francis and M.F. Hawthorne, Inorg. Chem., 1971, 10, 594
9. G. Ferguson, M.C. Jennings, A.J. Lough, S. Coughlan, T.R. Spalding, J.D. Kennedy, X.L.R. Fontaine and B. Stibr, J. Chem. Soc., Chem. Commun., 1990, 891
10. G. Ferguson, S. Coughlan, T.R. Spalding, X.L.R. Fontaine, J.D. Kennedy and B. Stibr, Acta. Cryst., 1990, C46, 1402
11. G. Ferguson, Faridoon and T.R. Spalding, Acta. Cryst., 1988, C44, 1368
12. J. Bould, P. Brint, X.L.R. Fontaine and M. Thornton-Pett, J. Chem. Soc., Chem. Commun., 1989, 1763
13. X.L.R. Fontaine, H. Fowkes, N.N. Greenwood, J.D. Kennedy and M. Thornton-Pett, J. Chem. Soc., Dalton Trans., 1987, 1431
14. E.J. Ditzel, X.L.R. Fontaine, N.N. Greenwood, J.D. Kennedy and M. Thornton-Pett, J. Chem. Soc., Chem. Commun., 1989, 1262

Novel *exo-nido*-Ruthena and *closo*-Rhodacarboranes. Chemistry, Structure, and Catalytic Activity in Organic Reactions

I. T. Chizhevsky[1], I. A. Lobanova[1], I. V. Pisareva[1], T. V. Zinevich[1], V. I. Bregadze[1], A. I. Yanovsky[1], Yu. T. Struchkov[1], C. B. Knobler[2], and M. F. Hawthorne[2]

[1]INSTITUTE OF ORGANO-ELEMENT COMPOUNDS, RUSSIAN ACADEMY OF SCIENCE, 28 VAVILOV STR., 117813, MOSCOW, RUSSIA
[2]DEPARTMENT OF CHEMISTRY AND BIOCHEMISTRY, UNIVERSITY OF CALIFORNIA, LOS ANGELES CA 90024, USA

Exo-nido-metallacarboranes of transition metals are of considerable synthetic, structural and catalytic interest. Only a few such compounds have been prepared and fully characterized by spectroscopic and x-ray diffraction methods.[1-2]

The first exo-nido ruthenacarboranes, 5,6,10-[Cl(Ph$_3$P)$_2$Ru]-5,6,10-μ-(H)$_3$-7-R-8-R'-10-H-7,8-C$_2$B$_9$H$_6$ (1a-d, a R=R'=H; b R=R'=Me; c R=H, R'=Me; d R=H, R'=PhCH$_2$), have been prepared and characterized (Figure 1) and their behavior in solution was investigated by ^{1}H, ^{11}B, ^{31}P and ^{1}H{^{11}B} NMR spectroscopy.[3] The solid state structure of cluster 1a has been elucidated by x-ray crystallography.[3] Unlike other known exo-nido species, in the solid state this compound contains three two-electron three-center B-H...M (metal) bonds which serve to connect the ruthenium atom with the nido carborane cage. An additional hydrogen atom in 1a was located exclusively at the boron 10 atom of the open face.

Exo-nido-ruthenacarboranes with nido-7,9-C$_2$B$_9$H$_{11}$ and nido-7-Ph-7,9-C$_2$B$_9$H$_{10}$ which we tried to use as ligands proved to be unstable in the solid state and were characterized only in solution by NMR methods. In the reaction of RuCl$_2$(PPh$_3$)$_3$ with [nido-7-Ph-7,9-C$_2$B$_9$H$_{10}$]$^-$K$^+$ we have separated two principal products: zwitterionic ruthenacarborane (2), proposed to appear from the rearrangement of an unstable exo-nido species, and nido-7-Ph-10 (or 11)-(PPh$_3$)-7,9-C$_2$B$_9$H$_{10}$ (3) as one of the side products of this rearrangement reaction (Figure 2). The structure of 2 was confirmed by x-ray diffraction and NMR data.

Due to the facile intramolecular oxidative addition processes which occur, exo-nido ruthenacarboranes 1a-d have been found to exist in solution as an equilibrium mixture of three isomeric exo-nido species. Each of the isomers

possesses only two bridging B-H...Ru bonds and one new B-Ru
sigma-bond and an additional terminal hydrogen atom at the
Ru^{IV} center.[3] Unlike the other known exo-nido

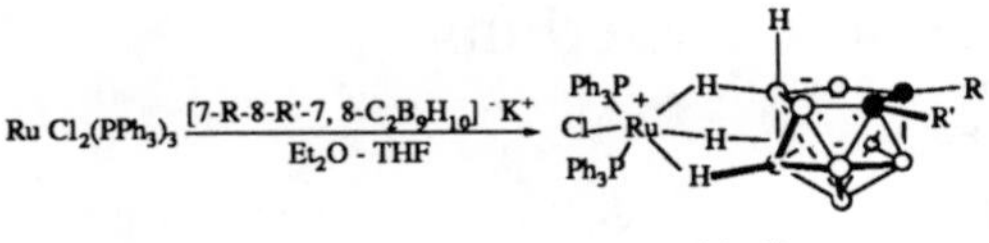

1(a-d)

Figure 1: Synthesis of the exo-nido-ruthena carboranes based on nido-7,8-carboranes

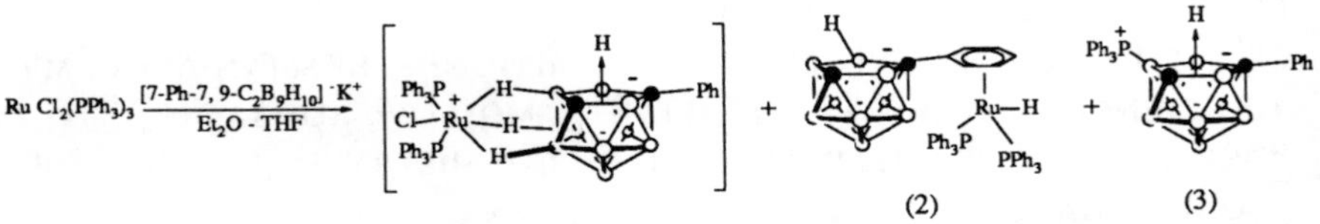

(2)   (3)

Figure 2: Attempted synthesis of exo-nido-ruthena carboranes based on Ph-substituted
nido-7,9-carboranes

species, the closo-tautomers of the compound obtained were
not detected in the ^1H or $^{11}B\{^1H\}$ NMR spectra of the above
exo-nido isomeric mixture. This suggests a high barrier
for the interconversion of these exo-nido and closo-
ruthenacarboranes. The unsubstituted closo-tautomer of 1a,
closo-3,3-$(PPh_3)_2$-3-Cl-3-H-3,1,2-$RuC_2B_9H_{11}$ (4) has been
prepared by thermal non-reversible rearrangement of 1a in
nearly quantitative yield. Its structure was confirmed by
x-ray diffraction[4] (Figure 3). No indications of such
rearrangement of any substituted exo-nido species 1b-d were
observed even at higher temperatures.

Exo-nido ruthenacarborane 1a can serve as a convenient
starting reagent for the synthesis of bimetallic clusters
containing the carborane moiety. The reaction between 1a
and dimeric diene-or dicarbonyl rhodium complexes, [(η^4-
diene)$RhCl]_2$ and [(CO)$_2RhCl]_2$, in EtOH in the presence of
KOH resulted in the mixed-metal complexes, [η^4-
diene)$RhRu(\mu$-H)(PPh_3)_2$($\eta^5$-$C_2B_9H_{11}$) (5a, diene=COD; b, diene
=NBD) and (CO)(Ph_3P)$RhRu(\mu$-H)(PPh_3)_2$($\eta^5$-$C_2B_9H_{11}$) (6) along
with mononuclear closo-3-PPh_3-3,3-(CO)$_2$-3,1,2-$RuC_2B_9H_{11}$ (7)
as a side product, respectively (Figure 4). The structure
of the Ru-Rh bimetallic clusters 5a and 6 have been
elucidated by x-ray diffraction. Because of the pure
square planar configuration around the rhodium atom, both
compounds studied by x-ray diffraction can be regarded[5] as
unsaturated 34e species with the rhodium atom having a
sixteen-electron shell.

Two alternative routes have been found for synthesis of the anionic closo-ruthenacarborane, $[closo-3,3-(PPh_3)_2-3-H-3,1,2-RuC_2B_9H_{11}]^-$ (8), which was proposed to be the key intermediate in the processes mentioned above. Both the reaction of $(Ph_3P)_3RuHCl$ with $Tl_2[7,8-C_2B_9H_{11}]$ in THF or reduction of $[Et_4N][closo-3,3-(PPh_3)_2-3-Cl-3,1,2-RuC_2B_9H_{11}]$ by $LiAlH_4$ in THF led to the anionic closo-hydridoruthenacarborane 8. The structure has been confirmed by x-ray diffraction.[6] As expected, anionic 8 as well as neutral closo-ruthenacarborane 4 reacted readily with dimeric complex $[(\eta^4-COD)RhCl]_2$ to produce the postulated mixed-metal cluster 5a under the same conditions used for its original preparation.

Ruthenacarboranes with closo-and exo-nido-structures were found to be exceptionally effective catalyst precursors for the cyclopropanation reaction between ethyl diazoacetate and some olefins (cyclooctene, 1-hexene, 2-methylstyrene, n-butyl vinyl ether, isoprene, etc.).[7] With all olefins tested, unsubstituted exo-nido-and closo-ruthenacarboranes, 1a and 4, respectively, gave comparable yields and stereoselectivities in the formation of the less stable endo-cyclopropane isomer. This suggests an exo-nido-closo rearrangement of the catalysts used during the

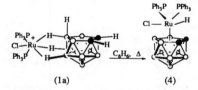

(1a)   (4)

Figure 3: Thermal rearrangement of exo-nido-ruthena carborane

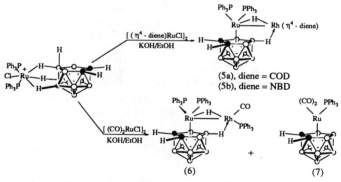

Figure 4: Synthesis of mixed-metal clusters

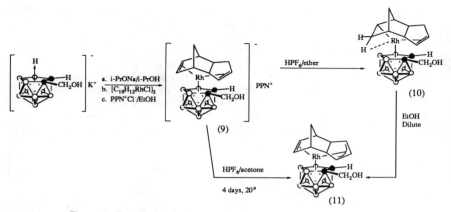

Figure 5: Reactions of anionic complex (9) under acidic conditions

catalytic process, in good agreement with the
chemical results obtained in the thermal reaction.

Upon protonation of the anionic closo-rhodacarborane
with a dicyclopentadiene ligand which also has a
hydroxymethyl function attached to the carborane cage
(complex 9, see Figure 5), we have observed high
regioselectivity in protonation at one of the carbon atoms
of the double bonds of the tricyclic ligand. The reaction
occurs smoothly in the presence of HPF6 in ether to give
stable neutral complex (10) with an "agostic" C-H...Rh
bond. In aqueous acetone solution the new allylolefinic
closo-rhodacarborane (11) was the main product. It was
evident from other experiments that complex 10 was observed
as an apparent intermediate in the latter reaction.
Moreover, complex 11 was obtained directly from compound 10
when its very dilute ethanol solution was stored for a week
and then separated by column chromatography on silica gel.
Both compounds were dimeric in the solid state (from x-ray
diffraction)[6] as well as in a concentrated solution of
nonpolar solvents, due to the presence of a strong O-H...O
hydrogen bond.

References

1. D. A. T. Young, R. J. Wiersema and M. F. Hawthorne, J. Am. Chem. Soc., 1971, 93, 5687.
2. J. A. Doi, R. G. Teller and M. F. Hawthorne, J. Chem. Soc., Chem. Commun., 1980, 80.
3. I. T. Chizhevsky, L. A. Lobanova, V. I. Bregadze, P. V. Petrovskii, V. A. Antonovich, A. V. Polyakov, A. I. Yanovsky and Yu. T. Struchkov, Mendeleev Comm., 1991, 47.
4. I. T. Chizhevsky, L. A. Lobanova, V. I. Bregadze, P. V. Petrovskii, A. V. Polyakov, A. I. Yanovsky and Yu. T. Struchkov, Metalloorganicheskaya Khim, 1991, 4, 957.
5. Y. C. Lin, C. B. Knobler and H. D. Kaesz, J. Am. Chem. Soc., 1981, 103, 1216.
6. Manuscript in preparation.
7. A. Demonceau, E. Saive, Y. de Froidmont, A. F. Noels, A. J. Hubert, I. T. Chizhevsky, I. A. Lobanova and V. I. Bregadze, Tetrahedron Lett., 1992, 33, 2009.

Modulation of the B(3)–H→Ru Distances in 7,8-Dicarba-*nido*-Undecarborate Derivatives

F. Teixidor[1], C. Viñas[1], and R. Kivekäs[2]

[1]INSTITUT DE CIÈNCIA DELS MATERIALS (CSIC), CAMPUS UNIVERSITAT AUTÒNOMA DE BARCELONA, BELLATERRA, 08193 BARCELONA, SPAIN
[2]DEPARTMENT OF CHEMISTRY, UNIVERSITY OF HELSINKI, VUORIKATU 20, SF-00100, FINLAND

It has been found that the length of the S,S'-connecting string modulates the B(3)...M distance in 7,8-dithio-7,8-dicarba-*nido*-undecaborate(1-) complexes[1] (see Fig.1); shorter strings produce shorter B(3)...M distances. With short S,S'-exocluster chains such as S-CH_2-CH_2-S (**L6**) the B(3)-H→M interaction could be found for the first time in $[7,8-C_2B_9H_{12}]^{1-}$ derivatives in $[RuCl(\textbf{L6})(PPh_3)_2]^2$ (see Fig.2), or a B(3)-Rh σ interaction had been proven in $[N(CH_3)_4][RhCl\{7,8-\mu-S(CH_2CH_2)S-C_2B_9H_{10}\}\{\sigma-7,8-\mu-S(CH_2CH_2)S-C_2B_9H_9\}]^3$.

In general, the B-H→M or B-M bonds for derivatives of $[7,8-C_2B_9H_{12}]^{1-}$ involve at least one boron atom of the open pentagonal C_2B_3 face, e.g. when *exo-nido* coordination takes place, there are two B-H→M interactions, the first one being with an open face boron atom while the second is with a boron atom of the second layer[4]. The B(3)-H→M contact described here is unique in the sense that it involves B(3) which is a boron atom of the second layer, and which is the only boron atom in the cage connected to both carborane carbon atoms. Due to this especial position distinctive properties could be foreseen[5]. Here we discuss how the B(3)-H→M interaction can be modulated from non-coordinating distance to coordinating, which is achieved by the macro-cyclic/carborane mutual influence. To prove this B-H→Ru modulation, the set of chemicals listed in Table 1 have been examined. The ligands' re-presentations are schematically indicated in Fig. 1. The compounds have been produced following similar reactions to this described in Eq. 1 for the synthesis of $[RuCl(\textbf{L6})(PPh_3)_2]$.

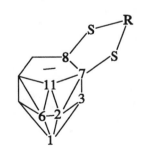

Fig.1 .-

R= $-(CH_2-)_n$ n= 1,2,3 (L5, L6, L7);
R= $-CH_2-(CH_2OCH_2-)_n CH_2$ n=1,2 (L12,L15)

The B(3)-H→Ru interaction has been studied by ^{11}B-NMR, ^{1}H-NMR and X-ray crystallography.

$$[NMe_4](L6) + RuCl_2(PPh_3)_3 \xrightarrow[\text{ethanol/N}_2]{} [RuCl(L6)(PPh_3)_2] + \text{other}$$

Eq. (1)

The ^{1}H-NMR spectra evidence the existence of a modulation in the B(3)-H→Ru interaction. Resonances near -18 ppm are displayed for almost every compound presented (Table 1), which are associated to B-**H**→Ru. Exceptions to the -18 ppm value are found in [RuH(**Ln**)(PPh_3)_2] (n= 6-8) with hydride ^{1}H-NMR resonances at -4.3; -3.4 and -2.1 ppm, respectively; [RuCl(**Ln**)(dmso)_2] (n=6,7) with ^{1}H-NMR resonances at -14.7 and -14.0 ppm, respectively, and [RuCl(**L5**)(PPh_3)_2] at -14.2. This broad field of B-H→M resonances ranging from -2.1 to -18.2 in a series of very related compounds suggests that different sorts of the B-H→M interactions are present, which tentatively are drawn as B-H→M (form A) and B-H→M (form B) (Fig.3). The interaction B-H→M (B) being in the less negative δ ^{1}H-NMR values, while B-H→M (A) is in the more negative δ ^{1}H-NMR values. The pure Ru-H bonds found in the hydride series have this resonance at -21 ppm, and this value could be taken as a M-H non-boron interacting hydride. The only ^{1}H-NMR spectra that have shown two sets of signals at the hydride region correspond to the hydride series

Table 1.- Displacement of the B(3) ^{11}B-NMR (ΔδB(3)) and δH(3) ^{1}H-NMR resonances.

compound	(ΔδB(3)) (ppm)	δH(3) ^{1}H-RMN (ppm)
[RuCl(**L5**)(PPh_3)_2]	10	-14.2
[RuCl(**L6**)(PPh_3)_2]	20	-17.4
[RuCl(**L7**)(PPh_3)_2]	20	-17.3
[RuCl(**L8**)(PPh_3)_2]	16	-18.2
[RuCl(**L6**)(PMePh_2)_2]	20	-17.4
[RuCl(**L7**)(PMePh_2)_2]	18	-17.1
[NMe_4][RuCl_2(**L0**)(PMePh_2)_2]	1	-
[RuH(**L6**)(PPh_3)_2]	15	-4.3
[RuH(**L7**)(PPh_3)_2]	10	-3.4
[RuH(**L8**)(PPh_3)_2]	6	-2.1
[RuCl(**L6**)(dmso)_2]	20	-14.7
[RuCl(**L7**)(dmso)_2]		-14.0
[NMe_4][RuCl(**L6**)_2]	20	-17.5
[NMe_4][RuCl(**L7**)_2]	13	-7.2

[RuH(**Ln**)(PPh$_3$)$_2$] (n= 6-8). Resonances near -3 ppm and -21 ppm (1:1) are observed in these compounds which are assigned to B-H→Ru(B) and Ru-H. Those NMR values can, as a first approximation, tentatively be used as "pure" for B-H→Ru(B) and Ru-H. Consequently, the observed chemical shifts for the other Ru complexes are hypothesized to be due to intermediate states between the B-H→Ru(B) and B-H→Ru(A) forms. The ^{11}B-NMR spectra of the complexes are very reminiscent of those of the respective parent **Ln**$^-$ ligands; however, the resonance assigned to B(3) in the free ligand is displaced to higher field in the complexes. The B(3) ^{11}B-NMR resonance assignment has been corroborated by a heteronuclear correlation (HETCOR) 2D NMR spectrum of [RuH(**L7**)(PPh$_3$)$_2$]. The displacement, $\Delta\delta$B(3), calculated according to Eq. 2, varies considerably from one compound to another.

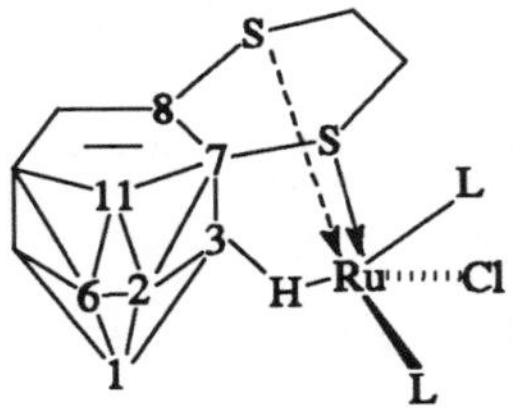

Fig.2

$$\Delta\delta B(3) = \delta(B(3))_{complex} - \delta(B(3))_{free\ ligand} \qquad \text{Eq. (2)}$$

The $\Delta\delta$B(3) value is upfield, and near 20 ppm for most of the compounds. Table 1 presents the $\Delta\delta$B(3) for the different complexes. Smaller $\Delta\delta$B(3) values are found for the hydride series [RuH(**Ln**)(PPh$_3$)$_2$] decreasing from 15 (**L6**) to 10 (**L7**) to 6 (**L8**). Other values out of the 15/20 ppm range are for [RuCl(**L5**)(PPh$_3$)$_2$] ($\Delta\delta$B(3)=10) and [(NMe$_4$)][RuCl(**L7**)$_2$] ($\Delta\delta$B(3)=13).

With the common exception of the particularly strained **L5**, the hydride series trend **"The longer the external chain the smaller the $\Delta\delta$B(3) value"**, is extensible to the other series. The gradation of $\Delta\delta$B(3) values (Table 1) implies a gradation of Ru-B interactions in them. The largest $\Delta\delta$B(3) would imply more Ru-B σ direct bond contribution in the B-H→M bond, as it is expressed in B-H→M(A); on the contrary the smallest $\Delta\delta$B(3)^3 value suggests more B-H σ contribution in the Ru-H→B interaction (B-H→M)(B). It is striking the small $\Delta\delta$B(3) value for [NMe$_4$][RuCl$_2$(**Lo**)(PMePh$_2$)$_2$]. This insignificant displacement for $\Delta\delta$B(3) was expected since **L$_o$** is an open chain ligand, which is equivalent to an infinite S,S' connecting string.

Following the discussion on [NMe$_4$][RuCl$_2$(**Lo**)(PMePh$_2$)$_2$], the ^{1}H-NMR spectrum does not display any hydride signal in this compound, which proves the absolute lack of B-H→Ru interaction in an infinite S,S'- connecting string. The X-ray crystal structures of the chemicals listed in Table 2 prove the above discussion on

B⁻H⁻Ru (A)   B⁻H⁻Ru (B)

Fig.3

the B-H→M modulation. In this case a stronger B-H→M interaction will imply a shorter B...M distance and a smaller w angle (Fig.4). Emphasizing the [RuCl(**Ln**)(PPh$_3$)$_2$] series, it is observed in Table 2 that the w angle decreases as the ring size decreases; also the B(3)...M distance decreases along with the decrease in size of the ring. As a result there is a gradation of stability of the B...M interaction from zero (**Lo**), open chain, to B-H→M forms (A) and (B), depending on the length of the chain. As a result, a way of modulating the B(3)-H→M interaction has been made possible.

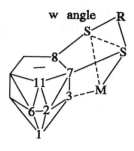

Fig.4 .- The w angle concept

Table 2.-

Chemical	Ring size	w (°)	B3-M (Å)	S-S (Å)
Pd(**L15**)$_2$	15	156.8	3.55	3.23
Rh(**L12**)(PPh$_3$)$_2$	12	154.2	3.55	3.22
PdCl(**L6**)(PPh$_3$)	6	131.0	3.27	2.97
Rh(**L6**)(PPh$_3$)$_2$	6	129.4	3.30	2.96
[NMe$_4$][RuCl(**L6**)$_2$]	6	117.3	2.58	2.96
[NMe$_4$][RhCl(**L6**)(σ-**L6**)]	6	108.0	2.12	2.97
RuCl(**L8**)(PPh$_3$)$_2$	8	120.4	2.48	3.21
RuCl(**L6**)(PPh$_3$)$_2$	6	110.0	2.41	3.01
RuCl(**L5**)(PPh$_3$)$_2$	5	102.7	2.39	2.78

REFERENCES

a) F. Teixidor and R.W. Rudolph, J.Organomet.Chem.,1983, 241, 301. b) C. Viñas, W.M. Butler, F. Teixidor and R.W. Rudolph, Organometallics, 1984, 3, 503. c) C. Viñas, W.M. Butler, F. Teixidor and R.W. Rudolph Inorg.Chem., 1986, 25, 4369. d) F. Teixidor, A.M. Romerosa, J. Rius, C. Miravitlles, J. Casabó, C. Viñas and E. Sánchez, J.Chem.Soc.,Dalton Trans. 1990, 525. e) F. Teixidor, C. Viñas,J. Rius, C. Miravitlles and J. Casabó, J.Inorg.Chem. 1990, 29, 149.
F. Teixidor, J.A. Ayllon, C. Viñas, R. Kivekäs, R. Sillanpää, J. Casabó J.Chem.Soc., Chem.Comm.,1992, 1281.
F. Teixidor, A. Romerosa, C. Viñas, J. Rius, C. Miravitlles and J. Casabó, J.Chem.Soc., Chem.Commun., 1991, 192.
J.D. Hewes, C.W.Kreimendhal, T.B. Marder and M.F. Hawthorne, J.Am.Chem.Soc. 1984, 106, 5757.
Extracted from the paper by F. Teixidor, J.A. Ayllon, C. Viñas, R. Kivekkäs, R. Sillanpää and J. Casabó. To be published.

Triphenylstannyl Derivatives of Pentaborane(9) and Hexaborane(10)

D. K. Srivastava, H. Fang, N. P. Rath, and L. Barton

DEPARTMENT OF CHEMISTRY, UNIVERSITY OF MISSOURI-ST. LOUIS, ST. LOUIS MO 63121, USA

Group IV derivatives of *nido*-pyramidal boranes have been known since the pioneering studies of Gaines and others[1]. Alkyl derivatives of Si, Ge, Sn, and Pb have been inserted into a basal B-B bond in the anion $[B_5H_8]^-$.[1] The Si and Ge derivatives are known to undergo rearrangements such that the MR_3 substituent migrates from the bridging position to the terminal, basal B(2) position, ultimately to the terminal apical B(1) position, to afford equilibrium mixtures in which product with substituent in the 1-position predominates.[2] More recently analogous Ph_3Sn derivatives were prepared in our laboratory and we demonstrated, using ^{119}Sn and ^{11}B NMR spectroscopy, that the species isomerize exclusively to the 1-isomer in coordinating solvents but that the 2,3-μ-isomer is obtained if reactions are carried out in non-coordinating solvents.[3] Examples of Group IV-substituted hexaboranes are very rare. The only known Group IV derivatives of B_6H_{10} are the species 1-$Me_3MB_6H_9$,[4a] (M = Si, Ge) which are formed in the reaction between salts of either $[2-Me_3MB_5H_7]^-$ or $[1-Me_3MB_5H_7]^-$ and H_2BCl. The related 2-$[Me_3Si]-\mu-[Me_2B]B_5H_7$[4b] is also known but this latter formal hexaboranyl-Group IV system is really a B_5H_9 derivative. This paper extends this previously described work and also reports on the syntheses of the first organotin derivative of hexaborane(10).

When $K[B_5H_8]$ and $SnCl_2Ph_2$ are stirred for 16h at 25°C in THF, the species 1-$(SnClPh_2)B_5H_8$(**I**) is formed in 56% yield. **I** is identified by NMR spectroscopy and X-ray crystallography. Figure 1 shows the crystal structure of one of the two crystallo-graphically unique molecules in the unit cell of **I**. Also shown are the ^{11}B and ^{119}Sn NMR spectra. Of particular note is the ^{11}B-^{119}Sn coupling ($J_{11B-119Sn}$ = 1272 Hz) resulting in the 1:1:1:1 quartet in the ^{119}Sn spectrum, compared to the single broad resonance, fwhm = 108 Hz, for the bridging isomer, **II**, described below. Apparently the $SnClPh_2$ moiety is sufficiently sterically hindered such that the 1-isomer is the exclusive product. Although the bridging isomer, 2,3-μ-$(SnPh_3)B_5H_8$(**II**), shown in Figure 2 below along with its ^{11}B and ^{119}Sn NMR spectra, is the exclusive product when the reaction between $SnClPh_3$ and $K[B_5H_8]$ is carried out in a non-coordinating solvent such as CH_2Cl_2, the same reaction using $SnCl_2Ph_2$ always results in the formation of complex mixtures.

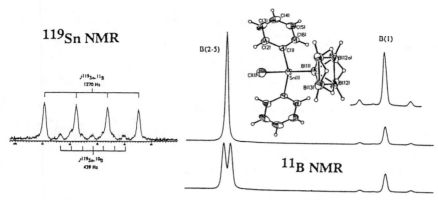

<u>Figure 1</u> Crystal structure, 96.3 MHz ^{11}B and $^{11}B\{^1H\}$ NMR spectra, and 111.7 MHz ^{119}Sn NMR spectra of 1-$(SnClPh_2)B_5H_8$ (**I**).

The reaction between $SnCl_2Ph_2$ and $K[B_5H_8]$ in CH_2Cl_2, in 1:2 mole ratio, forms exclusively the species μ,μ'-$SnPh_2(B_5H_8)_2$ (**III**). It is identified by ^{11}B, ^{119}Sn, and ^1H NMR spectra, and by its mass and infrared spectrum. Elemental analysis indicate some decomposition but observed mass spectra correlate well with calculated ones. Attempts to grow crystals suitable for X-ray analysis have been unsuccessful to date. If μ,μ'-$SnPh_2(B_5H_8)_2$ is stored at 25°C for several months in $CDCl_3$, and its NMR spectra

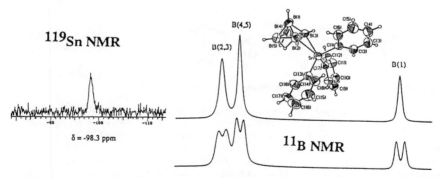

<u>Figure 2</u> Crystal structure, 96.3 MHz ^{11}B and $^{11}B\{^1H\}$ NMR spectra, and 111.7 MHz ^{119}Sn NMR spectrum of 2,3-μ-$(SnPh_3)B_5H_8$ (**II**).

monitored periodically, rearrangement is clearly observed. After one month, a second isomer, which we tentatively assign as $\mu,2'$-$SnPh_2(B_5H_8)_2$ (**IV**), is present in about equal concentration as **III**, and after four months a third isomer, assigned as $\mu,1'$-$SnPh_2(B_5H_8)_2$ (**V**), is observed such that the **III**, **IV**, and **V** are present in approximately 1:2:1 mole ratio, respectively.

When μ,μ'-$SnPh_2(B_5H_8)_2$ is treated with Et_2O, the species rearranges rapidly to the $\mu,2'$-isomer, but when treated with THF it forms the $\mu,1'$-isomer. If the initial reaction is carried out in THF, the product appears to consist of two isomers of $\mu,1'$-

$SnPh_2(B_5H_8)_2$. The $\mu,2'$-isomer appears to be the least stable and, when treated with THF, it degrades rather than rearrange to the $\mu,1'$-isomer. The μ,μ'-isomer is apparently much more stable and rearrangement to the $\mu,2'$-, and $\mu,1'$-isomers occurs rather than degradation. The $\mu,1'$-isomer is prepared independently in the reaction between $1-(SnClPh_2)B_5H_8$ and $K[B_5H_8]$ in CH_2Cl_2. This was expected from our earlier observations that the insertion of the Ph_3Sn moiety into a basal B-B bond in $[B_5H_8]^-$ stops at the formation of the bridging isomer and that isomerization to the 2- and 1- isomer is quite slow in this solvent.[3] NMR data for the product of the reaction between $1-(SnClPh_2)B_5H_8$ and $K[B_5H_8]$ are identical to those obtained for the ultimate product of the reaction between two moles of $[B_5H_8]^-$ and $SnPh_2Cl_2$, that is **V**. The formation of tin-bridged bipentaboranes is summarized in Figure 3. below.

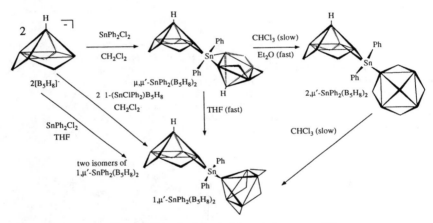

<u>Figure 3</u> Formation of tin-bridged bipentaboranes(9).

In the case of hexaborane(10), insertion of the $SnPh_3$ moiety into a basal B-B bond, by the reaction of $SnClPh_3$ with $K[B_6H_9]$ in CH_2Cl_2, results in the formation of $2,3-\mu-SnPh_3B_6H_9$ (**VI**). The species is identified by elemental analysis, ^{11}B, ^{119}Sn and ^1H NMR spectra, mass and infrared spectra. **VI** forms B_6H_{10} on protonation suggesting that the Sn group occupies a bridging position. The ^{11}B NMR spectrum and the proposed structure are given in Figures 4. and 5. respectively. Low temperature NMR spectra suggest a pyramidal structure with four different boron environments. The boron-11 decoupled proton spectrum gives two resonances for the bridging H atoms and four for the terminal ones, at -105°C, but at room temperature only a single resonance is observed for the basal terminal H atoms. To account for these observations, we assume that the bridging H atoms in **VI** are fluxional on both the boron-11 and proton NMR time-scale but that the motion of the Hμ atoms is partially quenched, as observed for other $2,3-\mu$-metalladerivatives of hexaboranes.[5] At temperatures above about -10°C some simple motion of the cage relative to the SnR_3 group is invoked to account for the spectra of **VI**. Processes equivalent to those given in Figure 5; account for the observed spectra. The

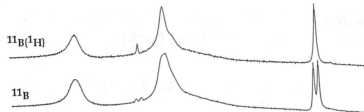

Figure. 4 96.3 MHz ^{11}B NMR spectra of 2,3-μ-(SnPh$_3$)B$_6$H$_9$(VI) in CDCl$_3$/CHClF$_2$ at -40°C.

stannylhexaborane, **VI**, is much less stable than its pentaboranyl

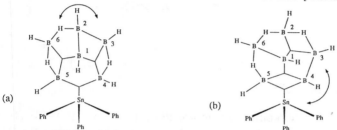

Figure. 5 Proposed fluxional motion of (a) H atoms, and (b) SnPh$_3$ group in **VI**.

congeners. Rearrangement does not occur in the presence of bases; rather degradation of the cage occurs.

REFERENCES

1.   (a). D. F. Gaines, in "Boron Chemistry 4, Plenary Lectures at the 4th International Meeting on Boron Chemistry, Salt Lake City, 1979", R. W. Parry, G. J. Kodama, Ed.; Pergamon Press: Oxford. 1980. (b) T. C. Geisler and A. D. Norman, <u>Inorg. Chem</u>. 1970, <u>9</u>, 2167. (c) A. B. Burg, <u>Inorg</u>. <u>Chem</u>. 1974, <u>13</u>, 1010.

2.   D. F. Gaines and T. V. Iorns, <u>Inorg. Chem</u>. 1971, <u>10</u> 1094.

3.   (a) L. Barton and D. K. Srivastava, <u>J. Chem. Soc., Dalton Trans</u>. <u>1992</u>, 1327. (b) D. K. Srivastava, N. P. Rath, and L. Barton, <u>Organometallics</u>, <u>1992</u>, <u>11</u>, 2263.

4.   (a) D. F. Gaines, S. Hildenbrandt and J. Ulman, <u>Inorg. Chem</u>. <u>1974</u>, <u>13</u>, 2792. (b) Gaines, D. F.; Ulman, J. <u>J. Organomet. Chem</u>. <u>1975</u>, <u>93</u>, 281.

5.   (a) D. L. Denton, W. R. Clayton, M. Mangion, S. G. Shore and E. A. Meyers, <u>Inorg. Chem</u>. 1976, <u>15</u>, 541. (b) R. J. Remmel, D. L. Denton, J. B. Leach, M. A. Toft and S. G. Shore, <u>Inorg. Chem</u>. 1981, <u>20</u>, 1270. (c) D. K. Srivastava, N. P. Rath, L. Barton, J. D. Ragaini, O. Hollander, R. Godfroid and S. G. Shore, <u>Organometallics</u> 1993, <u>12</u>, 2017.

Icosahedral Bis(phosphine) Metalladiarsa- and Distibaboranes Containing Nickel and Palladium

S. A. Jasper Jr., S. Roach, J. N. Stipp, J. C. Huffman, and L. J. Todd

DEPARTMENT OF CHEMISTRY AND MOLECULAR STRUCTURE CENTER, INDIANA UNIVERSITY, BLOOMINGTON, IN 47405, USA

1 INTRODUCTION

We have previously reported an improved synthesis of $1,2-As_2B_{10}H_{10}$ with $B_{10}H_{14}$ as the starting material.[2] Treatment of this icosahedral diarsaborane with excess piperidine forms the $7,8-As_2B_9H_{10}^-$ ion in good yield.[3] This arsenic-containing monoanion which is isoelectronic with the $7,8-C_2B_9H_{12}^-$ ion may form an extensive series of metal complexes similar to known metallacarboranes. In previous studies several metalladiarsaborane complexes such as $(C_5H_5)Co(B_9H_9As_2)$, $(Ph_2PCH_2CH_2PPh_2)Ni(B_9H_9As_2)$, $L_2Pd(B_9H_9As_2)$, $L_2Pt(B_9H_9As_2)$, $LClPd(5-L-B_9H_8As_2)$ [L=PMe_2Ph or PPh_3], $(PPh_3)_2HRh(B_9H_9As_2)$, and $(C_5Me_5)Rh(B_9H_9As_2)$ have been reported.[2,4-6]

Substitution of a neutral ligand, such as a phosphine, for H$^-$ on a B atom in a borane cage results in a 'charge-compensated'[7] cage; that is, a cage formally one unit of charge more positive. In the specific case for arsaboranes, this corresponds to changing $B_9H_9As_2^{2-}$ to $B_9H_8(L)As_2^-$ (L=neutral 2 e$^-$ donor). Several metallaheteroborane complexes reported in the literature contain charge-compensating ligands. Methods of preparation include formal ligand rearrangement from the metal onto the cage, reduction of a metallacarborane complex by a Lewis base, addition of R_2S to a protonated metallocene-type sandwich complex, and metallation of a charge-compensated carborane ligand.[7]

2 SYNTHESES AND DISCUSSION

The reaction of $(Me_2PPh)_2MCl_2$ species (M=Ni or Pd) with the *nido*-$7,8-As_2B_9H_9^{2-}$ ion and $(Me_2PPh)PdCl_2$ with the *nido*-$7,8-Sb_2B_9H_9^{2-}$ ion at room temperature led to the formation of icosahedral bis(phosphine) metalladiheteroboranes in low to moderate yields for M=Ni (**1** for As) or M=Pd (**3** for As and **5** for Sb), equation 1.

$$nido\text{-}7,8\text{-}X_2B_9H_{10}^- + n\text{-BuLi} + (PMe_2Ph)_2MCl_2 \text{ -----> } \quad (1)$$
$$closo\text{-}1,1\text{-}(PMe_2Ph)_2\text{-}1,2,3\text{-}MX_2B_9H_9 + n\text{-BuH} + LiCl + Cl^-$$
$$M=Ni, Pd, or Pt; X=As or Sb$$

The numbering system for these icosahedral metallaheteroboranes is as illustrated in Figure 1.

Structural Considerations

The most striking structural feature of the $(PMe_2Ph)_2Pd(B_9H_9As_2)$ (3) and $(PMe_2Ph)_2Pd(B_9H_9Sb_2)$ (5) molecules is the distortion of the 12 membered cage that results from the inclusion of the relatively large As, Sb, and Pd atoms. As may be expected, the 2.7996 Å Sb(2)-Sb(3) distance in (5) is much greater than a typical 1.61 Å C-C distance in the isoelectronic $B_9H_9(CH)_2M$ cage, and even greater than the 2.488 Å As-As distance in (3). Since these cages are isoelectronic, the cage distortion due to the arsenic and antimony atoms is primarily a steric effect, and not an electronic one.

The X-ray crystal structures of (3) and (5) show that the Pd-As and Pd-Sb distances are not equal: Pd-As(2) is 2.6835(13) Å and Pd-As(3) is 2.5304(13) Å; similarly, Pd-Sb(2) is 2.7074 Å and Pd-Sb(3) is 2.7865 Å. Also, the P-Pd-P plane is nearly, but not exactly, parallel to the X-X vector. There is significant twisting of the $(R_3P)_2Pd^{2+}$ fragment relative to the B_3X_2 face such that one R_3P is closer to the X atoms than the other. This asymmetry is consistently observed in crystal structures of the type $(R_3P)_2M(B_9H_9X_2)$ (M=Pd or Pt; X=CH, As, or Sb). This means that the $[(R_3P)_2Pd]$ HOMO is twisting such that one lobe is closer to, and thus has better overlap with, the same X atom. Thus, there are two different Pd-X distances, and the shorter is always on the side where the R_3P is closer to the X atoms. The twisting of the $(R_3P)_2M^{2+}$ fragment may be due to crystal packing forces, as all of the aforementioned crystals have the same space group.

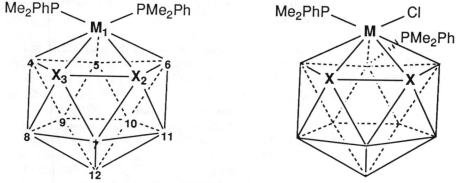

M=Ni, Pd; X=As, Sb

Figure 1 Numbering system and geometry for icosahedral metallaheteroboranes (1, 3, 5, left) and (2, 4, 6, right)

Derivatives with a phosphine or amine bonded to boron

In the reaction of $(PMe_2Ph)_2NiCl_2$ with $B_9H_9As_2^{2-}$, one other compound in addition to $(PMe_2Ph)_2Ni(B_9H_9As_2)$ was formed in low yield. This second compound (2) is believed to be formed by a phosphine-hydride interchange to give $(PMe_2Ph)HNi(5-PMe_2Ph-B_9H_8As_2)$ which then undergoes hydrogen-chlorine exchange to generate (2), $(PMe_2Ph)ClNi(5-PMe_2Ph-B_9H_8As_2)$. In the reaction with the analogous palladium reagent, the same phosphine-hydride interchange is observed to occur, to form $(PMe_2Ph)ClPd(5-PMe_2Ph-B_9H_8As_2)$, which has been reported.[6] However, under suitable conditions, the palladium complex can be further converted to a 6-chloro substituted derivative (4), $(PMe_2Ph)ClPd(6-Cl-5-PMe_2Ph-B_9H_7As_2)$. This type of phosphine-hydride interchange reaction has been reported for a number of (phosphine)metallacarborane complexes containing nickel,[8,9] platinum,[10] rhodium,[11] and ruthenium.[12,13] The metal-hydride products were not isolated but reacted with chlorinating agents to form the chlorometal compounds described above.

The inclusion of $(PMe_2Ph)_2MCl_2$ (M=Ni or Pd) in THF solution allows phosphine-hydride interchange to occur in $(PMe_2Ph)_2-M(B_9H_9As_2)$ species at room temperature. We therefore suggest that in our system there may be an intermolecular process operating, which involves the metal center of a reactive intermediate derived from $(PMe_2Ph)_2MCl_2$.

The factors influencing the distribution of $(PMe_2Ph)_2Pd(B_9H_9As_2)$ versus the rearranged products, $(PMe_2Ph)ClPd(5-PMe_2Ph-B_9H_8As_2)$ or (4), have not been fully investigated; however, some general conclusions seem evident. In the room temperature synthesis of the platinum analogue, no rearrangement product such as $(PMe_2Ph)ClPt(PMe_2Ph-B_9H_8As_2)$ was observed, although attempts to synthesize this by several methods are in progress. For the palladium case, there is a distribution of $(PMe_2Ph)_2-Pd(B_9H_9As_2)$ and $[(PMe_2Ph)ClPd(5-PMe_2Ph-B_9H_8As_2)$ or (4)] which favors $(PMe_2Ph)_2Pd(B_9H_9As_2)$ at room temperature.

In the reaction of $7,8-B_9H_{10}Sb_2^-$ with $(PMe_2Ph)_2PdCl_2$, inclusion of piperidine, $C_5H_{10}NH$, in the reaction mixture results in the formation of *closo*-1-Cl-1-PMe_2Ph-5-C_5H_{10}NH-1,2,3-PdSb_2B_9H_8 (6), in low (<10 %) yield. In the event, stirring $1,2-B_{10}H_{10}Sb_2$ (362 mg, 1.00 mmol) with excess piperidine generates the $7,8-B_9H_{10}Sb_2^-$ ion in 2 hrs. After addition of THF (60 mL), treatment with BuLi (1.0 mL of a 2.1 M solution in hexanes), followed by $(PMe_2Ph)_2PdCl_2$ (454 mg, 1.00 mmol) results in a dark solution. After stirring 4 hrs, the THF was removed, and the solids chromatographed (on silica gel) with toluene. The resulting purple band (R_f=0.95 in CH_2Cl_2) formed bright purple crystals from slow evaporation of pentane. The spectroscopic data[14] are consistent with the formulation indicated as (6).

This fairly nonpolar crystalline compound may be formed by initial exchange of the free piperidine with phosphine in $(PMe_2Ph)_2PdCl_2$, to form $(PMe_2Ph)(C_5H_{10}NH)PdCl_2$; this would then be followed by amine-hydride interchange by an intermolecular process, in a similar manner to phosphine-hydride exchange discussed above.

3 REFERENCES AND NOTES

1. S. A. Jasper Jr., S. Roach, J. N. Stipp, J. C. Huffman, and L. J. Todd, <u>Inorg. Chem.</u>, 1993, <u>32</u>, in press, and refs. therein.
2. T. P. Hanusa, N. Roig de Parisi, J. G. Kester, A. Arafat, and L. J. Todd, <u>Inorg. Chem.</u>, 1987, <u>26</u>, 4100.
3. J. L. Little, S. S. Pao, and K. K. Sugathan, <u>Inorg. Chem.</u>, 1974, <u>13</u>, 1752.
4. J. L. Little and S. S. Pao, <u>Inorg. Chem.</u>, 1978, <u>17</u>, 584.
5. X. L. R. Fontaine, J. D. Kennedy, M. McGrath, and T. R. Spalding, <u>Magn. Res. Chem.</u>, 1991, <u>29</u>, 711.
6. M. McGrath, T. R. Spalding, X. L. R. Fontaine, J. D. Kennedy, and M. Thornton-Pett, <u>J. Chem. Soc. Dalton Trans.</u>, 1991, 3223.
7. H. C. Kang, S. S. Lee, C. B. Knobler, and M. F. Hawthorne, <u>Inorg. Chem.</u>, 1991, <u>30</u>, 2024.
8. S. B. Miller and M. F. Hawthorne, <u>J. Chem. Soc. Chem. Commun.</u>, 1976, 787.
9. R. E. King III, S. B. Miller, C. B. Knobler, and M. F. Hawthorne, <u>Inorg. Chem.</u>, 1983, <u>22</u>, 3548.
10. G. K. Barker, M. Green, F. G. A. Stone, A. J. Welch, and W. C. Wolsey, <u>J.Chem. Soc. Chem. Commun.</u>, 1980, 627.
11. R. T. Baker, M. S. Delaney, R. E. King III, C. B. Knobler, J. A. Long, T. B. Marder, T. E. Paxson, R. G. Teller, and M. F. Hawthorne, <u>J. Am. Chem. Soc.</u>, 1984, <u>106</u>, 2965.
12. C. W. Jung and M. F. Hawthorne, <u>J. Am. Chem. Soc.</u>, 1980, <u>102</u>, 3024.
13. C. W. Jung, R. T. Baker, and M. F. Hawthorne, <u>J. Am. Chem. Soc.</u>, 1981, <u>103</u>, 810.
14. ^{11}B NMR (CH_2Cl_2): ∂ 43.4 (1B, s), 29.0 (1B, d, J_{BH}=141 Hz), 18.9 (2B, d, J_{BH}=130 Hz), -0.7 (2B, d, J_{BH}=135 Hz), -12.6 (1B, d), and -14.0 (2B, d).
$^{31}P\{^1H\}$ NMR $(CDCl_3)$: ∂ 2.3 (s). ^1H NMR $(CDCl_3)$: ∂ 1.40-1.58 (6H, mult, β,γ-$C_5H_{10}NH$), 3.05-3.32 (4H, mult, α-$C_5H_{10}NH$), 1.878 (6H, d, J_{HP}=10.99 Hz, P*Me*), 7.39-7.42, 7.65-7.72 (5H, mult, P*Ph*).

Electrochemistry of Tetradecker Metallacarboranes

J. R. Pipal* and R. N. Grimes

DEPARTMENT OF CHEMISTRY, UNIVERSITY OF VIRGINIA,
CHARLOTTESVILLE VA 22901, USA

1 INTRODUCTION

As recently reported[2] our group has prepared a series of tetradecker metallacarboranes of general formula $\{Cp^*Co(Et_2C_2B_3H_2X)\}_2M$ where $Cp^* = C_5Me_5$, M = Ni, Co, Ru and X = Cl, Br, I, Me, $COCH_3$, butynyl. In addition $\{Cp^*Co(Et_2C_2B_3HCl_2)\}_2Ni$ was made. Synthesis of the compounds from double deck precursors is possible only if X is somewhat electron withdrawing. A general view of these tetradeckers is shown in Figure 1.

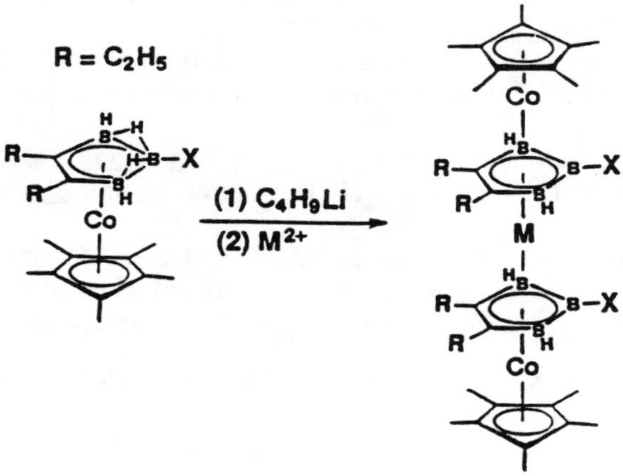

Figure 1   Synthesis and structure of tetradecker
metallacarboranes

The preparation of these compounds (hereafter referred to by $CoMCoX_2$ or $CoNiCoCl_4$) was undertaken as part of an effort to synthesize molecular compounds which might be conductors or serve as precursors to larger multideck conductive polymers. The electrochemical studies were done to assess the degree of electron delocalization within the compounds and to observe the effect of the X substituent. Electron delocalization is essential if these compounds are likely to be molecular conductors or if polymeric molecules made by linking these tetradeckers are to be molecular conductors

2 RESULTS

All of these compounds exhibit two reversible one electron reductions, one reversible one electron oxidation (except for X = $COCH_3$), and a second irreversible oxidation. For the M = Ni compounds the reversible reductions are separated by 0.78 v on average and for the M = Co compounds they are separated by 1.59 V on average. Controlled potential electrolysis was used to demonstrate that all the reversible processes were one electron transfers. Tables 1 and 2 give the results of the cyclic voltammetry studies for these compounds in 1,2-dimethoxyethane (DME) and dichloromethane DCM.

3 DISCUSSION

One immediately can see the effect of the substituent X especially for the $CoNiCoX_2$. As X becomes more electron withdrawing, the reduction potentials become more positive indicating a greater ease of reduction and difficulty in oxidizing as one would expect. Since tetradeckers could be synthesized only if X is electron withdrawing, the stability of these compounds requires

Table 1 Cyclic voltammetry results ($E_{1/2}$ in volts) for reversible processes in DME.

Compound	+/0	0/-	-/2-
$CoNiCoMe_2$	-0.08	-1.84	-2.64
$CoNiCo(CH_2CCMe)_2$	0.04	-1.72	-2.56
$CoNiCo[C(O)Me]_2$	0.33 [a]	-1.48	-2.32
$CoNiCoCl_2$	0.18	-1.44	-2.17
$CiNiCoBr_2$	0.22	-1.37	-2.10
$CoNiCoCl_4$	0.30	-1.18	-1.96 [b]
$CoNiCoCl_4F_2$	0.36	-1.04	-1.76
$CoCoCoMe_2$	-0.16	-1.48	-3.36 [a]
$CoCoCo[C(O)Me]_2$	0.15 [a]	-1.08	-2.77
$CoCoCoCl_2$	0.07	-1.13	-2.62
$CoCoCoI_2$	0.18	-1.03	-2.53[a,c]
$CoRuCoCl_3$	0.48 [a]	-1.09	-2.23

[a] irreversible
[b] 2-/3- (irreversible) at -3.42V
[c] 2-/3- (irreversible) at -3.10V

Table 2   Cyclic Voltammetry results ($E_{1/2}$ in volts) for reversible processes in DCM.

Compound	+/0	0/-	-/2-
CoNiCoMe$_2$	-0.21	-1.91	
CoNiCo(CH$_2$CCMe)$_2$	-0.06	-1.81	
CoNiCo[C(O)Me]$_2$	0.24[a]	-1.57	-2.31
CoNiCoCl$_2$	0.06	-1.54	-2.26
CoNiCoBr$_2$	0.15	-1.44	-2.14
CoNiCoCl$_4$	0.23	-1.24	-1.97
CoNiCoCl$_4$F$_2$	0.31	-1.11	-1.78
Unknown Rose Red[b]	1.12	-0.24	-1.02
CoCoCoMe$_2$	-0.30	-1.55	
CoCoCo[C(O)Me]$_2$	0.07[a]	-1.19	
CoCoCoCl$_2$	0.00	-1.19	
CoCoCoI$_2$	Not done		
CoRuCoCl$_3$	0.39[a]	-1.16	-2.27

[a] irreversible   [b] see Discussion

removal of electron density from the metallacarborane cluster per se to the substituent.

The results also strongly suggest that there is electronic communication within the tetradecker. The molecules are symmetrical. If there were no communication, then one would expect either one electron reduction of the central metal followed by a two electron reduction of the equivalent terminal cobalts or a two electron reduction followed by a one electron reduction. That two widely separated one electron processes are seen instead strongly supports electron delocalization. The results seen could be consistent with no communication only if both reductions were localized on the central metal which we think unlikely given that all three central metals studied showed similar results.

An unexpected result was the electrochemical synthesis of {Cp*Co(Et$_2$C$_2$B$_3$Cl$_2$F)}$_2$Ni from {Cp*Co(Et$_2$C$_2$B$_3$HCl$_2$)}$_2$Ni. Oxidation of the latter to a dication is irreversible, but in the presence of Bu$_4$NPF$_6$ supporting electrolyte the dication is fluorinated to produce a tetradecker of unknown structure which upon isolation is converted to {Cp*Co(Et$_2$C$_2$B$_3$Cl$_2$F)}$_2$Ni. The tetradecker of unknown structure is listed in Table 2 as "unknown rose red" to reflect the color of the solution after oxidation. Attempts to isolate it were unsuccessful; only CoNiCoCl$_4$F$_2$ was found. The cyclic voltammogram is consistent with those seen for all the CoNiCo tetradeckers. In addition, the isolation of CoNiCoCl$_4$F$_2$ in good yield suggests that the unknown rose red compound retains the features of a tetradecker. The positive shift in $E_{1/2}$ value compared even to CoNiCoCl$_4$F$_2$ is rather remarkable and remains to be explained.

4   ACKNOWLEDGMENTS

The work was supported by the U. S. Army Research Office and the National Science Foundation. JRP acknowledges the sabbatical leave provided by Alfred University.

5   REFERENCES

1.   On sabbatical from Alfred University
2.   Piepgrass, K.W.; Meng, X.; Hölscher, M.; Sabat, M.; Grimes, R.N. *Inorganic Chemistry* **1992**, *31*, 5202
3.   Merkert, J.W.; Davis, J.H.; Geiger, W.E.; Grimes, R.N. *J. Am. Chem Society* **1992**, *114*, 9846

The Photochemistry of Borane and Metallaborane Cluster Complexes

Craig M. Davis, Bruce H. Goodreau, Rona A. Nardone, Lianna R. Orlando, and James T. Spencer*

DEPARTMENT OF CHEMISTRY AND THE W. M. KECK CENTER FOR MOLECULAR ELECTRONICS, CENTER FOR SCIENCE AND TECHNOLOGY, SYRACUSE UNIVERSITY, SYRACUSE, NY 13244, USA

1. Introduction

The photochemistry of a relatively large number of organometallic species has been intensively investigated and has proven to be a very fertile area of study. Photochemical processes in both inorganic and organic chemistry have frequently been found to provide very efficient methods for the preparation of otherwise synthetically inaccessible compounds. In contrast, the photochemistry of borane and metallaborane compounds has not been explored in detail.

The photochemistry of boron hydride compounds has been limited to relatively few examples. These reactions can be generally divided into two types; those in which the borane or metallaborane compound undergoes photochemically induced intramolecular elimination and/or rearrangement processes and those in which intermolecular reactions between two or more reactants are photochemically promoted. Several examples are known, outside of our work, of photochemical transformations involving clearly intramolecular processes.[1-4] An example of this type of process is the reversible decarbonylation of $Mn(CO)_4B_3H_8$ shown in Figure 1.[1] Irradiation of this butterfly-shaped structural analog of $arachno$-B_4H_{10} results in the photolytic cleavage of one carbonyl-metal bond, subsequent loss of the carbonyl ligand and the formation of a M-H-B bond, thus transforming the triborane unit into a tridentate ligand. The CO loss was found to be a reversible process with the addition of CO to the solution regenerating the original bidentate triborane system. The photochemical reactions

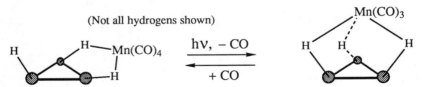

Figure 1. Photochemical interconversion of $MnCO)_4B_3H_8$ and $Mn(CO)_3B_3H_8$.

of polyhedral boranes with other reagents (intermolecular processes) is limited to four cases in the literature.[5-8] As an example of this reaction type, the cophotolysis of three organometallic compounds, $[(CO)_4ReBr]$, $[(\eta^5\text{-}C_5H_5)(CO)_2FeI]$, and $[(\eta^5\text{-}C_5H_5)(CO)_3WCl]$, with the octahydrotriborate(1-) anion was found to produce the $[(CO)_4ReB_3H_8]$, $[(\eta^5\text{-}C_5H_5)(CO)FeB_3H_8]$, and

[η^5-C$_5$H$_5$)(CO)$_3$WB$_3$H$_8$] complexes, respectively. The photo-formation of the rhenium based [(CO)$_4$ReB$_3$H$_8$] complex is shown in Figure 2. In these reactions, each metal center was found to lose its halide and one carbonyl ligand with the ultimate formation of two three-center two-electron M-H-B bonds.

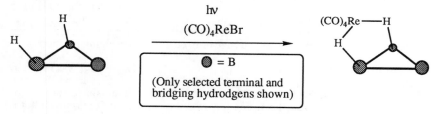

Figure 2. Intermolecular photochemical reaction of boron hydrides.

Our initial work in the photochemistry of boranes and metallaboranes involved the photolysis of [Fe(η^5-C$_5$H$_5$)(CO)$_2$B$_5$H$_8$(μ-P(C$_6$H$_5$)$_2$)] to sequentially form the tetraborane [Fe(η^5-C$_5$H$_5$)(CO)B$_4$H$_6$(P(C$_6$H$_5$)$_2$)] complex and, after passage down a silica gel column, the triborane [Fe(η^5-C$_5$H$_5$)(CO)B$_3$H$_7$-(P(C$_6$H$_5$)$_2$)] complex, as shown in Figure 3.[9] This photochemical process was found to yield a single photoproduct in a remarkably clean reaction. Based on these initial studies, we have thus begun a systematic investigation of the photochemistry and rearrangement chemistry of borane and metallaborane species.

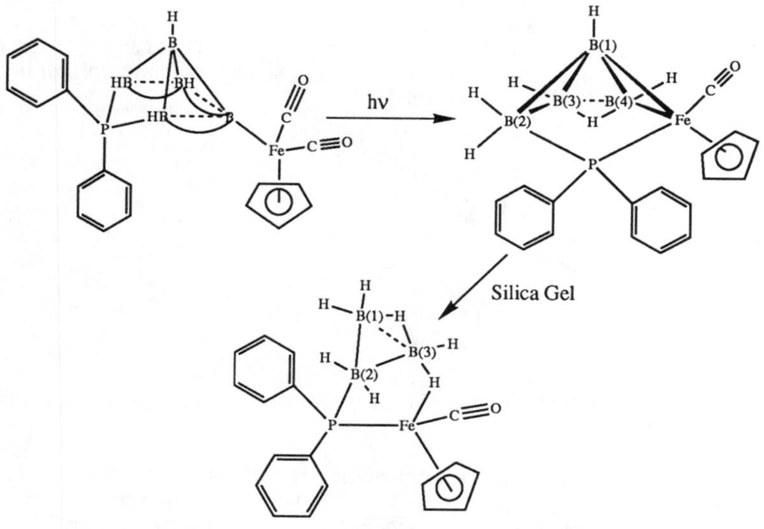

Figure 3. Photolysis of [Fe(η^5-C$_5$H$_5$)(CO)$_2$B$_5$H$_8$(μ-P(C$_6$H$_5$)$_2$)].

In this lecture, we describe the results of our photochemical investigations with two polyhedral borane species, [2-(Fe(η^5-C$_5$H$_5$)(CO)$_2$)B$_5$H$_8$] and the octahydrotriborate(1-) anion, [B$_3$H$_8$]⁻.

2. Results and Discussion

In our work, we have initially studied the effect of a photochemically generated unsaturated iron center from a $Fe(\eta^5\text{-}C_5H_5)(CO)_2$ group variously substituted on a pentaborane(9) cage system. This $Fe(\eta^5\text{-}C_5H_5)(CO)_2$ organometallic moiety is readily decarbonylated upon UV irradiation, generating an unsaturated 16 electron iron center. Borane cluster ligands, such as pentaborane(9), offer the possibility that a photochemically generated unsaturated metal center could insert into borane cage or isomerize to a bridging position.

The photochemical irradiation of $[2\text{-}(Fe(\eta^5\text{-}C_5H_5)(CO)_2)B_5H_8]$ was found to yield the cage inserted $[2\text{-}(\eta^5\text{-}C_5H_5)(CO)\text{-}2\text{-}FeB_5H_8]$ complex in 63 % yield. This rearrangement is shown in Figure 4. ^{11}B NMR analysis of the photo-reaction mixture showed the formation of only one boron-containing photo-product, $[2\text{-}(\eta^5\text{-}C_5H_5)(CO)\text{-}2\text{-}FeB_5H_8]$. Characterization of this new complex was by ^{1}H and ^{11}B NMR, infrared, mass spectral, and single-crystal X-ray analyses. The reaction appears to be non-reversible and no reaction was observed when the $[2\text{-}(\eta^5\text{-}C_5H_5)(CO)\text{-}2\text{-}FeB_5H_8]$ complex was treated with either CO or PPh3 ligands. In the attempted photochemical preparation of $[2\text{-}(\eta^5\text{-}C_5H_5)(CO)\text{-}2\text{-}FeB_5H_8]$ in pentane rather than THF, $[(\eta^5\text{-}C_5H_5)(CO)_2Fe]_2$ was the only identifiable product. This result suggests that homolysis of the Fe-B bond occurred rather than decarbonylation of the $(\eta^5\text{-}C_5H_5)(CO)_2Fe$ group. This suggestion is consistent with the known homolysis pathway reported for the benzyl substituted analog, $[(\eta^5\text{-}C_5H_5)(CO)_2FeCH_2C_6H_5]$, which was reported to undergo both CO dissociation and homolytic fission of the M-C bond upon irradiation.[10] The homolysis products, $(\eta^5\text{-}C_5H_5)(CO)_2Fe\cdot$ and $\cdot B_5H_8$, could then undergo recombination to give $[(\eta^5\text{-}C_5H_5)(CO)_2Fe]_2$ and $(B_5H_8)_2$. Observing only the $[(\eta^5\text{-}C_5H_5)(CO)_2Fe]_2$ complex indicates that the borane radical is probably highly unstable and decomposes before it can be observed. This reaction also suggests that THF solvent is in some fashion important in the formation of $[2\text{-}(\eta^5\text{-}C_5H_5)(CO)\text{-}2\text{-}FeB_5H_8]$.

(Hydrogen atoms not shown)

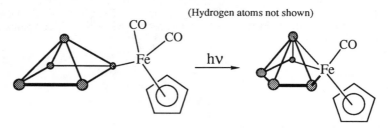

Figure 4. Photochemical expansion of the cluster framework of $[2\text{-}(Fe(\eta^5\text{-}C_5H_5)(CO)_2)B_5H_8]$.

We have also begun photochemical investigations of the di-substituted pentaborane complex, $[2,4\text{-}(Fe(\eta^5\text{-}C_5H_5)(CO)_2)_2B_5H_7]$. Upon irradiation, this complex was found to cleanly lose one carbonyl from one of the two $[(Fe(\eta^5\text{-}C_5H_5)(CO)_2)]$ units terminally bound to the cage. Similar behavior to that observed for the mono-substituted $[2\text{-}(Fe(\eta^5\text{-}C_5H_5)(CO)_2)B_5H_8]$ complex was found for the di-substituted cage after CO loss in which the iron center which loses the carbonyl ligand inserts into the pentaborane(9) cage framework in a cage expansion reaction to form $[2\text{-}(\eta^5\text{-}C_5H_5)(CO)\text{-}4\text{-}(Fe(\eta^5\text{-}C_5H_5)(CO)_2)\text{-}2\text{-}FeB_5H_8]$. Further irradiation of this intermediate diiron product was found to generate only $[(\eta^5\text{-}C_5H_5)(CO)_2Fe]_2$ and the mono-iron substituted $[2\text{-}(Fe(\eta^5\text{-}C_5H_5)(CO)_2)B_5H_8]$ complex previously described. This result also suggests that homolysis of the Fe-B bond on the remaining terminally bound $[(\eta^5\text{-}C_5H_5)(CO)_2Fe]$ unit occurred rather than decarbonylation. As observed in the photolysis of $[2\text{-}(Fe(\eta^5\text{-}C_5H_5)(CO)_2)B_5H_8]$ in pentane, the homolysis product $(\eta^5\text{-}C_5H_5)(CO)_2Fe\cdot$ could then undergo recombination to give the observed $[(\eta^5\text{-}C_5H_5)(CO)_2Fe]_2$ complex. The $\cdot[2\text{-}(Fe(\eta^5\text{-}C_5H_5)(CO)_2)B_5H_8]$ radical presumably formed must abstraction a hydrogen to form the observed $[2\text{-}(Fe(\eta^5\text{-}C_5H_5)(CO))B_5H_8]$ complex. The source of the hydrogen atom could be starting material, decomposition products or solvent. It seemed possible, however, that the $\cdot[2\text{-}(Fe(\eta^5\text{-}C_5H_5)(CO)_2)B_5H_8]$ radical might be stable enough to exist long enough to be experimentally observed. Stabilization might occur because of the electron delocalization within the cluster bonding. ESR spectra of the THF solution immediately after irradiation of $[2\text{-}(\eta^5\text{-}C_5H_5)(CO)\text{-}4\text{-}(Fe(\eta^5\text{-}C_5H_5)(CO)_2)\text{-}2\text{-}FeB_5H_8]$ showed no such radical.

We have also begun a systematic investigation into the photochemistry of non-metal borane species.[11] Our initial work has been focused on the octahydrotriborate(1-) anion, which displays exceptional oxidative and hydrolytic stabilities. The photochemical irradition of the octahydrotriborate(1-) anion with alcohols has been found to provide clean, rapid, and total alcoholysis of this robust boron hydride to form $HB(OC_2H_5)_2$, which ultimately forms $B(OC_2H_5)_3$. Deuterated ethanol (EtOD) was employed to probe the mechanism of this phenomenon. In contrast, the photochemical irradition of the octahydrotriborate(1-) anion in a wide variety of chlorohydrocarbons has been found to lead to the sequential chlorination of the cluster without cage degradation. Mechanistic studies involving quantum yield measurements and reactive intermediate trapping experiments indicate that these photochemical processes probably proceed through radical chain mechanisms.

3. Acknowledgments

We wish to thank the National Science Foundation (Grant No. MSS-89-09793), the Wright-Patterson Laboratory (Award No. F33615-90-C-5291), and the Industrial Affiliates Program of the Center for Molecular Electronics for support of this work.

4. References

1.   D.F. Gaines and S. J. Hildebrandt, <u>Inorg. Chem.</u>, 1978, <u>14</u>, 794.
2.   R. N. Leyden, B. P. Sullivan, R. T. Baker, and M. F. Hawthorne, <u>J. Am. Chem. Soc.</u>, 1978, <u>100</u>, 3758.
3.   L. G. Sneddon, D. C. Beer, and R. N. Grimes, <u>J. Am. Chem. Soc.</u>, 1973, <u>95</u>, 6623.
4.   R. V. Schultz, F. Sato, and L. J. Todd, <u>J. Organomet. Chem.</u>, 1977, <u>125</u>, 115.
5.   D.W. Lowman, P.D. Ellis, and J.D. Odom, <u>Inorg. Chem.</u>, 1973, <u>12</u>, 681.
6.   D.F. Gaines and S.J. Hildebrandt, <u>Inorg. Chem.</u>, 1978, <u>14</u>, 794.
7.   R.J. Astheimer and L.G. Sneddon, <u>Inorg. Chem.</u>, 1984, <u>23</u>, 3207.
8.   (a) S. Trofimenko, <u>J. Am. Chem. Soc.</u>, 1966, <u>88</u>, 1899; (b) S. Trofimenko and H.N. Cripps, <u>J. Am.Chem. Soc.</u>, 1965, <u>87</u>, 653.
9.   B. H. Goodreau, L. R. Orlando, and J. T. Spencer, <u>J. Am. Chem. Soc.</u>, 1992, <u>114</u>, 3827.
10. R. J. Kazlauskas and M. S. Wrighton, <u>Organometallics</u>, 1982, <u>1</u>, 602.
11. C. M. Davis and J. T. Spencer, <u>Inorg. Chim. Acta</u> 1993, in press.

Paramagnetic Metallacarborane Complexes. Synthesis, Electrochemistry, ESR, and P-NMR Studies

M. Stephan[1,2], P. Mueller[1,2], U. Zenneck*[,3], R. N. Grimes*[,2], and W. Siebert*[,1]

[1]ANORGANISCH-CHEMISCHES INSTITUT DER UNIVERSITÄT, 6900 HEIDELBERG, GERMANY
[2]DEPARTMENT OF CHEMISTRY, UNIVERSITY OF VIRGINIA, CHARLOTTESVILLE, VA 22901, USA
[3]INSTITUT FÜR ANORGANISCHE CHEMIE DER UNIVERSITÄT ERLANGEN-NÜRNBERG, 8520 ERLANGEN, GERMANY

1 SUMMARY

In our joint effort to develop a more detailed picture of the electronic structure and properties of multi-decker systems[1] we have performed the synthesis of the air-stable, paramagnetic 29-electron iron-cobalt triple-decker complexes (2a,b,c) in very high yield. We studied the redox properties of these species by combined use of cyclic voltammetry, ESR and NMR in comparison to earlier investigations on mononuclear 7-vertex closo-ferra- and cobaltacarborane clusters (3) and (4)[2], and more recently to the triple-decker complexes $(C_5Me_5)Fe(Et_2C_2B_3H_3)Co(C_5H_5)$ (5) and $(C_5Me_5)Co(Et_2C_2B_3H_3)Co(C_5Me_5)$ (6).

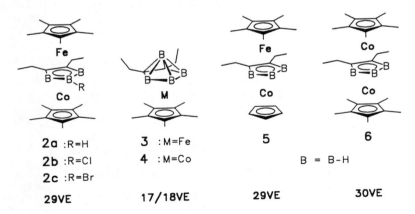

2a :R=H	3 :M=Fe	5	6
2b :R=Cl	4 :M=Co		
2c :R=Br		B = B-H	
29VE	17/18VE	29VE	30VE

The ^1H NMR spectra of the paramagnetic species $2a^+$, 2a, $2a^{2-}$, as well as 5^+, 5, 6^+ and 6^- were assigned by recording series of spectra during quantitative stepwise reduction of the neutral complexes in THF-d_8 using a potassium mirror in a sealed NMR tube[2,3] or recording spectra of mixtures of the chemically oxidized cations in various concentrations with the neutral species in $CDCl_3$.

2 RESULTS AND DISCUSSION

Whereas the strategy of the synthesis of the first neutral paramagnetic $(C_5H_5)Fe(Et_2C_2B_4H_3)Co(C_5H_5)$ (29 electrons) triple-decker complex containing the C_2B_3 bridging ligand[4] was not generally applicable for stacking sandwich systems, the reaction of pentamethylcyclopentadienide anion $(C_5Me_5^-)$ with anhydrous ferrous chloride and the $(R_2C_2B_3H_5)Co(C_5Me_5)^-$ (R = H,Cl,Br) anions (1a,b,c) in THF with subsequent isolation via column chromatography afforded black-brown and airstable crystals of 2a,b,c in yields between 62% and 92%[5] (Figure 1).

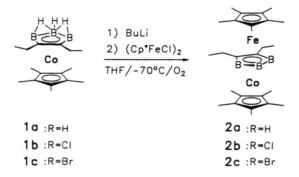

1) BuLi
2) (Cp*FeCl)₂
THF/-70°C/O₂

1a :R=H	**2a** :R=H
1b :R=Cl	**2b** :R=Cl
1c :R=Br	**2c** :R=Br

<u>**Figure 1**</u> Synthesis of the triple-decker complexes 2a,b,c

A substituent dependent redox behavior was observed for 2a,b,c encompassing an electron-transfer series with reversible one-electron steps from the 28-electron cation to the paramagnetic 31-electron dianion. For 2a we measured the redox potentials in DME/Bu_4NPF_6 (E' in V vs. SCE): $2a^{2+/+}$: +1.36V (irr.); $2a^{+/0}$: 0.35V; $2a^{0/-}$: -1.51V; $2a^{-/2-}$: $E_p^{red} < -3.1V$.[5] 2a was readily oxidized with 1 equivalent $AgBF_4$ in ether to form the purple 28-electron cation $2a^+$ in quantitative yield.

The 29-electron complexes were examined via ESR spectroscopy conducted in toluene glass between 300 and 110K (Table 1). Showing no ^{59}Co hyperfine structure the measurements are consistent with the presence of one unpaired electron in the vicinity of the iron atom supporting the designation of a formal metal oxidation state of +3. The axial symmetric ESR spectra can be compared to data of isoelectronic species such as the diborolyl-carboranyl "hybrid" complex $(Et_2C_2B_4H_4)Fe(Et_2MeC_3B_2Et_2)Co(C_5H_5)$[6] (7), and the diiron complex $(C_5H_5)Fe(Et_2MeC_3B_2Et_2)Fe(C_5H_5)$[7] (8) measured under the same conditions, as well as that of $(Et_2C_2B_4H_4)Fe(C_5Me_5)^2$ (3) and $(\eta^6-C_6Me_6)Fe(Et_2C_2B_4H_4)^+$ cation[8] (9^+). The striking similarity in these spectra supports the presumption of comparable electronic structures among different families of Fe(III)-carborane complexes and is further supported by NMR measurements.

<u>Table 1</u>   X-band ESR data of 2a,b,c at 110K in toluene
          glass (standard, LiTCNQ: g = 2.0025).

Comp.	<g>	$g_{\parallel}$	$g_{\perp}$	Ref.
2a	2.120	2.400	2.035	[5]
2b	2.160	2.455	2.031	[5]
2c	2.160	2.500	2.030	[5]
3	2.220	2.689	1.979	[2]
7	2.194	2.583	2.000	[6]
8	2.280	2.930	1.980	[7]
9+a)	-	2.486	2.002	[8]

a) generated by coulometric oxidation of 9;
   measured in CH_2Cl_2 at T=77K

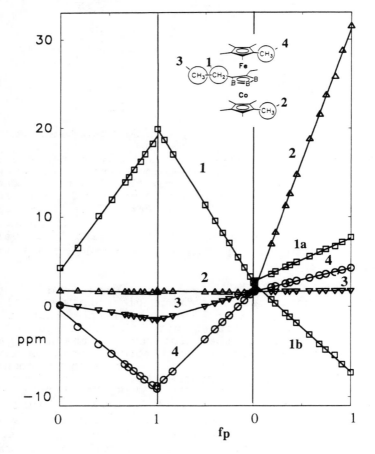

<u>Figure 2</u>   Correlation diagram for ^1H NMR signals of 2a$^+$/2a,
          2a/2a$^-$ and 2a$^-$/2a^{2-} mixtures (tabulated data are
          available as supplementary material).

The ^{1}H-NMR spectra of paramagnetic 2a and the dianion 2a[2-] were completely assigned to the green 30-electron diamagnetic species 2a[-] via stepwise reduction in THF-d$_8$ in a sealed NMR tube which contained a potassium mirror well above the solution surface.[3] Monitoring the NMR-spectrum at each stage of the reduction we observed an averaged set of ^{1}H signals indicating rapid electron transfer between the neutral and 2a[-], and the redox species 2a[-] and 2a[2-]. The ^{1}H NMR spectrum of the apparently paramagnetic but ESR inactive 28-electron cation 2a[+] was correlated to 2a by taking spectra of variously concentrated mixtures of 2a and chemically oxidized 2a[+] in CDCl$_3$.

We obtained correlation diagrams which in addition to the analysis of the designation of the signals gave insight into the spin density distribution, the extent of electron delocalization and metal-metal communication in this FeCo-dinuclear complex. In accordance with ESR studies of both 2a and 2a[2-] we observe the paramagnetism to be largely located in Fe- and Co-centered highest occupied molecular orbitals. These results can be compared to investigations done on the redox species of the triple-decker species 5+/5, 5/5-, 6/6- and the metallcarboranes 3/3[-] and 4/4[-]. It shows that common electronic properties do exist for the 17-electron iron complex 3 and 2a, and the 19-electron cobalt complex 4[-] and 2a[2-].

Acknowledgement. This work was supported by NATO International Research Grant 0196/85, by the Deutsche Forschungsgemeinschaft (SFB 247), U.S. Army Research Office and by NSF (Grant CHE 8721657).

3 REFERENCES

1. a) A. Feßenbecker, M.D. Attwood, R.F. Bryan, R.N. Grimes, M.K. Woode, M. Stephan, U. Zenneck, W. Siebert, Inorg.Chem. 1990, 29, 5157-5163; b) A. Feßenbecker, M.D. Attwood, R.N. Grimes, M. Stephan, H. Pritzkow, U. Zenneck, W. Siebert, Inorg.Chem. 1990, 29, 5164-5168
2. M. Stephan, J.H. Davis, X. Meng, K.J. Chase, J. Hauß, U. Zenneck, H. Pritzkow, W. Siebert, R.N. Grimes, J.Am.Chem.Soc. 1992, 114, 5214-5221
3. F.H. Köhler, U. Zenneck, J. Edwin, W. Siebert, J.Organomet.Chem. 1981, 208, 137-144
4. J.H. Davis, E.Sinn, R.N. Grimes, J.Am.Chem.Soc. 1989, 111, 4776-4784
5. M. Stephan, Ph.D. Thesis, Universität Heidelberg, 1992
6. M.D. Attwood, K.K. Fonda, R.N. Grimes, G. Brodt, D. Hu, U. Zenneck, W. Siebert, Organometallics 1989, 8, 1300-1303
7. G. Brodt, Ph. D. Thesis, Universität Heidelberg, 1988
8. J.M. Merkert, W.E. Geiger, M.D. Attwood, R.N. Grimes, Organometallics 1991, 10, 3545-3549

Transition Metal-mediated Reactions of Boron Hydride

Hee-Joo Jeon and Sang Ook Kang

DEPARTMENT OF CHEMISTRY, COLLEGE OF NATURAL SCIENCES,
KOREA UNIVERSITY, CHOONG-NAM 339-700, KOREA

1 INTRODUCTION

Transition-metal reagents are widely used in polyhedral borane chemistry to catalyze or promote a variety of transformations. We previously reported our results[1] concerning the metal-mediated alkylation reaction of a boron hydride, in which Fischer-type carbene complexes were found to promote the conversion of arachno-$S_2B_7H_8^-$ to the alkylated cage compound, arachno-$RCH_2S_2B_7H_8$, in essentially quantitative yields under mild conditions. In this article we report details of these and related reactions of a variety of Fischer-type carbene complexes with arachno–$S_2B_7H_8^-$.

2 RESULTS AND DISCUSSION

The reaction of a variety of Fischer-type carbene complexes with arachno-$S_2B_7H_8^-$ were explored, and all reactions were found to proceed at moderate temperature, be highly selective, and give good yields of alkylated cage products. Thus, the reaction of the arachno-$S_2B_7H_8^-$ anion with the Fischer-type carbene complexes, $(CO)_5M\{C(R_1)(R_2)\}$(M= Cr, W; R_1= CH_3, C_6H_5; R_2= OCH_3, SC_6H_5), followed by protonation of the resulting anion **I** with the anhydrous HCl resulted in the formation of compound **II**, which was isolated as an air-sensitive, white crystalline product in good yield:

$$arachno–S_2B_7H_8^- \ + \ (CO)_5M\{C(R_1)(R_2)\} \ ---->$$
$$[(CO)_5M\{C(R_1)(R_2)(S_2B_7H_8)\}]^- \ \mathbf{I} \qquad (1)$$

$$[(CO)_5M\{C(R_1)(R_2)(S_2B_7H_8)\}]^- \ \mathbf{I} \ + \ H^+ \ ---->$$
$$arachno–R_1CH_2S_2B_7H_8 \ \mathbf{II} \qquad (2)$$
$$(M= Cr, W; \ R_1= CH_3, \ C_6H_5; \ R_2= OCH_3, \ SC_6H_5)$$

The reaction was found to proceed at room temperature to give the unstable metallathiaborane intermediate **I**. These compounds can be obtained quantitatively in all reactions described above but decompose rapidly in solution above -30 °C. The new compounds would be $[(CO)_5M\{C(R_1)R_2(S_2B_7H_8)\}]^-$ supported by ^{11}B NMR data. The ^{11}B NMR spectrum consists of seven resonances of equal intensity. All the resonances are split into B-H coupled doublets except for the resonance at -21.5 ppm which appears as a singlet in the proton coupled spectrum. A complex ^{1}H NMR spectrum of **I** is obtained at room temperature. The extremely complex and broad spectrum (25 °C) precludes structural assignment for **I**. However, broad high-field absorption observed at -10.3 ppm may indicate the presence of the metal hydride complex. Thus, the proposed structure for complex **I** is that of a substituted *arachno*-6,8-$S_2B_7H_9$ system in which the carbene complex fragment has substituted for a boron vertex in a 4-boron atom. We had hoped that the addition of cationic carbene complexes[2,3] to *arachno*-$S_2B_7H_8^-$ would lead to the corresponding neutral metal complexes. However, we found that the addition of $[Cp(CO)_2Fe\{C(CH_3)(OCH_3)\}]^+BF_4^-$ to *arachno*-$S_2B_7H_8^-$ did not give the desired product but rather decomposed material.

It was also found that *in situ* reaction of the anion **I** with anhydrous HCl in methylene chloride resulted in the good yield formation of the alkylated derivatives of *arachno*-6,8-$S_2B_7H_9$. Exact mass measurements support the proposed composition of $CH_3CH_2S_2B_7H_8$ **IIa** and $C_6H_5CH_2S_2B_7H_8$ **IIb**. Thiaboranes of the formula $RCH_2S_2B_7H_8$ (R = CH_3, **IIa**; C_6H_5, **IIb**) would be arachno skeletal electron systems (9 cage atoms and 12 skeletal electron pairs) and would be expected to adopt the open-cage geometry found in *arachno*-6,8-$S_2B_7H_9$.

The ^{11}B NMR spectra of **IIa** and **IIb** have several similar features and support the structures proposed in Figure 1. The ^{1}H NMR spectrum of **IIa** strongly supports the proposed formulation, showing two alkyl CHs in a relative ratio of 2:3 and one distinct type of bridging hydrogens. Upon boron decoupling these broad CH resonances collapse to a triplet and quartet arising from an ethyl group. Also, in agreement with the proposed structure, the ^{1}H NMR spectrum of **IIb** exhibits broad resonance for B-H protons as well as resonances expected for the benzylic protons.

It has been found that insertion reaction is one of the characteristic features of Fischer-type carbene complexes. So far, carbene ligands insert into the silicon-,[4] germanium-,[5] and tin-hydrogen bonds.[6] In contrast, there is no known example for the insertion of a metal coordinated carbene into a B-H bond. It should be noted that a typical B-H bond energy is similar to that for an analogous Si-H bond.[7] This means that insertion reactions involving boron hydrides occur under similar conditions. Previously, *arachno*-$S_2B_7H_8^-$ was found to insert into a variety of polarizable organic compounds[8] such as nitriles and ketones to generate the corresponding *hypho*-$CH_3CNS_2B_7H_8^-$ [9] and *hypho*-$S_2B_6H_9^-$ [10] as shown in equations 3 and 4, respectively.

$$S_2B_7H_8^- + CH_3CN \longrightarrow \textit{hypho}\text{-}CH_3CNS_2B_7H_8^- \qquad (3)$$
$$S_2B_7H_8^- + (CH_3)_2CO \longrightarrow \textit{hypho}\text{-}S_2B_6H_9^- \qquad (4)$$

The result of the reactions above suggests that *arachno*-$S_2B_7H_8^-$ anion might also readily attack other polarized multiple bonds. Indeed, we have found that *arachno*-$S_2B_7H_8^-$ anion readily reacts with Fischer-type carbene complexes at room temperature. Similar to the reactions with nitriles and ketones, cage B-H insertion results in the production of new alkyl substituted thiaboranes, *arachno*-4-RCH_2-6,8-$S_2B_7H_8$ (R = CH_3, **IIa**; C_6H_5, **IIb**), in good yield.

All of the insertion reactions studied make use of electrophilic carbene complexes $(CO)_5M\{C(R_1)(R_2)\}$(M= Cr, W; R_1= CH_3, C_6H_5; R_2= OCH_3, SC_6H_5). These species react with *arachno*-$S_2B_7H_8^-$ at room temperature, below those required for ligand dissociation. In addition, no boron containing side products are observed. All available evidence points to direct reaction of these metal carbene complexes with *arachno*-$S_2B_7H_8^-$ without prior formation of carbene-borane complexes via ligand substitution.

Although the mechanisms of the reactions reported herein have not been determined, the above observations suggest that the reaction with *arachno*-$S_2B_7H_8^-$ may involve steps analogous to those given in equations 3 and 4 to yield a metallathiaborane complex, which could then undergo protonation to produce alkylated products.

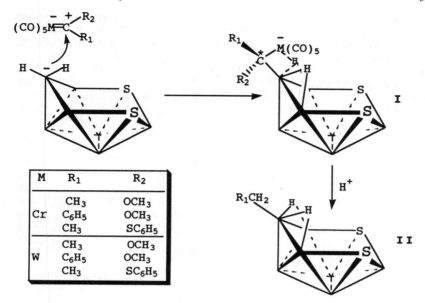

M	R_1	R_2
Cr	CH_3	OCH_3
	C_6H_5	OCH_3
	CH_3	SC_6H_5
W	CH_3	OCH_3
	C_6H_5	OCH_3
	CH_3	SC_6H_5

Figure 1   Possible reaction sequence leading to the formation of *arachno*-4-$R_1CH_2S_2B_7H_8$

3 ACKNOWLEDGMENT

The present studies were supported by the Basic Science Research Institute program, Ministry of Education and the Korea Science and Engineering Foundation (KOSEF 921-0300-030-2). We thank Professor Larry G. Sneddon for his help in obtaining NMR results and for his useful discussions and suggestions.

REFERENCES

1.   H. J. Jeon, J. J. Ko, S. J. Kim, D. S. Shin, and S. O. Kang, <u>Bull. Korean Chem. Soc.</u>, 1992, <u>13</u>, 220.
2.   C. P. Casey, W. H. Miles, and H. Tukada, <u>J. Am. Chem. Soc.</u>, 1985, <u>107</u>, 2924.
3.   M. Brookhart, J. R. Tucker, and G. R. Husk, <u>J. Am. Chem. Soc.</u>, 1983, <u>105</u>, 258.
4.   J. A. Connor, and P. D. Rose, <u>J. Organomet. Chem.</u>, 1970, <u>24</u>, C45.
5.   E. O. Fischer, and K. H. Dotz, <u>J. Organomet. Chem.</u>, 1972, <u>36</u>, C4.
6.   J. A. Connor, P. D. Rose, and R. M. Turner, <u>J. Organomet. Chem.</u>, 1973, <u>55</u>, 111.
7.   W. L. Jolly, 'The Principle of Inorganic Chemistry', McGraw-Hill, New York, 1976, p41.

8.   S. O. Kang and L. G. Sneddon, In Electron Deficient
      Boron and Carbon Clusters; G. A. Olah, K. Wade, and
      R. E. Williams Eds.; John Wiley and Sons: New York,
      1990; p 195.
9.   S. O. Kang, G. T. Furst and L. G. Sneddon,
      <u>Inorg. Chem.</u>, 1989, <u>28</u>, 2339.
10.   S. O. Kang and L. G. Sneddon, <u>J. Am. Chem. Soc.</u>,
      1989, <u>111</u>, 3281.

HETEROBORANE DERIVATIVES AND
COMPLEX BOROHYDRIDES

Aza-*closo*-dodecaborane(12): The Story of Six-coordinate Nitrogen

P. Paetzold*, J. Müller, F. Meyer, H.-P. Hansen, and L. Schneider

INSTITUT FÜR ANORGANISCHE CHEMIE, TECHNISCHE HOCHSCHULE AACHEN, D-52056 AACHEN, GERMANY

1 INTRODUCTION

Thirty years ago, the discovery of six-coordinate carbon in dicarba-*closo*-dodecaborane, $C_2B_{10}H_{12}$,[1] made the chemists finally become aware that carbon will not under all circumstances be a good boy who strictly obeys classical valence bond rules. Carbon once being debunked as a brother of the rascal boron, we became interested in the question, whether nitrogen could also behave in such an unorthodox manner. With the electronic equivalency of a CC and a BN unit in mind, we tried to synthesize the analogues of the *closo*-species $C_2B_{10}H_{12}$ and $MC_2B_9H_{11}$ (M stands for a metal-complex fragment which contributes two cluster electrons) and also of the *nido*-species $C_2B_9H_{13}$, $C_2B_9H_{12}^-$, and $C_2B_9H_{11}^{2-}$. The formulae of the expected analogues would be $NB_{11}H_{12}$, $MNB_{10}H_{11}$, $NB_{10}H_{13}$, $NB_{10}H_{12}^-$, and $NB_{10}H_{11}^{2-}$, respectively. It is well known that carbon has a greater tendency than boron to reduce its connectivity in cluster opening or degradation reactions. One might expect this tendency to continue in the series of increasing electronegativity from boron via carbon to nitrogen. Anyhow, four-fold connectivity of nitrogen within the skeleton, i.e. non-classical five-fold coordination with inclusion of the terminally bonded hydrogen, had been realized years ago in the cluster molecules *nido*-$NC_2B_8H_{11}$ [2] and *closo*-NB_9H_{10};[3] in both molecules the positions of the higher connectivity five are reserved to boron. What about such higher cluster connectivity with nitrogen, including a sixth coordination place for either an electron pair or an *exo*-ligand?

REFERENCES

1. T.L. Heying, J.W. Ager, S.L.Clark, D.J. Mangold, H.L. Goldstein, M. Hillman, R.J. Polak, J.W. Szymanski, *Inorg. Chem.* 1963, *2*, 1089; M.M. Fein, J. Bobinski, N. Mayes, N. Schwartz, M.S. Cohen, *Inorg. Chem.* 1963, *2*, 1111; L.I. Zakharkin, V.I. Stanko, V.A. Brattsev, Y.A. Chapovskii, Y.T. Struchkov, *Izw. Akad. Nauk SSSR, Ser. Khim.* 1963, 2069.
2. J. Plešek, S. Heřmánek, J.C. Huffman, P. Ragatz, R. Schaeffer, *J. Chem. Soc. Chem. Commun.* 1975, 935; J. Plešek, B. Štibr, S. Heřmánek, *Chem. Ind.* 1974, 662.
3. A. Arafat, J. Baer, J.C. Huffman, L.J. Todd, *Inorg. Chem.* 1986, *25*, 3757.

2 INCORPORATION OF NITROGEN IN B_9 AND B_{10} SKELETONS

closo-, *nido-*, and *arachno-*Azadecaboranes have been the first representatives of larger azaborane clusters. The starting material is the classical $B_{10}H_{14}$ which loses a B atom when treated either with Me_2N-NS [1] or with $NaNO_2$.[2] Poor yields of derivatives of *arachno-*NB_9H_{14} [Eq.(1)] or a 55% yield of *nido-*NB_9H_{12} [Eq. (2)] are obtained, respectively. Concerning the mechanism of these reactions, there is a lack of transparency, whereas the thermal elimination of H_2 from *nido-*NB_9H_{12} gives *closo-*NB_9H_{10} in a most transparent way [Eq. (3)].

$$B_{10}H_{14} \xrightarrow{+ Me_2N-NS} (Me_2N)NB_9H_{12}^- \xrightarrow{Na} NB_9H_{13}^- \xrightarrow[+ MeCN]{Br_2} NB_9H_{12}(NCMe) \quad (1)$$

$$B_{10}H_{14} \xrightarrow[- H_2]{+ NaNO_2} \xrightarrow[\substack{- HBO_2 \\ - NaHSO_4}]{+ H_2SO_4} NB_9H_{12} \quad (2)$$

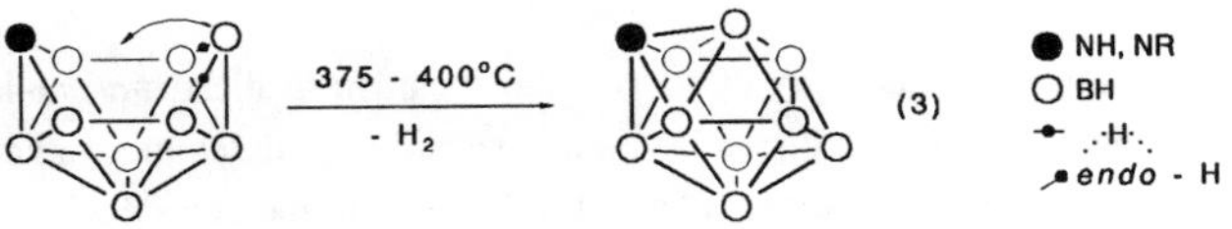

(3)

● NH, NR
○ BH
→ ‥H‥
↗ *endo* - H

nido   *closo*

We started from ethyl instead of sodium nitrite and synthesized NB_9H_{12} in a one-step reaction, which is not yet mechanistically but at least stoichiometrically clear [Eq. (4)].[3]

$$B_{10}H_{14} + 3\ EtONO \longrightarrow NB_9H_{12} + B(OEt)_3 + NO + NO_2 + H_2 \quad (4)$$

A mechanistically lucid access to NB_9H_{12} is achieved by the thermolysis of the well known *arachno-*$B_9H_{13}(NH_3)$.[4] We suggest an *exo/endo* exchange of the ligands H and NH_3 as the first step, followed by the elimination of H_2 [Eq. (5)].[5]

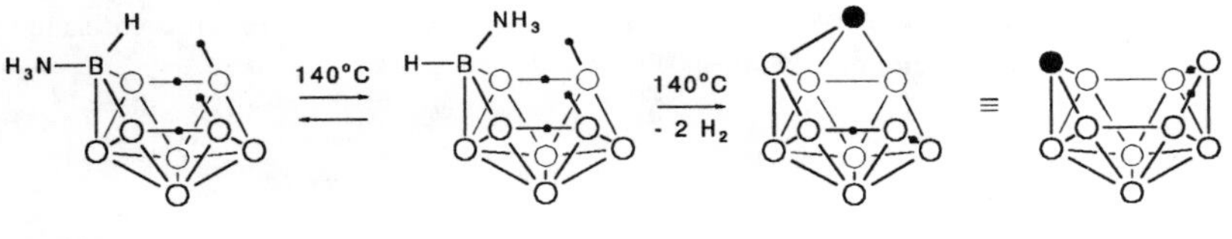

(5)

N-Alkyl derivatives RNB_9H_{11} are available in a similar way by applying RNH_2 instead of NH_3. Starting from *arachno-*$B_9H_{13}(SMe_2)$,[6] two steps are necessary [Eq. (6)].[7]

$$B_9H_{13}(SMe_2) \xrightarrow[- SMe_2]{+ RNH_2} B_9H_{13}(NRH_2) \xrightarrow[- 2\ H_2]{140°C} RNB_9H_{11} \quad (6)$$

In order to gain an NB_{10} skeleton, we had planned to decompose 6-(azido)-decaborane(14), $B_{10}H_{13}(N_3)$, expecting the nitrene intermediate to be arranged to aza-*nido*-un-decaborane, $NB_{10}H_{13}$. With the synthesis of $B_{10}H_{13}Cl$ from $B_{10}H_{12}(SMe_2)_2$ and HCl in mind,[8] we tried to synthesize $B_{10}H_{13}(N_3)$ from $B_{10}H_{12}(SMe_2)_2$ and HN_3. Unexpectedly, we observed a 1:2 reaction with the evolution of 1 mole of N_2. A potential nitrene intermediate added two H atoms and entered into a bridge position, giving the decaborane $B_{10}H_{12}(N_3)(NH_2)$. The thermolysis of this product in boiling xylene yielded the desired aza-*nido*-borane $NB_{10}H_{13}$ besides as much of the gases N_2 and NH_3 as is equivalent to the decomposition of 1 mole of HN_3. Apparently, the bridging amino-group had been incorporated into the borane skeleton as an NH unit [Eq. (7)]. Unfortunately, the yield of $NB_{10}H_{13}$ was only 5%.[9] Nevertheless, *nido*-$NB_{10}H_{13}$ after Eq. (7) had opened the synthesis of *closo*-$NB_{11}H_{12}$ (see below).

$$+ 2\ HN_3 \quad \underset{-\ N_2}{\overset{-\ 2\ SMe_2}{\longrightarrow}} \qquad \overset{140°C}{\underset{-\ "HN_3"}{\longrightarrow}} \qquad (7)$$

REFERENCES

1. W.R. Hertler, F. Klanberg, E.L. Muetterties, *Inorg. Chem.* 1967, *6*, 1696.
2. K. Baše, F. Hanousek, J. Plešek, B. Štibr, A. Lycka, *J. Chem. Soc. Chem. Commun.* 1981, 1162; K. Baše, *Coll. Czech. Chem. Soc.* 1983, *48*, 2593.
3. P. Paetzold, F. Meyer, H.-P. Hansen, L. Schneider, Technische Hochschule Aachen, unpublished.
4. E.L. Muetterties, F. Klanberg, *Inorg. Chem.* 1966, *5*, 315; G.M. Bodner, F.R. Scholer, L.J. Todd, L.E. Senor, J.C. Carter, *Inorg. Chem.* 1971, *10*, 942.
5. J. Müller, P. Paetzold, U. Englert, J. Runsink, *Chem. Ber.* 1992, *125*, 97.
6. S. Heřmánek, J. Plešek, B. Štibr, F. Hanousek, *Coll. Czech. Chem. Commun.* 1968, *33*, 2177.
7. F. Meyer, J. Müller, M.U. Schmidt, P. Paetzold, *Inorg. Chem.*, submitted.
8. J. Plešek, B. Štibr, S. Hermánek, *Coll. Czech. Chem. Commun.* 1966, *31*, 4744.
9. J. Müller, P. Paetzold, R. Boese, *Heteroatom Chem.* 1990, *1*, 461.

3 INCORPORATION OF BORON IN THE NB_9 SKELETON

The deprotonation of *nido*-NB_9H_{12} with Li*t*Bu and subsequent addition of thf·BH_3 yields the *arachno*-species $Li[NB_{10}H_{14}]$ whose structure is suggested from 2D-NMR ^{11}B-^{11}B evidence. The anion loses a molecule of H_2 upon protonation, allowing quite a convenient access to *nido*-$NB_{10}H_{13}$ [Eq. (1)].[1]

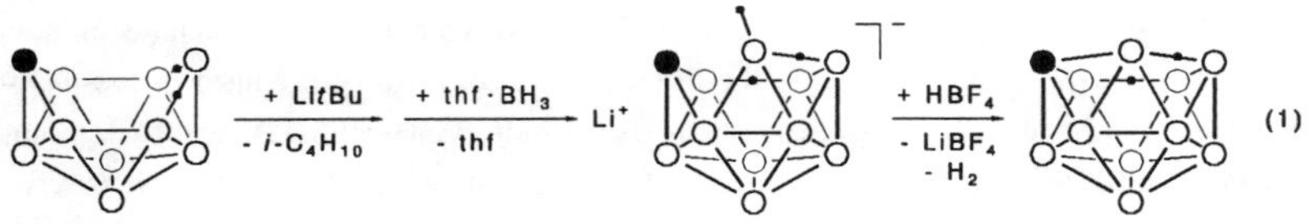

$$(1)$$

The successfull application of $Et_3N \cdot BH_3$ in order to transform $NB_{10}H_{13}$ into $NB_{11}H_{12}$ [2] (see below) prompted us to apply the same borane in order to incorporate boron into the skeleton of *nido*-NB_9H_{12}. Instead of an incorporation, we observed a simple *exo*-borylation of nitrogen at 120°C.[3] In a melt of excess $Et_3N \cdot BH_3$ at 200°C, however, the NB_9 skeleton adopts even two BH units, thus making the *closo*-NB_{11} skeleton available. The N-borylation remains, and one of the ethyl groups of NEt_3 closes a six-membered ring by ethylating a neighboring B atom [Eq. (2)]. The product crystallizes in the triclinic space group $P1$.[1]

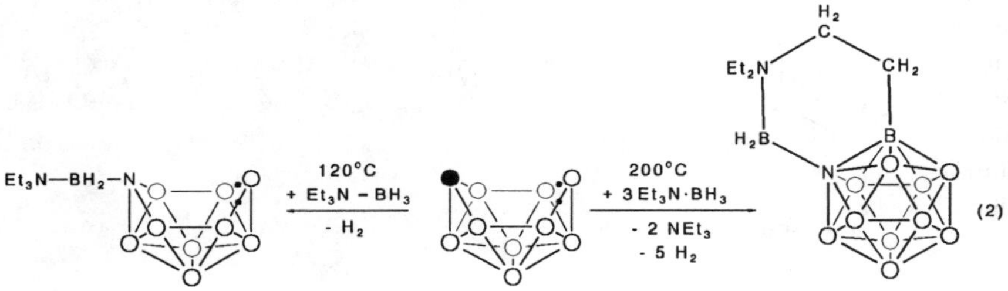

$$(2)$$

Could the ring-closure of reaction (2) be avoided by starting from $D \cdot BH_3$ with bases D different from NEt_3? We found out that Me_2S and NMe_3 are such bases, which permit the closure to the *closo*-NB_{11} skeleton already at 140°C. The N-borylation cannot be avoided [Eq. (3)].[1]

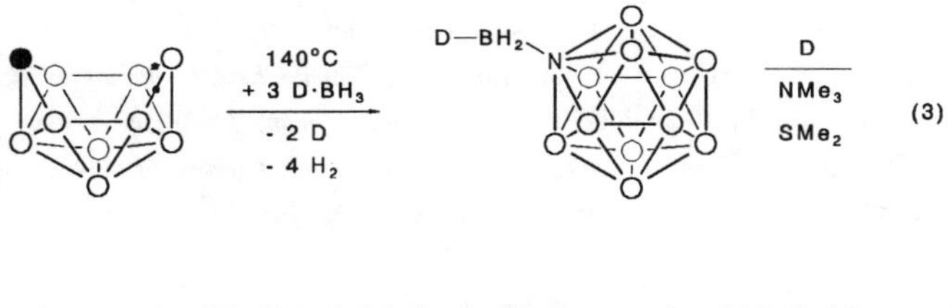

$$(3)$$

In order to make the N-borylation impossible, we employed N-alkyl boranes RNB_9H_{11}. Two BH units are inserted when an excess of $Me_2S \cdot BH_3$ is applied at 170°C in decalin [Eq. (4)].[4]

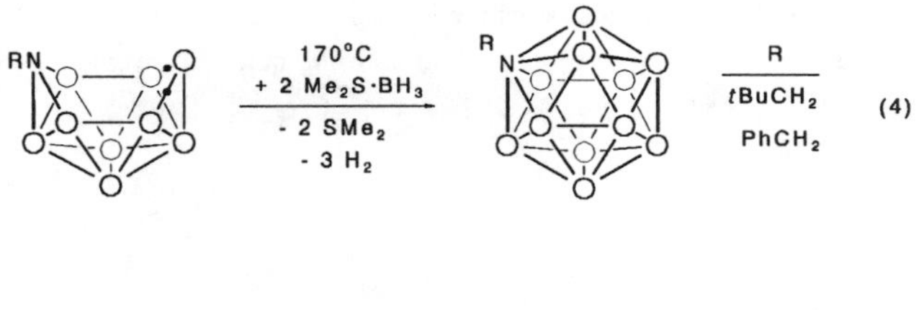

$$(4)$$

REFERENCES

1. P. Paetzold, F. Meyer, H.-P. Hansen, L. Schneider, Technische Hochschule Aachen, unpublished.
2. J. Müller, J. Runsink, P. Paetzold, *Angew. Chem. Int. Ed. Engl.* 1991, *30*, 175.
3. J. Müller, P. Paetzold, U. Englert, J. Runsink, *Chem. Ber.* 1992, *125*, 97.
4. F. Meyer, J. Müller, M.U. Schmidt, P. Paetzold, *Inorg. Chem.*, submitted.

4 INCORPORATION OF BORON OR METAL IN THE NB_{10} SKELETON

When we had that 5% yield of *nido*-$NB_{10}H_{13}$ [Eq. (7) in Section 2] in our hands in 1989, we immediately tried to close it by the action of $Et_3N \cdot BH_3$. We identified the *closo*-anion $NB_{11}H_{11}^-$ and the corresponding cation $NHEt_3^+$ as the products, from which the neutral *closo*-species $NB_{11}H_{12}$ was available on protolysis with HBF_4 in an overall-yield of 47% [Eq. (1)]. Apparently, $NB_{11}H_{12}$ is a stronger Brönsted acid than $NHEt_3^+$ or $NB_{11}H_{11}^-$ a weaker base than NEt_3. A 5:5:1 set of ^{11}B-NMR signals clearly indicated the C_{5v} structure of $NB_{11}H_{12}$; the ^1H-NMR triplet [$^1J(HN) = 62.5$ Hz] for the N-bonded proton is in accord with the symmetrical charge distribution around the N atom on the C_5 axis.[1] $NB_{11}H_{12}$ is the first example of six-coordinate nitrogen incorporated in a cluster skeleton with five-fold connectivity.

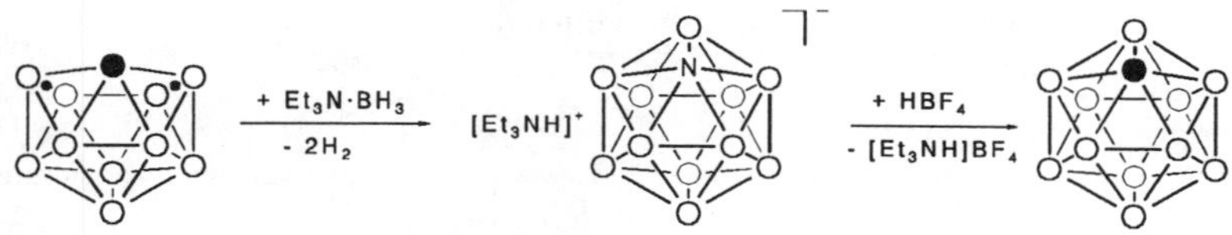

One usually starts from the *nido*-anions $C_2B_9H_{12}^-$ or $C_2B_9H_{11}^{2-}$ in order to close the nido-C_2B_9 skeleton with a metal-complex fragment. The analogous *nido*-anions $NB_{10}H_{12}^-$ and $NB_{10}H_{11}^{2-}$ are easily formed by the action of an alkali triethylborate on $NB_{10}H_{13}$ in a 1:1 or 2:1 ratio, respectively [Eq. (2)]. The extra-hydrogen atom in $NB_{10}H_{12}^-$ is found in a bridging position on the mirror plane through the molecule, according to NMR evidence. The position of the analogous H atom in $C_2B_9H_{12}^-$ had been discussed for years, but is now confirmed to be chiefly in an *endo*-position, again on a mirror plane.[2]

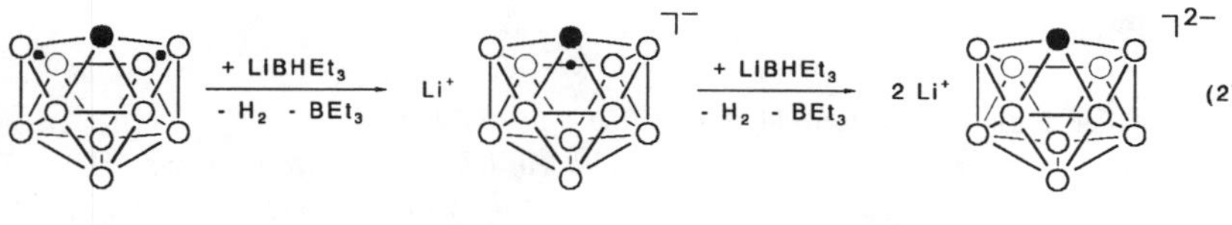

Chlorotris(triphenylphosphane)rhodium closes $nido\text{-}NB_{10}H_{12}^-$ in the same way as had been found for $nido\text{-}C_2B_9H_{12}^-$.[3] The closure is accompanied by the loss of one phosphane molecule and the migration of the extra-H atom from boron to the metal [Eq. (3a)].[4] The known metallation of the $nido$-anion $C_2B_9H_{11}^{2-}$ by the dimeric dichloro-η^6-(hexamethylbenzene)-ruthenium[5] can as well be applied to $NB_{10}H_{11}^{2-}$ [Eq. (3b)].[6] The N atom adopts a five-fold connectivcity in both azametalladodecaborane skeletons, well established by X-ray analyses. The metallation of $NB_{10}H_{11}^{2-}$ seems to be quite a general reaction. We were able to close $NB_{10}H_{11}^{2-}$ with an $[Ni(PPh_3)_2]^{2+}$ unit [Eq. (3c)] and also to bind two anions $NB_{10}H_{11}^{2-}$ to a Co^{3+} cation in a sandwich-type product [Eq. (3d)].

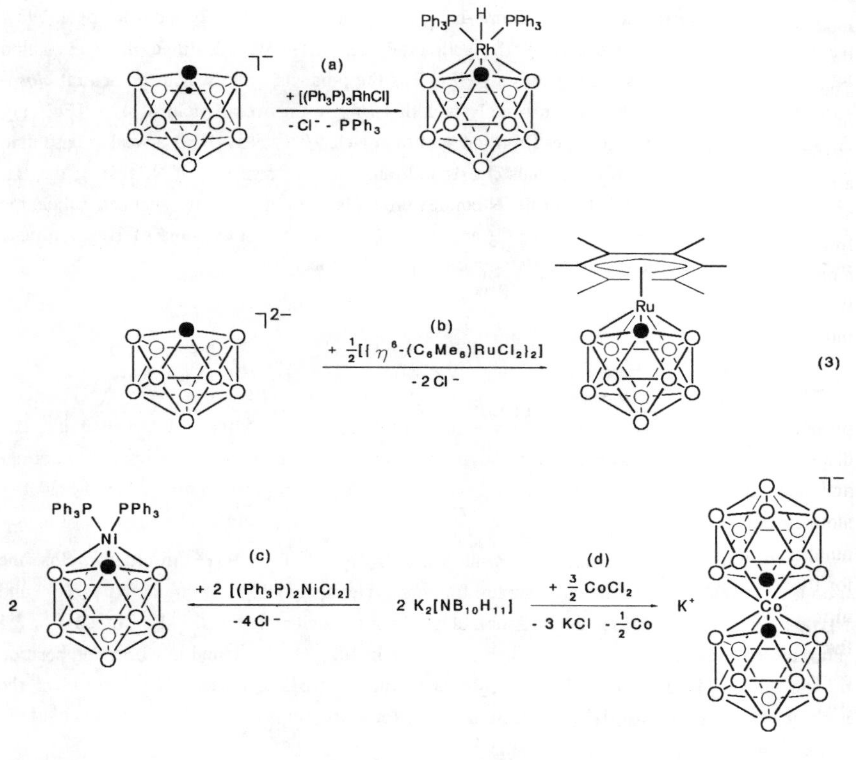

REFERENCES

1.   J. Müller, J. Runsink, P. Paetzold, *Angew. Chem. Int. Ed. Engl.* 1991, *30*, 175.

2.   J. Buchanan, E.J.M. Hamilton, D. Reed, A.J. Welch, *J. Chem. Soc. Dalton Trans.* 1990, 677.

3.   T.E. Paxson, M.F. Hawthorne, *J. Am. Chem. Soc.* 1974, *96*, 4674; G.E. Hardy, K.P. Kallahan, C.E. Strouse, M.F. Hawthorne, *Acta Crystallogr. Sect.* 1976, *B 32*, 264.

4.   H.-P. Hansen, J. Müller, U. Englert, P. Paetzold, *Angew. Chem. Int. Ed. Engl.* 1991, *30*, 1377.

5. M. Bown, J. Plešek, K. Baše, B. Štíbr, X.L.R. Fontaine, N.N. Greenwood, J.D. Kennedy, *Magn. Reson. Chem.* 1989, *27*, 947.

6. P. Paetzold, F. Meyer, H.-P. Hansen, L. Schneider, Technische Hochschule Aachen, unpublished.

5 COMMENTS ON THE STRUCTURE OF AZADODECABORANE

Structural evidence comes from an ab initio calculation on $NB_{11}H_{12}$ at an HF/6-31G* level,[1] from a gas-phase electron diffraction (GED) study on $NB_{11}H_{12}$,[2] and from a crystal structure determination of $(PhCH_2)NB_{11}H_{11}$.[3] The GED data could be equally well fit by four models with C_{5v} symmetry. As the best model the one was taken which allowed a calculation of the ^{11}B NMR shifts by the IGLO method in the best agreement with our experimental values.[4] The thirty skeletal bond lengths can be arranged in five zones of equal distances in the gas phase molecule of C_{5v} symmetry. This symmetry is still present in solutions of $(PhCH_2)NB_{11}H_{11}$ as far as the characteristic 5:5:1 set of ^{11}B NMR chemical shifts is concerned, which is apparently due to a rapidly rotating benzyl group. In crystalline $(PhCH_2)NB_{11}H_{11}$, however, the rigid benzyl group as well as the monoclinic lattice make all thirty edges of the icosahedron-type skeleton different from each other. Nevertheless, the five zones remain such as each zone contains bond distances in a narrow range. In Table 1 the NB_{11} skeleton is looked at as a bicapped pentagonal antiprism with the N atom as the upper capping apex. The X-ray data are mean values from the five or ten individual bond distances, respectively, in each zone.

The BB bond distances in the crystal are smaller than in the free molecule, a general phenomenon. Unexpectedly, the situation is reversed for the BN bond distances. Anyhow, the BN distances are the shortest ones in the skeleton, though they are rather long, if compared to normal aminoboranes (≈ 1.40 Å) or amine-boranes (≈ 1.60 Å). They seem to be determined by the smaller atomic radius of nitrogen, as compared to boron, on the one hand and by the high coordination number of nitrogen on the other hand. Recently, we found a mean bond distance of only 1.516 Å for the BN bonds in the crystal structure of *closo*-NB_9H_{10} in which the boron is six-, but the nitrogen atom five-coordinated.[5] Apparently, the increase of the coordination number of N from five to six makes the BN bonds significantly longer. The average BN bond length in the metal complex $[(Ph_3P)_2RhH]NB_{10}H_{11}$ of 1.694 Å with again six-coordinate nitrogen is close to that of $(PhCH_2)NB_{11}H_{11}$.[6]

The largest BB bond lengths are those in the upper pentagonal zone thus illustrating the chief distortion of the NB_{11} skeleton from icosahedral geometry: The N atom penetrates towards the center of the polyhedron and the adjacent B atoms make way by forming a larger pentagon. Parallel to this situation, the five BH bonds at this pentagon are shifted out of radial direction towards the N atom, thereby shielding the N atom considerably. This shielding effect may be the reason why we were unable to bind larger alkyl groups, e.g. the *tert*-butyl group, to the N atom of aza-*closo*-dodecaborane.

Table 1.   Skeletal bond distances (Å) of $NB_{11}H_{12}$ (calculated and from gas-phase electron diffraction) and of $PhCH_2)NB_{11}H_{11}$ (averaged from X-ray diffraction)

	$NB_{11}H_{12}$ Calcd.	$NB_{11}H_{12}$ GED	$(PhCH_2)NB_{11}H_{11}$ X-ray
Upper pyramidal zone	1.710	1.716	1.719
Upper pentagonal zone	1.821	1.825	1.796
Antiprismatic zone	1.767	1.791[a]	1.745
Lower pentagonal zone	1.808	1.791[a]	1.773
Lower pyramidal zone	1.798	1.791[a]	1.770

[a] Mean value of 20 edges in 3 zones.

REFERENCES

1.   R. Zahradnik, V. Balaji, J. Michl, *J. Comput. Chem.* 1991, *12*, 1147.
2.   D. Hnyk, H. Bühl, P. von Ragué Schleyer, H.V. Volden, S. Gundersen, J. Müller, P. Paetzold, *Inorg. Chem.*, 1993, *32*,.
3.   F. Meyer, J. Müller, M.U. Schmidt, P. Paetzold, *Inorg. Chem.*, submitted.
4.   J. Müller, J. Runsink, P. Paetzold, *Angew. Chem. Int. Ed. Engl.* 1991, *30*, 175.
5.   P. Paetzold, F. Meyer, H.-P. Hansen, L. Schneider, Technische Hochschule Aachen, unpublished.
6.   J. Müller, P. Paetzold, U. Englert, J. Runsink, *Chem. Ber.* 1992, *125*, 97.

6 OPENING OF THE CLOSED AZADODECABORANE SKELETON

The 1,2-dicarba-*closo*-dodecaborane $C_2B_{10}H_{12}$ can be opened by the attack of bases like NaOEt in EtOH; this opening process is accompanied by the loss of H_2 and of one B atom as $B(OEt)_3$, giving the *nido*-anion $C_2B_9H_{12}^-$.[1] We found out that aza-*closo*-dodecaborane is opened by the attack of mere methanol without degradation. We first started from the *N*-methyl derivative in a solution of CH_2Cl_2 in the presence of an excess of MeOH. We identified the *nido*-anion $[MeNB_{11}H_{11}(OMe)]^-$ as the product in solution, which could be crystallized as $[N(PPh_3)_2][MeNB_{11}H_{11}(OMe)]$ [Eq. (1)].[2] The anion in the crystal exhibited the same structure that had been concluded for the anion in solution from NMR data. The corresponding cation in the primary methanolysis process must be H^+, solvated by excess MeOH or perhaps CH_2Cl_2. It could not be detected by spectroscopic methods. All efforts to isolate the free acid $H[MeNB_{11}H_{11}(OMe)]$, apparently quite a strong acid, ended with the decomposition of the cluster.

The structure of the anion can be described to contain an open, non-planar pentagon with nitrogen of three-fold connectivity on a mirror plane and with the methoxy group bonded terminally to the neighboring B atom on the mirror plane. Two B atoms of connectivity 4 are bridged by an H atom. This H atom is the one that had been exchanged by the methoxy group in its original terminal position. Methanol can be excluded as its source since no deuterium is transferred to the bridging position upon methanolysis with CD_3OD, and there is no exchange of that H atom on standing neither with an excess of methanol nor with protons in the medium in the case of any protic solvent. The opening of $MeNB_{11}H_{11}$ apparently proceeds through the attack of the base at a B atom close to the N apex, the opening of two BN bonds, the migration of the terminal H atom to a bridging position and the dissociation at the methanolic OH bond.

The *nido*-structure of the anion $[MeNB_{11}H_{11}(OMe)]^-$ is related to a hypothetical *closo*-$NB_{12}H_{13}$, which is isoelectronic with the hypothetical anion $B_{13}H_{13}{}^{2-}$. A favorable structure of the closed 13-vertex polyhedron may be derived from the icosahedron by opening an edge and inserting a vertex with the connectivity 4; this vertex would be the NH unit in the case of $NB_{12}H_{13}$. Such an inserting process lets two vertices have the extreme connectivity 6. The resulting structure of C_{2v} symmetry had been claimed by theory.[3] Another way to look at this structure is to have a non-planar hexagon and a pentagon fused together in a kind of antiprismatic manner and to cap both polygons. The capping apex over the hexagon and one vertex in the pentagon adopt the connectivity 6, one in the hexagon 4, all others 5. *closo*-Structures of this type are realized in metallaboranes, e.g. $[(CpCo)C_2B_{10}H_{12}]$,[4] where the three positions with the connectivities 6, 6 and 4 are occupied by Co, B, and one of the C atoms, respectively. Going from *closo*-$NB_{12}H_{13}$ to *nido*-$NB_{11}H_{12}{}^{2-}$, such a B atom ought to be removed that reduces the extra-connectivities from 6 to 5 and from 4 to 3 and, consequently, the connectivity of two B atoms from 5 to 4. This is exactly the structure formed in the opening process of $MeNB_{11}H_{11}$ with MeOH. *nido*-Structures of that type are known in several heteroboranes. In *nido*-$Et_4C_4B_8H_8$,[5] the connectivity of the four electronegative C atoms can additionally be reduced to values of 3 and 4 by opening a bond. In *nido*-$[(CpCo)Se_2B_9H_9]$,[6] the rather electronegative Se atoms are found in the positions with the low connectivities 3 and 4, and the electropositive Co atom in a position with the higher connectivity 5, coordinated by both the Se atoms.

Current Topics in the Chemistry of Boron

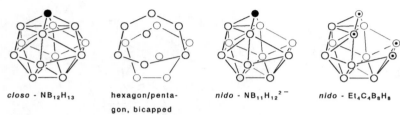

| *closo* - $NB_{12}H_{13}$ | hexagon/penta-
gon, bicapped | *nido* - $NB_{11}H_{12}{}^{2-}$ | *nido* - $Et_4C_4B_8H_8$ |

The opening process according to Eq. (1) can be generalized. The azaboranes $RNB_{11}H_{11}$ may be attacked by $LiBHEt_3$, LiMe, KO*t*Bu, or $[S(NMe_2)_3][Me_3SiF_2]$ instead of MeOH [Eq. (2)].[7]

REFERENCES

1. R.A. Wiesboeck, M.F. Hawthorne, *J. Am. Chem. Soc.* 1964, *103*, 1642.
2. F. Meyer, J. Müller, P. Paetzold, R. Boese, *Angew. Chem. Int. Ed. Engl.* 1992, *31*, 1227.
3. L.D. Brown, W.N. Lipscomb, *Inorg. Chem.* 1977, *16*, 2989.
4. G.B. Dunks, M.M. McKown, M.F. Hawthorne, *J. Am. Chem. Soc.* 1971, *93*, 2541; E.R. Churchill, B.G. DeBoer, *Inorg. Chem.* 1974, *13*, 1411.
5. T.L. Venable, R.B. Maynard, R.N. Grimes, *J. Am. Chem. Soc.* 1984, *106*, 6187.
6. G.D. Friesen, A. Barriola, P. Daluga, P. Ragatz, J.C. Huffman, L.J. Todd, *Inorg. Chem.* 1980, *19*, 458.
7. P. Paetzold, F. Meyer, H.-P. Hansen, L. Schneider, Technische Hochschule Aachen, unpublished.

Acknowledgement We gratefully acknowledge the support of this work by the Deutsche Forschungsgemeinschaft and by the Fonds der Chemischen Industrie.

Crowded Heteroboranes

J. Cowie, D. J. Donohoe, N. L. Douek, G. O. Kyd, Z. G. Lewis, T. D. McGrath, J. M. S. Watmough, and A. J. Welch

DEPARTMENT OF CHEMISTRY, UNIVERSITY OF EDINBURGH, EDINBURGH EH9 3JJ, UK

Heteroboranes which are deliberately overcrowded are potentially useful and interesting species. We have focussed recent attention on formally *closo* icosahedral carbametallaboranes of general formula $3\text{-}L_x\text{-}1\text{-}R\text{-}2\text{-}R'\text{-}3,1,2\text{-}MC_2B_9H_9$ where L_x (the *exo*-metal ligand set), R and R' are sterically-demanding substituents. With this class of compound intramolecular crowding is maximised simply because the three cluster heteroatoms are mutually adjacent (Figure 1).

Hawthorne *et al* [1] have already demonstrated the utility of crowding in heteroboranes. Species $3,3\text{-}(PPh_3)_2\text{-}3\text{-}H\text{-}1\text{-}R\text{-}2\text{-}R'\text{-}3,1,2\text{-}closo\text{-}Rh^{III}C_2B_9H_9$ exist in solution in equilibrium with a tautomeric <u>vertex extruded</u> *exo-nido* Rh^I form which is a homogeneous catalyst precursor for a range of organic transformations. In these systems the proportion of the *exo-nido* form is directly governed by relief of intramolecular crowding between metal-bound phosphine ligands and C_{cage}-bound substituents.

Metal vertex extrusion is clearly a fairly effective way for a heteroborane to separate its bulky substituents, but it is not the only way by which an overcrowded heteroborane can "react" to its situation. Recently, our studies have concentrated on the sterically-induced <u>polyhedral deformations</u> and <u>facile polyhedral isomerisations</u> of carbametallaboranes, focussing particularly on complexes of the mono and diphenyl carbaborane ligands $[7\text{-}Ph\text{-}7,8\text{-}nido\text{-}C_2B_9H_{10}]^{2-}$ and $[7,8\text{-}Ph_2\text{-}7,8\text{-}nido\text{-}C_2B_9H_9]^{2-}$. In such systems the anisotropic steric requirement of the phenyl group introduces an additional degree of subtlety, necessitating specification of the conformation of that group relative to the cage; to this end the angle θ is defined as the complement of the average C_{cage}-C_{cage}-C-C torsion angle.

Theoretical [2] and experimental [3] study of 1-Ph-1,2-*closo*-$C_2B_{10}H_{11}$ shows that the optimum molecular conformation is when $\theta=90°$ since this allows maximum electronic conjugation between ring and cage. In 1,2-Ph$_2$-1,2-*closo*-$C_2B_{10}H_{10}$ mutual crowding between the rings means that a similar conformation is not possible, and very low θ values are observed. This situation is broadly maintained in the diphenylcarbaborane in ligand form [4] (Figure 2).

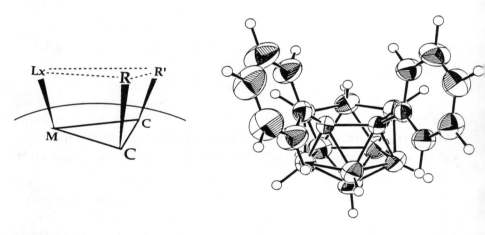

Figure 1   Figure 2

Serically-Induced   Polyhedral   Deformation

In 3,1,2-MC_2B_9 carbametallaboranes the degree of crowding betwen *exo* metal ligands and C_{cage}-bound substituents will be affected by any lateral slip (Δ) of the metal fragment across the carbaborane ligand face, and several years ago model studies on uncrowded carbametallaboranes[($C_2B_9H_{11}$) ligand] showed that the optimum position of the metal could be traced to the <u>orbital number</u> (o.n.) of the metal fragment (the number of available cluster bonding orbitals).

We have found that complexes of diphenylcarbaborane with {Ph$_3$PHg} (o.n.=1) and {codPd} (o.n.=2) fragments [5,6] display clear signs of substantially enhanced slipping because of steric crowding (Table1). Moreover, in these complexes the cage Ph groups are substantially and sequentially twisted about the C$_{cage}$-C axis relative to their conformation in the free ligand.

Table 1

	Δ/ Å	<θ>/°
Ph$_3$PHgPh$_2$C$_2$B$_9$H$_9$	1.10	29.1
Ph$_3$PHgC$_2$B$_9$H$_{11}$	0.92	-
codPdPh$_2$C$_2$B$_9$H$_9$	0.52	48.4
codPdC$_2$B$_9$H$_{11}$	0.24	-

The {Ph$_3$PCu} fragment has an o.n. of 3 and consequently should sit fairly centrally above a C$_2$B$_3$ ligand face. In (Ph$_3$PCu)$_2$C$_2$B$_9$H$_{11}$ [7] one {Ph$_3$PCu} occupies vertex 3 of a closed icosahedron (Δ=0.25Å) with the other *exo* capping the Cu(3)B(4)B(8) face, Cu-Cu=2.58Å. In the C-diphenyl analogue of this species [8] (Figure 3) the vertex Cu is substantially slipped (Δ=0.61Å) to relieve crowding between its PPh$_3$ ligand and the cage-bound Ph groups, with the result that the *exo* {Ph$_3$PCu} fragment is now required to cap the B(8)B(9)B(11) face, there being no Cu-Cu bond. This complex thus represents an example of crowding within a heteroborane that results in a truly different stereochemistry.

The {Cp*Rh} fragment also has an o.n. of 3 but a substantially greater steric demand than {Ph$_3$PCu}. Consequently, in Cp*RhPh$_2$C$_2$B$_9$H$_9$ [9] the Ph groups are required to adopt conformations with high θ values (<θ>=81.1°) even though this results in their substantial mutual crowding. To relieve this the Ph substituents are pushed apart, prising open the C(1)-C(2) connectivity to *ca*. 2.5Å (Figure 4).

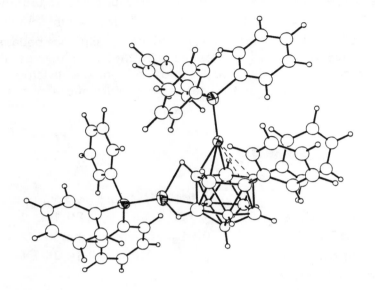

Figure 3

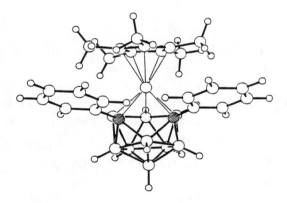

Figure 4

The open RhCB(6)C face thus generated engenders the description *pseudocloso*. Fully analogous *pseudocloso* carbaruthenaboranes have also been prepared and characterised [8] with various {(arene)Ru} fragments. RMS misfit calculations [10] show that, geometrically, *pseudocloso* carbametallaboranes appear to lie about midway between true *closo* species and *hypercloso* systems recently described by Stone *et al*[11]. *Hypercloso* heteroboranes are characterised by only n skeletal electron pairs, and it is of considerable interest to note that our *pseudocloso* carbarhoda- and carbaruthenaboranes all show consistent and substantial high frequency shifts in their ^{11}B NMR spectra relative to *closo* species, supporting the idea of a measurable loss of cluster electron density from *closo* to *pseudocloso*.

Serically-Induced Facile Polyhedral Isomerisation

The polytopal isomerisation of heteroboranes and the possible mechanism(s) by which such rearrangements occur continue to be the subject of considerable interest. Generally such processes require substantially elevated temperatures, making experimental assessment of possible mechanistic pathways difficult.

We have found [12] that a family of carbaplatinaboranes, $3-3-(PR_3)_2-1-Ph-3,1,2-closo-PtC_2B_9H_{10}$ [$R_3=Me_2Ph$, $(OMe)_3$, Et_3, Ph_3, *etc.*] undergo smooth and complete isomerisation at temperatures generally between 40° and 70°C to yield both $3,1,11-PtC_2B_9$ isomers (Figure 5). Attempts to synthesise $3,1,2-PtC_2B_9$ species with the diphenylcarbaborane ligand by slowly warming a frozen mixture (CH_2Cl_2) of $Tl[TlPh_2C_2B_9H_9]$ and $cis-PtCl_2(PR_3)_2$ lead only to the 3,1,11-isomer, *i.e.* this isomerisation occurs even below room temperature.

Clearly, the diphenylcarbaplatinaboranes undergo such facile rearrangements to obviate severe overcrowding, but for the monophenyl analogues the situation is not so simple - indeed, for $PR_3=PMe_2Ph$ the precursor and one of the rearranged products are crystallographically isomorphous. However, inspection of the conformation adopted by the $\{PtP_2\}$ fragment relative to the C_2B_3 face of the C-adjacent carbaborane ligand clearly shows that the monophenyl $3,1,2-PtC_2B_9$ species is conformationally destabilised as the result of intramolecular crowding (its isomorphous rearrangement product is not) and presumably therefore experiences a reduced activation barrier to isomerisation.

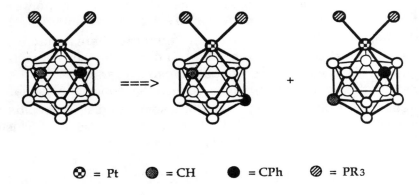

⊗ = Pt   ◉ = CH   ● = CPh   ⊘ = PR₃

Figure 5

Studies [13] are currently underway to utilise these low temperature, sterically-induced, rearrangements to gain useful information on the isomerisation mechanism(s).

1. Hawthorne, M. F. *et al, J. Am Chem. Soc.*, 1984, **106**, 2965; 2979; 2990
2. Lewis, Z. G. and Welch, A. J., *Acta Crystallogr.*, 1993, **C49**, 705.
3. Donohoe, D. J. and Welch, A. J. Unpublished results.
4. Cowie, J., Donohoe, D. J., Douek, N. L., and Welch, A. J., *Acta Crystallogr.*, 1993, **C49**, 710.
5. Lewis, Z. G. and Welch, A. J., *Acta Crystallogr.*, 1993, **C49**, 715.
6. Kyd, G. O. and Welch, A. J. Unpublished results.
7. Kang, H. C., Do, Y., Knobler, C. B., and Hawthorne, M. F., *Inorg. Chem.*, 1988, **27**, 1716.
8. Cowie, J. and Welch, A. J. Unpublished results.
9. Lewis, Z. G. and Welch, A. J. *J. Organomet. Chem.*, 1992, **430**, C45.
10. Macgregor, S. A., Wynd, A. J., Gould, R. O., Moulden, N., Taylor, P., Yellowlees, L. J., and Welch, A. J., *J. Chem. Soc., Dalton Trans.*, 1991, 3317.
11. Attfield, M. J., Howard, J. A. K., Jelfs, A. N. de M., Nunn, C. M., and Stone, F. G. A., *J. Chem. Soc., Dalton Trans.*, 1987, 2219.
12. Baghurst, D. R. *et al, J. Organomet. Chem.*, 1993, **447**, C14
13. McGrath, T. D. and Welch, A. J. Unpublished results.

Perturbation of Boron Clusters by Heteroatoms or by Substituents

S. Heřmánek and J. Plešek

INSTITUTE OF INORGANIC CHEMISTRY, ACADEMY OF SCIENCES OF THE CZECH REPUBLIC, 259 68 ŘEŽ NEAR PRAGUE, CZECH REPUBLIC

1 INTRODUCTION

Distribution of electrons within borane skeletons and, especially, the charges on individual vertices have been for long time a matter of interest to both theoretical and practical boron chemists. In deltahedral boron clusters, the electron density on individual skeletal atoms is influenced, in the first approximation, by:
1) the number of neighboring vertices in the given skeleton, decreasing from three to five B neighbors[1]
2) the number of skeletal electrons, increasing in the order: *closo* ($2n+2$) < *nido* ($2n+4$) < *arachno* ($2n+6$) < *hypho* ($2n+8$)[2]
3) the polarizability of the B-H_{exo} bond
4) the perturbation exerted by heteroatom/s if present
5) the perturbation exerted by substituent/s if present
    Transmissions of electrons through particular skeletons brought by the effects under 4) and 5) deserve special attention. Many findings exist which indicate that an "aromaticity" of boron clusters is distinctly different from that of organic compounds. The very simple transfer of I and M effects from benzene to icosahedral compounds seems to be, therefore, questionable. A stepwise solving of these problems is the matter of this work.
    As ^{11}B NMR characteristics indicate,[3] the greatest perturbance is brought by heterovertices. While in organic molecules an electronegative heteroatom E (E = N, P, O, S, etc.) is the center of enhanced electron density, quite opposite situation is with heteroborane clusters[4] in which the heteroatom seems to be the more plus charged the higher its electronegativity (C < N < S). The reason is that the main-group heteroatom substitutes a skeletal boron atom of the same formal number of skeletal electrons. Thus :CH or :NO are equivalent to a formal vertex :BH^-, and :NH or :S to a formal vertex :BH^{2-}. Such electron-rich boron vertices are in reality not existing in cluster hydroborates since these "surplus" electrons are dissipated throughout the whole skeleton. The same fate meets the skeletal electrons brought in by an electronegative heteroatom: they are dissipated through the skeleton, changing the heteroatom to the electropositive center.

2   ESCA MEASUREMENTS OF CLUSTER THIA-BORANES

To confirm the above hypotheses, and to determine the real charge on heteroatoms of various connectivities and distributions of bonds/electrons, the ESCA measurements of S(2p) on a group of thiaboranes and heterothiaboranes have been performed.[5] This group is rich in the number of compounds, and sulfur is of high electronegativity, offering thus an optimum series for the determination of changes brought by the perturbing heteroatom.

Approximate estimation of electron deficiency on the sulfur atom can be made by means of Lipscomb's topological formulae involving both the classical two-center two-electron (2c2e) and three-center two-electron (3c2e) bonds. For three (1,2), four (3,4) and five (5,6) adjacent vertices, always two bonding variants are available. Of these, the one is preferred, in which the sulfur atom is bound to the neighbors by higher number of classical bonds, i.e. sulfur loses lesser part of electrons. This series also indicates that the electropositivity of the sulfur vertex should decrease in the order: *closo-* > *nido-* > *arachno-* > *hypho-*, and with increasing connectivity: five > four > three.

closo, nido

| 1 | 2 | 3 | 4 |

closo, nido   *arachno*

| 5 | 6 | 7 | 8 |

Table I shows that the simple predictions based on topological formulae are in full agreement with the ESCA results. Somewhat lower $S(\delta+)$ charges are a result of a back donation from the BH vertices to strongly plus charged S atom. This study and the preliminary results obtained with carbaboranes indicate that the **positive charge on a heterovertex is the greater, the more electronegative is the heteroatom.** The ESCA results thus point out the incorrectness in transferring mechanically the common knowledge from organic to borane chemistry, e.g. to consider the electronegative heterovertices negatively charged in heteroboranes.[6]

The *hypho*-compounds have still not been measured, but the above indication is in agreement with the fact that the anion $S_2B_6H_9^-$ **18** was successfully alkylated,[7] while our attempts to alkylate *closo-*, *nido-* and *arachno-* species failed.

Anion **18**, its C- analog **19**,[8] $Me_3N-CB_5H_{11}$ **20**,[9] and other heteroboranes with the heteroatom bound to two

Table I. Binding Energy Differences ΔE_b of B(1s) – S(2p) Electrons (in eV) with Selected *closo-*, *nido-* and *arachno-* Thia-boron Clusters. Reference level: C(1s) E_b of adventitious carbon (284.8 eV).
[a] Charge $Q = (E_b - 163.8)/3.38$ (Ref.[10])

closo 2n+2	*nido* 2n+4	*arachno* 2n+6

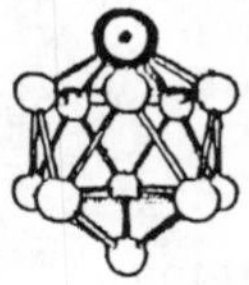

SB$_{11}$H$_{11}$ **9**
23.8 +0.68[a]

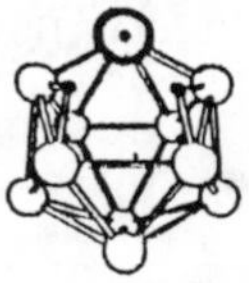

7–SB$_{10}$H$_{12}$ **10**
24.5 +0.18[a]

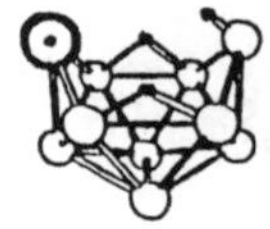

6–SB$_9$H$_{12}^-$ **11**
25.14 0.0[a]

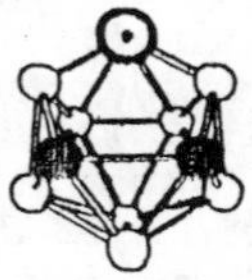

7,8,10–C$_2$SB$_8$H$_{10}$ **12**
24.8

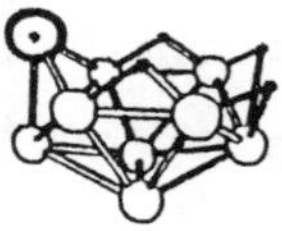

4–SB$_8$H$_{12}$ **13**
25.3

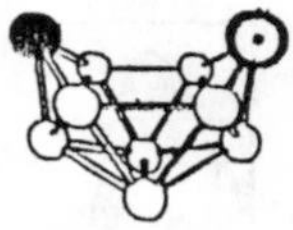

6,9–CSB$_8$H$_9^-$ **14**
24.8

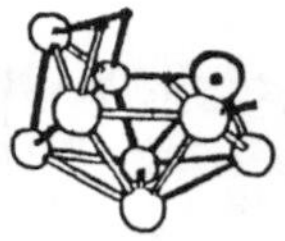

6–SMe$_2$–4–SB$_8$H$_{10}$ **15**
25.6; (23.8)

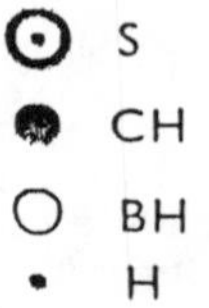

⊙ S
● CH
○ BH
• H

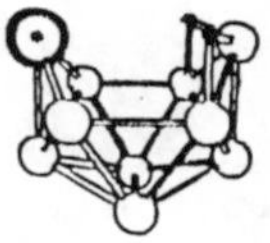

6–SB$_9$H$_{11}$ **16**
25.0

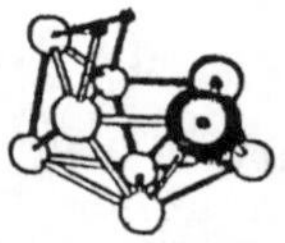

4,6–S$_2$B$_7$H$_9$ **17**
25.6

boron atoms by two classical 2c2e bonds can be considered either for *hypho-* species of eight, eight and six vertices, respectively, or for heteroatom bridged *nido-*B$_6$ (18,19), and *arachno-*B$_5$ (20) boranes. A decision, which of these possibilities is valid, is very important both for their chemistry and, especially, for nomenclature reasons based on the number of vertices.

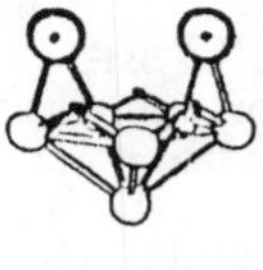

18

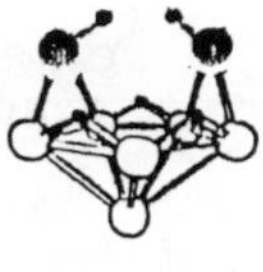

19

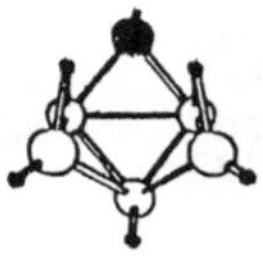

20

Assignment of a given skeleton to a right class (i.e.
closo-, *nido-*, etc.) can be realized using two notional
approaches, the debor-[4] and seco-[3] concepts. While the
former procedure removes one, two and three vertices
from the "optimum" *closo-* skeleton, the latter one
removes two, three and four edges from the *closo-*
skeletons of the same number of vertices, transforming
them to *nido-*, *arachno-* and *hypho-* types, respectively.

At present, however, no objective method for such assignment exists. In the case that ESCA measurements will
find these heteroatom bridges positively charged, a partial delocalization of their electrons and, consequently,
classifying as skeletal atoms could be considered.

3   MCD MEASUREMENTS OF BORON CLUSTER COMPOUNDS

Another method which could help us solve this
problem is MCD spectroscopy. Our preliminary results[11]
have indicated that while *nido-* B-cluster compounds
exhibit + - character of the spectrum, *arachno-* species
show the - + character (Fig. 1), irrespective of the
presence of heteroatoms or substituents. In case that
MCD spectra of *closo-*, *nido-*, *arachno-* and *hypho-*
compounds will have alternating character of the beginning of the spectra, a significant tool for determining
the class of B-clusters will be found.

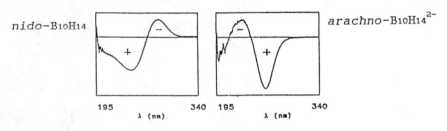

Figure 1. MCD Spectra of *nido-*$B_{10}H_{14}$ (**21**) in cyclohexane
and *arachno-*$B_{10}H_{14}{}^{2-}$ (**22**) in acetonitrile at room temp.

4   QUANTUM CHEMICAL CALCULATIONS ON $B_6H_6{}^{2-}$, $F\text{-}B_6H_6{}^{2-}$, $NC\text{-}B_6H_6{}^{2-}$ AND SB_5H_5 COMPOUNDS

Distribution of electrons, i.e. electron densities
on individual atoms, vertices, edges and faces of basic
B-cluster skeletons were repeatedly studied since the
time of Lipscomb's basic works.[1] The present state of
affairs shows that in order to get reliable results, population analysis must be based on a high quality *ab
initio* calculations and precisely optimized structures
(e.g. accuracy of bond distances <0.1 pm). In order to
meet these strict conditions, very symmetrical skeletons
$B_6H_6{}^{2-}$ (**21**), $FB_6H_5{}^{2-}$ (**22**), $NC\text{-}B_6H_5{}^{2-}$ (**23**) and SB_5H_5 (**24**)
were selected, and their geometry, electron densities
(e.d.) on all atoms (Figure 3), as well as energies and
e.d. maps of individual MO's were optimized[12] at the SCF
level using TZ + P + D , [i.e. better than $6\text{-}3111G^{**}{}_{++}$]
(Figure 2).

NC-B₆H₅²⁻ (23)   B₆H₆²⁻ (21)   FB₆H₅²⁻ (22)

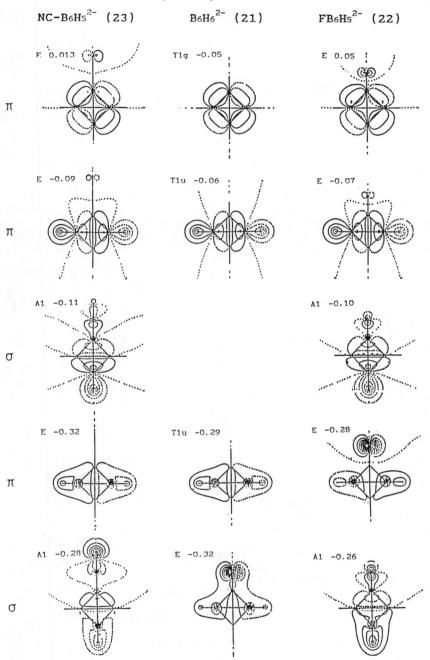

Figure 2 Electron density maps of MO's of B₆H₆²⁻ (21), FB₆H₅²⁻ (22) and NC-B₆H₅²⁻ (23) involved in the "Π" and σ substituent-cluster interactions.[12]

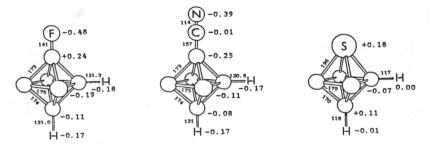

Figure 3 Structures, bond distances (petit) and e.d. on individual atoms in F–B$_6$H$_5$$^{2-}$, NC–B$_6H_5$$^{2-}$ and SB$_5$H$_5$

Using the above computing procedure and e.d. maps, all MO's were reliably assigned, and correlation diagram of FB$_6$H$_5$$^{2-}$, NCB$_6H_5$$^{2-}$ and parent B$_6$H$_6$$^{2-}$ was drawn.

A comparison of the skeletal MO's of substituted compounds with the parent ones (Fig. 2) allows to state:
1) both +M (F) and –M (CN) substituents show only anti-bonding orbital interactions of a "π" character (with the exception of one p(F) bonding interaction with the low lying T1u (–0.29) MO)
2) main transfers of electrons take place in σ orbitals
3) changes in e.d. have character of a polarization of original MO's; in the case of F, an insignificant trans-fer of electrons from the skeleton to F occurs.

E.d. on individual atoms at 22–24 (Fig. 3) and at the basic anion B$_6$H$_6$$^{2-}$ (B: –0.15, H: –0.19, B–B 174, B–H 121 pm) show minimum changes on H atoms in vicinal and antipodal positions brought by F, CN and hetero-S, but significant changes on B atoms, especially on the sub-stituted B atoms.

REFERENCES

1. W.N. Lipscomb, *Boron Hydrides*, W.A. Benjamin, Inc., New York, 1963; W.A. Dixon, D.A. Kleier, J.H. Halgren, W.N. Lipscomb, *J. Am. Chem. Soc.* 1977, *99*, 6226.
2. K. Wade, *Chem. Commun.* 1971, 792; *Adv. Inorg. Radiochem.* 1976, *18*, 1.
3. S. Heřmánek, *Chem. Revs* 1992, *92*, 325.
4. R.E. Williams, *Adv. Inorg. Radiochem.* 1976, *18*, 67 (129)
5. Z. Bastl, S. Heřmánek, J. Plešek, Z. Janoušek, J. Holub, in preparation.
6. E.D. Jemmis, G. Subramanian, L. Radom, *J. Am. Chem. Soc.* 1992, *114*, 1481.
7. S.O. Kang, L.G. Sneddon, *J. Am. Chem. Soc.* 1989, *111*, 3281.
8. T. Jelínek, J. Plešek, S. Heřmánek, B. Štíbr, *Main Group Met. Chem.* 1987, *10*, 397.
9. J. Duben, S. Heřmánek, B. Štíbr, *J. Chem. Soc., Chem. Commun.* 1978, 287.
10. B. Folkesson, R. Larsson, *J. Electron Spectrosc. Relat. Phenomena* 1990, *50*, 251.
11. J. Michl, K. Baše, T. Jelínek, S. Heřmánek, in prep.
12. Z. Havlas, S. Heřmánek, in preparation.

A New Series of Thia-, Carbaaza-, Carbathia-, and Diazaboranes with Eight, Nine, and Ten Cluster Vertices

J. Holub[1], T. Jelínek[1], B. Štíbr[1], J. D. Kennedy[2], and
M. Thornton-Pett[2]

[1]INSTITUTE OF INORGANIC CHEMISTRY, ACADEMY OF SCIENCES OF
THE CZECH REPUBLIC, 259 68 ŘEŽ NEAR PRAGUE, CZECH REPUBLIC
[2]SCHOOL OF CHEMISTRY, UNIVERSITY OF LEEDS, LEEDS LS2 9JT, UK

1 INTRODUCTION

Since the early beginning of cluster borane chemistry,
there has been considerable interest in the development
of experimental methods for the synthesis of main-group
heteroboranes, i.e. compounds in which one or more cage
vertices have been replaced by an isolobal heteroatomic
group. Our two groups have been for a longer time
interested in the synthesis, reactivity, structure, and
NMR behaviour of medium-sized heteroborane cages that
contain nitrogen or sulphur as typical cluster
constituents.[1] The aim of this contribution is to report
our most recent results in this rapidly developing area
of inorganic-cluster chemistry.

2 NEW TYPES OF EIGHT TO ELEVEN-VERTEX MONO- AND DIHETEROBORANES WITH CARBON, NITROGEN, AND SULPHUR VERTICES

These types of compounds are usually available from
reactions between open-cage borane species and suitable
heteroatom-insertion reagents, such as sodium nitrite,
polysulphides and sulphites (see refs. 1 and 2, for
example). Depending on reaction conditions, such
reactions typically result in partial removal of one or
several boron vertices under concomitant insertion of
{NH} or {S} units into the cage. This
cluster-degradation/insertion approach was employed
earlier by our and other groups for the preparation of a
series of essential heteroboranes and carbahetero-
boranes. We continue contributing to this concept with
the aim of the isolation of new cluster types and
establishment of NMR and structural systematics in this
area.

All compounds discussed below were fully
characterized by ^{1}H and ^{11}B NMR spectroscopy and mass
spectrometry. The NMR measurements were completed by
[^{11}B-[11]B]-COSY and [1]H-{[11]B(selective)} experiments, and
the [^{1}H-^{1}H]-COSY technique was employed whenever these

techniques led to ambiguous results. These methods permitted unambiguous assignments of all ^{11}B and ^{1}H resonances to individual {BH} cluster units and comparison of the NMR patterns with those for structurally related compounds.

Nine-Vertex Thiaboranes

Reactions between *arachno*-4-SB8H12 and selected Lewis bases (for L = SMe2, MeCN, NMe3, pyridine, and urotropine) in dichloromethane (2-24 h at room temperature) have produced a long series of compounds of the 6(*exo*)-L-*arachno*-4-SB8H10 (1) constitution[3] in yields 20 - 90 %. An alternative approach to compounds **I** consists in the replacement of the SMe2 ligand in structure (1) by stronger bases, such as pyridine and urotropine. Some derivatives (for L = MeNC, PPh3, quin, and *iso*-quin) have also been isolated as side products from reactions between *nido*-6-SB9H11 and corresponding ligands. Compounds of type (1) are isostructural with the monocarbaborane analogues 6(*exo*)-L-*arachno*-4-CB8H12,[4] as documented by intercomparison of corresponding NMR data and by X-ray diffraction studies on derivatives (1) with L = MeNC, uro, and 1/2 uro.[3]

(1)   (2)

Treatment of the SMe2 derivative of (1) with NaH in boiling THF gave a good yield (72%) of the first nine-vertex *nido* thiaborane, the anionic [*nido*-9-SB8H9]$^-$ (2).[5] This can be reconverted into compounds of type (1) (for L = SMe2 and py) by reactions with the corresponding base L in dichloromethane, followed by acidification. As demonstrated by comparative NMR studies, anion (2) is structurally related, though not isostructural, to other parent representatives of the nine-vertex *nido* series, such as [B9H12]$^-$ and 1-CB8H12.

Eight- to Ten-Vertex Diheteroboranes

A series of the ten-vertex parent diheteroboranes of the general *arachno*-6,9-E^{1}E^{2}B8H10 constitution (3), where E^1 = E^2 = NH ; E^1 = CH2, E^2 = NH ; E^1 = CH2, E^2 = S, and E^1 = NH, E^2 = S has been prepared[6-8] from reactions involving *nido*-6-EB9H12 and *arachno*-4-EB8H12 (where E = CH2, NH or S) as starting boron-cluster reagents.

Thus, the reaction between *nido*-6-NB9H12 and NaNO2 in THF at room temperature has resulted in the isolation of the first parent diaazaborane *arachno*-6,9-N₂B₈H₁₂ (yield 4%),[6] an improved yield (*ca.* 10%) being obtained using BuONO in benzene as an alternative nitrogen source. An analogous compound, *arachno*-6,9-CNB₈H₁₃, was isolated in 15% yield as the main product from the reaction between 4-CB₈H₁₄ and NaNO₂ in THF.[7] The same compound, accompanied by the formation of the first nine-vertex carbaazaborane *arachno*-4,6-CNB₇H₁₂ (4) (yield 5%), can also be obtained in 10% yield from a similar BuONO reaction in benzene.[8] An analogous reaction between BuONO and *nido*-6-SB₉H₁₁ in ether yielded the first azathiaborane, *arachno*-6,9-NSB₈H₁₁, along with its 6-HO- derivative (yields 15 and 35%, respectively); lower yields of the former compound (10%) were obtained in a similar way, using *arachno*-4-SB₈H₁₂ as a boron-cluster reagent.[8]

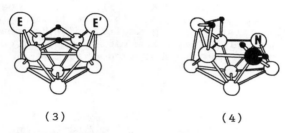

(3)   (4)

Much more developed has been the chemistry of *arachno*-6,9-CSB₈H₁₂[9] which can be obtained in a reasonable yield (48%) by treatment of 4-CB₈H₁₄ with elemental sulphur in the presence of triethylamine in chloroform.[7] Bridge-deprotonation of this carbathiaborane with NaH leads to the [*arachno*-6,9-CSB₈H₁₁]⁻

(5)   (6)

anion. This can be oxidized by acetone (ambient temperature, 24 h, yield 38%) to give the anionic *nido* carbathiaborane congener, [6,9-CSB₈H₉]⁻ (5). The reaction of this anion with hydrochloric acid at room temperature resulted in one-boron cluster degradation to afford the previously reported[10] nine-vertex species *arachno*-4,6-CSB₇H₁₁ (yield 55%). This was deprotonated and converted in turn to the eight-vertex *hypho* anion [7,8-CSB₆H₁₁]⁻ (6), a carbathiaborane analogue of [7,8-S₂B₆H₉]⁻,[11] in 65% yield by the action of aqueous acetone (reflux, 5h). Methylation of (6) with methyl iodide in THF at room temperature has produced the neutral 8-Me-*hypho*-7,8-CSB₆H₁₁ in 75% yield. The [6,9-CSB₈H₁₁]⁻ anion was also converted to *nido*-7,9-CSB₉H₁₁ (7) by heating at 120° C.

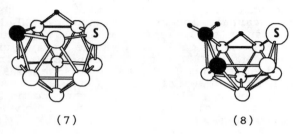

(7)   (8)

A new ten-vertex *arachno* dicarbathiaborane, formulated as 5,6,9-C₂SB₇H₁₁ (8), was prepared from *nido*-5,6-C₂B₈H₁₂ in 20% yield by treatment with sulfur and triethylamine in chloroform. NMR measurements suggest a straightforward similarity of this compound to the ligand derivatives 9-L-*arachno*-5,6-C₂B₈H₁₂.

REFERENCES

1.   B. Štíbr, J. Plešek and S. Heřmánek, in 'Molecular Structure and Energetics', Vol. 5, Advances in Boron and the Boranes; J. F. Liebman, A. Greenberg and R. E. Williams, Eds., VCH Publishers, Inc.: New York, 1988, Chapter 3, pp. 35-70, and references therein.

2.   P. A. Wegner, in 'Boron Hydride Chemistry', E. L. Muetterties, Ed., New York, 1975, Chapter 12, and references therein.

3.   J. Holub, B. Štíbr, J. D. Kennedy, M. Thornton-Pett and T. Jelínek, Inorg. Chem., submitted.

4.   J. Plešek, T. Jelínek, B. Štíbr and S. Heřmánek, J. Chem. Soc.,Chem. Commun, 1988, 348.

5.   J. Holub, B. Štíbr and J. D. Kennedy, Collect. Czech Chem. Commun., 1993, in press.

6.   B. Štíbr, J. D. Kennedy and T. Jelínek, J. Chem. Soc., Chem. Commun., 1990, 1309.

7.   J. Holub, T. Jelínek, J. Plešek, B. Štíbr, S. Heřmánek and J. D. Kennedy, J. Chem. Soc., Chem. Commun., 1991, 1389.

8.   T. Jelínek, J. D. Kennedy and B. Štíbr, J. Chem. Soc., Chem. Commun., manuscript in preparation.

9.   J. Holub, J. D. Kennedy, T. Jelínek and B. Štíbr, J. Chem. Soc., Dalton Trans., submitted.

10.   J. Plešek, S. Heřmánek and Z. Janoušek, Collect. Czech. Chem. Commun., 1977, 42, 785.

11.   S. O.Kang and L. G. Sneddon, J. Am. Chem. Soc., 1989, 111, 3281-3289.

The First Oxaborane Species: 12-Vertex-28e-*nido* $B_{11}H_{12}O^-$

A. Ouassas[1], B. Fenet[2], H. Mongeot[3], and B. Frange[3]

[1]LABORATOIRE DE CHIMIE DE COORDINATION, FACULTÉ DES
SCIENCES SEMLALIA, BOULEVARD DE SAFI, MARRAKECH, MAROC
[2]CENTRE DE RMN, 43 BOULEVARD DU 11 NOVEMBRE 1918, 69622
VILLEURBANNE CÉDEX, FRANCE
[3]LABORATOIRE DE PHYSICOCHIMIE MINÉRALE 1, 43 BOULEVARD DU
11 NOVEMBRE 1918, 69622 VILLEURBANNE CÉDEX, FRANCE

1 INTRODUCTION

In spite of its ready availibility, starting either from $NaBH_4$ and $BF_3.Et_2O$ (or any Lewis acid)[1] or B_5H_9 and KH_2, the chemistry of the nido-$B_{11}H_{14}^-$ still remains quite unexplored. Thus it could either lose or add a proton according to both following reactions :

$$B_{11}H_{14}^- \rightarrow B_{11}H_{13}^{2-} + H^+ \qquad (1)$$

$$B_{11}H_{14}^- + H^+ \rightarrow B_{11}H_{15} \qquad (2)$$

Furthermore, $B_{11}H_{13}^{2-}$ gives rise to the closo $B_{11}H_{11}^{2-}$ by pyrolysis. However, the structures of these eleven-boron species have been established with varying degrees of certainty. Thus, $B_{11}H_{14}^-$ itself was only recently characterized in the solid state by Shore et al.[3] whereas Lipscomb's hypothesis of an H_3^+ triangle interacting with a $B_{11}H_{11}^{2-}$ moiety still prevails in solution[4]. On the other hand, if the crystallographic structure of $B_{11}H_{13}^{2-}$ has been known for a long time[5], neither $B_{11}H_{15}$ nor $B_{11}H_{11}^{2-}$ are known although the latter is supposed to display a C_{2v} symmetry. Anyhow, most of the chemistry devoted to these compounds aimed at including a twelfth atom within the boron framework in order to achieve an icosahedral geometry. Besides Hawthorne's pioneering work with nickellocene[6], such an approach is mainly developed in the States by two different groups : Shore et al. on one hand[7] (Al, P), Little and Todd on the other[8] (Se, Te, As, Sb, Bi, Sn, Ge, Pb) leading to inclusion of a metal or more commonly a non-metallic atom. Hereafter, we are only interested with those works where an eleven-boron species is employed as starting material although others routes could be used instead. As far as we are concerned, we first duplicated previous results of one of this group by means of Phase Transfer Catalysis Methods. Thus, the reaction $B_{11}H_{14}^- + M_2O_3$ (M = As, Sb, Bi) was investigated in the biphasic NaOH 10N/CH_2Cl_2 system leading to the expected icosahedral species $B_{11}H_{11}M^-$ with good yields. During these reactions, a by-product was also obtained[9] first supposed to be a B-hydroxy derivative $B_{11}H_{13}OH^-$, i.e. belonging to the family $B_{11}H_{13}.L$ (L = OH^-). A related derivative (L = Me_2S) has been claimed for in earlier literature[10] while $B_{11}H_{14}^-$ itself can formally be considered as the parent compound (L = H^-). The following paper is mainly concerned with the reinvestigation of the structure of this by-product by means of NMR techniques since our former formulation was at variance with later results. Hereafter, the so-obtained compound is shown to be an oxaborane anion formulated as $B_{11}H_{12}O^-$, M^+ (M = Na, NEt_4, $C_6H_5CH_2NEt_3$).

2 RESULTS

The ^{11}B NMR spectrum of this derivative displays the typical pattern for an eleven-boron framework with a symmetry plane(2/1/1/2/2/2/1) with an hydrogen atom attached to each boron. The ^1H NMR spectrum by careful integration reveals the presence of one bridge hydrogen per cage (0.89 to 0.99 according to M) and a charge close to one The 2D ^1H-^{11}B spectrum allows the exact location of the bridge hydrogen, in agreement with selective irradiation of the bridge proton that results in an enhancement of the signal of the related boron atoms. The 2D ^{11}B-^{11}B spectrum enables the straightforward assignment of all boron atoms; two points deserve further comments : (i) no correlation could be observed between B10 and B9,11 in spite of repeated attempts (ii) with proper choice of NMR parameters, a very small cross-peak was indeed observed between B8, B9 and B7, B11. These two facts, in addition to the deshielding observed for the ^{11}B chemical shifts and the T_1 measurements for the relevant atoms B9, B10 and B11 strongly support the presence of a strong electronegative atom in capping position. From elemental analysis and accurate mass measurement of the anion using Fast Atom Bombardment techniques, this atom ought to be oxygen. Thus, the reported compound is $B_{11}H_{12}O^-$, M^+ (M = Na, NEt_4, $C_6H_5CH_2NEt_3$), the cation being exchangeable by usual metathesis reactions (Figure 1). This is a nido-12-vertex-28 electrons oxaborane (assuming four electrons for the oxygen atom according to classical electron counting rules) and it is the first example of oxaborane, i.e. of a cluster with boron, hydrogen and oxygen. The related examples so-far reported always contained metallic moieties within the boron framework such as h^6-arene-iron in $C_6Me_3H_3FeOB_8H_{10}$ prepared by Sneddon et al.[11] or h^5-RhCp in $C_5Me_5RhCl(PMe_2Ph)OB_{10}H_9$[12] and $C_5Me_5Rh(NEt_3)OB_{10}H_{10}$ obtained by Greenwood, Kennedy et al.[13] While the former compound is isoelectronic with nido-$B_{10}H_{14}$ the two other, whose solid state structures have been established (Figure 2), are isoelectronic with the above reported $B_{11}H_{12}O^-$ using the isolobal analogy. An other closely related anionic species $(MeO)B_{11}H_{11}NMe^-$ has been reported quite recently by Paetzold et al.[14] (Figure 3) which was readily obtained by methanolysis of the closo-$B_{11}H_{11}NMe$. All these literature reports, together with our spectroscopic results bring compelling evidence to the proposed structure for $B_{11}H_{12}O^-$, M^+. Unfortunately, no crystallographic determination could be performed so far because of disorder at ambient temperature. However one should be careful since other structures (at least seven) were indeed observed for closely related 12-vertex-28-electrons metallacarboranes species[15].

Our next challenge is to obtain the closo-$B_{11}H_{11}O$ 26-electrons derivative, with a formally pentacoordinated oxygen atom. Such a target seems to be attainable since the closely related, air-stable closo-$B_{11}H_{11}NH$ (Figure 4), with an isoelectrreonic NH group instead of oxygen, has recently been prepared by Paetzold et al.[16]

3 CONCLUSION

A Boron-Hydrogen cluster being considered as an electron-deficient moiety, the addition of such an electonegative atom as oxygen would usually result in the destruction of the cage and, indeed, examples of B, H, O derivatives used to be scarce. However, a somewhat different approach can be derived from Pearson's

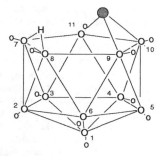

<u>Figure 1</u> Structure of $B_{11}H_{12}O^-$ in solution from NMR data.

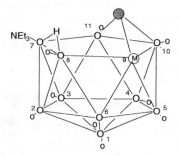

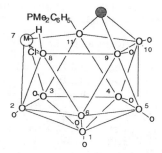

<u>Figure 2</u>. Structure of the two oxarhodaboranes prepared by Greenwood, Kennedy et al. (M = RhC_5Me_5, O is the shaded atom). The numbering is ours.

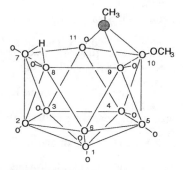

<u>Figure 3</u> Structure of $(MeO)B_{11}H_{11}NMe^-$ (N is the shaded atom). The numbering is ours.

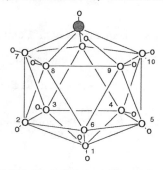

<u>Figure 4</u> Structure of the closo $B_{11}H_{11}NH$ (N is the shaded atom). The numbering is ours.

HSAB theory, the boron behaving as a soft center and the oxygen atom as the hard one. Thus, the presence of an hydroxy substituent in a boron-hydrogen cluster is rather unlikely but quite a different situation is observed with a bridging oxygen since this allows the electronic density to be shared by two different boron atoms: the first example[17], discovered by the Czech group, was $(B_{10}H_{13})_2O$, an isoelectronic oxo-bridged-rhodaborane being reported later on[13]. With a triply coordinated oxygen, in a capping position, further delocalisation of the electronic density could be obtained: three neutral compounds were quoted above prepared by Sneddon et al[11] and Greenwood, Kennedy et al.[12,13]; the anionic species $B_{11}H_{12}O^-$ belongs to the same family. Finally, the existence of a formally pentacoordinated oxygen does not appear as unrealistic and we are presently investigating in our group oxidation reactions of $B_{11}H_{12}O^-$ as possible route to $B_{11}H_{11}O$.

The same behavior could apply, to a lesser extent, to the nitrogen atom and may explain the unexpected stability of clusters with bridging nitrogen. Furthermore, fluorine could possibly be included within a boron framework.

4 REFERENCES

1. G. B. Dunks and K. P. Ordonez, Inorg. Chem., 1978, 17, 1514.
   G. B. Dunks, K. Barker, E. Hedaya, C. Hefner, K. Palmer-Ordonez and P.Remec, Inorg. Chem., 1981, 20, 1692.
2. N. S. Hosmane, J. R. Wermer, Z. Hong, T. D. Getman and S. G. Shore, Inorg.Chem., 1987, 26, 3638.
3. T. D. Getman, J. A. Krause and S. G. Shore, Inorg. Chem., 1988, 27, 2398.
4. G. R. Eaton and W. N. Lipscomb, 'NMR Studies of Boron Hydrides and Related Compounds', W. A. Benjamin, Inc., New York, 1969.
5. C. J. Fritchie Jr., Inorg. Chem., 1967, 6, 1199.
6. R. N. Leyden, B. P. Sullivan, R. T. Baker and M. F. Hawthorne, J. Amer. Chem. Soc., 1978, 100, 3758.
7. T. D. Getman and S. G. Shore, Inorg. Chem., 1988, 27, 3439.
   28, 3612.
8. G. D. Friesen and L. J. Todd, J. Chem. Soc., Chem. Comm., 1978, 349.
   J. L. Little, Inorg. Chem., 1979, 18, 1598.
   T. P. Hanusa, N. R. de Parisi, J. G. Kester, A. Arafat and L. J. Todd, Inorg. Chem. 1987, 26, 4100.
   J. L. Little, M. A. Whitesell, J. G. Kester, K. Folting and L. J. Todd, Inorg. Chem., 1990, 29, 804.
   R. W. Chapman, J. G. Kester, K. Folting, W. E. Streib and L. J. Todd, Inorg. Chem., 1992, 31, 979.
9. A. Ouassas, C. R'Kha, H. Mongeot and B. Frange, Inorg. Chim. Acta, 1991, 180, 257.
10. L. J. Edwards and J. M. Makhlouf, J. Amer. Chem. Soc., 1966, 88, 4728.
11. R. P. Micciche, J. J. Briguglio and L. G. Sneddon, Inorg. Chem., 1984, 23, 3992.
12. X. L. R. Fontaine, H. Fowkes, N. N. Greenwood, J. D. Kennedy and M. Thornton-Pett, J. Chem. Soc., Chem. Comm., 1985, 1722.
13. E. J. Ditzel, X. L. R. Fontaine, H. Fowkes, N. N. Greenwood, J. D. Kennedy, P. Mackinnon, Z. Sisan and M. Thornton-Pett, J. Chem. Soc., Chem.Comm., 1990, 1692.
14. F. Meyer, J. Müller, P. Paetzold and R. Boese, Angew. Chem., 1992, 104, 1221.
15. R. B. Maynard, E. Sinn and R. N. Grimes, Inorg. Chem., 1981, 20, 1201.
16. J. Müller, J. Runsink and P. Paetzold, Angew. Chem., 1991, 103, 201.
17. J. Plesek, S. Hermanek and B. Stibr, Coll. Czech. Chem. Comm., 1968, 33, 691.

Specific Cage Expansion of the Decaborane(14) System

Donald F. Gaines and Adam N. Bridges

DEPARTMENT OF CHEMISTRY, UNIVERSITY OF WISCONSIN-MADISON, MADISON, WI 53706, USA

1 INTRODUCTION

High-yield preparations of the anion $B_{11}H_{14}{}^{-1}$ using lower boron hydrides have been developed and optimized over the last 15 years[1,2] Althought the mechanisms for these reactions remain unclear, they have eliminated the need to prepare this parent compound from decaborane, a more obvious and systematic route. There are certain instances where it is necessary to replace optimum routes with more systematic routes. Our interest in regiospecific [11]B isotopic labeling and substitution of the nido-$B_{11}H_{14}{}^{-1}$ anion has led us to develope rational undecaborane syntheses from common decaborane precursors using clues from the original reaction between decaborane(14) and borohydride anion.[3] Herein we report several new methods for the preparation of $B_{11}H_{14}{}^{-1}$ derivatives and the development of specific alkyl sustitution of the B_{11} cage.

2 SYNTHESIS

Preparation of the Parent Anion $B_{11}H_{14}{}^{-1}$

Work was first conducted on preparation of the parent $B_{11}H_{14}{}^{-1}$ anion, and then extended to accommodate substitution chemistry. It is known that decaborane(14) reacts with borohydride at elevated temperatures to produce $B_{11}H_{14}{}^{-1}$. This reaction is postulated to proceed via $B_{10}H_{13}{}^{-1}$ and BH_3, which react at elevated temperature to produce $B_{11}H_{14}{}^{-1}$ and H_2. We report below rational syntheses of $B_{11}H_{14}{}^{-1}$ salts from $B_{10}H_{13}{}^{-1}$, $B_{10}H_{12}{}^{-2}$ and $B_{10}H_{12} \bullet 2L$ under mild conditions.

Reaction of $B_{10}H_{13}{}^{-1}$ with excess $Cl\text{-}BH_2 \bullet SMe_2$ forms two products, $B_{11}H_{14}{}^{-1}$ and $B_{10}H_{14}$, in a ratio of 1:1. The stoichiometric production of

$B_{10}H_{14}$ suggests that the $B_{10}H_{13}{}^{-1}$ anion acts as a proton acceptor in the formation of $B_{11}H_{14}{}^{-1}$. Since $B_{10}H_{13}{}^{-1}$ does not deprotonate itself and Cl-$BH_2 \cdot SMe_2$ is not readily deprotonated by other common bases, it is probable that $B_{10}H_{13}{}^{-1}$ accepts a proton from an activated complex, perhaps a form of $B_{11}H_{15}$. The reaction of $B_{10}H_{13}{}^{-1}$ with 0.5 molar equivalents Cl-$BH_2 \cdot SMe_2$ forms three products: $B_{10}H_{14}$, $B_{11}H_{14}{}^{-1}$, and a species similar in structure to $B_{11}H_{14}{}^{-1}$, but soluble in benzene and thermally stable. We postulate that this latter species could be the protonated intermediate from which $B_{11}H_{14}{}^{-1}$ is formed.

It was then surmised that an alternate proton acceptor could be found that would take the place of $B_{10}H_{13}{}^{-1}$, thus increasing the yield of $B_{11}H_{14}{}^{-1}$. An excellent candidate was the so-called "proton sponge", 1,8-bis(dimethylamino)naphthalene (PS). When $B_{10}H_{13}{}^{-1}$ was reacted with Cl-$BH_2 \cdot SMe_2$ in the presence of 1 equivalent of proton sponge, the sole product was $B_{11}H_{14}{}^{-1}$.

$$Na[B_{10}H_{13}] + Cl\text{-}BH_2 \cdot SMe_2 + PS \xrightarrow{\ R_2O\ } PSH^+[B_{11}H_{14}] + NaCl \qquad (1)$$

There are several possible mechanistic explanations for the success of this reaction. One assumes that the reaction proceeds as if the proton sponge were not present, and the decaborane formed is deprotonated to regenerate starting material. The other assumes that the proton sponge reacts more rapidly than $B_{10}H_{13}{}^{-1}$ with the B_{11} intermediate. The production of only NaCl precipitate indirectly supports the latter postulate.

Sodium hydride was also found to be a suitable proton acceptor, thus providing a more synthetically useful cation and an inexpensive substitute for proton sponge.

$$Na[B_{10}H_{13}] + Cl\text{-}BH_2 \cdot SMe_2 + NaH \xrightarrow{\ THF\ } Na[B_{11}H_{14}] + NaCl + H_2 \qquad (2)$$

When reaction **2** is conducted in Et_2O it proceeds only to the mixture of $B_{11}H_{14}{}^{-1}$ and $B_{10}H_{14}$. When the Et_2O is removed and replaced with THF, the reaction then proceeds quickly to completion. This suggests that NaH is not competing with $B_{10}H_{13}{}^{-1}$, but rather it is simply regenerating $B_{10}H_{13}{}^{-1}$. The low solubility of NaH is possibly a factor in its sluggish reactivity and inability to compete with $B_{10}H_{13}{}^{-1}$ in Et_2O. The $B_{11}H_{14}{}^{-1}$ synthesis can also be effected directly from $B_{10}H_{14}$ via reaction **3**.

$$B_{10}H_{14} + CL\text{-}BH_2 \cdot SMe_2 + 2\,NaH \xrightarrow{\ THF\ } Na[B_{11}H_{14}] + NaCl + 2\,H_2 \qquad (3)$$

The two other decaborane(14) systems have found limited use in the production of $B_{11}H_{14}{}^{-1}$ in fair to high yields. $Na_2[B_{10}H_{12}] \cdot THF$ reacts with Cl-$BH_2 \cdot SMe_2$ to produce $B_{11}H_{14}{}^{-1}$ exclusively as shown in reaction **4**.

$$Na_2[B_{10}H_{12}] \cdot THF + Cl\text{-}BH_2 \cdot SMe_2 \xrightarrow{\text{THF}} Na[B_{11}H_{14}] + NaCl \tag{4}$$

Low solubility and fluctionality of the $B_{10}H_{12}^{-2}$ cage may preclude its use in derivative syntheses. The bis-adduct of decaborane(14), $B_{10}H_{12} \cdot 2L$ (L=MeCN, Me_2S) has also been shown to produce significant amounts of $B_{11}H_{14}^{-1}$, as shown in reaction 5, although the cation in this case has yet to be identified.

$$B_{10}H_{12} \cdot 2L + BH_3 \cdot THF \xrightarrow{\text{THF}} B_{11}H_{14}^{-1} + \dots \tag{5}$$

Synthesis of Monoalkyl Undecaboranes

Several of the above synthetic methods have been employed in the production of the first alkyl derivatives of $B_{11}H_{14}^{-1}$. Systematic construction of monoalkyl derivatives of $B_{11}H_{14}^{-1}$ via boron insertion can utilize two synthetic strategies. One involves the addition of an unsubstituted monoborane to an alkyl decaborane (equ. 6 and 7), the other involves the addition of an alkylmonoborane to an unsubstituted decaborane (equ. 8 and 9). Our development of high-yield preparations of monoalkyldecaboranes[4] has made the former strategy readily accessible.

$$Na[6\text{-}R\text{-}B_{10}H_{12}] + Cl\text{-}BH_2 \cdot SMe_2 + PS \xrightarrow{\text{Et}_2\text{O}} PS\overset{+}{H}[7\text{-}R\text{-}B_{11}H_{13}] + NaCl \tag{6}$$

$$6\text{-}R\text{-}B_{10}H_{13} + Cl\text{-}BH_2 \cdot SMe_2 + 2\,NaH \xrightarrow{\text{THF}} Na[7\text{-}R\text{-}B_{11}H_{13}] + NaCl + 2H_2 \tag{7}$$

The second strategy has yielded two useful syntheses, although this approach may be limited by the availability of monoalky-monoboranes. The thexyl group (Thx) has been a very convenient substituent in this regard.

$$Na[B_{10}H_{13}] + Thx\text{-}BHCl \cdot SMe_2 + PS \xrightarrow{\text{Et}_2\text{O}} PS\overset{+}{H}[7\text{-}Thx\text{-}B_{11}H_{13}] + NaCl \tag{8}$$

$$B_{10}H_{12} \cdot 2SMe_2 + Thx\text{-}BH_2 \cdot THF + PS \xrightarrow[\text{3 days}]{\text{THF}} PS\overset{+}{H}[7\text{-}Thx\text{-}B_{11}H_{13}] \tag{9}$$

3 CHARACTERIZATION OF $R\text{-}B_{11}H_{13}^{-1}$ COMPOUNDS

Salts of the anions $7\text{-}R\text{-}B_{11}H_{13}^{-1}$ appear to be fairly air stable solids. The ^{11}B NMR of $7\text{-}Thx\text{-}B_{11}H_{13}^{-1}$ displays a spectrum of six doublets and one singlet: $\delta = +9.6$ s (1B); -12.3 d 2(B); -13.6 d (3B); -19.1 d (2B); -22.0 d (2B); -27.1 d (1B). $^{11}B\text{-}^{11}B$ 2-D COSY NMR provides an unambiguous assignment of all boron resonances However, the cross peak coupling between the substituted boron B(7) and B(8,11) is conspicuously absent, possibly suggesting that bridging hydrogens may occupy static positions spanning these sites on the nido cage face.

The crystal structure of 7-Thx-$B_{11}H_{13}^{-1}$, determined as the triphenyl(methyl)phosphonium salt (fig. 1), is very similar to that of the parent compound $B_{11}H_{14}^{-1}$ as determined by Shore et al.[5] However, the B-B distances between the substituted boron B(7) and its nearest neighbors in the alkyl derivative are longer compared to the parent, indicating that the alkyl group imparts some distortion to the cage. The exoskeletal hydrogens occupy three bridging positions around the face of the cage. This is in contrast to the parent structure, in which one of the hydrogens occupies an *endo* position. As expected, two of the hydrogens are situated between B(7)-B(8) and B(7)-B(11) connectivities which correlates well with the ^{11}B-^{11}B COSY results. Bridging hydrogen bonding is often suspected of distorting the magnetization along the B-B vector, thus producing an absence of cross-coupling in the 2-D array.

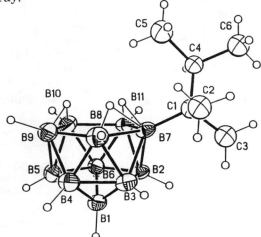

Figure 1   X-ray Structure of $Ph_3PMe[7\text{-Thx-}B_{11}H_{13}]$

This research was supported by the National Science Foundation

REFERENCES

1. Dunks, G. B.; Palmer-Ordonez, K. *Inorg. Chem.* **1978,** *17,* 1514.
2. Hosmane, N. S.; Wermer, J. R.; Hong, Z.; Getman, T. D.; Shore, S. G. *Inorg. Chem.* **1987,** *26,* 3638.
3. Aftandilian,, V. D.; Miller, H. C.; Parshall, G. W.; Muetterties, E. I. *Inorg. Chem.* **1962,** *1,* 734.
4. Bridges, A. N.; Gaines, D. F. *Organometallics* **1993,** *12,* 2015.
5. Getman, T. D.; Krause, J. A.; Shore, S. G. *Inorg. Chem.* **1988,** *27,* 2398.

The Dawn of Chiral Deltahedral Boranes

B. Grüner* and J. Plešek

INSTITUTE OF INORGANIC CHEMISTRY, ACADEMY OF SCIENCES OF
THE CZECH REPUBLIC, 250 68 ŘEŽ NEAR PRAGUE, CZECH REPUBLIC

1. INTRODUCTION

Substitution at most cage boron skeletons or skeletal distortion may lead to chirality. The first example of such chiral species was reported by Hertler[1] as early as 1964. However, up to the end of 1990 only 19 chiral cage boron compounds were known[2]. The main reason that these asymmetrical species had received only marginal attention lies probably in the fact that all chiral compounds were resolved by tortuous classical methods, mainly *via* diastereomeric intermediates. Several excellent results[2] were achieved in this pioneering era. However, it became apparent that further research in this interesting and promising field have strongly needed more effective resolution methods, which are rapid, reliable, sensitive and inexpensive. Such methods, HPLC on on Chiral Stationary Phases (CSPs), have been used by organic chemists for nearly two decades[3]. Therefore we decided to develop new methods enabling resolution of enantiomers of chiral cage boron compounds on the basis of CSPs available, wondering that nobody realized this obvious approach before us.

2. RESULTS

Within the two last years we have succeeded in resolution of more than 35 racemic species into opticaly pure enantiomers using β-cyclodextrin (β-CD) CSPs, the inexpensive and widely used CSPs with intercalation separation mechanism and a broad range of enantio-selectivities. Sufficient amounts of enantiomers were obtained for circular dichroism spectra (CD) and for crystal growth for X-ray study using these methods. Two native and one modified β-CD CSPs were tested throught this study: β-CD CYCLOBOND I (A) with β-CD molecule attached *via* 6-8 atomic spacer, TESSEK β-CD with directly bonded β-CD (B) and laboratory made directly bonded acetyl-β-CD column (C). However, the intercalation mechanism on these CSPs requires the use

of aqueous-organic mobile phases. Therefore, a selection
of hydrolytically resistant and sufficiently soluble
protochiral cage boron compounds was essential for
success. The selected systems had to be synthetized
for this purpose, prepared in extreme purity and well
characterized by other methods (^{1}H, ^{11}B NMR, GC-MS, HPLC).

Separation of carboranes

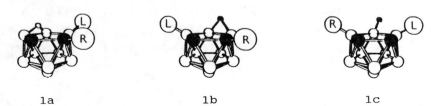

1a   1b   1c

Eleven vertex series[4] The first attempts of chiral
resolution were made using mono and disubstituted
zwitterionic species R,L-7,8-$C_2B_9H_{11}$ (1a, b: R= H, Me,
Ph, L= Me_2S, Py, $CH_2NC_5H_5$; 1c: R= Me, L= Py). From comparison of retention behaviour of 11 enantiomeric pairs,
several observations made during this study were important for understanding the separation mechanism: 1) One
substituent on the skeleton $[C_2B_9H_{11}]^-$ is sufficient for
successful enantiomeric discrimination. 2) In all cases,
the compounds enter the β-CD cavity by their more hydrophobic carborane part. Substituents (including Ph-,
Py-) give interactions (steric, hydrogen bonding) with
hydroxyls on the rim of the β-CD cone. 3) Distinct
differences were observed between two native β-CD CSPs
CYCLOBOND I (C) and direcly bonded β-CD (D). While the
former material has proven to be efficient in the separation of most enantiomeric pairs under study, the
latter CSPs was effective only in the resolution of
'*meta-*' substituted species (1 b, c)[4].

Ten vertex series. Most recently we have succeeded in
separation of incredibly stable amine substituted
zwitterionic *arachno-* compounds in the ten vertex
series (2a: L= H_3N, $BuNH_2$, t-$BuNH_2$, Piper., Et_2NH;
2b: L= Bu_2NH, Et_3N). However, application of native β-CD
CSPs (A and B) allowed for the resolution of only two
compounds, Et_3N- 5,10-$C_2B_8H_{12}$ and $C_5H_{10}N$-5,10-$C_2B_8H_{12}$.
Surprisingly, members of both series differ in positions

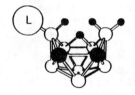

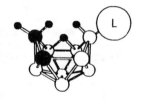

2a   2b

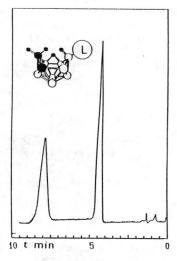

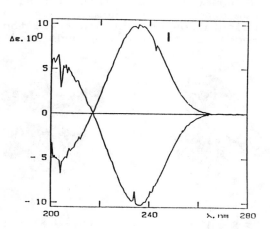

Figure 1: **a)** Separation of enantiomers of the t-BuNH$_2$-5,6-C$_2$B$_8$H$_{12}$ on acetyl β-CD column

b): CD spectra of the 6-Et$_3$N-5,6-C$_2$B$_8$H$_{12}$ enantiomers (I- first eluting)

of skeletal carbons. Six other compounds were resolved using the modified acetyl-β-CD support (C). Example is given in Figure 1a. CD spectra of carboranes not bearing chromophoric groups exhibited simple pattern in CD. One example is given in Fig. 1b.

Metallaboranes.

A variety of metallaboranes were resolved on β-CD CSPs[4-8]. In opposite to the carborane series, the directly bonded β-CD material (B) has proven to be more efficient. Relatively good resolution values were obtained for both mixed sandwich series having eleven or ten vertex carborane ligands (3= 4-MeS-3-Co-1,2-C$_2$B$_9$H$_{10}$, 4= Me$_6$C$_6$-RuC$_2$B$_8$H$_9$Br). On the other hand, in the series of compounds with two carborane ligands,(e. g.4-Me$_2$S-4'-MeS-3-Co-(1,2-C$_2$B$_9$H$_{10}$)$_2$), a remarkable decrease of resolution was observed in comparison with similar mixed sandwich species. This loss of enantioselectivity is probably a consequence of averaging of interactions with both carborane parts of such molecules, and the relatively free rotation of ligands in solution around the central atom.

Bridged metallaboranes[6,7] The good resolution values in the series of the bridged cobaltacarboranes [6,6'-RE (1,7- C$_2$B$_9$H$_{10}$)$_2$-2-Co] (E= S<, R= none, Me, Et, Pr, i-Pr, Bu, Hex, CH$_3$OOCCH$_2$; E= O<, R=none; E= N<, R=H, Me) with rigid helicaly twisted structures on the native β-CD CSPs confirmed the above assumption. The resolution values decrease monotonously with increasing size of bridge substituent, showing almost no dependence on substituent polarity. However for the substituent size

up to n-butyl, nearly baseline resolutions can be obtained.
CD spectra of the first eluting enantiomers of all
compounds in this series have exhibited the identical
orientation of Cotton curves; therefore the separation
mechanism for all compounds should be the same.

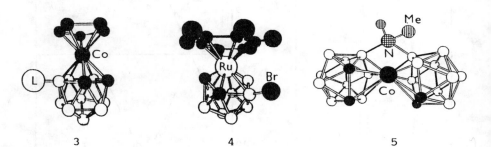

3   4   5

       More recently, three closely related compounds(5)
$[4,8'-R_2N(1,2- C_2B_9H_{10})_2-3-Co]$ (R= H, Me) were resolved with even better results[8]. We suspect the larger
dipole moments are responsible for enhancement in
resolution. All metallaborane enantiomers exhibit very
complex patterns in their Cotton curves. However, CD
spectra are by far more informative than UV spectra
and may help future theoretical chemists calculate
subtle differences in excitation levels.
       The samples of crystals of pure enantiomers and
inclusion complexes with β-CD were recently submitted
for X-ray diffraction study. We hope these results
will help solve the mystery of separation mechanism.
       We believe the methods outlined in this article will
open new possibilities in mechanistic and stereochemical
studies and potential applications for these stereochemically unique species in which chirality arises
from twisted structures rather than from distinct
chiral centers.

REFERENCES

1. W.R. Hertler, *J.Am. Chem. Soc.*, 1964, *86*, 2949.
2. J. Plešek, S. Heřmánek, B. Štíbr, *Pure Appl. Chem.*,
   1991, *63*, 399.
3. S. Ahuja,'Chiral Separations by Liquid Chromatography'Am. Chem. Soc., Washington D.C. 1991.
4. J. Plešek, B. Grüner, P. Maloň, *J. Chromatogr.*,1992,
   626, 197.
5. J. Plešek, B. Grüner, T. Vaněk, H. Votavová,
   J. Chromatogr., 1993, *633*, 73.
6. J. Plešek, B. Grüner, P. Maloň *Collect. Czech. Chem.*
   Commun., in press
7. J. Plešek, B. Grüner, J. Fusek, H. Votavová, *Collect.*
   Czech. Chem. Commun., in press
8. J. Plešek, B. Grüner, S. Heřmánek, J. Fusek,
   H. Votavová, *Collect. Czech. Chem. Commun.*, in press

TlB$_{11}$H$_{14}$ – A Highly Versatile Synthon for Twelve Vertex Cluster Chemistry

Bernd Brellochs

INSTITUT FÜR ANORGANISCHE CHEMIE DER UNIVERSITÄT MÜNCHEN,
MEISERSTRASSE I, D-8000 MÜNCHEN 2, GERMANY

In continuation of our studies of the chemistry of nido-$B_{11}H_{14}^{-}$ [1] we focus on the synthesis and characterisation of new heterododecaboranes $EB_{11}H_{11}^{n-}$.

In recent years major developments in the introduction of main group elements appeared in the literature[2,3] ($BiB_{11}H_{11}^{-}$, $NB_{11}H_{12}$).

In general these reactions proceed in consecutive steps until the complete incorporation of the heteroatom E into the cluster. A more simple view may be the interaction of $B_{11}H_{11}^{4-}$ and the heteroatom in an appropriate oxidation state.

$$B_{11}H_{11}^{4-} + \overset{+y}{E}_{(HG)}^{y+} \longrightarrow \overset{+y}{E}_{(HG)}B_{11}H_{11}^{4-(+y)} \qquad (1)$$

The wish to get a deeper insight into the reaction path (distribution of electrons and bonding situations)

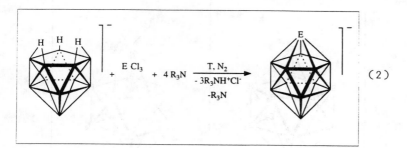

$$\text{+ } E\,Cl_3 \text{ + 4 } R_3N \xrightarrow[\substack{-3R_3NH^+Cl^- \\ -R_3N}]{T, N_2} \qquad (2)$$

leading to closo-heterododecaboranes has led us to examine the formation and the chemistry of TlB$_{11}$H$_{14}$. In combination with a large variety of substituted elemental halides RmEHaln gives us the

opportunity to characterise the species involved on their
way to $closo-EB_{11}H_{11}$ systems.

$TlB_{11}H_{14}$ is a bright-orange coloured thermo- and photo-
labile[14] compound which can serve as a valuable synthon
for the buildup of higher boron clusters. Though it is
sparingly soluble only in DMSO, it can be reacted with
various halides in toluene heterogeneously.
Reactions with halides EX (E= R_2B, RBX, alkyl, RCO, R_3Si,
R_3Sn, R_2P, RPX) yield a wide range of neutral twelve
vertex boranes $EB_{11}H_{14}$.

Group III halides ($GaCl_3$, $InCl_3$) react with $B_{11}H_{14}^-$ under
base-assistance (amine) to two different stages[14] of
the system $B_{11}H_{14}^-/EHal_n$ (cf. scheme 1).

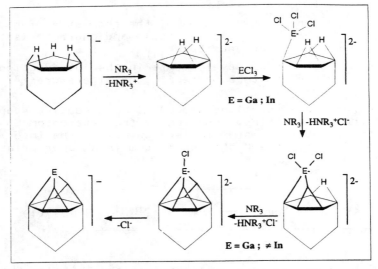

Scheme 1 Reaction paths $Ga/InCl_3 //B_{11}H_{14}^-$

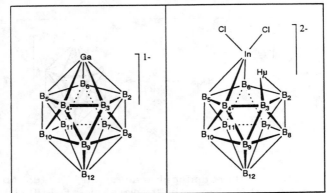

Figure 1 Molecular structures $GaB_{11}H_{11}^-$, $Cl_2InB_{11}H_{12}^{2-}$

The NMR-data indicate that the Cl_2In-unit and the μ-H (changing its habit between exo- and bridging-type) are both involved in an exchanging process on the pentagonal face of the B_{11}-unit (Table 1).

NMR- Daten	1-			2-		
Position	2-6	7-11	12	2-6	7-11	12
$\delta^{11}B$ / [ppm]	-20.3	-21.3	-25.4	-18.2	-16.0	-18.2
$^1J(^{11}B,H_t)$/[Hz]	135	142	130	124	132	124
Position	2-6	7-11	12	2-6	7-11	12
δ^1H / [ppm]	0.63		0.78	1.47	1.32	1.75
				$H\mu = -5.42$		

Table 1 ^{11}B-, ^1H-NMR-data of $GaB_{11}H_{11}^-$ and $Cl_2InB_{11}H_{12}^{2-}$

$POCl_3$ or PCl_5 do react with $B_{11}H_{14}^-$ to $PB_{11}H_{11}^-$.
The NMR-spectroscopic data of the whole series of Group V-compounds $EB_{11}H_{11}^-$ are compared in Table 2.
Additional data for $PB_{11}H_{11}^-$ are : ^{31}P -91.6ppm, $^1J(B_{2-6},P)= 43Hz$, $^3J(B_{12},P)= 28Hz$, $^2J(H_{2-6},P)= 20Hz$, $^4J(H_{12},P) < 2.5Hz$.

References
1. Imeboron VII, 1987, Torun, Poland
2. J.L.Little, M.A.Whitesell, J.G.Kester, K.Folting, L.J.Todd, Inorg.Chem.,1990, 29, 804
3. J.Müller, J.Runsink, P.Paetzold, Angew.Chem., 1991, 103, 201 .

NMR DATEN	(2-)			(1- N)			(1- P)		
Position	2-6	7-11	12	2-6	7-11	12	2-6	7-11	12
$\delta^{11}B/(J_{BH})$ ppm/(Hz)				^{11}B in CD_2Cl_2					
CH_2CL_2		-15.8(124)		-10.3(147)		-1.0(137)	-11.3(145)	-7.0(140)	4.5(130)
THF		-17.3(123)		-		-	-11.4(144)	-6.8(140)	4.9(131)
$C_2D_2Cl_4$		-16.3(122)		-		-	-11.1(148)	-7.0(141)	4.5(135)
δ^1H/ppm									
CD_2Cl_2		-		1.79		1.98	-	-	-
$C_2D_2Cl_2$		1.09		-		-	1.52	1.84	2.27

NMR DATEN	(1- As)			(1- Sb)			(1- Bi)		
Position	2-6	7-11	12	2-6	7-11	12	2-6	7-11	12
$\delta^{11}B/(J_{BH})$ ppm/(Hz)									
CH_2CL_2		-8.9(144)	6.6(133)	-9.7(137)	-8.7(132)	8.2(133)	-6.6(135)	-8.1(137)	10.4(138)
THF	-9.4(160)	-8.3(139)	7.4(133)	-9.6(146)	-8.2(139)	9.5(135)	-7.1(142)	-8.1(134)	11.2(135)
$C_2D_2Cl_2$		-8.9(136)	6.6(131)	-10.0(142)	-9.0(138)	7.9(146)	-6.9(165)	-8.4(154)	10.1(159)
δ^1H/ppm									
$C_2D_2Cl_2$	1.53	1.88	2.92	1.65	2.38	4.43	2.09	4.33	9.61

Table 2 $^{11}B-$, ^1H-NMR-data of $EB_{11}H_{11}^{-}$ (E= BH^-, N^3), P, As, Sb, Bi)

Topological and Geometrical Aspects of the Icosahedral Structural Units in Elemental Boron and Boron-rich Borides

DEPARTMENT OF CHEMISTRY, UNIVERSITY OF GEORGIA, ATHENS, GEORGIA 30602, USA

The icosahedron is well-known as a building block for discrete boron-rich ions and molecules such as $B_{12}H_{12}^{2-}$ and $C_2B_{10}H_{12}$.[1,2] Such species are characterized by exceptional chemical stability and have been described as three-dimensional aromatic systems analogous to planar polygonal two-dimensional aromatic systems of which benzene is the prototype.[3,4] An important feature of these delocalized icosahedral molecules and ions is the presence of a 12-center core bond formed by overlap of the unique internal or radial orbitals of the vertex atoms, which are all directed towards the center of the icosahedron.

This paper summarizes the extension of this topological approach to the study of the chemical bonding in infinite solid state structures constructed from boron icosahedra, namely allotropes of elemental boron and boron-rich metal borides. Further details of these ideas are provided elsewhere.[5,6,7] These ideas on boron-rich species are also useful for the development of chemically based structural models of icosahedral quasicrystals formed by alloys rich in aluminum, the heavier congener of boron.[5]

Packing icosahedra into three-dimensional space requires distortion of their symmetry from the ideal I_h to D_{3h}. The 12 vertices are partitioned into two types, namely six rhombohedral vertices (R in Figure 1), which define the directions of the rhombohedral axes and form a prolate (elongated) trigonal antiprism and six equatorial vertices (E in Figure 1), which lie in a staggered belt around the equator and form an oblate (flattened) trigonal antiprism.

In the simple (α) rhombohedral allotrope of boron[8] all boron atoms are part of discrete icosahedra. In a given B_{12} icosahedron the external orbitals of the rhombohedral borons are each used to form a two-center bond with a rhombohedral boron of an adjacent B_{12} icosahedron and the external orbitals of the equatorial borons are each used to form a three-center bond with equatorial borons of two adjacent B_{12} icosahedra. The available $(12)(3) = 36$ electrons from an individual B_{12} icosahedron are fully used as follows:

Skeletal bonding	
12-center core bond	2 electrons
12 2-center surface bonds: (12)(2) =	24 electrons
External bonding	
(a) Rhombohedral borons:	
1/2 of six 2-center bonds: (6/2)(2) =	6 electrons
(b) Equatorial borons:	
1/3 of six 3-center bonds: (6/3)(2) =	4 electrons
Total electrons required	*36 electrons*

α-Rhombohedral boron thus has a closed-shell electronic configuration.

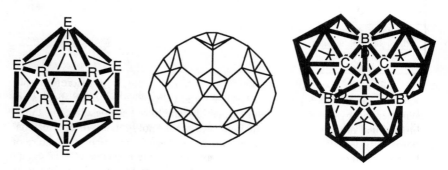

Figure 1 The icosahedron with its rhombohedral (R) and equatorial (E) vertices (left), the Samson complex (center), and the 28-vertex polyhedron linking three Samson complexes in β-rhombohedral boron (right).

The structure of β-rhombohedral boron is considerably more complicated than that of α-rhombohedral boron. The basic building block is a B_{84} Samson complex (Figure 1). Within the B_{84} Samson complex, the external orbital of each of the twelve boron atoms of a central B_{12} icosahedron forms a two-center bond with the external orbital of an apical boron atom of a B_6 pentagonal pyramid ("half-icosahedron") leading to the $B_{12}(B_6)_{12} = B_{84}$ stoichiometry of the Samson complex. The external surface of a B_{84} Samson complex is a B_{60} truncated icosahedron similar to the famous C_{60} truncated icosahedron of fullerene. The ideal I_h symmetry of the Samson complex is distorted to D_{3h} in the rhombohedral lattice environment similar to the B_{12} icosahedron in α-rhombohedral boron discussed above.

Despite the considerably greater complexity in the structure of β-rhombohedral boron relative to α-rhombohedral boron the structural principles for both of these boron allotropes are quite similar in that the rhombohedral linkages between adjacent polyhedral building blocks are two-center linkages and the equatorial linkages are three-center linkages (Table 1). β-Rhombohedral boron[9] has a rhombohedral packing of B_{84} Samson complexes linked by B_{10}

polyhedra and an interstitial boron atom so that the fundamental structural unit is $B_{84}(B_{10})_{6/2}B = B_{105}$. The rhombohedral B_6 pentagonal pyramids each overlap with analogous B_6 rhombohedral pentagonal pyramids of adjacent B_{84} Samson complexes to form six new B_{12} icosahedral cavities. The equatorial B_6 pentagonal pyramids each overlap with analogous B_6 equatorial pentagonal pyramids of two adjacent B_{84} Samson complexes by means of an additional B_{10} unit to form new B_{28} polyhedra (Figure 1) serving as a "glue."

Table 1 Analogy between the packing of units of ideal icosahedral symmetry in α- and β-rhombohedral boron

	α-Rhombohedral	*β-Rhombohedral*
Icosahedral unit	Icosahedron	Truncated icosahedron
Number of vertices	12	60
Rhombohedral linkages	2-center bonds	Icosahedral cavities
Equatorial linkages	3-center bonds	Ideal B_{28} polyhedra from three fused icosahedra

The B_{28} "glue" in β-rhombohedral boron has local C_{3v} symmetry and is formed by fusion of three icosahedra. One vertex (A in Figure 1) is shared by all three icosahedra and four vertices (B and D in Figure 1) are shared by two icosahedra leading to the $3(B_7B_{4/2}B_{1/3}) = B_{28}$ stoichiometry. The 28 boron atoms furnish a total of $(4)(28) = 112$ valence orbitals of which 24 (i.e., one orbital/boron atom except for borons A and B in Figure 1) are required for external bonding. This leads to the following counting scheme for electrons and orbitals:

24 external 2-center bonds:	24 orbitals	24 electrons
3 × 12-center core bonds:	36 orbitals	6 electrons
26 surface bonds	52 orbitals	52 electrons
Total required	*112 orbitals*	*82 electrons*

The closed shell configuration is thus B_{28}^{2+} with $(28)(3) - 2 = 82$ valence electrons.

A complicating feature of β-rhombohedral boron is the partial occupancies of some of the sites and the presence of interstitial atoms as follows:
(1) Three of the boron vertex sites of the B_{10} unit linking three B_6 pentagonal pyramids to form the B_{28} polyhedron (Figure 1), namely the D vertices, are only partially occupied (73.4%).
(2) An interstitial boron atom is located within bonding distance of six of the partially occupied D vertices corresponding approximately to an isolated tetracoordinate boron (i.e., $(0.734)(6) = 4.4$). These interstitial boron atoms each have a B^- closed shell configuration similar to the boron atoms in BH_4^- and $B(C_6H_5)_4^-$.
(3) Each B_{84} Samson complex has a partially occupied (24.8%) interstitial site.

With these considerations in mind the following scheme can be used to count borons and electrons in β-rhombohedral boron:

	Boron Atoms	Net Charge
Central B_{12} icosahedron	12	–2
6/2 Rhombohedrally located peripheral B_{12} icosahedra:		
(6/2)(12) =	36	
(6/2)(–2) =		–6
6/3 Equatorially located peripheral B_{27} polyhedra:		
(6/3)(27) =	54	
(6/3)(+2) =		+4
1 B(15) interstitial boron atom:	1	–1
(0.25)(6) = 1.5 B(16) interstitial boron atoms:		
(1.5)(1) =	1.5	
(1.5)(+3) =		+4.5
Total boron atoms and overall net charge:	*104.5*	*–0.5*

The net charge of –0.5 for a 313.5 (= (3)(104.5)) valence electron unit can be assumed to be zero within experimental error for partial occupancies, etc. Thus β-rhombohedral boron has a closed shell electronic configuration like α-rhombohedral boron.

Boron icosahedra are also found in boron-rich metal borides. In borides of electropositive metals such as Mg_2B_{14}, $LiAlB_{14}$, $MgAl_{2/3}B_{14}$, and $NaB_{0.8}B_{14}$ the electropositive metals form cations such as Li^+, Na^+, Mg^{2+}, etc. The boron subnetwork then has a corresponding negative charge, e.g. $B_{14}^{2-} = (B_{12}^{2-})(B^-)_2$. Lanthanides form very boron-rich borides of the stoichiometry LnB_{66} with very complicated structures. Thus the unit cell of YB_{66} has approximately 24 Y atoms and 1584 B atoms and contains an "icosahedron of icosahedra," i.e., $B_{12}(B_{12})_{12} = B_{156}$. The remaining boron atoms in YB_{66} are statistically distributed in channels that result from packing the $B_{12}(B_{12})_{12}$ units and form non-icosahedral cages which are not readily characterized.

REFERENCES

1.   E. L. Muetterties, ed., *Boron Hydride Chemistry*, Academic Press, New York, 1975.
2.   R. N. Grimes, *Carboranes*, Academic Press, New York, 1970.
3.   J.-I. Aihara, *J. Am. Chem. Soc.*, 1978, *100*, 3339.
4.   R. B. King and D. H. Rouvray, *J. Am. Chem. Soc.*, 1977, *99*, 7834.
5.   R. B. King, *Inorg. Chim. Acta*, 1991, *181*, 217.
6.   R. B. King, *Inorg. Chim. Acta*, 1992, *198–200*, 841.
7.   R. B. King, *Applications of Graph Theory and Topology in Inorganic Chemistry and Coordination Chemistry*, CRC Press, Boca Raton, Florida, 1993.
8.   B. F. Decker and J. S. Kasper, *Acta Crystallogr.*, 1959, *12*, 503.
9.   J. L. Hoard, D. B. Sullenger, C. H. L Kennard, and R. E. Hughes, *J. Solid State Chem.*, 1970, *1*, 268.

Kinetic for the Reaction of Polyhedral Boron Compounds with Alkynyl Pyrimidine Derivatives

Dae Dong Sung and Tae Seop Uhm

DEPARTMENT OF CHEMISTRY, DONG-A UNIVERSITY, PUSAN 604-714, KOREA

1. INTRODUCTION

Dicarba-*closo*-dodecaborane (carborane) has the geometry of a regular icosahedron with two carbon atoms at adjacent vertices. Among the icosahedral cage compounds of carborane, *o*-carboranes are prepared easily by reaction of an acetylene with decaborane in the presence of an appropriate Lewis base.

Recently Reynolds has reported that *ortho*-carboranyl pyrimidine derivatives has been synthesized by reaction of decaborane with pentanoate derivatives. In earlier works Hyatt and Todd accomplished a nitrile insertion reaction to neutral boron hydrides. The reactions of neutral boron hydrides with nitriles have usually involved initial electrophilic addition at nitrogen and have resulted in the production of simple base adducts.

The process of the reaction of decaborane with alkyne in the presence of nitrile solvent might be shown a competitive reaction mechanism. The present work investigates the possibility to occur a competitive reaction by the two nitriles.

2. Result and Discussion

The reactions of decaborane with alkynyl pyrimidines in acetonitrile, toluene and acetonitrile-toluene binary solvent mixtures are shown to be catalyzed by a factor of base-dependent of acetonitrile concentration. The parallel second order reaction was observed by reaction of *nido*-$B_{10}H_{14}$ with alkynyl pyrimidine in acetonitrile and generated *nido*-$B_{10}H_{12}C(MeNH_2)$ and *closo*-$B_{10}H_{10}CHCR$ competitively from the reaction.

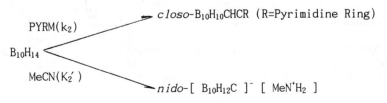

The reaction mechanism of a fair correlation with a parallel second order is explained as the following reaction steps.

$nido$-B$_{10}$H$_{14}$ + 2CH$_3$CN $\longrightarrow$ $arachno$-B$_{10}$H$_{12}$(CH$_3$CN)$_2$ + H$_2$

$arachno$-B$_{10}$H$_{12}$(CH$_3$CN)$_2$ $\longrightarrow$ $nido$-B$_{10}$H$_{12}$(CH$_3$CN) + CH$_3$CN

$nido$-B$_{10}$H$_{12}$(CH$_3$CN) $\longrightarrow$ $nido$-B$_{10}$H$_{12}$C$^-$(MeNH$_2^+$)

$nido$-B$_{10}$H$_{12}$(CH$_3$CN) + R-C≡C-R $\longrightarrow$ $closo$-B$_{10}$H$_{10}$CHCR

The rate constants of the reaction of alkynyl pyrimidine showed higher than those of the reaction of acetonitrile in parallel second order reactions.   Reactivity of alkynyl pyrimidines for decaborane showed a tendency ACMP > AMPP > DMPP > MPPD and this tendency was in accord with the basicity of the alkynyl pyrimidines.

Table 1. Rate constants(k_2) and activation parameters for the reaction of alkynyl pyrimidines with decaborane in acetonitrile

Alkynyl	Rate Constant(10^3 M^{-1}s^{-1})[a]			$\Delta H^{\neq}$	$\Delta S^{\neq}$	$\Delta G^{\neq}$
Primidine	333K	338K	343K	(Kcalmol^{-1})	(e.u)	(Kcalmol^{-1})
ACMP	4.53	15.1	66.3	60.2	111.0	22.7
AMPP	4.27	12.8	50.1	55.2	95.8	22.8
DMPP	4.09	12.0	42.4	52.4	87.4	22.8
MPPD	3.87	10.9	38.2	51.3	83.9	22.9

[a]Calculated from the equation, k_{obs} = k_0 + k_2[PYRM] (1) substituted the pseudo-first order rate constants(k_{obs}) which were obtained from the data of time-resolved UV/vis spectra which measured at 252nm (= λ_{max}) into the equation(1).

Table 2. Rate constants(k_2) and activation parameters for the reaction of alkynyl pyrimidines with decaborane in toluene

Alkynyl	Rate Constant(10^3 M^{-1}s^{-1})[a]			$\Delta H^{\neq}$	$\Delta S^{\neq}$	$\Delta G^{\neq}$
Primidine	333K	338K	343K	(Kcalmol^{-1})	(e.u)	(Kcalmol^{-1})
ACMP	3.74	10.9	29.5	46.2	68.9	22.9
AMPP	3.67	9.50	26.2	43.9	61.9	23.0
DMPP	3.42	8.85	23.3	42.9	58.6	23.1
MPPD	3.19	8.09	20.0	41.0	52.9	23.1

[a]Calculated from the equation, k_{obs} = k_0 + k_2[PYRM] (1) substituted the pseudo-first order rate constants(k_{obs}) which were obtained from the data of time-resolved UV/vis spectra which measured at 292nm (= λ_{max}) into the equation(1).

Table 3. Rate constants(k_2) and activation parameters for the reaction of alkynyl pyrimidines with decaborane in toluene-acetonitrile binary solvent mixtures

Alkynyl Primidine	MeCN content (v/v)%	Rate Constant(10^3 M^{-1}s^{-1})^a			$\Delta H^{\neq}$ (Kcalmol^{-1})	$\Delta S^{\neq}$ (e.u)	$\Delta G^{\neq}$ (Kcalmol^{-1})
		333K	338K	343K			
ACMP	30	3.92	12.0	41.7	53.0	89.1	22.8
	50	4.11	12.7	49.3	55.7	97.3	22.8
	70	4.28	13.9	58.2	58.5	105.9	22.7
AMMP	30	3.84	10.7	37.4	51.0	82.9	22.9
	50	3.93	11.0	40.8	52.4	87.3	22.9
	70	4.05	12.1	45.6	51.7	85.5	22.8
DMPP	30	3.65	9.23	30.9	47.8	73.2	23.0
	50	3.72	10.5	34.1	49.6	78.9	22.9
	70	3.79	11.4	38.6	52.0	86.1	22.9
MPPD	30	3.34	8.11	22.5	42.6	57.7	23.1
	50	3.49	8.90	26.6	45.4	66.1	23.0
	70	3.67	9.75	31.1	47.8	73.4	23.0

^{a}Calculated from the equation, $k_{obs} = k_0 + k_2[PYRM]$ (1) substituted the pseudo-first order rate constants(k_{obs}) which were obtained from the data of time-resolved UV/vis spectra which measured at 242nm($= \lambda$ max in 70(v/v)% MeCN content), 267nm($= \lambda$ max in 50(v/v)% MeCN content) and 285nm ($= \lambda$ max in 30(v/v)% MeCN content) into the equation(1).

Table 4. Rate constants(k_2') and activation parameters for the reaction of decaborane with acetonitrile

Rate constant(10^3 M^{-1}s^{-1})^a			$\Delta H^{\neq}$ (Kcalmol^{-1})	$\Delta S^{\neq}$ (e.u.)	$\Delta G^{\neq}$ (Kcalmol^{-1})
333K	338K	343K			
2.94	7.61	21.0	45.3	63.4	23.5

^{a}Calculated from the equation, $k_{obs} = k_0 + k_2[PYRM]$ (1) substituted the pseudo-first order rate constants(k_{obs}) which were obtained from the data of time-resolved UV/vis spectra which measured at 250 nm ($= \lambda$ max) into the equation(1).

REFERENCES
1. (a)H.C.Brown, "Boranes in Organic Chemistry" , Cornell University Press., Ithaca, New York(1972); (b)E.L.Muetterties, "The Chemistry of Boron and Its Compounds" Wiley, New York (1967).
2. R.C.Reynolds, T.W.Trask and W.D.Sedwick, *J. Org. Chem.*, 56, 2391 (1991).

3.   D. E. Hyatt, D. A. Owen and L. J. Todd, *Inorg. Chem.*, 5, 1749(1966).
4.   M. Yamamoto, *J. Chem. Soc.*, Perkin Trans., 1, 582(1981).
5.   E. Kuss and J. Moos, *Naturwissenschaften*, 48, 73(1961).
6.   M. Walter and L. Ramaley, *Anal. Chem.*, 45, 165(1973).
7.   E. A. Guggenheim, *Philos. Mag.*, 2, 538(1926).
8.   (a)J. W. Moore and R. G. Pearson, "Kinetics and Mechanism", John
      Wiley & Sons, New York, 300(1980); (b)C. H. Bamford and
      C. F. H. Tipper, "Chemical Kinetics", Elsevier, Amsterdam,
      Vol. 2, 14(1969).
9.   F. Nagy, "The Kinetics of Contact Catalytic Reactions",
      Z. G. Szabo (Ed), Hungarian Academy of Sciences, Budapest, 387
      (1966).
10.   H. Ketz, W. T. Tjarks and D. Gabel, *Tetrahedron Lett.*, 31, 4003
      (1990).
11.   T. L. Heying, J. W. Ager Jr., S. L. Clark, D. J. Mangold, H. L.
      Goldstein, M. Hillman, R. J. Polak and J. W. Szymanski, *Inorg.
      Chem.*, 2, 1089(1963).
12.   H. Shroeder, T. L. Heying and J. R. Reiner, *Inorg. Chem.*, 2, 1092
      (1963).
13.   R. Hoffman and W. N. Lipscomb, *J. Chem. Phys.*, 36, 3489(1962).
14.   F. E. Wang, R. Lewin and W. N. Lipscomb, *Proc. Natl. Acad. Sci. U. S.*,
      47, 996(1961).
15.   M. M. Fein, J. Bobinski, N. Mayers, N. Schwartz and M. S. Cohen,
      Inorg. Chem., 2, 1111(1963).
16.   M. M. Fein, D. Grafstein, J. E. Paustian, J. Bobinski, B. M.
      Lichstein, N. Mayers, N. N. Schwartz and M. S. Cohen,
      Inorg. Chem., 1115(1963).
17.   A. Arafat, G. D. Friesen and L. J. Todd, *Inorg. Chem.*, 22, 3721
      (1983).
18.   G. A. Olah, K. Wade and R. E. Williams, " Electron Defficient Boron
      and Carbon Clusters", John Wiley, New York, Chap. 2, (1991).

Structure and Reactions of Tetrahydroborates: New Results with a Fascinating Ligand

Heinrich Nöth*, Martina Thomann, Mathias Bremer, and Gerhard Wagner

INSTITUTE FOR INORGANIC CHEMISTRY, UNIVERSITY OF MUNICH, MEISERSTR, I, 80333 MUNICH, GERMANY

Tetrahydroborates, especially $NaBH_4$ and $LiBH_4$, are important reagents of significant industrial value [1]. They span the range from ionic compounds such as $CsBH_4$ to highly volatile covalent species such as $Al(BH_4)_3$ or $Zr(BH_4)_4$. The bonding in the latter compounds is characterized by electron deficient, multicenter bonding. In this respect the BH_4^- anion is an unusual ligand since it has no pair of free electrons available as provided for bonding by most other ligands.

In mononuclear tetrahydroborates $M(BH_4)_n$ and $L_mM(BH_4)_n$ the BH_4^- ligand can establish a bond to the metal center either via one, two or three of its hydrogen atoms (μ_1-, μ_2- and μ_3-BH_4). In di-, tri- and multinuclear complexes even all four hydrogens of the BH_4^- ligand may be involved in bonding. Although the bonding pattern can be established from IR data, especially when only one kind of bonding is present, this becomes increasingly difficult when different types of tetrahydroborate groups are present. Therefore, other methods must be employed to ascertain the bonding of the BH_4 groups in compounds under consideration. ^{11}B NMR is usually not very helpful, but X-ray diffraction proved to be most important, although the bonding may be different in solution especially in coordinating solvents. We report here new results in the field of titanium and alkaline earth tetrahydroborates.

Titanium Tetrahydroborates and Amido Titanium Hydrides
It is well known[2] that titanium(IV) alkoxides react with diborane in ether solvents to produce finally titanium(III) tetrahydroborate as ether solvates. Following the reaction in tetrahydrofurane, several distinct steps can be distinguished as shown in scheme I.

$$Ti(OR)_4 \rightarrow Ti(OR)_3BH_4 \rightarrow Ti(OR)_2(BH_4)_2 \rightarrow Ti(OR)(BH_4)_2 \rightarrow Ti(BH_4)_3 \quad (I)$$

The rate of $Ti(BH_4)_3$ formation depends on the nature of the groups R and increases in the series $Ph_3C < Me_3C < Me_2HC < Et$. Compounds of type $Ti(OR)_3BH_4$ and $Ti(OR)_2(BH_4)_2$ may also be prepared from the corresponding chloride by metathesis with $LiBH_4$ or $NaBH_4$ as pure, solvent free compounds.

$Ti(OR)(BH_4)_2$ was obtained as $Me_3COTi(BH_4)_2 \cdot 2THF$ as well as $[Me_3COTi(BH_4)_2]_2$. In each case μ_2-BH_4 groups are present. The molecular structure of the former is depicted together with the structure of $Ti(BH_4)_3 \cdot 2THF$ in Figure 1 which also features μ_2-BH_4 groups. In each case the B and O atoms form a trigonal bipyramidal array around the Ti center.

Amido titanium(IV) tetrahydroborates are readily accessible via metathesis from $(R_2N)_3TiCl$ or $(R_2N)_2TiCl_2$ and $LiBH_4$ in ether. $(R_2N)_3TiBH_4$ compounds contain μ_3-BH_4 groups. On pyrolysis they produce $(R_2N)_2BH$ and $R_2NH\cdot BH_3$ besides hydrogen. The stoichiometry of the reaction suggests the formation of amido titanium hydrides. This is further supported by reactions of these tetrahydroborates with amines or phosphines. At the present time, no well defined amido titanium hydride could be prepared by this route.

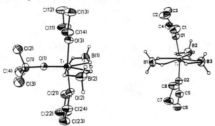

Figure 1: Molecular structure of $Me_3COTi(BH_4)_2\cdot 2THF$ and $Ti(BH_4)_3\cdot 2\ THF$.

However, when $(Et_2N)_3TiBH_4$ was used to reduce benzophenone a reaction according to scheme II was observed. Tris(diethylamino)titaniumhydride is an orange

$$(Et_2N)_3TiBH_4 + 2\ Ph_2CO \rightarrow (Et_2N)_3TiH + HB(OCHPh_2)_2 \quad (II)$$

colored oil and shows a ^{1}H-NMR signal at 0.2 ppm for its TiH proton. Similar reactions with zirconium tetrahydroborates are more complex and lead to multinuclear zirconium complexes as has been shown recently[3]. Reactions of this kind open a new area of research in transition metal hydride chemistry.

Tetrahydroborate Complexes of Calcium, Strontium and Barium
Very little is known about the structures of the alkaline earth metal tetrahydroborates. There are only two examples, $Mg(BH_4)_2\cdot 3THF$[4] and $Ca(BH_4)_2\cdot 2$ monoglym[5]. The former contains μ_2-, the latter μ_3-BH_4 groups. Preliminary results for $Sr(BH_4)_2\cdot 2THF$ indicate a linear array of Sr centers with bridging BH_4 groups and THF in *trans* positions whereby the Sr atoms are hexacoordinated by B and O atoms.

Treating $M(BH_4)_2$ (M = Ca, Sr, Ba) in THF with diglyme or 18-crown-6 (= L) the tetrahydroborates $M(BH_4)_2\cdot 2$ diglyme and $M(BH_4)_2\cdot L$ were obtained. Inspite of the

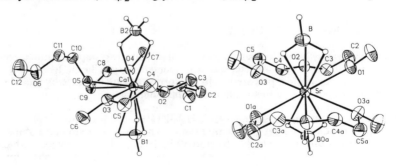

Figure 2: Molecular structures of $M(BH_4)_2\cdot 2$ diglyme (M = Ca, Sr). H atoms on carbon atoms omitted.

identical composition of the $M(BH_4)_2 \cdot 2$ diglyme the structure of the Ca complex is different from the Sr and Ba complexes, which are isostructural. Figure 2 shows two μ_3-BH_4 groups for the Sr complex with all oxygen atoms of the coordinated diglyme molecules bonding to the metal center. In contrast, only five oxygen atoms coordinate in the Ca case. More importantly, there are two types of BH_4 groups present in $Ca(BH_4)_2 \cdot 2$ diglyme, one μ_2- and one μ_3-BH_4 group. μ_3-BH_4 groups are also present in the crown complexes which have C. All these compounds have to be regarded as molecular since C_{3v} point group symmetry. They dissolve in nonpolar solvents. In contrast, diethylenetriamine reacts with $M(BH_4)_2$ in THF to give the salts $[M(trien)_2](BH_4)_2$.

Tetrahydroborate Complexes of Magnesium

In spite of many attempts we have so far been unable to obtain single crystals of $Mg(BH_4)_2$. This would be a particularly interesting example because it should be one of the borderline cases between ionic and covalent tetrahydroborates like $Be(BH_4)_2$. However, we have studied several magnesium tetrahydroborates of type $XMg(BH_4) \cdot nD$ (X = OR, NR_2, R; L = OEt_2, THF, glyme).

Alkylmagnesium tetrahydroborates crystallize as dimers from ether solutions, and the molecular structure of $MeMg(BH_4) \cdot OEt_2$ is depicted in Figure 3. It shows the presence of a μ_3-BH_4 group. However, the most notable feature are the bridging methyl groups. Two of its H atoms show agostic interactions with the Mg centers. In solution an equilibrium as described by Scheme III is established as detected by ^{25}Mg NMR spectroscopy, a well known Schlenk equilibrium. Most likely this is the reason why $RMg(BH_4)$ compounds react with ketones preferably with alkylation and not with reduction.

$$2\ RMg(BH_4) \rightleftharpoons MgR_2 + Mg(BH_4)_2 \quad \text{(III)}$$

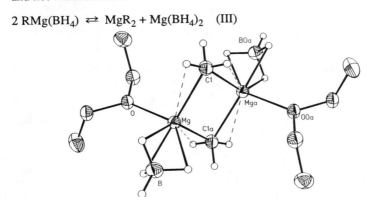

Figure 3: Molecular structure of dimeric $MeMg(BH_4) \cdot OEt_2$.

Organyloxo magnesium tetrahydroborates can be prepared by various methods; many of them crystallize well. Usually $ROMg(BH_4) \cdot nL$ compounds are dimerised via RO bridges, and μ_2- or μ_3-BH_4 groups are detected depending on the coordinated ether molecules. As a rule the coordination number of Mg is six, counting the H atoms of the BH_4 group as a ligand atom or four, if only the oxygen atoms (of RO and Ligand L) and the boron atom are taken into account. Mononuclear RO-$Mg(BH_4)$ compounds can be obtained with sterically demanding RO groups, such as the 2,4,6-tri-tert.-butylphenoxy group. In this case, the phenoxy group is linearily coordinated to the Mg atom, and the short MgO bond suggests a high degree of bond order.

Finally, two results are presented which we find difficulty in rationalizing. Firstly, the reaction of hexamethyldisilazane with $^iPrMgBH_4$ in diethylether gives access to two products, namely the expected $(Me_3Si)_2NMgBH_4 \cdot OEt_2$ but also the unexpected

$Me_3SiOMgBH_4 \cdot OEt_2$. The latter contains a μ_3-BH_4 group as shown in Figure 4. The amido magnesium tetrahydroborate did not crystallize well in contrast to its diglyme adduct which contains a μ_2 BH_4 ligand. Its structure is also represented in Figure 4.

The formation of the siloxy compound is not yet understood: it may result from ether cleavage. There is further evidence for this, since the same reaction, carried out at higher temperatures, produces single crystals which by X-ray structure analysis are shown to be a basic magnesium borohydride $Mg_4O(BH_4)_6 \cdot 4\ OEt_2$. The framework of this molecule is depicted in Figure 5.

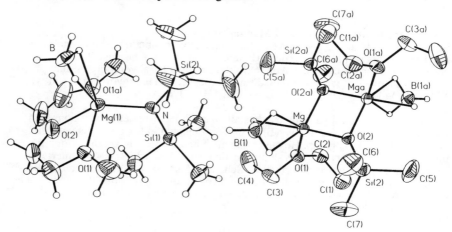

Figure 4: Molecular structures of $(Me_3Si)_2NMgBH_4 \cdot$diglyme and dimeric $Me_3SiOMgBH_4 \cdot OEt_2$.

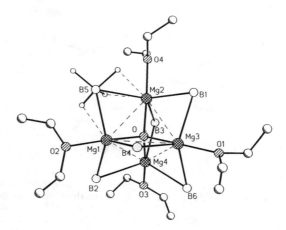

Figure 5: Core of the $Mg_4O(BH_4)_6 \cdot 4\ OEt_2$ molecule. Hydrogen atoms of only one BH_4 group are shown for clarity. Also C bonded H atoms are omitted.

Secondly, in attempts to synthesize alkoxymagnsium tetrahydroborates with an internal ether function methoxyethanol was reacted with $^iPrMgBH_4$ in ether and a reaction according to Scheme IV was expected.

$$^iPrMgBH_4 + MeOCH_2CH_2OH \rightarrow {}^iPrH + MeOCH_2CH_2OMgBH_4 \quad (IV)$$

However, very little isopropane was generated together with some hydrogen gas.The product did not contain a BH_4 group by ^{11}B NMR which, however, indicated the presence of a BR_4 group. X-Ray structure analysis revealed the structure of the compound which is formed in about 25 % yield. Its composition is $[Mg_7O(OCH_2-CH_2OCH_3)_{12}](BEt_4)_2$. The structure of the cation (Figure 6) shows a highly symmetric array of hexacoordinated Mg atoms. The MgO_6 octahedra share common edges.

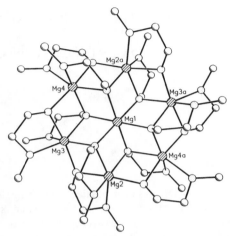

Figure 6: Structure of the cationic part of $[Mg_7(OCH_2CH_2OCH_3)](BEt_4)_2$.

The results described here as well as many newer reports in the literature demonstrate not only a rich a varied structural chemistry of metal tetrahydroborates but also that there must exist many polynuclear complexes with BH_4 groups as ligands. And it is to be expected that the BH_4 group in the latter compounds will act as bridging ligand involving two to four of its hydrogen atoms.

1. R. C. Wade, Speciality Inorganic Chemicals, R. Thompson, Editor, Special Publ. No. 40, Royal Soc. Chem., **1980**. B. D. James, M. G. H. Wallbridge, Prog. Inorg. Chem. **1970**, *11*, 99.
2. B. D. James, M. G. H. Wallbridge, J. Inorg. Nucl. Chem. **1966**, *28*, 2456.
3. J. E. Gozum, S. R. Wilson, G. S. Girolami, J. Am. Chem. Soc. **1992**, *114*, 9483.
4. H. Nöth, Z. Naturforsch. **1982**, *B 37*, 1493.
5. E. Hanecker, J. Moll, H. Nöth, Z. Naturforsch., **1984**, *B 39*, 424.

Kinetic and Mechanistic Studies of Gas-Phase Thermolysis and Cothermolysis Reactions of B_4H_8CO and of B_4H_{10}

S. J. Cranson, R. Greatrex, N. N. Greenwood, and M. Whitehouse

SCHOOL OF CHEMISTRY, UNIVERSITY OF LEEDS, LEEDS LS2 9JT, UK

1 INTRODUCTION

The thermal decomposition of B_4H_{10} has been studied experimentally in considerable detail,[1,2] and there is now general agreement that the initial step involves the reversible rate-determining elimination of di-hydrogen to generate the reactive intermediate $\{B_4H_8\}$ (1).

$$B_4H_{10} \overset{\text{slow}}{\rightleftharpoons} \{B_4H_8\} + H_2 \tag{1}$$

Results of recent *ab initio* calculations employed to evaluate the decomposition pathway are also consistent with this view.[3] Because B_4H_{10} is present in great excess, it has generally been assumed that the second step, which leads to the major initial product, B_5H_{11}, involves the reaction between $\{B_4H_8\}$ and B_4H_{10} (2), but this is by no means well established. Indeed, a simple kinetic analysis of this mechanism reveals an apparent inconsistency between the experimentally observed order of the reaction (unity), and the magnitude of the retardation in the presence of added H_2.[2] The possibility that the B_5H_{11} results from the reaction of $\{B_4H_8\}$ with itself (3) seemed to present a similar problem, despite the fact that this step has already been suggested as the route to B_5H_{11} in the low-temperature synthesis involving hydride-ion abstraction from $[B_4H_9]^-$.[4]

$$\{B_4H_8\} + B_4H_{10} \overset{\text{fast}}{\longrightarrow} B_5H_{11} + \{B_3H_7\} \tag{2}$$

$$\{B_4H_8\} + \{B_4H_8\} \overset{\text{fast}}{\longrightarrow} B_5H_{11} + \{B_3H_5\} \tag{3}$$

Even when the kinetic analysis is revised so that proper account is taken of the significant roles of H_2 and He in the energising and de-energising steps of the "unimolecular" decomposition stage we continue to find[5] that all the experimental observations (order in the presence of H_2, ratio of rates with and without H_2, very similar Arrhenius activation energy for the process with and without H_2) taken together cannot be made consistent with a unique set of model parameters - although each observation taken separately gives satisfactory agreement. This is true for a second step which is either linear [step (2)] or quadratic [step (3)] in $\{B_4H_8\}$. On the basis of the observed rate ratio the models all predict reaction orders in the presence of excess H_2 and He (present for calibration purposes) to be nearer to 2 than to the experimental value of 1. Since the chemical mechanisms seem to lack any plausible rivals, it may be that some aspect of the data is marred.

In seeking alternative ways of resolving this mechanistic problem we are confronted with the difficulty of devising definitive experiments to test the role of the reactive intermediates in this system. Apart from Fehlner's early mass spectrometric studies on reactions of $\{BH_3\}$, there has been very little work of this type.[6] An alternative approach is to study the macroscopic effect on the course of a reaction of adding either Lewis base "scavengers", such as CO, to trap the intermediates and prevent their further participation, or unstable adducts such as BH_3CO, B_3H_7CO and B_4H_8CO, which are thought to decompose readily to generate the intermediate in situ. We have recently probed details of the mechanism of decomposition of hexaborane(12) in this way,[7] and we now describe experiments with tetraborane(10) in which B_4H_8CO is used as a generator of $\{B_4H_8\}$, and CO is used as a scavenger. The thermolysis of B_4H_8CO has been studied qualitatively in the past,[8,9] but there has been no gas-phase kinetic work. In preparation for the cothermolysis work it was therefore important to study the kinetics of its own thermal decomposition to clarify the mechanism of its decomposition and to put this reaction on a more quantitative footing. The main aim of the work described below was to search for additional experimental clues that might illuminate the mechanistic problem outlined above.

2 RESULTS AND DISCUSSION

Thermolysis of B_4H_8CO Alone and in the Presence of Added Hydrogen

Reactions were carried out in the presence of a large background of an inert gas mixture comprising He, Ar and Kr in the ratio 98:1:1, and having a partial pressure of 100 mmHg. The changing composition of the gaseous mixture was monitored continuously by mass spectrometry as described elsewhere,[7] and a typical product analysis at 313 K is shown in Figure 1. The decomposition is seen to be complex, even in the early stages, the main products being B_5H_{11} and H_2, with B_2H_6, BH_3CO, and (in contrast to earlier work[8,9]) B_6H_{12} also being produced in appreciable amounts. $B_{10}H_{14}$ was observed in many of the runs, but the data could not be used to give a reliable quantitative analysis because the compound may not be completely volatile at 313 K. Initial-rate data for runs at 13 different initial pressures were recorded in the form of log-log plots and the slopes used to determine the order of the reaction at 313 K with respect to the concentration of B_4H_8CO. The values derived from the rate of consumption of B_4H_8CO and the rate of production of B_5H_{11} were both very close to unity (1.01 $\pm$ 0.05 and 1.2 $\pm$ 0.1, respectively). From measurements of the initial rate of consumption of B_4H_8CO over the pressure range

3.1 - 11.2 mmHg and at temperatures of 293 - 313 K, the activation energy was found to be 86 ± 3 kJ mol^{-1} and the pre-exponential factor 2.9×10^{11} s^{-1}, which is not unrealistic for a first order reaction. Measurements of the rate of production of B_5H_{11} gave values of 86 ± 6 kJ mol^{-1} and 6.6×10^{11} s^{-1}, respectively, for these Arrhenius parameters.

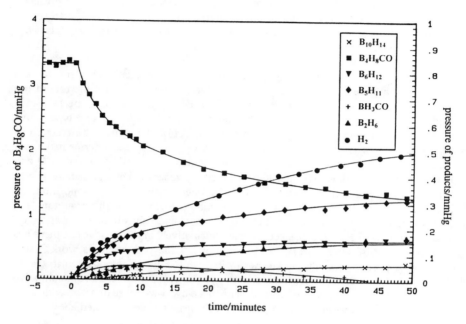

Figure 1 Pressure *vs.* time plots for the decomposition of B_4H_8CO at T = 313 K.

To study the effects of added H_2, runs were carried out at T = 313 K with H_2/B_4H_8CO pressure ratios of 4:1, 8:1 and 12:1 for an initial borane pressure of ca. 3.0 mmHg. As the hydrogen pressure was increased the initial rate of decomposition increased only marginally, and this effect became more pronounced towards the end of the reaction. From these observations it is clear that the order of the initial reaction with respect to H_2 is effectively zero, indicating that there is no direct reaction with the adduct and that any equilibrium involving elimination of H_2 is unimportant. Although a twelve-fold excess of added hydrogen has only a slight effect on the initial rate of decomposition of B_4H_8CO, the product analysis was altered significantly. There was a substantial build-up of B_4H_{10} (at the expense of B_5H_{11}), confirming that B_4H_8CO decomposes predominantly by elimination of CO and formation of $\{B_4H_8\}$ [step (4)], which is then converted to B_4H_{10} by step (-1). The fact that the initial rate of decomposition of B_4H_8CO is not dramatically increased by this scavenging process

$$B_4H_8CO \ \rightleftharpoons \ \{B_4H_8\} + CO \tag{4}$$

indicates that step (4) is probably rate controlling. On the other hand the observation that the initial rate is not substantially reduced is also significant. It suggests that the second step of the decomposition probably involves the interaction of $\{B_4H_8\}$ with itself, and not with B_4H_8CO. If this were the mechanism, the removal of $\{B_4H_8\}$ by added H_2 would compete with this second step and reduce the initial rate by a factor of two. The increase in the rate of consumption of B_4H_8CO over a longer period of time is attributed to the removal of the $\{B_4H_8\}$ generated in step (4) *via* step (-1). Step (-1) is in competition with step (-4) which, in the absence of added H_2, becomes progressively more important (with the build up of CO in the system) and leads to an apparent slowing of the rate of decomposition of B_4H_8CO.

A further interesting point is that only about 40% of the boron from B_4H_8CO ends up in volatile products, regardless of whether added H_2 is present or not. This is not usually the case with other borane decompositions, where the addition of H_2 tends to increase the conversion to volatiles.[4] In the B_4H_{10} decomposition, for example, this effect has been attributed to the interaction of H_2 with the reactive intermediate $\{B_3H_7\}$ to form B_2H_6. In the present reaction it seems likely that any reactive intermediates are scavenged by CO rather by H_2, and converted not to B_2H_6 but to non-volatile CO-containing polymeric solids.

Thermolysis of B_4H_8CO in the Presence of B_4H_{10}

It was hoped that by thermolysing B_4H_8CO in the presence of B_4H_{10}, at a temperature at which the latter is relatively stable, it might be possible to establish whether or not $\{B_4H_8\}$ reacts with B_4H_{10}, and thereby shed light on the precise sequence of events in the thermolysis of B_4H_{10} alone. Reactions were carried out at 40 ºC with B_4H_8CO at a pressure of 3.5 mmHg and B_4H_{10}/B_4H_8CO pressure-ratios of 1:1, 2:1 and 3:1. In all cases the B_4H_{10} remained essentially unaffected while the B_4H_8CO decomposed as though it were alone. This indicates that $\{B_4H_8\}$ does not react with B_4H_{10} under the conditions of this experiment, even when the latter is present in a three-fold excess. It therefore seems highly unlikely that $\{B_4H_8\}$ reacts with B_4H_{10} in the second step of the B_4H_{10} thermolysis. The conditions of this experiment do differ from those of the B_4H_{10} thermolysis in that the temperature is now lower, and an additional species, B_4H_8CO, is of course present. However, reactions involving highly reactive intermediates such as $\{B_4H_8\}$ are likely to be rapid and to have low activation energies; as such they might be expected to occur at relatively low temperatures. It would also be very surprising indeed if B_4H_8CO, in a 1:1 mixture with B_4H_{10}, were 100% efficient in competing for $\{B_4H_8\}$. The logical conclusion is that B_5H_{11} results from the reaction of $\{B_4H_8\}$ with itself.

Thermolysis of B_4H_{10} in the Presence of Excess CO

This particular reaction has been studied previously, but under conditions where the products were removed continuously from the system.[10] We have re-investigated the cothermolysis over the temperature range 45 - 75 ºC by continuous capillary sampling and mass spectrometric monitoring of the otherwise undisturbed reaction mixture. We have done this to generate a set of accurate rate data which could be compared directly with our previous measurements on the thermolysis of B_4H_{10} alone.[11] For a series of 11 measurements at 333 K, B_4H_8CO was observed to build up in the initial stages of the reaction, indicating that CO is very efficient in trapping $\{B_4H_8\}$. The mean ratio of production of B_4H_8CO to consumption of B_4H_{10} was found to be 0.85 ± 0.28, and the Arrhenius parameters calculated with respect to B_4H_{10} consumption were unchanged, within experimental error, from the values for the pure borane. These results confirm that under these conditions B_4H_{10} decomposes predominantly by the rate-determining elimination of H_2, and that the $\{B_4H_8\}$ does not then go on to react with a second mole of B_4H_{10}. If it did, the absolute rate of decomposition of B_4H_{10} would be halved in the presence of excess CO.

3. CONCLUSIONS

As a result of this work we conclude that the second step in the thermal decomposition of B_4H_{10} does not involve the reaction of $\{B_4H_8\}$ with B_4H_{10}. The intermediate, generated by the rate-controlling elimination of H_2 from B_4H_{10}, probably reacts with itself to give B_5H_{11} as the major product. It also seems likely that the same process occurs in the thermolysis of the adduct B_4H_8CO following the initial rate-determining elimination of CO.

4. ACKNOWLEDGEMENTS

We thank the SERC, the U.S. Army Research and Standardization Group (Europe) and Borax Consolidated Ltd. for financial support, and Mr Darshan Singh for experimental assistance.

5. REFERENCES

1.   N.N. Greenwood and R. Greatrex, Pure Appl. Chem., 1987, 59, 857, and references therein.
2.   M.D. Attwood, R. Greatrex and N.N. Greenwood, J. Chem. Soc., Dalton Trans., 1989, 391, and references therein.
3.   M.L McKee, J. Am. Chem. Soc., 1990, 112, 6753.
4.   M.A. Toft, J.B. Leach, F.L. Himpsl and S.G. Shore, Inorg. Chem., 1982, 21, 1952.
5.   T. Boddington and R. Greatrex, unpublished work.
6.   T.P. Fehlner, in "Boron Hydride Chemistry", E.L. Muetterties, ed., (Academic Press, New York, 1975), p.175.
7.   R. Greatrex, N.N. Greenwood and S.D. Waterworth, J. Chem. Soc., Dalton Trans., 1991, 643.
8.   J.R. Spielman and A.B. Burg, Inorg. Chem., 1963, 2, 1139.
9.   R.E. Hollins and F.E. Stafford, Inorg. Chem., 1970, 9, 877.
10.   G.L. Brennan and R. Schaeffer, J. Inorg. Nucl. Chem., 1961, 20, 205.
11.   R. Greatrex, N.N. Greenwood and C.D. Potter, J. Chem. Soc., Dalton Trans., 1986, 81.

The Chemistry of Elemental Sulfur and Selenium with the System THF–BH₃/BH₄⁻

H. Binder, K. Wolfer, H. Loos, H. Borrmann, A. Simon,
R. Ahlrichs, H.-J. Flad, and A. Savin

INSTITUT FÜR ANORGANISCHE CHEMIE DER UNIVERSITÄT, 70569
STUTTGART, GERMANY

Introduction

The reaction of $NaBH_4$/THF-BH_3 with elemental sulfur produces $Na[(BH_2)_5S_4]$ with hydrogen evolution. These nucleophilic degradation reactions proceed via $[H_3B-\mu_2-S(B_2H_5)]^-$, **1**, and the intermediate **3** by loss of BH_4^- and cyclization. **4** builds up a noradamantane skeleton B_5S_4. Thermal decomposition of **1** yields $Na_2[(BH_2)_6S_4]$, **5**, which possesses the adamantane skeleton B_6S_4. Furthermore **1** reacts with $NaBH_4$ in triglyme to form $[Na \cdot Triglyme]_2[S(BH_3)_4]$, **6**. The reaction of $NaBH_4$ with elemental selenium produces $Na_2[H_3B-Se-Se-BH_3]$, **7**, which on thermal decomposition forms $Na_2[(BH_2)_6Se_4]$, the Se analogue of **5** . **7** reacts with B_2H_6 quantitatively to yield $[H_3B-\mu_2-Se(B_2H_5)]^-$, **10**, which is the Se analogue of **1**. The Se analogue of **4** can be obtained by the reaction of $NaBH_4$/THF.BH_3 with elemental selenium, $[(BH_2)_5Se_4]^-$, **2**.

Whilst B_2H_6 is an electrophilic agent which preferentially attacks a molecule at a position of high electron density BH_4^- is nucleophilic. The sulfur atoms in elemental sulfur and in many chain-like sulfur compounds mostly react as electrophilic centers; they are susceptible to nucleophilic attack via a S_{N2} mechanism, scheme 1, Fig. 1.

The first step is a nucleophilic attack of the H^--ion on the S_8 ring. Thereby a S-S bond of the ring is broken. The second step is the addition of "BH_3" to the open sulfur chain anion. The following sequence of steps is an interaction between the SH-proton and "BH_3" forming H_2; further addition of "BH_3" to the same terminal S atom stabilizes this first ^{11}B-NMR spectroscopically detectable intermediate. According to scheme 1 eight molecules of type **1** arise from the degradation of the S_8 ring. **1** can be isolated as the

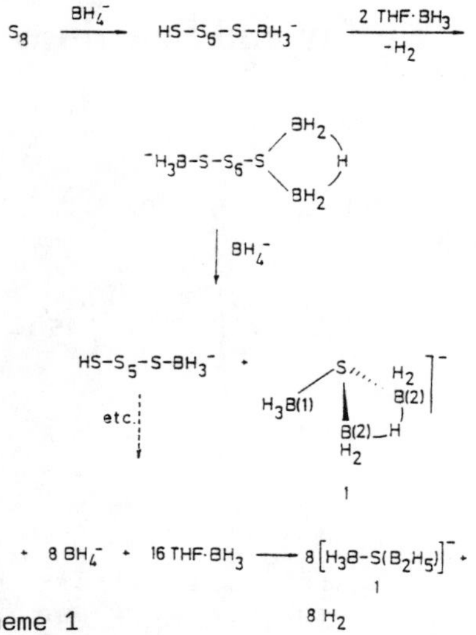

Scheme 1

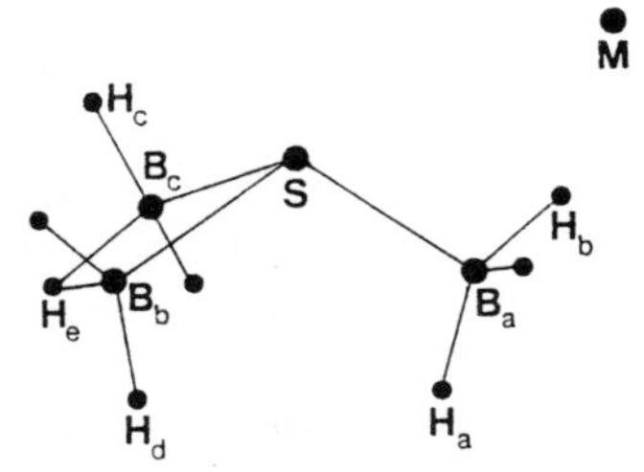

Figure 1 Calculated structure of $[H_3B\text{-}\mu_2\text{-}S(B_2H_5)]^-$, **1**

$(C_6H_5)_4P^+$-salt. If however excess sulfur is used a second type of degradation of the S_8 ring follows by nucleophilic attack of **1**, scheme 2. In contrast to scheme 1 the degradation of the S_8 ring is now confined to an intermediate with still one S-S bond **3**. The postulated compound **3** decomposes and undergoes an intramolecular cyclization thus forming a cage-like anion $[(BH_2)_5S_4]^-$ with the noradamantane skeleton B_5S_4, **4**,[1], Fig. 2. A new cage compound with the adamantane skeleton B_6S_4, **5**, is formed when **1** is heated in vacuum[2], Fig. 3.

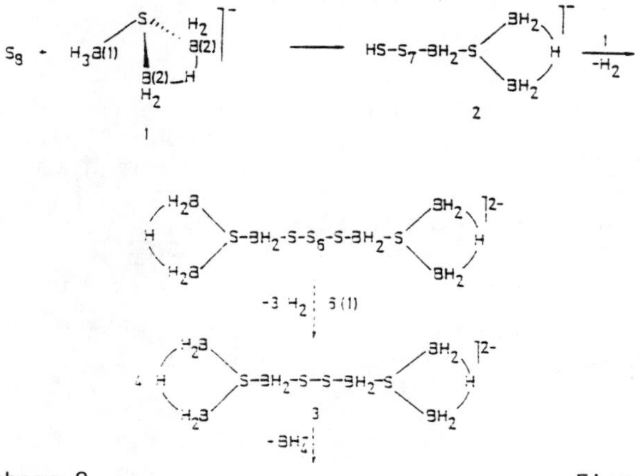

Scheme 2

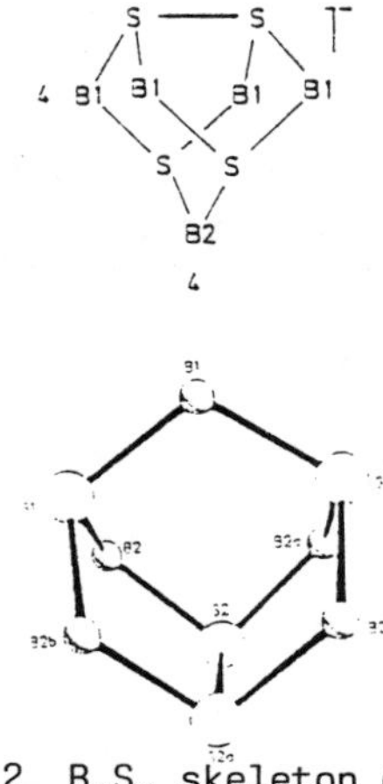

Figure 2 B_5S_4 skeleton of **4**

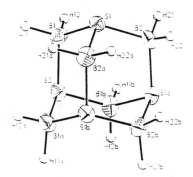

Figure 3   X-ray structure of $[(BH_2)_6S_4]^{2-}$, $\underline{5}$

Treatment of $\underline{1}$ with $NaBH_4$ in triglyme afforded a salt of the new anion $[Na \cdot Triglyme]_2[S(BH_3)_4]$ tetra(borane) sulfate(2-), $\underline{6}$. The anion of $\underline{6}$ may be viewed either as an adduct of B_2H_6 with S^{2-} or as a bridge substituted thia derivative of $B_2H_7^-$. Furthermore the anion of $\underline{6}$ is isoelectronic and isostructural with the SO_4^{2-} ion[3], Fig. 4.

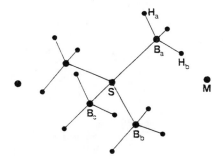

Figure 4   Calculated structure of $[S(BH_3)_4]^{2-}$, $\underline{6}$

The controlled nucleophilic degradation of elemental selenium with $NaBH_4$ proceeds only in triglyme according to scheme 3 forming $\underline{7}$.

$\underline{7}$ can be isolated but decomposes above R. T. to form $[(BH_2)_6Se_4]^{2-}$, $\underline{8}$,[2]. The anion of $\underline{8}$ consists of an adamantane skeleton which is isotypic with $\underline{5}$. A mixture of $THF \cdot BH_3 / BH_4^-$ reacts with elemental selenium to yield $[(BH_2)_5Se_4]^-$, $\underline{9}$, which is isotypic with $\underline{4}$, Fig. 5.

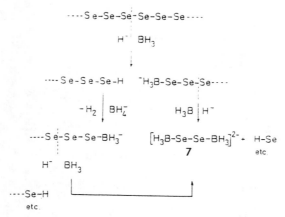

Scheme 3

Treatment of 7 with B_2H_6 forms $[H_3B-\mu_2-Se(B_2H_5)]^-$, 10, which is the selenium analogue of 1,Figure 6

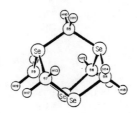

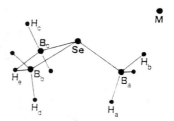

Figure 5
Proposed structure of
$[(BH_2)_5Se_4]^-$, 9

Figure 6
Calculated structure of
$[H_3B-\mu_2-Se(B_2H_5)]^-$, 10

Structural data of the compounds 1, 4, 6 and 10 have been calculated by SCF methods,[1,3]

REFERENCES

1.   H. Binder, K. Wolfer, W. Ehmann, W.-P. Pfeffer, K.Peters, H. Horn and R. Ahlrichs, *Chem. Ber.* 1992, 125, 651.

2.   H. Binder, H. Loos, K. Dermentzis, H. Bormann and A. Simon, *Chem. Ber.*, 1991, 124, 427.

3.   H. Binder, H. Loos, H. Bormann, A. Simon, H.-J. Flad and A. Simon *Z. anorg. allg. Chem.*, 1993, 619, 1.

Subject Index